T0318601

Mathematics in Science and Engineering

Algebraic and Combinatorial Computational Biology

Mathematics in Science and Engineering

Algebraic and Combinatorial Computational Biology

Edited by

Raina Robeva

Matthew Macauley

Series Editor

Goong Chen

Academic Press is an imprint of Elsevier
125 London Wall, London EC2Y 5AS, United Kingdom
525 B Street, Suite 1650, San Diego, CA 92101, United States
50 Hampshire Street, 5th Floor, Cambridge, MA 02139, United States
The Boulevard, Langford Lane, Kidlington, Oxford OX5 1GB, United Kingdom

Library of Congress Cataloging-in-Publication Data
A catalog record for this book is available from the Library of Congress

British Library Cataloguing-in-Publication Data
A catalogue record for this book is available from the British Library

ISBN 978-0-12-814066-6

For information on all Academic Press publications
visit our website at https://www.elsevier.com/books-and-journals

Publisher: Candice Janco
Acquisition Editor: Scott J. Bentley
Editorial Project Manager: Katerina Zaliva
Production Project Manager: Swapna Srinivasan
Cover Designer: Victoria Pearson

Typeset by SPi Global, India

Contents

Contributors

Numbers in parentheses indicate the pages on which the authors' contributions begin.

Boris Aguilar (147), Institute for Systems Biology, Seattle, WA, United States

Olcay Akman (351), Illinois State University, Normal, IL, United States

Robert Brijder (61), Department WET-INF, Hasselt University, Diepenbeek, Belgium

Timothy Comar (351), Benedictine University, Lisle, IL, United States

Carsten Conradi (279), Hochschule für Technik und Wirtschaft Berlin, Berlin, Germany

Carina Curto (213, 241), Department of Mathematics, The Pennsylvania State University, University Park, PA, United States

Robin Davies (89, 375), Biomedical Sciences, Jefferson College of Health Sciences, Roanoke, VA, United States

Joanna Ellis-Monaghan (35), Department of Mathematics, Saint Michael's College, Colchester, VT, United States

Stefan Forcey (319), Department of Mathematics, University of Akron, Akron, OH, United States

Urmi Ghosh-Dastidar (375), Department of Mathematics, New York City College of Technology, Brooklyn, NY, United States

Josselyn Gonzales (351), Illinois State University, Normal, IL, United States

Gabriela Hamerlinck (319), QUBES, BioQUEST Curriculum Consortium, Boyds, MD, United States

Hendrik Jan Hoogeboom (61), Department of Computer Science (LIACS), Leiden University, Leiden, The Netherlands

Daniel Hrozencik (351), Chicago State University, Chicago, IL, United States

Andy Jenkins (89), Department of Mathematics, University of Georgia, Athens, GA, United States

Nataša Jonoska (35, 61), Department of Mathematics and Statistics, University of South Florida, Tampa, FL, United States

John Jungck (1), University of Delaware, Newark, DE, United States

Logan Keefe (319), Department of Mathematics, Kent State University, Kent, OH, United States

Debra Knisley (1), Department of Mathematics and Statistics, East Tennessee State University, Johnson City, TN, United States

Jeff Knisley (375), Department of Mathematics and Statistics, East Tennessee State University, Johnson City, TN, United States

Matthew Macauley (89, 175), School of Mathematical and Statistical Sciences, Clemson University, Clemson, SC, United States

Katherine Morrison (241), School of Mathematical Sciences, University of Northern Colorado, Greeley, CO, United States

David Murrugarra (147), Department of Mathematics, University of Kentucky, Lexington, KY, United States

Greta Pangborn (1, 35), Department of Computer Science, Saint Michael's College, Colchester, VT, United States

Casian Pantea (279), West Virginia University, Morgantown, WV, United States

Manda Riehl (1), Department of Mathematics, Rose-Hulman Institute of Technology, Terre Haute, IN, United States

Masahico Saito (61), Department of Mathematics and Statistics, University of South Florida, Tampa, FL, United States

Widodo Samyono (375), Department of Mathematics, Jarvis Christian College, Charles A. Meyer Science and Mathematics Center, Hawkins, TX, United States

William Sands (319), Department of Computational Mathematics, Science, and Engineering, Michigan State University, MI, United States

Brandilyn Stigler (175), Department of Mathematics, Southern Methodist University, Dallas, TX, United States

Alan Veliz-Cuba (213), Department of Mathematics, University of Dayton, Dayton, OH, United States

Emilie Wiesner (1), Department of Mathematics, Ithaca College, Ithaca, NY, United States

Nora Youngs (213), Department of Mathematics and Statistics, Colby College, Waterville, ME, United States

Preface

When a mathematician or biologist hears the term "mathematical biology," the mental picture that comes to mind for many may be that of calculus-based techniques such as differential equations. There is, of course, much more of a diversity than this, though other types of mathematical biology often live under an umbrella with a different name. For example, many problems and techniques involving discrete mathematics have been relegated to the world of bioinformatics. Another large area of mathematical work in the life sciences is biostatistics, and yet another one emerging more recently is data science. Indeed, the lines between these fields are blurred and subjective. An area that involves mathematics and biology may be considered mathematical biology to some but not to others. Some research projects blend so many different fields that it is unnatural to separate into distinct silos such as "mathematics," "genomics," "computational biology," etc. Rather, they are true transdiscplinary *science* problems: a project on epidemiology might draw from applied mathematics, biology, public health, statistics and data science, computer science, network science, and economics; a project in phylogenetics might involve researchers from mathematics, computer science, a number of fields in biology, statistics, data science, and genomics; and a research group working on protein folding might consist of biologists, biochemists, biophysicists, mathematicians, statisticians, and computer scientists.

Early work involving discrete and algebraic methods to model biological systems can be traced back to (at least) the 1960s. In 1969, theoretical biologist Stuart Kauffman proposed modeling gene regulatory network with Boolean functions. Around the same time, biologist René Thomas pursued a similar modeling framework that he called "logical models." These types of models have been studied since under different names, such as Boolean networks, automata networks, generalized cellular automata, and others. In some cases, the models are not Boolean, but ternary, or feature a larger state space. If the state space is a finite field (if not, one can just expand it until it is), then the individual functions describing the model are polynomials. This opens a door to using the rich toolbox of computational algebra for analyzing such network models, leading to the province of *Algebraic Biology*. Among the many other examples where discrete mathematics and algebra facilitate progress in modern biology are the field of *Algebraic Statistics* that has proved instrumental for a number of problems in genomics and phylogenetics.

One hallmark of transdisciplinary research is that its results and subsequent publications could not have been produced only by expertise from a subset of the participating disciplines. This is a far cry from some multidisciplinary work where researchers from each discipline may work somewhat independently on individual "modules," then write separate sections for the project report and subsequent publication. Transdisciplinary research is also a powerful catalyst for accelerating advancement for each of the individual disciplines. In biology, the advent of high-throughput technology in the late 20th and early 21st century such as gene sequencers, RNA-Seq, and CRISPR, along with the rise of high-performance computing, has put this discipline firmly in the spotlight as a prime field to be transformed by mathematics and technology. In 2004, biologist Joel Cohen famously predicted that this is a two-way process when he published the paper titled "Mathematics is biology's next microscope, only better. Biology is mathematics' next physics, only better." The following year, mathematician Bernd Sturmfels asked in the title of a paper he wrote "Can biology lead to new theorems?," and then proceeded in the body of the paper to answer and support this claim in the affirmative.

The purpose of this book is to highlight some of the new areas of mathematical biology with combinatorial and algebraic flavors and a distinct computational/statistical component. It is in no way meant to be comprehensive, and reflects the personal preferences of the editors to highlight current trends in the discipline. Most importantly, the book reflects our efforts to address the urgent need to connect ongoing advances in discrete and algebraic mathematical biology with the academic curriculum where calculus-based methods still dominate the landscape. While the use of modern algebraic methods is now in the mainstream of mathematical biology research, this trend has been slow to influence the traditional mathematics and biology curricula. Students interested in mathematical biology have relatively easy access to courses that utilize classical analytic methods based on difference and differential equations. By contrast, students interested in algebraic and discrete computational approaches have fewer doors visibly open to them, and indeed may not even know that they exist. Several high-profile national reports have urged the mathematical biology community to enact steps to bridge this gap,[1] and since 2013, the editors have collaborated with groups of like-minded faculty to make headways in addressing this problem. Together, we have led several professional faculty development workshops—at the Mathematical Biosciences Institute at the Ohio State University (2013) and the National Institute for Mathematical and Biological Synthesis (NIMBioS) at the University of Tennessee (2014 and 2016)—focused on developing, disseminating, and classroom-testing novel educational materials based on cutting-edge research in discrete and combinatorial mathematical

1. The report *Vision and change in undergraduate biology education: a call to action* of American Association for the Advancement of Science (2011) and the National Research Council's report *The Mathematical Sciences in 2025* (2013) are just two examples.

biology. In fact, this book could be viewed as the third publication in a series that has been linked with those workshops.

The first book, titled *Mathematical Concepts and Methods in Modern Biology: Using Modern Discrete Models* and published in 2013, was edited by Raina Robeva and Terrell Hodge. Topics include Boolean networks, agent-based, and neuronal models, linear algebra models of populations and metabolic pathways, hidden Markov models in genetics, and geometric approaches in phylogenetics. The second publication, *Algebraic and Discrete Mathematical Methods for Modern Biology*, edited by Raina Robeva and published in 2015, covers topics from graph theory in systems biology, ecology, and evolution, more topics on Boolean networks, Petri nets, epidemiology on networks, linear algebraic approaches in genetics and metabolic analysis, computational phylogenetics, and RNA folding. Most of the material in these books is accessible to undergraduates who have not necessarily taken calculus. In addition to being ideal for undergraduates, these books can provide detailed introductions to the topics for biologists who have limited or even no calculus background. The current "Volume 3" explores a new set of topics with a distinct computational flavor, either not covered in the previous two, or topics that have emerged as fundamental to the field in the last few years. Although our target audience this time is primarily graduate students, we have made every effort to keep most of the topics accessible to advanced undergraduates as well. All three books are filled with examples and exercises to promote their use in the classroom, and feature notes on the use of specialized software for computation, analysis, and simulation. The chapters are designed to be largely independent from one another and can be viewed as starting points for undergraduate research projects or as entryways for graduate students and researchers new to the field of algebraic mathematical biology. They can also be used as "modules" for classroom use and independent studies. Solution guides containing the solutions to most exercises are also available.

The chapters of this volume are organized to highlight several common themes. We begin with a chapter on multiscale modeling, with a focus on the molecular level, followed by two chapters on the assembly of DNA. Chapters 4–6 involve topics on discrete models of the dynamics of molecular networks. More specifically, Chapter 4 introduces the local modeling framework, which attempts to clarify and unify a number of modeling paradigms, including Boolean networks, logical models, and automata networks. Chapter 5 considers these systems with stochastic features, which are sometimes called Stochastic Discrete Dynamical Systems.

Chapter 6 looks at the question of reverse engineering the wiring diagram, using techniques from combinatorial commutative algebra and algebraic geometry—namely Stanley-Reisner theory and the primary decomposition of square-free monomial and pseudomonomial ideals. Though Chapter 7 is on a problem from neuroscience, it also involves the same underlying algebraic framework as Chapter 6. The concept of a pseudomonomial ideal, as far as we can tell, had not been studied until it arose recently in several diverse areas in mathematical biology, from reverse engineering molecular networks to encoding the structure of place fields in neuroscience. Researchers are now studying and publishing on these objects and on so-called "neural ideals." This is a prime example of how biology is leading to new theorems, as predicted by Sturmfels. The neuroscience topic continues into Chapter 8 on threshold linear ODE models over graphs—a framework now used as a simple model of firing patterns in neurons. A central theme in this chapter is how to deduce the dynamics of the system from the structure of the underlying graph.

The focus of Chapter 9 is on multistationarity in biochemical reaction networks. Although this topic may appear unrelated, the aforementioned theme of connecting local network structure to global system dynamics emerges once again, after being introduced in Chapter 4, and being an underlying theme of Chapter 8. This question has appeared throughout the decades in different forms. Back in the 1980s, René Thomas posed these questions both in the context of logical models (recall, a variant of Boolean networks), which were popular models of gene networks, and in continuous differential equation frameworks. He observed that as a rule of thumb, positive feedback is a necessary condition for having multiple steady states (multistationarity), but negative feedback loops are necessary for cyclic attractors, and hence homeostasis. These conjectures have since been formalized and proven in a number of settings, from discrete models to differential equations.

Chapter 10 is on optimization and linear programming in phylogenetics, where the problem to infer and interpret a phylogenetic tree is useful in multiple contexts in biology and medicine. Finally, Chapters 11 and 12 examine classification in biology through clustering and machine learning, with examples ranging from protein families to environmental systems.

This book would not have been possible without the dedicated team of authors who felt passionately about the value of presenting their research results in a way that provides hands-on practical knowledge for readers ranging from advanced undergraduate students to researchers entering the field of algebraic and computational biology. We are grateful for their patience during the editing process and for their willingness to go through multiple revisions with us. We warmly appreciate the support of NIMBioS for the 2016 workshop *Discrete and Algebraic Mathematical Biology: Research and Education*. Work on many of the chapters in this volume started during this workshop and may not have materialized otherwise. Our personal thanks go to Katerina Zaliva, our Editorial Project Manager, who was gracious with her time, prompt to answer questions, and ready to adopt a cheerful attitude during some of the unavoidable challenges in the process. Finally, we thank our spouses, Catherine Gurri and Boris Kovatchev, for their patience and support throughout.

Matthew Macauley
Raina Robeva
August 27, 2018

Chapter 1

Multiscale Graph-Theoretic Modeling of Biomolecular Structures

John Jungck*, Debra Knisley†, Greta Pangborn‡, Manda Riehl§ and Emilie Wiesner¶

**University of Delaware, Newark, DE, United States, †Department of Mathematics and Statistics, East Tennessee State University, Johnson City, TN, United States, ‡Department of Computer Science, Saint Michael's College, Colchester, VT, United States, §Department of Mathematics, Rose-Hulman Institute of Technology, Terre Haute, IN, United States, ¶Department of Mathematics, Ithaca College, Ithaca, NY, United States*

1.1 INTRODUCTION

1.1.1 The Molecules of Life

Zuckerkandl and Pauling [1] identified three types of macromolecules within living systems as eusemantic, meaning that within their structure they carry evolutionary information from one generation to the next: DNA, RNA, and proteins. While DNA has received substantial attention because its three-dimensional structure was modeled by Watson and Crick in 1953, it is often viewed as a rather passive carrier of information from one generation to the next. By contrast, RNAs and proteins are considered to be active players within generations because they are also able to catalyze reactions, regulate metabolic functions, be synthesized and degraded many times over the life of an individual cell, and interact in complex RNA-protein assemblies. Furthermore, some viruses have RNA as their primary informational macromolecule from one generation to the next. Thus, the structures of RNA and protein macromolecules have been of substantial interest since their initial discovery.

RNA and protein molecules are both macromolecular polymers: that is, we can think of them as made up of a long sequence of a small number of molecular "subblocks." However, much of the function of these molecules is determined by interactions between these component pieces, as well as the three-dimensional arrangement of the macromolecule as a whole. Thus, RNA and proteins naturally lend themselves to analysis at a variety of different scales.

Algebraic and Combinatorial Computational Biology. https://doi.org/10.1016/B978-0-12-814066-6.00001-5

In this chapter, we describe how graph-theoretic approaches can be used to model and analyze RNA and protein structure at different scales. We begin with a brief primer on graph theory. Then we delve into a case study of RNA, describing its structure in more detail, giving a brief survey of the graph-related tools that biologists and mathematicians have used to analyze RNA structure, and then presenting a few graph-theoretic models of RNA structure in more detail. Finally, we consider a hierarchical network model of proteins that captures the relationships between atoms, amino acids, and molecular substructures.

1.2 GRAPH THEORY FUNDAMENTALS

Networks, or graphs as they are called in graph theory, are frequently used to model both RNA and protein structures. A graph is typically represented as a collection of points, called nodes in networks and vertices in graph theory, together with connecting lines. The lines are called edges in graphs and sometimes called links in networks. The edges may or may not have a direction assigned to them. If directions are assigned to the edges, then the graph is directed, otherwise it is undirected. More formally, an *undirected graph* G consists of a pair of sets (V, E) where V is a nonempty, finite set of vertices and $E \subset V \times V$ is a set of edges where edge $e = \{v_1, v_2\}$ connects vertices v_1 and v_2. Note that it is only the logical connections represented by the edges that matter, not the position of the vertices. See Fig. 1.1 for three drawings of the graph $G = (V = \{1, 2, 3, 4\}, E = \{\{1, 2\}, \{2, 3\}, \{3, 4\}, \{4, 1\}\})$. Some common graph models include: social networks (vertices correspond to people and edges to relationships), transportation networks (vertices correspond to cities and edges to flights or highways), computer networks (vertices correspond to machines and edges to cables), and biological networks (vertices may correspond to proteins and edges to interactions between proteins).

Edges frequently have costs or weights associated with them, such as cost or distance in transportation networks, bandwidth in computer networks, and extent of communication in a social network. Vertices may also have associated weights, such as the cost to locate a facility within a given town. In some cases edges may be directed; for example, in food webs the edges are directed from vertices corresponding to prey to vertices corresponding to predators.

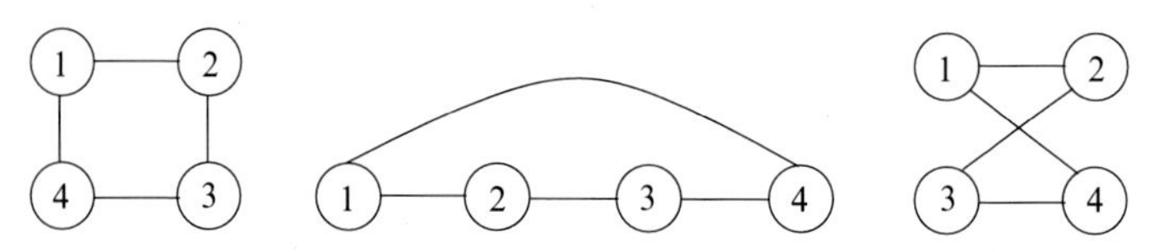

FIG. 1.1 Three drawings of the same graph.

- Two vertices v_1, v_2 are *adjacent* in G if there is an edge between them (i.e., $\{v_1, v_2\} \in E$).
- The *degree* of a vertex is the number of edges incident to it.
- A *length-k path* in G between vertices x and y is a sequence of vertices, $x = v_0, v_1, \ldots, v_k = y$, satisfying the property that v_i and v_{i+1} are adjacent in G for all i, $0 \leq i < k$ and each of the v_i are distinct.
- Two vertices are said to be *connected* if there is a path between them. G is connected if every pair of vertices in G is connected.
- A *cycle* in G is a path (with length at least three) where the starting and ending vertices are the same, but the remainder of the vertices are distinct.
- A *tree* T is a special type of graph in which all vertices are connected but there are no cycles. Any tree on n vertices will have $n - 1$ edges. A vertex with degree 1 is referred to as a leaf node. In some cases there is a designated root vertex r; the depth of a vertex v is the length of the unique path from r to v.
- A *graph invariant* is a property that depends only on the abstract structure of the graph (the vertex and edge sets), not on a particular labeling or embedding. Examples of graph invariants include: the maximum, minimum, and average vertex degrees; the size of the *maximum independent set* (the largest set of mutually nonadjacent vertices); and the *dominating number* (the minimum size of a dominating set S where every vertex in V is either in S or adjacent to a vertex in S).

1.3 MODELING RNA STRUCTURE

RNA stands for ribonucleic acid; an individual RNA is a macromolecular polymer, made up of repeated ribonucleotides. An individual ribonucleotide consists of a nucleobase, a sugar, and a phosphate. While many ribonucleotides exist in nature, we primarily focus on those RNAs that are composed of four bases: adenine, cytosine, guanine, and uracil, abbreviated as A, C, G, and U, respectively. Individual ribonucleotides are linked by bonds between the sugar and phosphate components. Unlike DNA, RNA is single stranded, which allows it to self-fold, forming double-stranded subregions. The double-stranded regions of RNA molecules are held together by hydrogen bonds between nucleobases of (nonadjacent) ribonucleotides. Finally, these macromolecules take on particular arrangements in space, (partially) controlled by the chemical bonds between ribonucleotides that determine their function.

These physicochemical features of RNA are traditionally separated into three structural levels. The *primary structure* of an RNA molecule is an abstraction of the sequence of phosphate-sugar bonds between individual nucleotides. Primary structure is described as a linear sequence that only uses the four-character nucleotide alphabet. Moreover, the sugar-phosphate bond is directional. This gives an orientation to the primary structure, indicated by $5'$ and $3'$ labels on the ends. The *secondary structure* refers to the pairing of nucleotides via bonds between their bases. *Tertiary structure* is the three-dimensional shape of an RNA molecule.

In this section, we focus on modeling RNA secondary structures. Scientifically, the problem of predicting the secondary structure of RNAs and proteins is crucial to understanding their function. When Holley's team [2] sequenced the first RNA (a transfer RNA) in 1965, they already recognized that this small molecule had a complex secondary and tertiary structure [2]. Even before scientists were able to sequence many of these macromolecules, they knew that determining their secondary structure from first principles is a very difficult problem. There are professional competitions for solving the folding of the primary structure of RNAs and proteins into secondary and tertiary structures (e.g., the BioVis Design Contests [3, 4] and CASP [5, 6]). There are also proofs that particular mathematical formulations of these processes are theoretically challenging, that is, NP-Hard [7, 8]. Citizen Science projects crowdsource these problems to see if human intuition and pattern recognition skills can improve upon foldings found by the best computer algorithms and heuristics (eteRNA [9, 10] for RNA folding and FoldIt [11] for proteins).

Classically, four planar graph representations of the secondary structure of RNAs have been used: Nussinov circles, airports, domes, and mountains (Fig. 1.2). In the first three cases, lines or curves connect pairs of nucleobases to represent a bond. The mountain diagram plots the number of base pairs enclosing a sequence position with the peak corresponding to a hairpin turn and the plateaus to unpaired bases. Later in Section 1.3, we will discuss airports and arc diagrams (close cousins to domes) in more detail. While these representations are the most widely used, new visualizations continue to be developed including space filling RNA curves [12], single-stranded RNA space filling curves [13], RNA bows [14], and probability airports [15].

The secondary structure of RNA molecules is constrained by a variety of factors. Bonds between nucleobases occur in particular patterns: The base A binds preferentially to U and G binds preferentially to C. The bases A and G have bigger bicyclic ring structures while C and U are monocyclic ring structures, and it takes less energy to break the two hydrogen bonds in A-U base pairs than three hydrogen bonds in G-C base pairs.

Thermodynamics affect the secondary structure globally, as well. A common criterion for determining whether an RNA secondary structural configuration is most plausible and stable is whether it has the lowest free energy compared with other potential foldings. Hydrogen bonds between nucleobases reduce free energy, and various configurations of unbonded regions of the RNA molecule may either raise or lower the free energy. For example, a region of six G-C base pairs would have a lower free energy than an equivalent region of six A-U base pairs. There usually needs to be three consecutive base pairs in a stack for an RNA secondary structure to be thermodynamically stable; similarly, loops of size one (which are also a type of bulge) are only stable if they are in midst of minimally five or six stacked nucleotide base pairs or a loop is longer than three

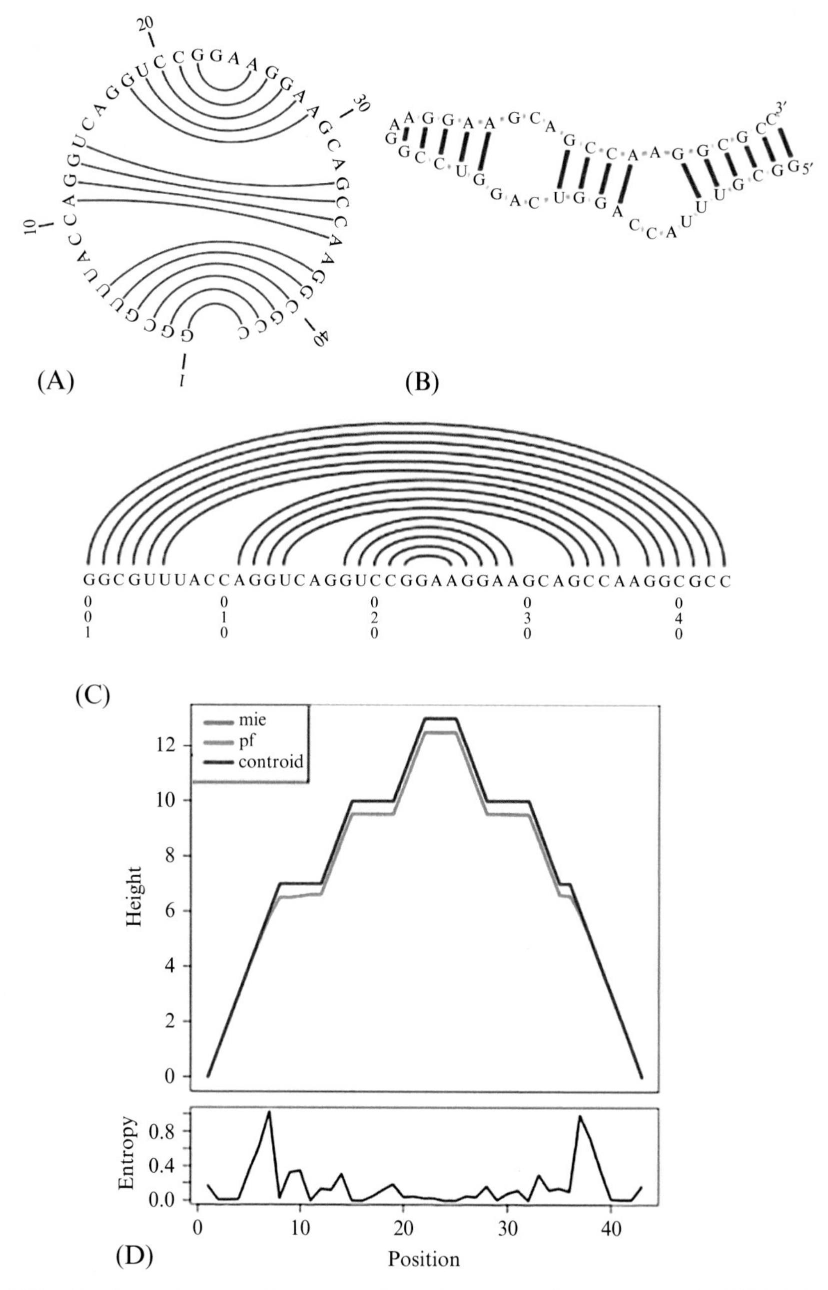

FIG. 1.2 Four planar graph representations of the secondary structure of *PDB_00032*. (A) Nussinov circle, (B) airport diagram, (C) dome plot, (D) mountain plot. *(Sequence information from M. Andronescu, V. Bereg, H.H. Hoos, A. Condon, RNA Strand: the RNA secondary structure and statistical analysis database, BMC Bioinf. 9 (1) (2008) 340, Figures A–C created with jVis.Rna, Available from: https://jviz.cs.sfu.ca/ (Accessed 20 October 2017), and Figure D created using the Vienna RNA secondary structure server, I.L. Hofacker, Vienna RNA secondary structure server, Nucleic Acids Res. 31 (13) (2003) 3429–3431.)*

unpaired nucleotides. Moreover, the interplay between regions of paired bases and unpaired regions contribute to the functionality of RNA. Thus, models of RNA secondary structure might be expected to represent both individual base pairs, as well as larger motifs in the molecule that result from such pairing.

There are a number of aspects of RNA secondary structure that are not well-reflected in the traditional models. The first is that there is not a well-defined mapping from the space of possible RNA primary structures to the space of secondary structures. It is the case that a given primary structure may fold into multiple secondary structures (possibly all reflecting relatively minimal free energy states); and that distinct primary structures may fold into similar secondary structures (due to the evolutionary feature of compensatory mutations that maintain positional base pairings without changing the secondary structure of RNA).

Three additional problems not included above are the existence of *pseudoknots*, *riboswitches*, and *circular RNAs*. Pseudoknots represent an interleaving of unbonded and bonded regions of the RNA sequence; they will be discussed below. Examples of ribozymes, programmed frame-shifting, and telomerase activity have pseudoknots essential to their functioning. Reidys et al. [17] describe a variety of algorithms and heuristics to predict RNA secondary structures that contain pseudoknots. Riboswitches violate the search for a singular thermodynamically stable RNA secondary structure because at least two different configurations are involved in their regulation of expression. Ritz et al. [18] discuss how evolution could simultaneously select for multiple forms. Kutchko and Laederach [19] go further and use riboswitches to critique the whole "prediction paradigm." Circular RNAs sometimes exist in thousands of copies and are expressed differentially in different human organs. Some of these differences have been used as molecular markers associated with particular diseases such as multiple sclerosis. Yet, because circular RNAs are more structurally constrained than linear RNAs, degrees of folding freedom are necessarily less. Cuesta and Manrubia [20] have used a variety of combinatorial approaches to estimate an asymptotic limit on the number of folds as the length of an RNA grows.

1.3.1 RNA Secondary Structure Features

In this section, we define important features of RNA secondary structures and use airports and arc diagrams to illustrate them. These features are the building blocks of the tree graph and dual graph models presented in Section 1.3.2.

Figs. 1.3 and 1.4 represent the secondary structure of an RNA molecule found in the organism *Bacillus subtilis*. Fig. 1.3 shows an airport, where the RNA backbone can be traced around the diagram, and base pair bonds cross the interior of the diagram (in blue). Fig. 1.4 shows an arc diagram for the same RNA structure; here, the backbone of the molecule appears along the bottom of the diagram, and base pair bonds are represents by arcs (in blue).

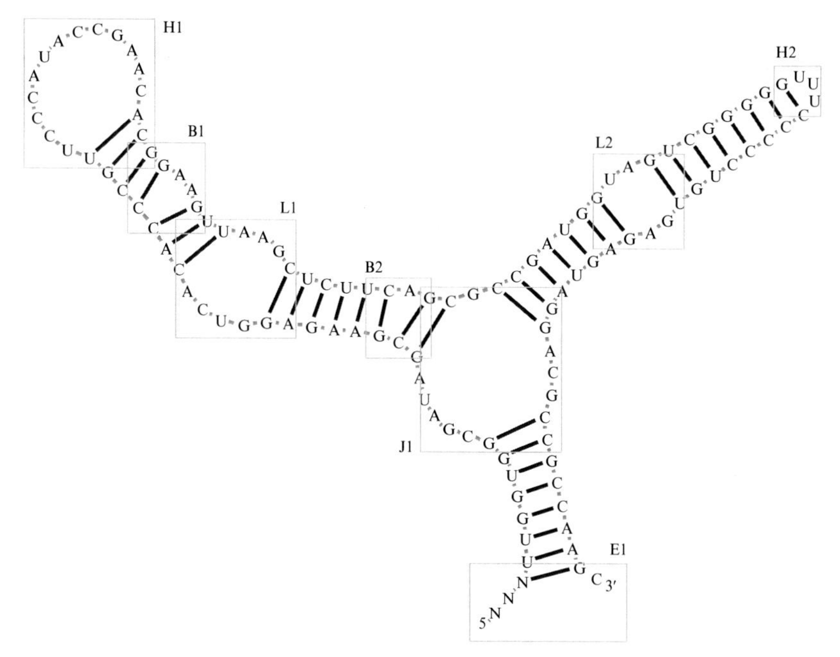

FIG. 1.3 Airport for 5S ribosomal RNA found in *Bacillus subtilis*. *B*, bulge; *E*, end; *H*, hairpin; *J*, junction; *L*, loop. *(RNA obtained from J.J. Cannone, S. Subramanian, M.N. Schnare, J.R. Collett, L.M. D'Souza, Y. Du, B. Feng, N. Lin, L.V. Madabusi, K.M. Müller, et al., The comparative RNA web (CRW) site: an online database of comparative sequence and structure information for ribosomal, intron, and other RNAs, BMC Bioinf. 3 (1) (2002) 2; visualization produced using JViz.Rna, K.C. Wiese, E. Glen, jViz.Rna—an interactive graphical tool for visualizing RNA secondary structure including pseudoknots, in: 19th IEEE International Symposium on Computer-Based Medical Systems, 2006. CBMS 2006, ISSN 1063-7125, 2006, pp. 659–664, https://doi.org/10.1109/CBMS.2006.104.)*

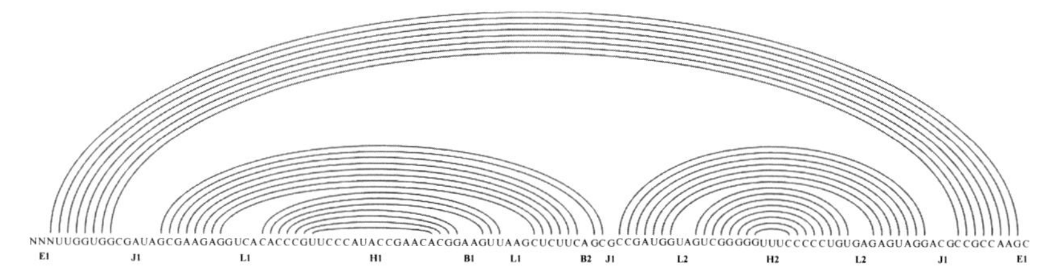

FIG. 1.4 Arc diagram for 5S ribosomal RNA found in *Bacillus subtilis*. *(RNA obtained from J.J. Cannone, S. Subramanian, M.N. Schnare, J.R. Collett, L.M. D'Souza, Y. Du, B. Feng, N. Lin, L.V. Madabusi, K.M. Müller, et al., The comparative RNA web (CRW) site: an online database of comparative sequence and structure information for ribosomal, intron, and other RNAs, BMC Bioinf. 3 (1) (2002) 2; visualization produced using JViz.Rna, K.C. Wiese, E. Glen, jViz.Rna—an interactive graphical tool for visualizing RNA secondary structure including pseudoknots, in: 19th IEEE International Symposium on Computer-Based Medical Systems, 2006. CBMS 2006, ISSN 1063-7125, 2006, pp. 659–664, https://doi.org/10.1109/CBMS.2006.104.)*

The formation of secondary structure in an RNA molecule naturally produces regions of paired and unpaired bases in recurring patterns. Researchers characterize these recurring substructures as follows:

- A *stem* refers to a set of consecutive base pairings, forming a double helix. Stems correspond to the sets of nested arcs, forming sets of matched pairs in the arc diagram.
- A *hairpin loop* is a region of unbonded bases formed when an RNA Strand folds back on itself to form a stem. The regions marked $H1$ and $H2$ in Figs. 1.3 and 1.4 show hairpin loops.
- An *internal loop* exists when two stems are interrupted by regions of unpaired bases on each strand. The regions $L1$ and $L2$ in Figs. 1.3 and 1.4 show internal loops.
- A *bulge* is formed when two stems are interrupted by regions of unpaired bases on one strand, as shown in regions $B1$ and $B2$ in Figs. 1.3 and 1.4.
- An *external loop* describes the region that includes the unpaired portion of the $3'$ and $5'$ ends of the RNA molecule. Region $E1$ in Figs. 1.3 and 1.4 forms an external loop.
- A *junction* (or *multibranch loop*) describes the meeting of three or more stems, which may or may not be separated by regions of unpaired bases. A junction is displayed in regions $J1$ in Figs. 1.3 and 1.4.
- Internal loops, bulges, and junctions arise when one of more sets of (noncrossing) nested base pairs form among the bases between another set of nested base pairs.
- A *psuedoknot* describes a structure in a sequence segment, located between two bonded regions, bonds with a region that is outside of these bonded regions. A pseudoknot appears in an arc diagram as crossing sets of nested arcs. Figs. 1.5 and 1.6 show an example pseudoknot.

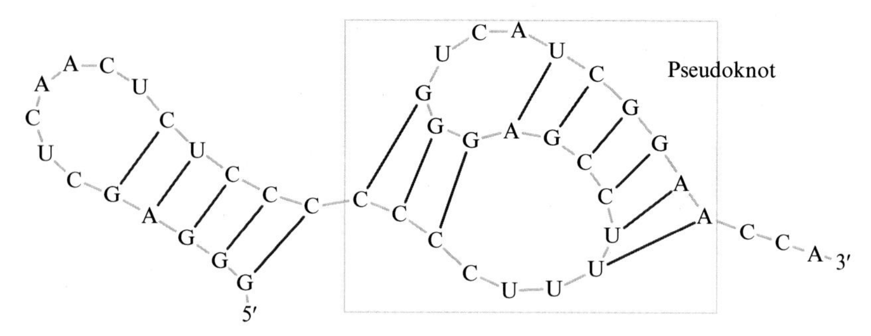

FIG. 1.5 Airport diagram of an RNA fragment from the turnip yellow mosaic virus. *(RNA obtained from M.H. Kolk, M. van der Graaf, S.S. Wijmenga, C.W.A. Pleij, H.A. Heus, C.W. Hilbers, NMR structure of a classical pseudoknot: interplay of single- and double-stranded RNA, Science 280 (5362) (1998) 434–438; visualization produced using JViz.Rna, K.C. Wiese, E. Glen, jViz.Rna—an interactive graphical tool for visualizing RNA secondary structure including pseudoknots, in: 19th IEEE International Symposium on Computer-Based Medical Systems, 2006. CBMS 2006, ISSN 1063-7125, 2006, pp. 659–664, https://doi.org/10.1109/CBMS.2006.104.)*

FIG. 1.6 Arc diagram of an RNA fragment from the turnip yellow mosaic virus. *(RNA obtained from M.H. Kolk, M. van der Graaf, S.S. Wijmenga, C.W.A. Pleij, H.A. Heus, C.W. Hilbers, NMR structure of a classical pseudoknot: interplay of single- and double-stranded RNA, Science 280 (5362) (1998) 434–438; visualization produced using JViz.Rna, K.C. Wiese, E. Glen, jViz.Rna—an interactive graphical tool for visualizing RNA secondary structure including pseudoknots, in: 19th IEEE International Symposium on Computer-Based Medical Systems, 2006. CBMS 2006, ISSN 1063-7125, 2006, pp. 659–664, https://doi.org/10.1109/CBMS.2006.104.)*

1.3.2 Tree and Dual Graph Models of RNA Secondary Structure

Given the network-like relationships between loops, stems, and other RNA structural elements, graph theory provides a natural tool for modeling RNA secondary structure. Here we present two modeling strategies, tree graphs and dual graphs. Then we discuss a few of the tools that researchers have used to analyze these models.

1.3.2.1 RNA Tree Graphs

When representing an RNA secondary structure as a tree graph, the general strategy is to represent regions of unbonded bases by vertices and to represent the paired regions that connect them by edges. In particular, we have the following rules for creating a tree graph, adopted from the models developed by Gan et al. [21]:

- A vertex represents any of the following: a hairpin loop, internal loop, or bulge containing more than one unbonded base pair corresponds to a vertex; a junction; the external loop, provided the $5'$ and $3'$ ends share a common stem.
- An edge connects two vertices when the corresponding elements of the RNA molecule are connected by a stem of at least two base pairings.

Fig. 1.7 shows the tree graph model for the RNA structure given in Fig. 1.3. Although tree graphs naturally capture important elements, this modeling scheme does not allow for the representation of all possible RNA secondary

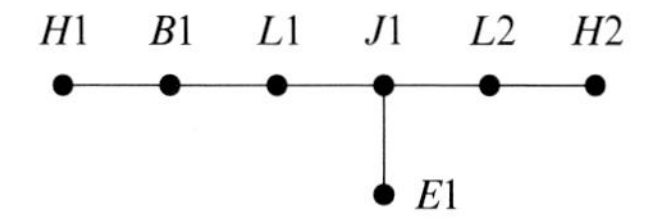

FIG. 1.7 Tree graph model of the 5S ribosomal RNA for *Bacillus subtilis*. Note that although this tree graph carries the labels of Figs. 1.3 and 1.4, we take tree graph models to be unlabeled.

structures. In particular, pseudoknots and $5'/3'$ ends that are not joined by a single stem cannot be represented. In Section 1.3.2.3, we introduce the dual graph model, which addresses these issues.

The tree graphs described here preserve information about the basic secondary structural elements on an RNA molecule, but they also discard considerable information, including: the type of each structural element, the number of bases or base pairs in each element, and the energy associated with each element (from a thermodynamic perspective). Other researchers have used similar models but enhanced their model with additional information about the RNA secondary structure. Le et al. [22] modeled RNA structure with labeled tree graphs, where the vertices are labeled by type (hairpin loop, bulge, etc.) and the number of nucleotides in the corresponding structural element, and the edges are labeled by the number of bases in the corresponding stem. A number of researchers (cf. [23]) have modeled RNA secondary structure using "rooted" tree graphs, where they distinguish the vertex corresponding to the $3'$ and $5'$ ends as a "root." The choices of each research team represent a trade-off between the fidelity and the simplicity of the model. Each of these variations of tree graphs (labeled, rooted, and unrooted and unlabeled) corresponds to important objects of study in graph theory; and thus the different model choices that researchers have made tap into different mathematical tools and questions.

1.3.2.2 Using Graph Statistics to Understand RNA Secondary Structure

The tree graph and dual graph representations of RNA secondary structure suggest many questions about the interplay between the mathematical and biological sides of these models: Which tree/dual graphs represent actual RNA secondary structures? Can the mathematical models be used to suggest undiscovered or newly designed RNA secondary structures? What biologically significant aspects of RNA secondary structure can be characterized mathematically in terms of their graph-theoretic models? What mathematically significant features of the models have biological interpretations?

Much of the work addressing these questions has used graph invariants as a starting point. In this section, we present examples of some of the graph invariants as applied to tree graph models of RNA.

Among the most fundamental graph statistics are vertex and edge counts: the number of vertices, denoted $|V(G)|$, and the number of edges, denoted $|E(G)|$, in a graph G. For example, for the graph in Fig. 1.7, $|V(G)| = 7$ and $|E(G)| = 6$. In general, for a tree graph T, $|V(T)| - |E(T)| = 1$ (see Exercise 6).

Gan et al. [21] used vertex counts to estimate the number of possible RNA secondary structures. Based on a survey of experimental results, they found that, on average, each vertex in a tree graph model corresponds to 20 nucleotides (abbreviated *nt*). Thus, tree graphs with V vertices approximately correspond to RNA molecules of length approximately $20V$ nt; and so the number of RNA

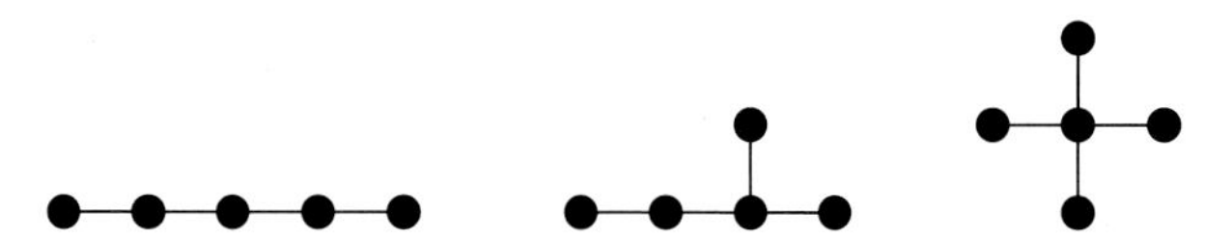

FIG. 1.8 Tree graphs with five vertices.

molecules of length $20V$ can be estimated by the number of distinct tree graphs on V vertices. For example, there are three trees on five vertices, shown in Fig. 1.8. Thus, this approach estimates that RNA molecules of length 100 nt will form one of only three distinct secondary structures.

To understand the significance of this estimate, consider the number of possible RNA primary structures of a given length. Since each nucleotide may be one of four possible options, there are 4^N possible RNA molecules of length N nt. Thus, there are $4^{100} \approx 1.6 \times 10^{60}$ theoretical RNA primary structures of 100 nt length.

There is not a known closed-form formula for the number of unlabeled trees with n vertices. However, there are implicit descriptions of the number of trees with n vertices via generating functions (cf. [24]); and the first several terms in this enumeration can be found at [25].

Another graph invariant that researchers have used to study RNA secondary structure is graph diameter, which measures the compactness of the graph. The *diameter* of a graph G is the maximal distance between any two vertices in G. For example, in the tree graph in Fig. 1.7, the most distant vertices are on either end of the tree, separated by five edges; thus, the diameter of this graph is 5. The tree graphs in Fig. 1.7 have diameter 4, 3, and 2, respectively.

Gevertz et al. [26] used prediction algorithms to determine secondary structures for randomly generated RNA primary sequences; then they used graph diameter to analyze the resulting secondary structures. They found that, for a given sequence length, the secondary structure tended toward trees with large diameters; that is, large diameter trees occurred at significantly greater rates than small diameter trees. This suggests that RNA folding patterns favor simpler, less compact structures.

Researchers have also used a number of other graph invariants to understand RNA secondary structure. For example, Gan et al. [21] applied the spectrum of a graph (i.e., the eigenvalues of the Laplacian matrix associated with a graph) to RNA secondary structure. A graph's spectrum gives information about its connectivity. (In particular, the second largest eigenvalue is generally inversely proportional to graph diameter and thus provides another measure of graph compactness.) Haynes et al. [27] analyzed secondary structures using various domination numbers (which measure the minimal size of vertex sets with various connectivity properties). Both research teams used these statistics, along with machine learning techniques, to make predictions about which trees are most likely to correspond to actual RNA secondary structures.

1.3.2.3 RNA Dual Graphs

RNA dual graphs were introduced by Gan et al. [21]; they provide another graph-theoretic model of the relationships between RNA secondary structural elements, but which reverses the roles of vertices and edges used by tree graphs. In this model:

- Vertices represent stems made up of two or more base pairs.
- Two vertices are connected if the corresponding stems are connected via a single-strand region of the RNA molecule.

For example, Fig. 1.9 shows the dual graphs for Figs. 1.3 and 1.5, respectively.

Note that these graphs may have more than one edge connecting a pair of vertices; in graph theory, such structures are often referred to as multigraphs. As shown, dual graphs can represent pseudoknots. Also, dual graphs—which do not capture information about the 3′ and 5′ ends—are not constrained to secondary structures where the 3′ and 5′ ends are connected to the same stem. This allows dual graphs to model all possible RNA secondary structures. Thus, dual graphs are a more broadly applicable model of RNA structures than tree graphs. On the other hand, tree graphs have been more fully studied from a mathematical perspective and thus have a richer mathematical toolkit.

As with tree graphs, dual graphs preserve only information about the relationships between secondary structural elements; and researchers have considered variations of the dual graph to address this. For example, Gan et al. [21] labeled edges by arrows (creating digraphs) to indicate the 5′/3′ orientation of the RNA molecule and labeled vertices to distinguish the stems attached to the 3′ and 5′ ends. Karklin et al. [28] labeled vertices by the number of base pairs in the corresponding stem and edges by the number of bases in the corresponding single-strand region.

1.3.2.4 Online RNA Resources

As knowledge and data about RNA structure has grown, researchers have developed many different online tools for accessing RNA data, making structure predictions, and visualizing RNA structures. Here we highlight a few that are most relevant to this chapter's content.

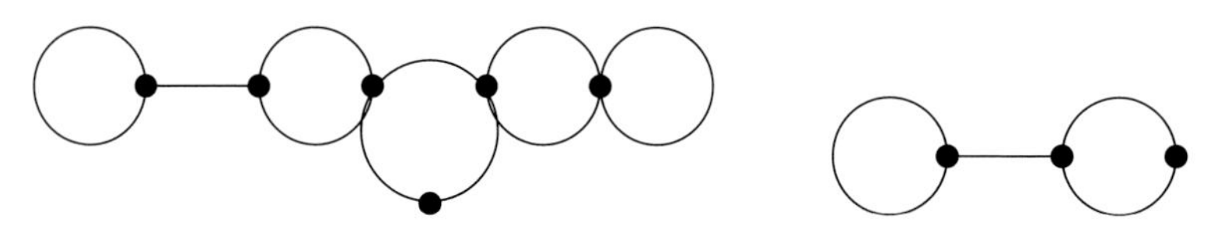

FIG. 1.9 Dual graphs for 5S ribosomal RNA for *Bacillus subtilis* (Fig. 1.3) and RNA fragment from the turnip yellow mosaic virus (Fig. 1.5).

RNA Strand [29] is a database of known RNA secondary structures. The entry for a given RNA secondary structure includes an airport diagram (as well as the underlying data describing the secondary structure, in a variety of formats), along with other biological information. One can also search the database by topological features of the secondary structure, specifying, for example, the number of hairpin loops. In connection with this chapter, one could use this database as a source of RNA secondary structures to model.

The RNA-As-Graphs [30, 31] database is specifically focused on graph-theoretic models of RNA secondary structure. In the tree graph database, tree graph models are organized by vertex count and information from the graph spectrum. Moreover, for each graph, the database includes information on known RNA with this secondary structure. The site also includes tools for computing the spectral data (Laplacian matrix and associated eigenvalues) for a graph associated with an RNA secondary structure, and for finding common subgraph structures for two RNA secondary structures.

1.3.3 Homework Problems and Projects

1. Create tree graphs for each airport diagram. (The RNA examples here have been retrieved from RNA Strand [29]. Citations indicate the original source of the molecule description. The visualizations were done with Jviz.Rna [16].)

   a. Yeast phenylalanine transfer RNA [32]

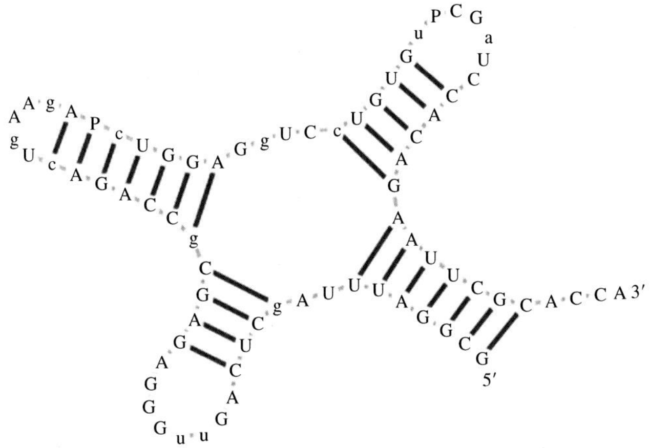

   b. RNase P RNA, *Schizosaccharomyces japonicus* [33]

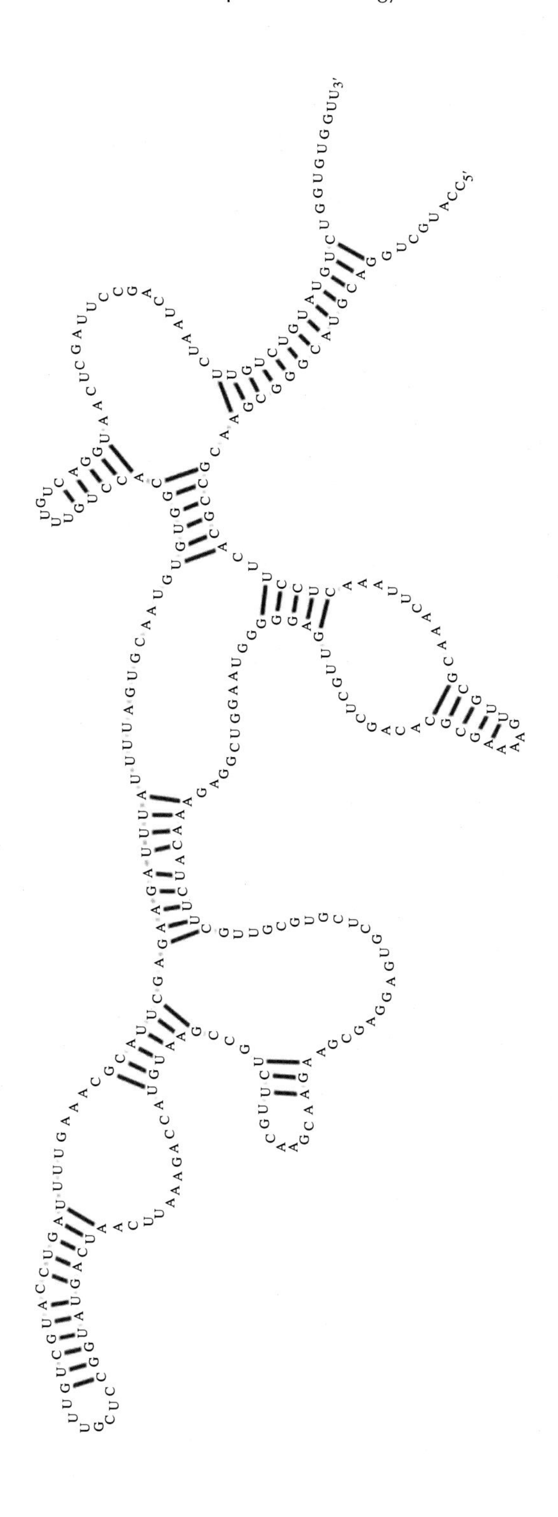

2. Create dual graphs for the RNA structures in Exercise 1.
3. Create dual graphs for each airport diagram. (The RNA examples here have been retrieved from RNA Strand [29]. Citations indicate the original source of the molecule description. The visualizations were done with Jviz.Rna [16].)
   a. RNA pseudoknot from simian retrovirus type-1 [34].

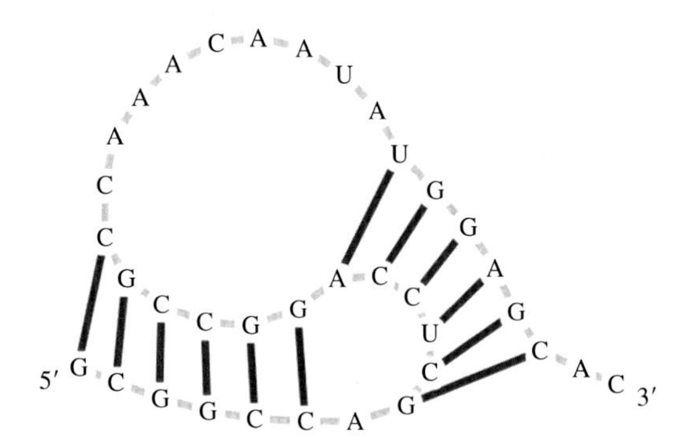

   b. A synthetic RNA [35]

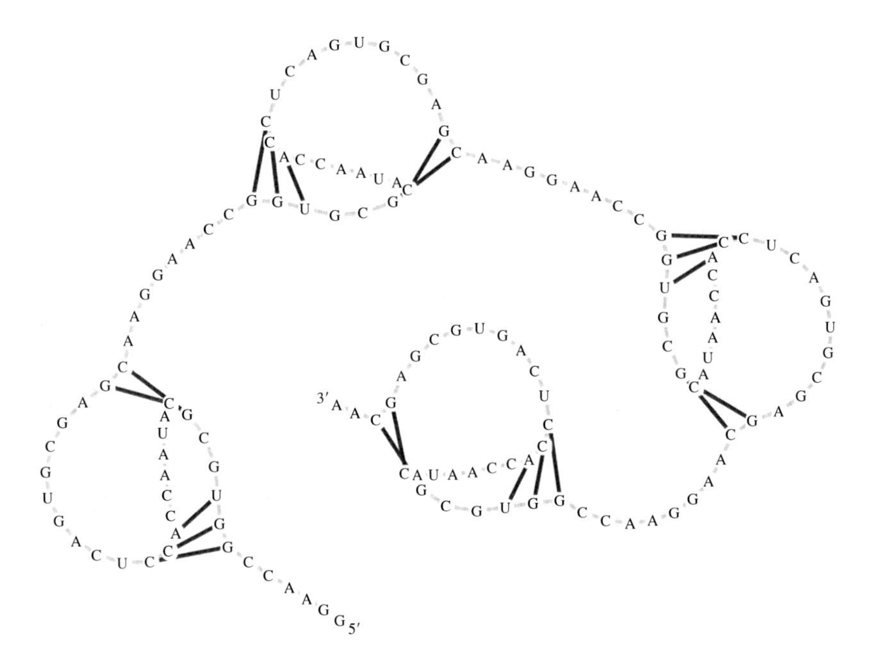

4. Determine all tree graphs with $n = 6$ vertices.
5. Determine all dual graphs with $n = 2$ vertices.

6. *Euler's formula*, a result in graph theory, relates the number of edges and vertices in a tree graph T: $|V(T)| - 1 = |E(T)|$.
   a. Verify this is true in the examples in Fig. 1.8.
   b. Give a general justification for this relationship.
   c. Interpret this in terms of RNA secondary structure: What is the relationship between stems, loops/bulges, and junctions?
7. In Section 1.3.2.2, we compared estimates for the number of secondary structures and primary structures for RNA molecules of length 80 nt. Generalize this by making a table comparing estimates for 20, 40, 60, 80, and 100 nt.
8. Determine the diameter of each tree in Exercise 4.
9. Consider a tree graph on n vertices. Determine formulas for the diameter of each class of tree.
   a. A path

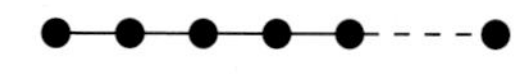

   b. A star graph

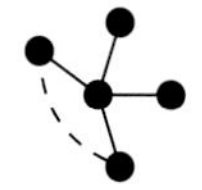

10. Project: Is it possible to determine which vertices in a tree graph correspond to hairpin loops? If it is, come up with a graph-theoretic description of them. If not, explain why and/or provide examples that illustrate this. Repeat for bulges, internal loops, junctions, and ends.
11. Project: Is it possible to detect a hairpin loop in a dual graph? If it is, come up with a graph-theoretic description. If not, explain why and/or provide examples that illustrate this. Repeat for bulges, internal loops, junctions, and pseudoknots.
12. Project: Develop a translation between dual graph and tree graph models for RNA secondary structures. Does a tree graph uniquely determine a dual graph, and vice versa? Develop a process for doing this, to the extent that it is possible, and discuss the limitations of such a translation.
13. Project: Section 1.3 focuses on using unlabeled, unrooted trees to model RNA secondary structure. However, rooted and/or labeled trees can also be used to model important aspects of RNA secondary structure. Explore this idea and how it affects the estimates of the number of possible RNA secondary structure techniques. As part of this exploration, consider the following:

a. Select biologically relevant information to use as vertex labels. What labels did you choose and why? What other choices might you have made?

b. Compare counts for trees, rooted trees, and labeled trees. (Work through some examples for small numbers of vertices. Research the formulas for these counts for large values.)

14. Project: Repeat the results of Gevertz et al. [26]: Use a random number generator to randomly generate RNA sequences of length 80 nt. Then use an online folding program such as RNAfold [36] to predict the secondary structure, create the appropriate tree models, and count the number of vertices. Examine the distribution you get.

For additional examples where these topics have been addressed in undergraduate research experience opportunities, please see [37, 38].

1.4 RNA STRUCTURE AND MATCHINGS

In Figs. 1.3 and 1.4, we saw that RNA secondary structures can be represented as arc diagrams. Informally, one can imagine pulling on the $5'$ and $3'$ ends in Fig. 1.3 and if the hydrogen bonds are stretchy, obtaining the arc diagram in Fig. 1.4. These arc diagrams are referred to as *partial matchings* by mathematicians. For simplicity, we will only consider *complete* matchings (i.e., matchings where all vertices are incident to an edge), and therefore contain n edges. We represent them as $2n$ points drawn along a horizontal line (the backbone) and arcs drawn between pairs of points represent the nucleotide bonds. Actual RNA strands always have nucleotides that are not bonded, but these structures can be reconstructed by adding any number of isolated vertices to a complete matching. We notate the set of complete matchings with n edges by $M(n)$. Note that the backbone of the RNA forms the backbone of the matching.

In a complete matching, the first vertex could be connected to any of the other $2n - 1$ vertices. Then the next leftmost vertex could be connected to any of the remaining $2n - 3$ vertices. In this manner, we see that $|M(n)| = (2n - 1)(2n-3)\cdots(3)(1)$. This is commonly denoted $(2n-1)!!$ But notice in a complete matching, we could have crossing arcs. Recall that in the example from Fig. 1.3, the arcs did not cross.

Matchings that contain no crossing pairs of edges are called *noncrossing matchings*. The number of noncrossing matchings with n edges is well-known to be counted by the nth Catalan number, $C_n = \frac{1}{n+1}\binom{2n}{n}$, and has been studied in several contexts, including pattern avoidance [39].

However, as noted above, one weakness of these noncrossing matchings is that they do not permit pseudoknots, which play critical roles in gene expression, and are of particular importance in viruses. For example, both a hepatitis B virus (HBV) and the SARS virus have catalytically active pseudoknots in their secondary structure [40].

In this section, we will develop terminology to describe pseudoknots as they appear in matchings in order to define several models for RNA secondary structure that allow, but severely limit, the number of pseudoknots present.

For a matching M, we denote the set of edges $E(M)$ and label these edges with $[n]$ in increasing order from left to right by their left endpoints. For the edge labeled i, we write $i = (i_1, i_2)$ where i represents the label of the edge and i_1, i_2 represent the position of the left and right endpoints of the edge, respectively.

A pair of edges $i = (i_1, i_2)$ and $j = (j_1, j_2)$ are said to be *nested* if $i_1 < j_1 < j_2 < i_2$ and *crossing* if $i_1 < j_1 < i_2 < j_2$. For a matching $M \in M(n)$, let $ne(M)$ and $cr(M)$ denote the number of pairs of nested edges and crossing edges in M, respectively.

Additionally, the edges $i = (i_1, i_2)$ and $j = (j_1, j_2)$ are said to form a *hairpin* if $j_1 = i_1 + 1$ and $j_2 = i_2 + 1$ (Fig. 1.10A). Note that this mathematical use of hairpin is unrelated to the biological definition of hairpin described earlier! In fact, a hairpin in a matching corresponds to a pseudoknot in an RNA secondary structure. For the rest of Section 1.4, when we use the term hairpin, we are referring to the mathematical usage. We will often place pseudoknot in parentheses after its use in order to help the reader. A nested sequence of edges that can be drawn above the backbone is called a *ladder* (Fig. 1.10B). Note that a ladder in a matching could arise as the result of a duplex, stem, junction, or branch in the RNA secondary structure.

Several matching models that allow pseudoknots have been developed for RNA secondary structures. They are often named after scientists who studied them, and include: Largest Hairpin Family (LHF), Dirks and Pierce (D&P), Reeder and Giegerich (R&G), Cao and Chen (C&C), and Lyngsø and Pedersen (L&P) [41]. They are built inductively from some small starting structure, with, in general, the two allowable operations being insertion of another matching within the current matching, or inflation of an edge by a matching. To inflate an edge by a ladder, we simply replace the current edge by a ladder of size $k > 1$. To insert a matching, we consider the position we would like to insert at, create the appropriate number of new vertices there, and reproduce the matching we are inserting beneath the matching we already had.

For example, the matching in Fig. 1.10B can be obtained from a single edge in two ways: we could have inflated the single edge by a ladder of four edges, or we could have inserted a ladder of three edges under the single edge

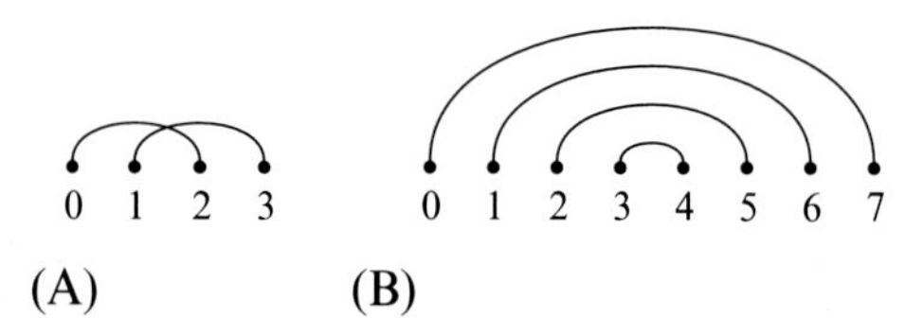

FIG. 1.10 (A) The matching is a hairpin (pseudoknot) and (B) the matching is a ladder of four edges.

we began with. Thus the same matching can be obtained by using the two operations in more than one way. On the other hand, the matching in Fig. 1.10A cannot be obtained from a single edge by inflating that edge by a ladder or by inserting a matching under a single edge. Thus, the two processes for creating new matchings limit the number of new pseudoknots.

For a more complicated example, examine Fig. 1.11A. This can be constructed by beginning with a hairpin (psuedoknot), and first inflating the left edge of the hairpin by a ladder of one edge. Next, a single edge can be inserted just after the left endpoint of the leftmost edge. Then another single edge can be inserted after the rightmost endpoint of the final edge. This inductive construction was not unique in this case, since the last two matching insertions could have been done in either order.

The matching insertion operation is biologically motivated. Inserting a matching below the current matching corresponds to breaking the RNA backbone between two nucleotides and inserting a strand between them—an *indel* operation. Inflating an edge by a ladder just increases the number of consecutively bonded base pairs in the corresponding stem, duplex, branch, etc. in the secondary structure, which only increases the stability of the structure.

We now describe two of these pseudoknot-allowing, inductively constructed families here, the two families with the most restrictive conditions on pseudoknots: the L&P family and the C&C family.

1.4.1 L&P Matchings

Matchings in the L&P (Lyngso and Pedersen) family can be constructed inductively by starting from either a hairpin or a single edge, and either (a) inflating an edge by a ladder on k edges or (b) inserting a noncrossing matching from below between two vertices of an L&P matching [41]. For an example of an L&P matching, see Fig. 1.12. This matching can be constructed as follows.

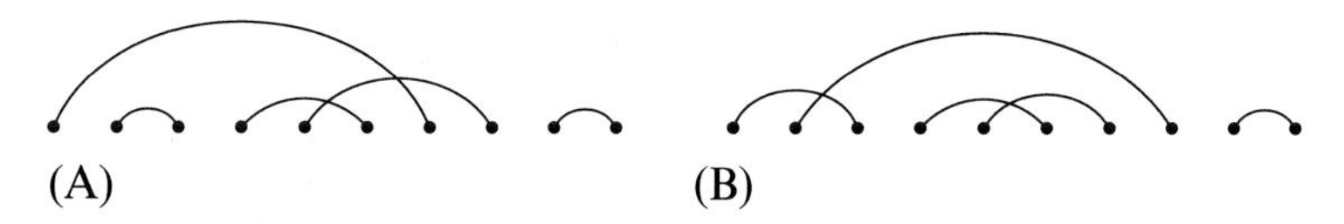

FIG. 1.11 Two matchings containing pseudoknots; (A) L&P and (B) not L&P. Both are C&C matchings.

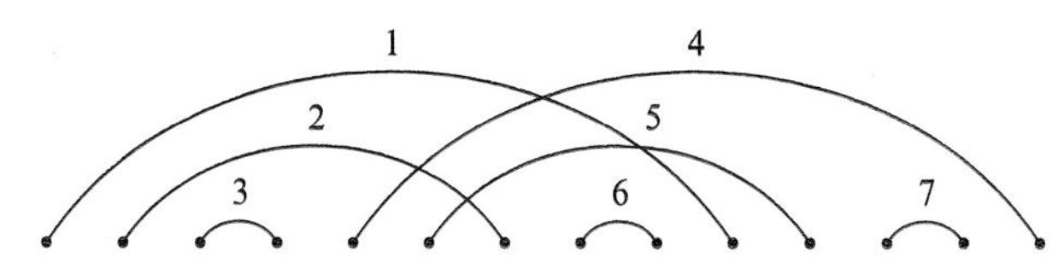

FIG. 1.12 An L&P matching.

1. Begin with a hairpin (pseudoknot) (edges 1 and 4)
2. Inflate each edge by a ladder (add edges 2 and 5)
3. Insert a single edge three times (add edges 3, 6, and 7)

The sequence for the number of L&P matchings with n edges begins 1, 3, 12, 51, 218, 926, 16,323, 67,866, 280,746 and is given by the formula $2 \cdot 4^{n-1} - \frac{3n-1}{2n+2}\binom{2n}{n}$ [41].

The process of constructing an L&P matching implies that such a matching contains a crossing precisely if the matching can be built inductively from a hairpin. As a result, any L&P matching that contains a crossing will have all crossings occur in an inflated hairpin. Given a matching M, we will label edges by left endpoint, as in Fig. 1.12. In this figure, edges 1 and 2 together with their crossing edges 4 and 5 comprise an inflated hairpin. Below we provide a precise definition of this structure, as given by Martinez and Riehl [41a].

Definition 1.1. A *maximal inflated hairpin* in an L&P matching is two sets of edges $A = \{a_1, \ldots, a_k\}$ and $B = \{b_1, \ldots, b_\ell\}$ such that

1. every pair of edges in A and every pair of edges in B is nested,
2. every $a_i \in A$ crosses every edge in B, and
3. every crossing in M occurs between edges in A and B.

Note that condition 3 ensures that the sets A and B are maximal. We let A be the set of edges with smaller labels (the left side of the inflated hairpin), and we say that M contains the inflated hairpin (A, B).

Any L&P matching consists of a maximal inflated hairpin with noncrossing matchings inserted below and/or on either side of the hairpin as in Fig. 1.13. It is possible for the inflated hairpin to be empty, yielding a noncrossing matching. For example, the matching in Fig. 1.11A is L&P: the first and third edges crossing the fourth edge form an inflated hairpin. However, the matching in Fig. 1.11B is not L&P, since the first four edges are all involved in crossings, but these four edges do not form an inflated hairpin (pseudoknot).

Notice that every edge not in the inflated hairpin of an L&P matching must begin and end between two vertices that are adjacent in the inflated hairpin,

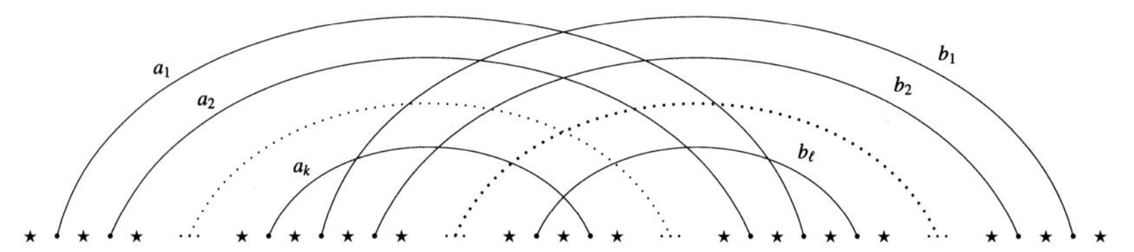

FIG. 1.13 The structure of an L&P matching: each $\star$ indicates a position where a noncrossing matching may be inserted. The edges $a_1, \ldots, a_k$ pairwise cross the edges $b_1, \ldots, b_\ell$ to form the inflated hairpin [41a].

otherwise it would be part of an inflated hairpin itself, and these in general arise from stems and hairpin loops (the biological kind) around the area of the pseudoknot.

1.4.2 The C&C Family

Starting from either a single edge or a hairpin, matchings of the *C&C*(Cao and Chen) *family* can be built inductively via the following operations: (1) insert C&C matchings in the allowable places (as shown in Fig. 1.14) and then (2) inflate the original edge(s) by ladders. The allowable places on a single edge include under the edge or to its right. The exclusion of insertion to the left is not biologically motivated, but instead helps us avoid creating the same matching multiple times. Any C&C matching that could be created by inserting to the left could also have been created by inserting to the right. The allowable insertion places if we begin with a hairpin excludes insertion to the left for the same reason as with a single edge, and also excludes insertion under the cross of the hairpin. This excludes the possibility of a pseudoknot within a pseudoknot, which has not been observed biologically. (Note that this "pseudoknot within a pseudoknot" is also not allowed in L&P matchings.)

An example of a C&C matching is shown in Fig. 1.15. To construct it, we start with a single edge and insert a hairpin (pseudoknot) under it. Two more complicated examples are shown in Fig. 1.11. The matching in Fig. 1.11A can be obtained by beginning with a hairpin, inserting a single edge between the two leftmost vertices, inserting a single edge after the rightmost vertex, and then inflating the leftmost original edge by a ladder.

The sequence for the number of C&C matchings on n edges begins 1, 3, 12, 51, 227, 1052, and 5030 [41].

Pseudoknots do appear in RNA secondary structure, and have important functional implications, but they are not exceedingly common. Both the L&P and C&C families allow pseudoknots, but their construction rules significantly limit the number there can be when compared with an arbitrary perfect matching on $2n$ vertices. Note also the differences between L&P and C&C: the former

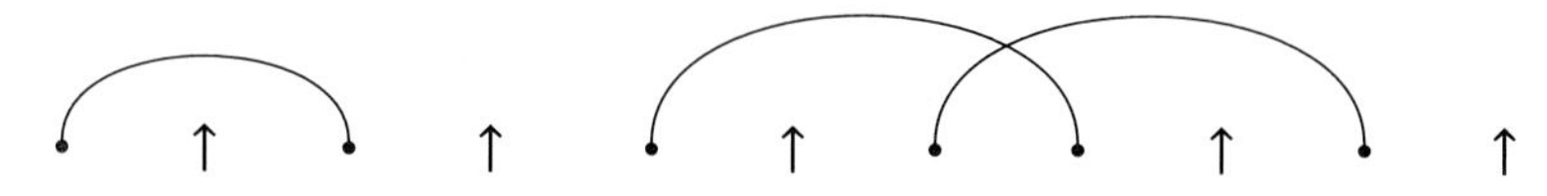

FIG. 1.14 The allowable places for insertion in the C&C family.

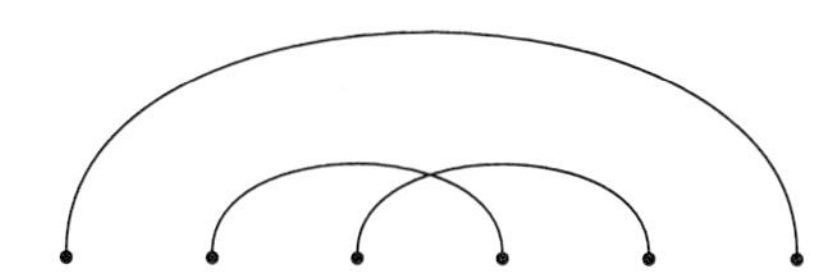

FIG. 1.15 A C&C matching on three edges.

allows only noncrossing matchings to be inserted, while the latter allows any C&C to be inserted. On the other hand, L&P contains more insertion sites than C&C. For small n, those two considerations exactly balance each other, and both sequences begin 1, 3, 12, and 51. However, for larger n, allowing more matchings to be inserted gives more freedom than having more insertion sites, and the C&C family grows faster [41].

1.4.3 Homework Problems and Projects

1. If we prohibit pseudoknots in our model, then the associated matchings will be noncrossing. How many matchings on three edges are noncrossing? How many on four edges? State and prove the recursive relationship that generates noncrossing matchings on n edges.
2. Describe the sequence of inductive steps required to create the C&C matching in Fig. 1.11B. Is this sequence unique?
3. Draw all L&P matchings with three edges. Draw all C&C matchings with three edges. How many matchings do they have in common?
4. The number of L&P matchings with n edges is given by $a(n) = 2 * 4^{(n-1)} - \binom{2n}{n}(3n-1)/(2n+2)$. The number of noncrossing matching with n edges is given by $(2n)!/(n!(n+1)!)$. How much quicker does the L&P family grow? (That is, what is the largest order term of the difference?)
5. Project: A number of statistics can be applied to matchings (e.g., crossing and nesting number). The crossing number $cr(M)$ counts the number of times a pair of edges in the matching cross. The nesting number for one edge counts the number of edges nested under it. The nesting number for a matching $ne(M)$ is the sum of the nesting numbers for each edge. Find the maximum possible crossing and nesting numbers for L&P and C&C matchings on n edges as a function of n. Compare this to the maximum crossing and nesting numbers for matchings which allow unlimited pseudoknots (called *perfect matchings*).
6. Project: We also define here a biologically motivated statistic called the pseudoknot number $pknot(M)$. A pseudoknot occurs in a strand of RNA when the strand folds on itself and forms secondary bonds between nucleotides, and then the same strand wraps around and forms secondary bonds again. However, when that pseudoknot has several nucleotides bonded in a row, we do not consider that a "new" pseudoknot. The pseudoknot number of a matching, $pknot(M)$, counts the number of pseudoknots on the RNA motif by deflating any ladders in the matching and then finding the crossing number on the resulting matching. For example in Fig. 1.16 we give two matchings containing hairpins (pseudoknots). Even though their crossing numbers both equal 6, we see that in Fig. 1.16A, these crossing arise from a single pseudoknot, and so their pknot number is 1, while in Fig. 1.16B, the pknot number is 3. Find the maximum pseudoknot number on the C&C matchings on n edges as a function of n. Compare this to the maximum pseudoknot number on all perfect matchings.

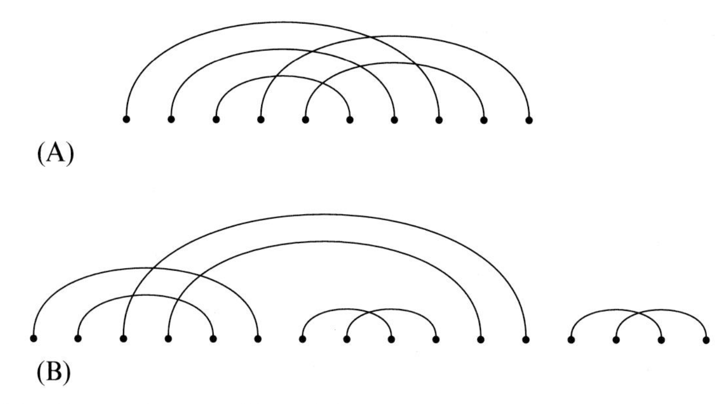

FIG. 1.16 Two matchings containing hairpins (pseudoknots), each with crossing numbers equal to 6, but (A) has a single pseudoknot while (B) has three.

7. Research question: The inductive process for generating L&P and C&C matchings uses insertion of matchings between two vertices since biologically this represents a strand of RNA being inserted into an existing RNA motif. Are there other biologically motivated methods for creating larger matchings from smaller matchings?

1.5 HIERARCHICAL PROTEIN MODELS

Proteins are sometimes referred to as macromolecules since they tend to be very large. Since one can describe a protein as being constructed by the bonding of a large number of smaller molecules, the structure of a protein molecule is frequently modeled as a hierarchical network or graph. In general, a *hierarchical network* consists of repeated copies of smaller networks that are connected to form a larger network. Given the hierarchical structure of a protein molecule, we begin with a discussion of the repeated smaller networks.

Every protein molecule is constructed by the molecular bonding of amino acids. There are 22 amino acids, but in nature we find that 20 of these are primarily used to build a protein molecule. Each amino acid can be modeled as a network, or graph. For example, the amino acid Leucine is shown in Fig. 1.17. Note that the atoms whose background has been shaded in yellow form what is known as the *residue* of the amino acid. The atoms above that are not shaded are called the backbone. The backbone structure of each of the amino acids is identical and therefore it is the residue structure that distinguishes them. A typical graph representation of Leucine is also shown in Fig. 1.17. In this model, the entire backbone structure is represented by a single node. One can consider this node as representing the central carbon atom of the backbone. The residue structure is modeled by the remaining nodes. The hydrogen atoms are typically not represented and hence this is called the hydrogen suppressed model.

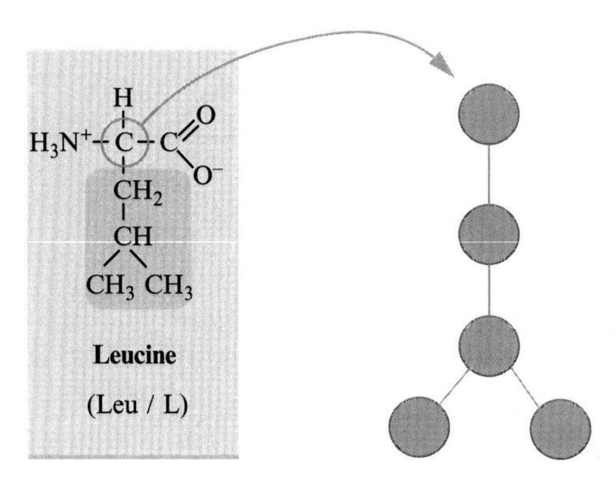

FIG. 1.17 Leucine.

Since there are 20 amino acids that are used to construct a protein and each amino acid has a single letter representation, the primary structure of a protein molecule can be given as a sequence of letters from an alphabet of size 20, called its FASTA sequence. For example, we might see:

MAGGTUVLLEFILILLVCAQPLKRLVRTQUIVNCSERKKLWTUI PYFKRRAQADE

The primary structures always begin with M. We can now see the natural hierarchical structure since we have the repeated use of smaller substructures in the molecule, namely the amino acid structures. However, a protein is not just a sequence of letters. The physical chain of amino acids bends and twists so that amino acids in the chain may form molecular bonds with other amino acids even though they are far apart on the chain. Since the chain of amino acids is the primary structure, let us now consider the secondary structure. There are several ways that amino acids can bond to other amino acids in the primary structure that are in relatively close proximity to form structures. These structures are called the secondary structures of a protein molecule (see Fig. 1.18).

Finally, the amino acids of one secondary structure may bond with the amino acids of a different substructure creating the final tertiary structure of the protein molecule. An example of a small protein molecule is given in Fig. 1.19. The secondary structures are each colored differently.

One of the earliest representations of a protein molecule by a graph was given by Koch et al. [42] who represented a β-strand as a vertex and bonds as edges. Later, Grigoriev and collaborators represented all α-helical structures in the form of connected graphs [43]. These early graph-theoretic approaches assigned nodes to represent secondary structures and the edges to represent some level of contact between them. Przytycka et al. [44] represented both α-helices and β-strands as vertices and edge connections were based on proximity in the tertiary structure. Others chose to let each atom be represented as a vertex and edges be based on proximity [45]. Each method has its advantages and disadvantages. One loses local information in the model where an entire

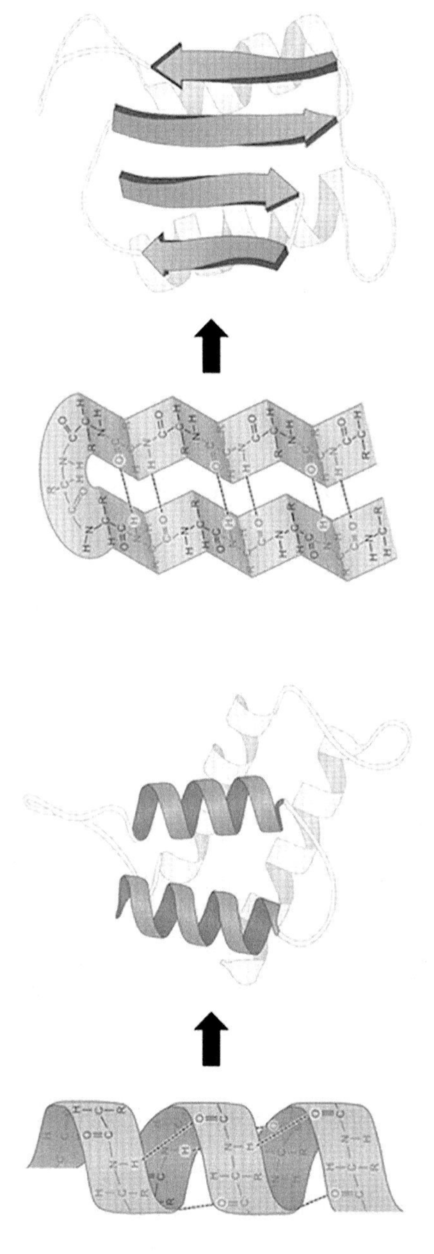

FIG. 1.18 Secondary structures: α-helix and β-sheet composed of four β-strands.

FIG. 1.19 Small protein.

substructure is represented as a vertex but gains a better global view. The converse is true for the alternative model.

In order to perform a biological function, a macromolecule must have a certain amount of flexibility. However, in order to maintain its structure, there must be a degree of rigidity. Researchers such a Jacobs have used graph-theoretic concepts of rigidity to quantify the rigidity of protein structures. With the emergence of network science, graph-theoretic models of protein structures, when viewed as networks, were found to have the characteristic properties of small-world networks [46, 47] including short average path lengths and high clustering coefficient values.

Although, the human genome has been fully sequenced, the corresponding three-dimensional structures of the proteins that each gene encodes are not all known. The Protein Data Bank (PDB) [48] is a repository of the known protein structures. For a given entry, one can retrieve the positioning of every atom in the three-dimensional structure. This information is stored and can be retrieved as a PDB file. The Protein Graph Repository [49] allows the user to submit a PDB file ID and returns the corresponding graph representation. The user may choose to let the vertices represent individual atoms or amino acids. An example is shown in Fig. 1.20 where each vertex represents an atom.

Graph representation can also be a visualization tool. Consider the protein in Fig. 1.21 and the two occurrences of amino acid K. One can see that the one in the center may be more critical to the protein's structure than the one shown at the top of the figure. Looking at the FASTA sequence, there is no indication that one of the Ks is more critical to the protein structure than another. This observation helps us understand why some amino acid changes due to a gene mutation are much more consequential than others, even though they may be seem to be equivalent changes in the primary structure.

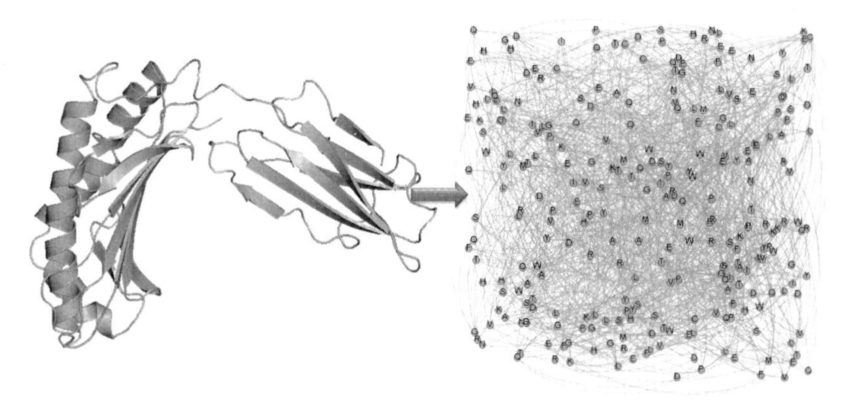

FIG. 1.20 Protein graph from the Protein Data Bank [49].

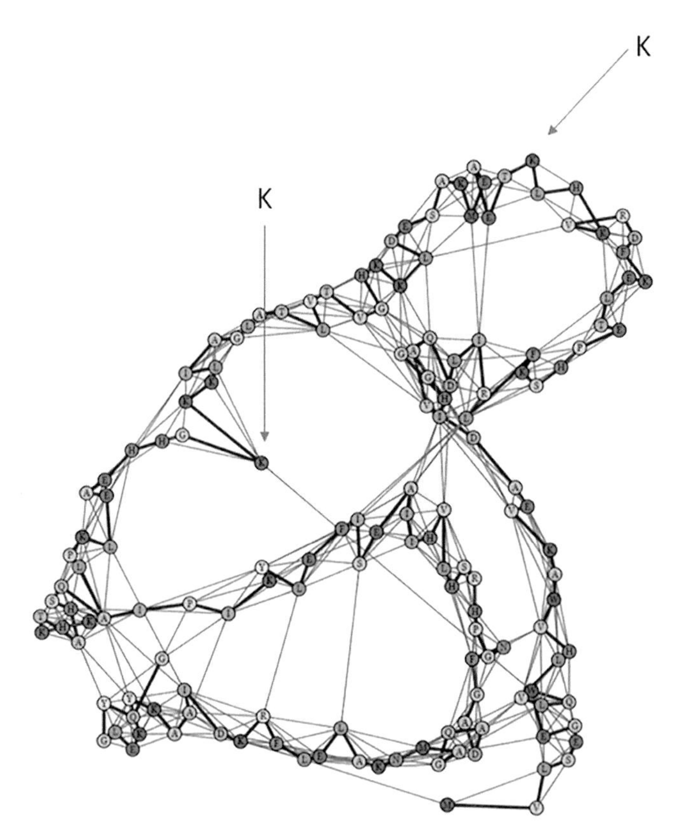

FIG. 1.21 Critical amino acid.

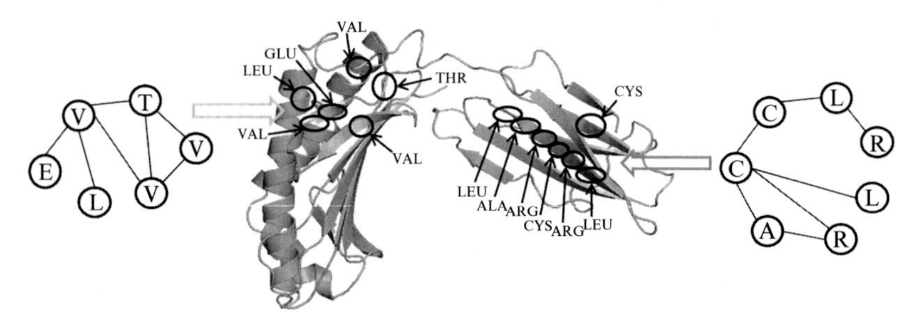

FIG. 1.22 Substructure graphs.

Using a graph to represent the molecule allows us to quantify it in numerous and meaningful ways. However, as previously discussed, we lose information about the local dynamics or the distribution of topological substructures. Regions of interest in the protein molecule can be studied as subgraphs of the protein graph, as seen in Fig. 1.22.

As previously discussed, given a large protein molecule, a change in a single amino acid can have negligible global consequences on the graphical representation. However, a change in a single amino acid can have very significant biological consequences. Therefore, to improve the graph-theoretic model, we need to create a method that is able to measure the significance of a local change on the global landscape. One approach to this is to create a hierarchical graph using graph nesting. In a nested graph, each vertex contains yet another graph. An example of a hierarchical graph with nested graphs is shown in Fig. 1.23.

At the top level, the graph is a single node. At the next level, the graph is a triangle. Embedded in each node of the triangle is a graph of an amino acid. The two amino acid graphs are representations of L (leucine) and V (valine). The graph of V appears twice. As another example, in Fig. 1.24 the top-level graph has eight nodes, $G_1, \ldots G_8$. Embedded in each node G_i is subgraphs of the protein graph. Each node of the embedded graphs is an amino acid. Of course, each amino acid is represented by a graph where each node is an atom. These repeated embeddings allow us to retain information at the atomic level within the graph that models the global interactions of substructures of the molecule.

In this case, the top-level graph is shown at the right-hand side of the figure where each vertex is labeled G_i. Notice that we also have assigned a vector of weights to each of these vertices. The weights correspond to the statistics of the corresponding nested graph. This method is iteratively applied. The vertices of the mid-level graphs each represent amino acids and the weights correspond to the nested graph of the nested amino acid graph. In order to explain how these weights are calculated and then applied, we need to discuss weighted graphical invariants.

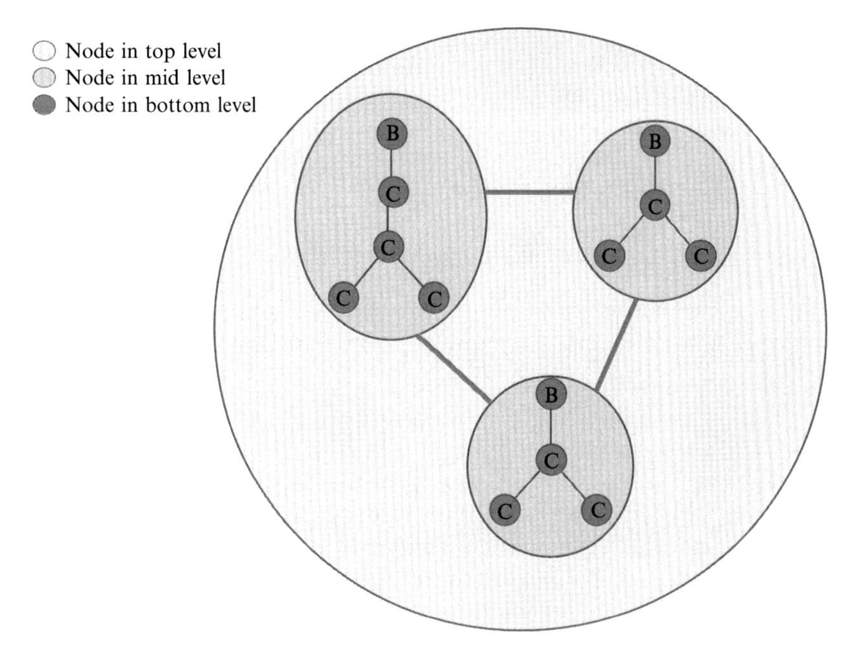

FIG. 1.23 Hierarchical graph.

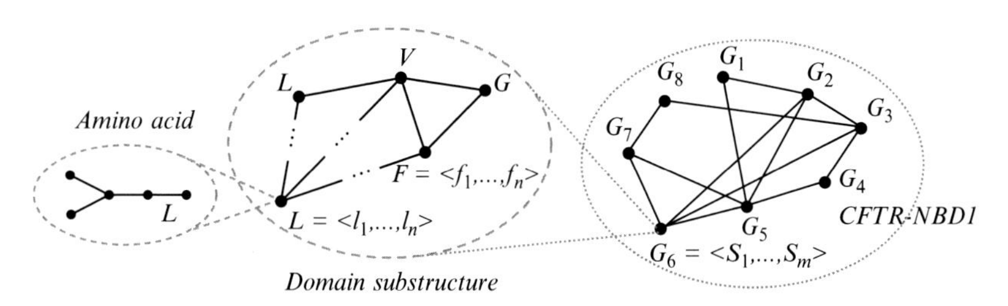

FIG. 1.24 Hierarchical graph levels.

1.5.1 Weighted Graph Invariants

In Fig. 1.25, we see the graphs of two amino acids, E and Q. Their residue structures are identical, but the atomic composition is slightly different. In particular, they differ by exactly one atom. The amino acid E has two oxygen atoms bound to its periphery whereas Q has an oxygen and a nitrogen. Therefore, to distinguish between the two residues, we must incorporate this difference into our model. We can make use of the fact that oxygen and carbon have different properties. For example, the values of their masses differ, so we can assign vertex weights to obtain different graphical invariant values.

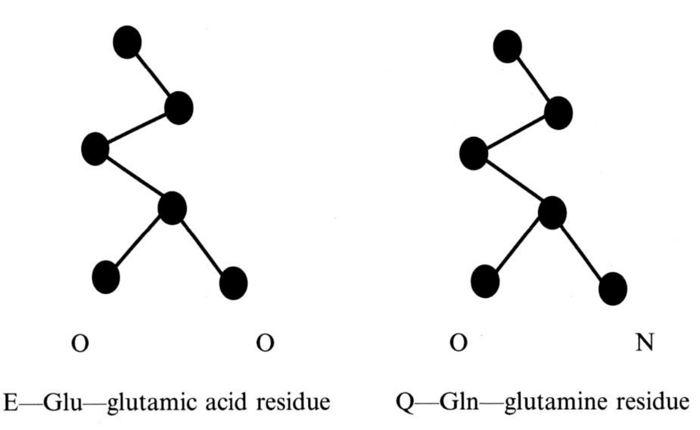

FIG. 1.25 Quantifying residue chains.

Recall that the degree of a vertex is the number of edges incident to it. Since every edge must have two vertices that define it, an equivalent definition for the degree of a vertex v is the number of neighbors or the number of vertices adjacent to it. By counting the number of neighbors, we are assuming each neighboring vertex contributes exactly 1 to the count. However, since vertices represent biomolecules in our model and these biomolecules are not equivalent, it is necessary to incorporate these differing properties by weighting the corresponding vertices. In order to quantify the biomolecule with standard graph-theoretic definitions of graph theory, we modify those definitions to include the vertex weights. For example, see Fig. 1.26.

We can also generalize the domination number to be the minimum weight of a dominating set among all dominating sets rather than the minimum cardinality. In the example in Fig. 1.27, the two black vertices are a minimum dominating set with a weight sum of 5. However, suppose that the remaining vertices each have a weight of 1. They too form a dominating set, and while they do not have minimum cardinality, their weight sum is 4. Consequently, the domination number is 4, not 5. Given that most mathematically defined graphical invariants utilize the cardinality of a set of vertices, a significant number of the invariants can be modified to incorporate weighted vertices. This allows us to quantify properties at the local level and store this information as vertex weights in the next level. For further discussions on this, see [50], and for an application, see [51].

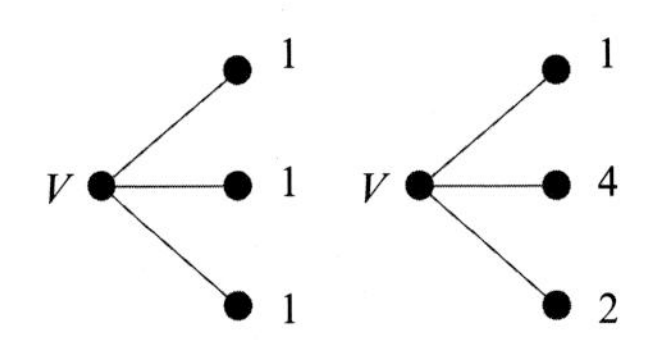

FIG. 1.26 When the vertices are not weighted we can assume the weights are 1. For example, the vertex v has degree 3 and weighted degree 7.

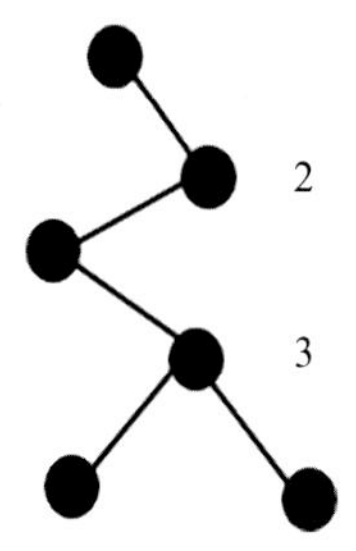

FIG. 1.27 If the vertices whose weights are not shown have an assumed weight of 1, then the weighted domination number is 4.

1.5.2 Homework Problems and Projects

1. Consider all of the distinct tree graphs on six vertices. For each tree, label the vertices using the integers n where $0 < n < 10$. Find the following:
   a. The maximum weighted degree of each tree
   b. The weighted domination number of each tree
2. Label the trees in problem 1 with labels 1–5. Construct a cycle on five vertices and label the vertices 1–5. Assign each tree in problem 1 to a vertex in the cycle and then connect at least two vertices that are not already connected in the cycle. Assign the maximum weighted degree measure of the corresponding tree to the vertex in this graph. Using this vertex weighted graph, find the weighted domination number.
3. Determine the graph of each of the 20 amino acids using the method shown in Fig. 1.25. Label the vertices with the corresponding atom and choose a method based on the atomic structure of the residue to weight the vertices. Find the weighted domination number of each amino acid graph.
4. Project: With input from a molecular biologist, select a set of 6–10 proteins that are similar in structure and function whose three-dimensional structure is known. For each protein, partition the molecule into domains. In each domain, determine the graph representation letting each amino acid be a vertex and edges for pairs of amino acids are within a distance of seven angstroms in the three-dimensional structure. Label the vertices in the domain graph with the corresponding value of the amino acid graph such as those found in problem 3. Using this nested graph approach, determine values for each of the proteins and use these values to reveal which of the proteins are most similar, based on these graph-theoretic measures.

REFERENCES

[1] E. Zuckerkandl, L. Pauling, Molecules as documents of evolutionary history, J. Theor. Biol. 8 (2) (1965) 357–366.

[2] R.W. Holley, G.A. Everett, J.T. Madison, A. Zamir, Nucleotide sequences in the yeast alanine transfer ribonucleic acid, J. Biol. Chem. 240 (5) (1965) 2122–2128.

[3] BioVis Design Contest, Available from: http://biovis.net/2017/index.html (Accessed 20 October 2017).
[4] W.C. Ray, R.W. Rumpf, B. Sullivan, N. Callahan, T. Magliery, R. Machiraju, B. Wong, M. Krzywinski, C.W. Bartlett, Understanding the sequence requirements of protein families: insights from the BioVis 2013 contests, in: BMC Proceedings, vol. 8, BioMed Central, 2014, p. S1.
[5] Protein Structure Prediction Center, Available from: http://www.predictioncenter.org/ (Accessed 23 July 2018).
[6] J. Moult, A decade of CASP: progress, bottlenecks and prognosis in protein structure prediction, Curr. Opin. Struct. Biol. 15 (3) (2005) 285–289.
[7] B. Berger, T. Leighton, Protein folding in the hydrophobic-hydrophilic (HP) model is NP-complete, J. Comput. Biol. 5 (1) (1998) 27–40.
[8] R.B. Lyngsø, C.N. Pedersen, RNA pseudoknot prediction in energy-based models, J. Comput. Biol. 7 (3–4) (2000) 409–427.
[9] eteRNA, Available from: http://www.eternagame.org/web/ (Accessed 20 October 2017).
[10] J. Anderson-Lee, E. Fisker, V.K.M. Wu, J. Kong, J. Lee, M. Lee, M. Zada, A. Treuille, R. Das, Principles for predicting RNA secondary structure design difficulty, J. Mol. Biol. 428 (5) (2016) 748–757.
[11] foldIT, Available from: https://fold.it/ (Accessed 20 October 2017).
[12] G.A. Pavlopoulos, P. Kumar, A. Sifrim, R. Sakai, M.L. Lin, T. Voet, Y. Moreau, J. Aerts, Meander: visually exploring the structural variome using space-filling curves, Nucleic Acids Res. 41 (11) (2013) e118.
[13] C.W. Geary, E.S. Andersen, Design principles for single-stranded RNA origami structures, in: International Workshop on DNA-Based Computers, Springer, 2014, pp. 1–19.
[14] D.P. Aalberts, W.K. Jannen, Visualizing RNA base-pairing probabilities with RNAbow diagrams, RNA 19 (4) (2013) 475–478.
[15] K. Sato, M. Hamada, K. Asai, T. Mituyama, CENTROIDFOLD: a web server for RNA secondary structure prediction, Nucleic Acids Res. 37 (suppl 2) (2009) W277–W280.
[16] jVis.Rna, Available from: https://jviz.cs.sfu.ca/ (Accessed 20 October 2017).
[17] C.M. Reidys, F.W. Huang, J.E. Andersen, R.C. Penner, P.F. Stadler, M.E. Nebel, Topology and prediction of RNA pseudoknots, Bioinformatics 27 (8) (2011) 1076–1085.
[18] J. Ritz, J.S. Martin, A. Laederach, Evolutionary evidence for alternative structure in RNA sequence co-variation, PLoS Comput. Biol. 9 (7) (2013) e1003152.
[19] K.M. Kutchko, A. Laederach, Transcending the prediction paradigm: novel applications of SHAPE to RNA function and evolution, Wiley Interdisciplinary Rev. RNA 8 (1) e1374, https://doi.org/10.1002/wrna.1374.
[20] J.A. Cuesta, S. Manrubia, Enumerating secondary structures and structural moieties for circular RNAs, J. Theor. Biol. 419 (2017) 375–382.
[21] H.H. Gan, S. Pasquali, T. Schlick, Exploring the repertoire of RNA secondary motifs using graph theory; implications for RNA design, Nucleic Acids Res. 31 (11) (2003) 2926–2943.
[22] S.-Y. Le, R. Nussinov, J.V. Maizel, Tree graphs of RNA secondary structures and their comparisons, Comput. Biomed. Res. 22 (5) (1989) 461–473.
[23] C. Heitsch, S. Poznanoviä, Combinatorial insights into RNA secondary structure, in: Discrete and Topological Models in Molecular Biology Natural Computing Series, 2013, p. 145166, https://doi.org/10.1007/978-3-642-40193-0_7.
[24] F. Harary, E.M. Palmer, Graphical Enumeration, Academic Press, New York, NY, 1973.
[25] Sequence A000055, Available from: http://oeis.org/A000055, 2018.

[26] J. Gevertz, H.H. Gan, T. Schlick, In vitro RNA random pools are not structurally diverse: a computational analysis, RNA 11 (6) (2005) 853–863.

[27] T. Haynes, D. Knisley, E. Seier, Y. Zou, A quantitative analysis of secondary RNA structure using domination based parameters on trees, BMC Bioinf. 7 (1) (2006) 108.

[28] Y. Karklin, R.F. Meraz, S. Holbrook, Classification of non-coding RNA using graph representations of secondary structure, in: R.B. Altman, T.A. Jung, T.E. Klein, A.K. Dunker, L. Hunter (Eds.), Biocomputing 2005: Proceedings of the Pacific Symposium, Hawaii, USA, January 4–8, 2005, 2005, pp. 4–15.

[29] M. Andronescu, V. Bereg, H.H. Hoos, A. Condon, RNA Strand: the RNA secondary structure and statistical analysis database, BMC Bioinf. 9 (1) (2008) 340.

[30] H.H. Gan, D. Fera, J. Zorn, N. Shiffeldrim, M. Tang, U. Laserson, N. Kim, T. Schlick, RAG: RNA-As-Graphs database—concepts, analysis, and features, Bioinformatics 20 (8) (2004) 1285–1291, https://doi.org/10.1093/bioinformatics/bth084.

[31] J.A. Izzo, N. Kim, S. Elmetwaly, T. Schlick, RAG: an update to the RNA-As-Graphs resource, BMC Bioinf. 12 (1) (2010) 219.

[32] L. Jovine, S. Djordjevic, D. Rhodes, The crystal structure of yeast phenylalanine tRNA at 2.0 Å resolution: cleavage by Mg^{2+} in 15-year old crystals, J. Mol. Biol. 301 (2) (2000) 401–414.

[33] J.W. Brown, The ribonuclease P database, Nucleic Acids Res. 27 (1) (1999) 314, https://doi.org/10.1093/nar/27.1.314.

[34] P.J.A. Michiels, A. Versleyen, C.W.A. Pleij, C.W. Hilbers, H.A. Heus, Solution structure of the pseudoknot of SRV-1 RNA, involved in ribosomal frameshifting, J. Mol. Biol. 310 (5) (2001) 1109–1123, https://doi.org/10.2210/pdb1e95/pdb.

[35] D. Sussman, J.C. Nix, C. Wilson, Molecular recognition by the vitamin B12 RNA aptamer, Nat. Struct. Mol. Biol. 7 (1) (2000) 53, https://doi.org/10.2210/pdb1ddy/pdb.

[36] A.R. Gruber, R. Lorenz, S.H. Bernhart, R. Neuböck, I.L. Hofacker, The vienna RNA website, Nucleic Acids Res. 36 (suppl 2) (2008) W70–W74.

[37] R. Ellington, J. Wachira, A. Nkwanta, RNA secondary structure prediction by using discrete mathematics: an interdisciplinary research experience for undergraduate students, CBE Life Sci. Educ. 9 (3) (2010) 348–356.

[38] R. Willenbring, RNA secondary structure, permutations, and statistics, Discret. Appl. Math. 157 (7) (2009) 1607–1614.

[39] J. Bloom, S. Elizalde, Pattern avoidance in matchings and partitions, 2012, arXiv preprint arXiv:1211.3442.

[40] I. Brierley, S. Pennell, R.J.C. Gilbert, Viral RNA pseudoknots: versatile motifs in gene expression and replication, Nat. Rev. Microbiol. 5 (8) (2007) 598–611.

[41] A.F. Jefferson, The Substitution Decomposition of Matchings and RNA Secondary Structures (Ph.D. thesis), University of Florida 2015, https://search.proquest.com/docview/1726910322 (Accessed 23 July 2018).

[41a] M. Martinez, M. Riehl, A bijection between the set of nesting-similarity classes and L&P matchings. Discret. Math. Theor. Comput. Sci. 19 (2), (2018).

[42] I. Koch, F. Kaden, J. Selbig, Analysis of protein sheet topologies by graph theoretical methods, Proteins Struct. Funct. Bioinf. 12 (4) (1992) 314–323.

[43] I.V. Grigoriev, A.A. Mironov, A.B. Rakhmaninova, Interhelical contacts determining the architecture of alpha-helical globular proteins, J. Biomol. Struct. Dyn. 12 (3) (1994) 559–572.

[44] T. Przytycka, R. Srinivasan, G.D. Rose, Recursive domains in proteins, Protein Sci. 11 (2) (2002) 409–417.

[45] D.J. Jacobs, A.J. Rader, L.A. Kuhn, M.F. Thorpe, Protein flexibility predictions using graph theory, Proteins Struct. Funct. Bioinf. 44 (2) (2001) 150–165.

[46] A.R. Atilgan, P. Akan, C. Baysal, Small-world communication of residues and significance for protein dynamics, Biophys. J. 86 (1) (2004) 85–91.
[47] M. Vendruscolo, N.V. Dokholyan, E. Paci, M. Karplus, Small-world view of the amino acids that play a key role in protein folding, Phys. Rev. E 65 (6) (2002) 061910.
[48] Protein Data Bank, Available from: https://www.pdb.org (Accessed 31 October 2017).
[49] W. Dhifli, A.B. Diallo, ProtNN: fast and accurate protein 3D-structure classification in structural and topological space, BioData Mining 9 (1) (2016) 30.
[50] D.J. Knisley, J.R. Knisley, Seeing the results of a mutation with a vertex weighted hierarchical graph, in: BMC Proc., vol. 8, 2014, p. S7.
[51] D. Knisley, J. Knisley, Predicting protein-protein interactions using graph invariants and a neural network, Comput. Biol. Chem. 35 (2) (2011) 108–113.

FURTHER READING

[52] I.L. Hofacker, Vienna RNA secondary structure server, Nucleic Acids Res. 31 (13) (2003) 3429–3431.
[53] J.J. Cannone, S. Subramanian, M.N. Schnare, J.R. Collett, L.M. D'Souza, Y. Du, B. Feng, N. Lin, L.V. Madabusi, K.M. Müller, et al., The comparative RNA web (CRW) site: an online database of comparative sequence and structure information for ribosomal, intron, and other RNAs, BMC Bioinf. 3 (1) (2002) 2.
[54] K.C. Wiese, E. Glen, jViz.Rna—an interactive graphical tool for visualizing RNA secondary structure including pseudoknots, in: 19th IEEE International Symposium on Computer-Based Medical Systems, 2006. CBMS 2006, ISSN 1063-7125, 2006, pp. 659–664, https://doi.org/10.1109/CBMS.2006.104.
[55] M.H. Kolk, M. van der Graaf, S.S. Wijmenga, C.W.A. Pleij, H.A. Heus, C.W. Hilbers, NMR structure of a classical pseudoknot: interplay of single- and double-stranded RNA, Science 280 (5362) (1998) 434–438.

Chapter 2

Tile-Based DNA Nanostructures

Mathematical Design and Problem Encoding

Joanna Ellis-Monaghan*, **Nataša Jonoska**† **and Greta Pangborn**‡
**Department of Mathematics, Saint Michael's College, Colchester, VT, United States, †Department of Mathematics and Statistics, University of South Florida, Tampa, FL, United States, ‡Department of Computer Science, Saint Michael's College, Colchester, VT, United States*

2.1 INTRODUCTION

DNA is becoming an exciting new nanoscale building material due to the same chemical and physical properties that allow it to self-replicate and encode genetic information. About 35 years ago, Nadrian C. Seeman invented DNA-based bottom-up assembly [1], which uses the Watson-Crick complementarity of the nucleotides comprising DNA, to control the shape of the assembled structures. Such nano-scale constructions are obtained by designing DNA molecules with unsatisfied sequences of bases at prescribed locations, which then adhere to one another to form the targeted structures. This has led to molecular scaffoldings made of DNA that have potential for a wide range of applications such as biomolecular computing, drug delivery, nano-robotics, and biosensors (see [2–11]).

Various laboratories around the world have designed and produced synthetic DNA molecules that assemble into preprogrammed nanostructures, starting with the first breakthrough with branched DNA molecules [12, 13], nanoscale arrays [14, 15], and even a variety of DNA and RNA knots [16–18]. In this chapter, we focus on the tile- or block-based self-assembly of targets that have the underlying structure of graphs, including many polyhedral skeletons [19, 20, 20, 21, 21–23], crystalline lattices [24], and numerous common graph classes [25–27].

In all self-assembly processes, an essential step in achieving the nanostructure is designing the molecular building blocks, and moreover, determining how they should adhere to one another and where in the final structure they will appear. Since many targets of DNA self-assembly have the structure of graphs, these design strategy problems fall naturally into the realm of graph theory. For the methods discussed here, the reader should be familiar with basic graph

Algebraic and Combinatorial Computational Biology. https://doi.org/10.1016/B978-0-12-814066-6.00002-7

theory (e.g., Eulerian graphs, augmenting edges, degree sequences, directed graphs, graph covering, chromatic number, common classes of graphs, etc.), as well as elementary linear algebra (using a matrix to encode and solve a system of equations, linear independence, etc.). Some references that provide this background are included in Section 2.7.

2.2 LABORATORY PROCESS

The earliest self-assembly process, pioneered by Seeman [28], builds target structures from modular units called *branched junction molecules*. These asterisk-shaped molecules have adhesion sites at the ends of their arms, allowing them to join in a prescribed way to form larger molecules with the desired structures.

Recall that a strand of DNA consists of a sequence of nucleotides that contain one of the four bases: cytosine [C], guanine [G], adenine [A], and thymine [T]. These bases come in complementary pairs, with adenine bonding to thymine (A $\leftrightarrow$ T) and cytosine with guanine (C $\leftrightarrow$ G). A sequence of nucleotides is also oriented, one side (phosphorylated side) denoted as the $5'$-end (the "beginning") and the other (ending with a hydroxyl group) as the $3'$-end (the "end"). Two oppositely oriented strands containing complementary sequences of nucleotides bond together to form the well-known double helical structure, referred to as duplex DNA, as shown in the bottom of Fig. 2.1. A sequence of nucleotides is *unsatisfied* if it is not paired with a complementary sequence. A *cohesive-end* is a sequence of unsatisfied bases extending out of a duplex DNA molecule. Cohesive-ends form the fundamental mechanism of modular self-assembly because molecules with complementary cohesive-ends can bond to each other, thus attaching the modular units to one another to form a new molecule as in Fig. 2.1.

The mathematical models in this chapter identify complementary cohesive-ends, without specifying the exact sequence of bases. Thus, we generally represent the whole sequence corresponding to a cohesive-end with a single

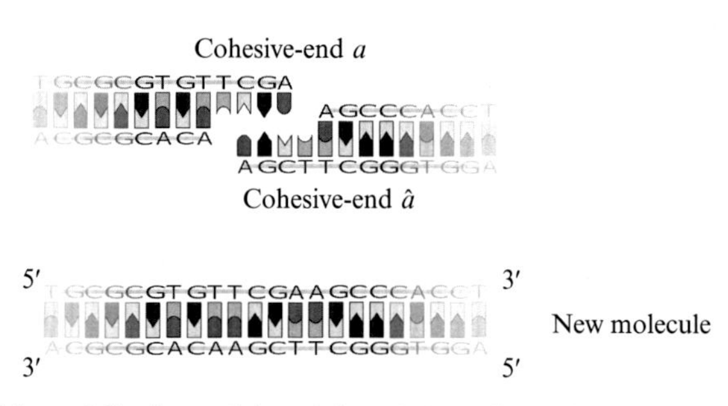

FIG. 2.1 Sticky end ligation. *(Adapted from M.P. Ball, Available from: https://commons.wikimedia.org/wiki/File:Ligation.svg.)*

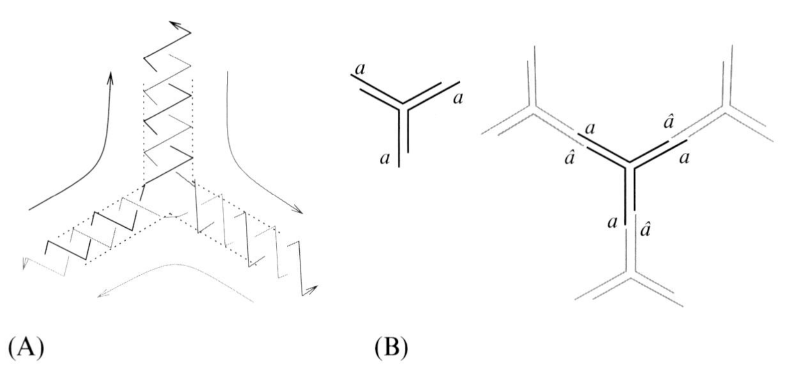

FIG. 2.2 Modular components and their assembly. (A) Three-branched junction molecule whose arms are duplex DNA that have extended single-stranded cohesive-ends. (B) Schematics of branched junction molecules joining together through cohesive-ends.

alphabet symbol (e.g., one of $a, b, c, \ldots$) and its complementary sequence as a hatted letter (correspondingly, $\hat{a}, \hat{b}, \hat{c}, \ldots$). In Fig. 2.1, the cohesive-ends are denoted with a and $\hat{a}$, respectively.

The building blocks, which we also call modular components, may be simple motifs, such as the branched junction molecules used in [21, 29–34]. These branched junction molecules have some number of arms (generally between 2 and 12), each with a cohesive-end at the terminus, emerging from a central point. Fig. 2.2A gives an example of three-armed branched molecule, with a single-strand extending beyond the duplex to form a cohesive-end at the terminus of each arm. The arrows indicate the polarity of the DNA strands, with the arrows pointing from the 5′-end to the 3′-end. This three-armed junction molecule consists of three strands of DNA, as indicated by the colors red, blue, and black within the helical structure. Each strand has a portion that is complementary to each of the other two strands, and all strands end with a single-stranded cohesive-end extension (usually 2–10 nucleotides). This single-stranded part can join its Watson-Crick complement when the branched junction molecule is placed in a test tube with other molecules containing cohesive-ends with the complementary sequence. The arms of the branched junction molecules may be flexible, formed, for example, from longer double-stranded DNA with one strand extending beyond the other to form cohesive-ends. On the other side, the arms may be more complex structures (several duplex DNA molecules bundled together), and may be reinforced with additional molecular structures for greater rigidity as in [23].

Fig. 2.2B is a schematic depiction of three-junction molecules annealing together. The left-hand side of Fig. 2.2B shows a three-junction molecule with its cohesive-ends labeled with a and $\hat{a}$. The right-hand side shows a possible structure formed by multiple copies of the same unit by allowing the complementary cohesive-ends to join.

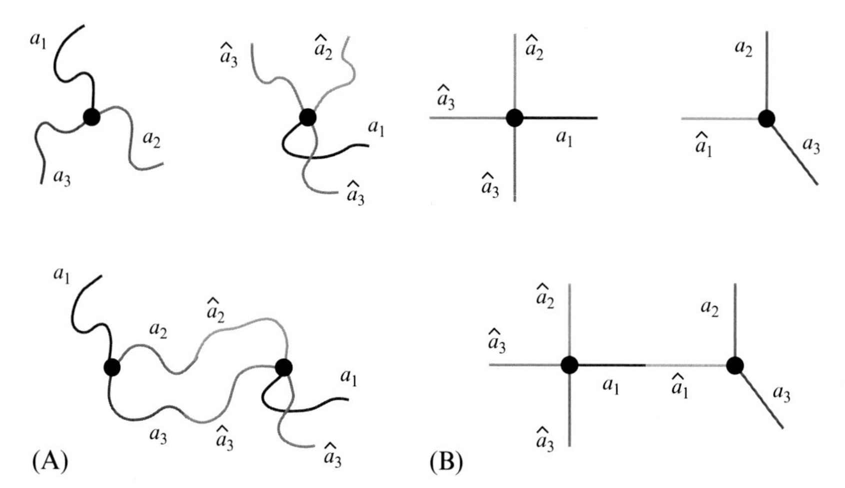

FIG. 2.3 Different annealing of flexible and rigid tiles. (A) Flexible tiles have arms that can bend and reach in different directions, hence two pairs of arms with complementary cohesive-ends can bind. (B) Rigid tiles have geometric constraints and due to their structure, only one pair of arms with complementary cohesive-ends can bind.

Fig. 2.3 shows a further simplification of k-armed branched junction molecules. Here, each molecule is depicted with a central vertex (small black circle) and number of lines corresponding to the arms of the molecule attached to it. If we allow each molecular arm to be 200–300 base pairs long, or add unpaired bulged sections [35], then the arms become rather flexible too and we deliberately show them curved (e.g., Fig. 2.3A). When the arms are shorter, or if they have been reinforced in some way, the modular components are more highly structured for greater rigidity and geometric control, as in the DNA tripods of [23]. Rigid molecules carry additional geometric information, as the fixed configuration of the edges about the vertices constrains the assembly. Fig. 2.3A and B shows examples of assemblies with flexible and rigid modular components, respectively. Although the cohesive-end of the modular components are the same in both the rigid and the flexible case, due to the flexibility of the arms, only three cohesive-ends remain unpaired in the left construct, while five such unpaired cohesive-ends remain to the right. The geometric constraint of the building blocks to the right do not allow for a_2 to bond with $\hat{a}_2$, nor for a_3 to bond with $\hat{a}_3$.

In most settings, the assembly process happens with all modular units combined in a single step (*one-pot assembly*), allowing all molecules to take part in the assembly of the desired structure [15, 24], and in general, we can assume that there is an arbitrarily large number of modular units of each type present. Thus, there are two common design objectives. One is to minimize the number of different units and cohesive-end types needed to assemble a given target graph-like structure. The other is to determine what graph-like structures might self-assemble from a given set of modular units.

2.3 GRAPH THEORETICAL FORMALISM AND TOOLS

Recall that a graph G consists of a set $V = V(G)$ of vertices, a set $E = E(G)$ of edges, and a map $\mu: E \to V^{(2)}$ where $V^{(2)}$ is the set of (not necessarily distinct) unordered pairs of elements of V. If $\mu(e) = \{u, v\}$, then u and v are the vertices incident with e. An edge e is a *loop* if $u = v$, that is, it is incident with only one vertex. A *multiple edge* occurs if $\mu(e) = \mu(f)$ for some $e, f \in E$. Additionally, we say that (v, e) is a *half-edge* of G if $v \in \mu(e)$, and we write H for the set of half-edges of G.

In the simplest setting, a k-armed branched junction molecule, and its k arms with cohesive-ends, corresponds to a k-degree vertex with its k incident half-edges. Below, we introduce the main notions that we use in our model similarly as in [30, 33, 36, 37].

Definition 2.1. Following [30, 33, 36, 37], we use the combinatorial objects listed here for tile-based assembly design.

- *Cohesive-end type.* Branched k-armed junction molecules have arms formed from DNA strands, possibly multiple strands, especially in the case of rigid molecules. Given a finite set of symbols Σ, called an alphabet, the extended cohesive-ends on the arms are denoted by an "un-hatted" letter in Σ and its complement by the same letter, but "hatted," that is $\hat{\Sigma} = \{\hat{x} \mid x \in \Sigma\}$. We do not allow palindromic cohesive-ends, so $a \neq \hat{a}$ for all $a \in \Sigma$. Moreover, $\hat{\hat{a}} = a$.
- *Bond-edge type.* A cohesive-end type joined to its complement forms a bond-edge type, which we identify by the un-hatted letter label, so, for example, cohesive-ends a and $\hat{a}$ will join to form a bond edge of type a.
- *Tile.* The combinatorial abstraction of a branched junction molecule is called a *tile*. It consists of a vertex with half-edges labeled by the cohesive-end types on the arms of the branched junction molecule the tile represents, and is denoted by a multiset of its cohesive-end types whose multiple entries of the same cohesive-end type are indicated by the exponent to the corresponding symbol. The number of arms of a tile t is indicated with $\#t$. For example, Fig. 2.3 shows two tiles $t_1 = \{a_1, a_2, a_3\}$ with $\#t_1 = 3$ and $t_2 = \{a_1, \hat{a}_2, \hat{a}_3^2\}$ with $\#t_2 = 4$. The rigid versions of the tiles require additional information about the position of the arms relative the center vertex, which we address later.
- *Pot.* A pot is a collection of tiles such that for any cohesive-end type that appears on any tile in the pot, its complement also appears on some tile in the pot. More precisely, a *pot* is a set $P = \{t_1, \ldots, t_k\}$ where each t_i is a tile $(i = 1, \ldots, k)$ and for all $a \in \Sigma$, if there is i such that $a \in t_i$, then there is $j \in \{1, \ldots, k\}$ such that $\hat{a} \in t_j$. The set of bond-edge types that appear in tiles of P is denoted with $\Sigma(P)$, and we write $\#P$ to denote the number of distinct tile types in P.

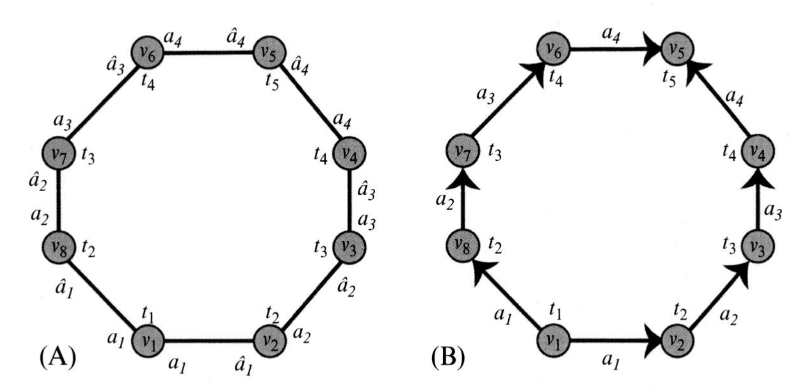

FIG. 2.4 (A) An assembly design of an 8-cycle. (B) The assembly design can be depicted through labeled edge orientations from a_i to $\hat{a}_i$.

- *Assembly design.* An assembly design is a labeling λ: $H \to \Sigma \cup \hat{\Sigma}$ of the half-edges of a graph G with the elements of Σ and $\hat{\Sigma}$ such that if $e \in E(G)$ and $\mu(e) = \{u, v\}$, then $\widehat{\lambda(v, e)} = \lambda(u, e)$. This means that each edge receives both a hatted and an un-hatted symbol on its half-edges. We use the convention that λ provides each edge with an orientation that starts from the un-hatted half-edge to the hatted half-edge, as in Fig. 2.4.
- The set of *tiles associated with an assembly design* λ of a graph G is the set $P_\lambda(G) = \{ t_v \mid v \in V(G) \}$ where $t_v = \{ \lambda(v, e) \mid v \in \mu(e), e \in E(G) \}$. This means that for each vertex v of G, the assembly design specifies a tile t_v whose multiset is the set of labels of half-edges incident to v. We expand the labeling λ to a labeling on vertices with λ: $V \to P_\lambda(G)$ such that $\lambda(v) = t_v$.

A target graph-like molecule assembles from branched molecules, with arms with complementary cohesive-ends joining to form the final structure. The centers of the branched junction molecules correspond to the vertices of the underlying graph, and the joined arms correspond to its edges.

Abstractly, we think of a graph assembling from a set of tiles, with a tile placed at each vertex of the graph, and each half-edge of the graph corresponds to a cohesive-end type so that the edges of the graph correspond to complementary cohesive-end types joined together to form a bond edge. We note that in an experiment, there may be as many as 10^{12} to 10^{15} copies of each molecule. Thus, the model assumes that an arbitrarily large number of each branched junction molecule is available during the assembly process, so we allow each tile type to be used repeatedly.

More precisely, we say that a pot P *realizes* a graph G if we can assign a tile type in P to each vertex and its incident half-edges (the labels of a tile t assigned to a vertex v must be in bijection with the half-edges of v) such that each edge of G is assigned a complementary pair of cohesive-end labels. This leads to the following definition:

Definition 2.2. Let G be a graph with vertices V, edges E, and half-edges H. A pot P *realizes the graph* G if there exists an assembly design $\lambda \colon H \to \Sigma \cup \hat{\Sigma}$ such that $P_\lambda(G) \subseteq P$.

It follows from the definition that for every assembly design λ of G, the pot $P_\lambda(G)$ realizes G.

Example 2.1. Consider the set of tiles $P = \{t_1, t_2, t_3, t_4, t_5\}$ where $t_1 = \{a_1^2\}$, $t_2 = \{\hat{a}_1, a_2\}$, $t_3 = \{\hat{a}_2, a_3\}$, $t_4 = \{\hat{a}_3, a_4\}$, and $t_5 = \{\hat{a}_4^2\}$. Because for each $i = 1, 2, 3, 4$ both cohesive-ends a_i and $\hat{a}_i$ appear in a tile of P, the set P is a pot. For a graph that is a cycle on eight vertices, $V = \{v_1, \ldots, v_8\}$, and eight edges, $E = \{e_1, \ldots, e_8\}$ with $\mu(e_i) = \{v_i, v_{i+1 \bmod 8}\}$, there is an assembly design with labels depicted in Fig. 2.4A. We can read off the tiles associated with this assembly design through the labels of the vertices. This pot realizes the 8-cycle because all tiles that are associated with the design are in P. Note that P realizes a 16-cycle as well with an assembly design obtained from two copies of the 8-cycle, such that the bonds a_4 at the two copies of tiles t_5 and t_4 swap their corresponding complementary pairs. Fig. 2.4B shows a convention of indicating an assembly design by labeling oriented edges where the orientation is taken from un-hatted a_i to hatted a_i.

Thus, a design strategy for a graph-like molecule with the structure of a graph G is obtained if G is realized by a pot P. We can then propose an experiment to assemble the target graph-like molecule from branched junction molecules whose arms have cohesive-end configurations corresponding to the tile types listed in the pot. However, there may be many graphs that can be realized by a given pot P.

Definition 2.3. The set of graphs realized by a pot P is called the *output* of P and is denoted $\mathcal{O}(P)$.

We say that a DNA complex (a graph-like molecule) is *complete* if it has no unmatched cohesive-ends. In general (except for various lattices), as follows from Definition 2.2, we assume that the final realization of a target graph is a complete complex. Given a target graph-like structure, the challenge is to determine a *minimal* pot from which the target will assemble under various laboratory settings. Alternatively, given a pot, we may seek to determine what graphs might be realized by it, that is, to determine the output of the pot.

There are many different laboratory settings depending, for example, on how highly structured the branched junction molecules are, how symmetric the target is, how the final product will be recovered, and even what ancillary molecules might be involved (nanocargo, attachment sites, etc.). Here we will consider design strategies for both flexible and rigid tiles under two different yield constraints.

2.3.1 Flexible Tiles

The simplest branched junction molecules have arms consisting of double-stranded DNA, with one strand extending beyond the other at the end of each arm to form the cohesive-ends, and have no reinforcing structure at the junction, as in Fig. 2.2. Because lengthy double-stranded DNA is quite flexible, the model assumes that the arms may bend and change position relative to each other about the center of the junction. Because the arms are flexible, we do not need to capture any additional geometric information about the tile or distinguish between permutations of the cohesive-ends about a vertex, so we need only specify the multiset of cohesive-end types to describe each tile.

Fig. 2.5 shows a pot $P = \{t_1, t_2\}$ containing two flexible tiles, $t_1 = \{a^3\}$ and $t_2 = \{a, \hat{a}^2\}$, as well as a possible graph that can be realized with this pot. We can introduce new graph invariants in this setting, namely the minimum number of tile types needed to realize a graph G, which we denote by $T(G) = \min\{\#(P) \mid P \text{ realizes } G\}$, and the minimum number of bond-edge types, which we denote by $B(G) = \min\{\#\Sigma(P) \mid P \text{ realizes } G\}$. With this formalism, the basic design problem for flexible tiles with one-pot assembly becomes:

Statement 2.1 (Flexible Tiles Problem)**.** Given a graph G, determine a pot P with minimum number of tiles and/or bond-edge types required to construct a DNA complex with the structure of G. More precisely, determine $T(G)$ and $B(G)$. Furthermore, provide an assembly design for G and the resulting specifications of the tiles in a pot P realizing G.

This problem statement can be reformulated as the following purely combinatorial problem by thinking of the edges of a graph being directed from unhatted to hatted labels as in Fig. 2.4B. For example, the figure indicates that vertex v_1 can be labeled by a tile $\{a_1^2\}$ while vertex v_2 can be labeled with a tile

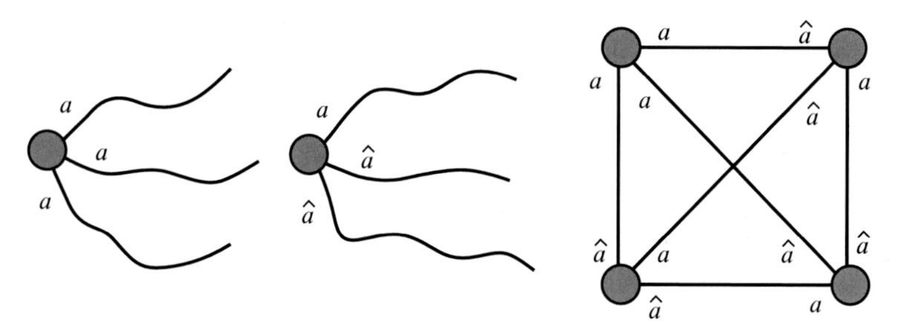

FIG. 2.5 A pot containing two tiles, and a graph realized by the pot.

$\{\hat{a}_1, a_2\}$. The tiles in $P_\lambda(G)$ can be "read off" from the vertices by listing the multisets of labels of the incident edges, incoming arcs with a hatted symbol, and outgoing arcs with un-hatted symbol.

Statement 2.2 (Graph Theoretical Flexible Tiles Problem)**.** Find an edge-labeled orientation of a graph G using a minimal alphabet for the labels, with as few different multisets labeling of the half-edges about the vertices as possible. That is: (a) Given a graph G, determine an assembly design λ_t such that $\#P_{\lambda_t}(G) = T(G)$ and an assembly design λ_b that provides $B(G)$ labels of the edges. (b) Characterize graphs for which $\lambda_t = \lambda_b$. (c) For each graph G, determine the minimal gap between the number of edge labels provided by λ_t and $B(G)$.

Another significant problem is to determine the output of a pot P. Although the design strategies provide a set of tiles that can realize a given graph structure, such a set may realize many other structures, and determining all possible outputs helps in filtering out the undesired constructs.

Statement 2.3 (The Pot Output)**.** Given a pot P with a finite set of tile types, determine the set $\mathcal{O}(P)$ of graphs realized by P.

We discuss some combinatorial approaches to these problems under two different laboratory settings. We also provide an algorithm for obtaining an unconstrained assembly design for a given graph and give a linear algebra strategy for determining $\mathcal{O}(P)$.

2.3.2 Flexible Tiles, Unconstrained Case

For experimental purposes, typically we want the target graph to be the "smallest" (fewest vertices) complete complex realized by a pot, but if we are not constrained by this, then there is a fast algorithm leading to a design solution. In fact, if we do not have any restrictions on the incidental creation of smaller complete complexes, then any graph can be realized by a pot with just one bond-edge type, and the number of different tile types depends only on the number of different vertex degrees appearing in the graph. We use superscript $T^u(G)$ to indicate that $T(G)$ is considered in an unconstrained case.

Fig. 2.5 shows a pair of three-arm tiles that realize the clique K_4. All vertices in K_4 have degree 3, so there must be two tile types (since there must be the same number of hatted vs. un-hatted labels). So $T^u(K_4) = 2$. The following algorithm determines an assembly design for an input graph G, providing a pot that for every vertex of degree k contains one tile type if k is even, and at most two tile types if k is odd. We use an augmented Eulerian graph. Recall that a graph is *Eulerian* if it contains an Eulerian circuit, and this is equivalent to it having no vertices of odd degree. Additionally, every graph has an even number of odd degree vertices. Together, this means that if there are odd degree vertices in the graph, we can add edges connecting pairs of odd degree vertices to obtain an Eulerian graph. This addition of new edges forms an *augmented Eulerian graph*.

Algorithm 2.1.

1. Create an augmented Eulerian graph G' from G by partitioning the odd degree vertices into pairs and then adding edges between pairs of odd degree vertices.
2. Construct an Eulerian circuit for G'.
3. Choose a direction to traverse the Eulerian circuit, and then record the corresponding orientation of each edge.
4. Delete the augmented edges, leaving an orientation $\overrightarrow{G}$ of the original graph G, and then label all the edges a to yield an assembly design for G.
5. Label each outgoing half-edge with a and each incoming half-edge with $\hat{a}$.
6. Add a tile type in the pot with j cohesive-ends of type a and k cohesive-ends of type $\hat{a}$ whenever there is a vertex of $\overrightarrow{G}$ with out-degree j and in-degree k.

Since the edge orientations come from an Eulerian circuit, every even degree vertex has an equal number of incoming and outgoing edges (and hence the corresponding tile has an equal number of hatted and un-hatted ends). Once the augmenting edges are deleted, the number of incoming and outgoing edges adjacent to an odd degree vertex differs by one, so there are at most two tile types corresponding to each odd vertex degree. This algorithm provides an upper bound for $T^u(G)$ for a graph G.

Fig. 2.6 illustrates the above algorithm. For the construction of a cube, four augmenting edges must be added to ensure that all vertices have even degree (Fig. 2.6, left). An Eulerian circuit results in an orientation of the edges of the augmented graph with two incoming and two outgoing edges at each vertex (Fig. 2.6, middle). When the augmenting edges are removed the resulting pot can be taken $P = \{t_1 = \{a^2, \hat{a}\}, t_2 = \{a, \hat{a}^2\}\}$ (Fig. 2.6, right). Note that, as was the case for K_4, the cube cannot be constructed from a single tile type since any

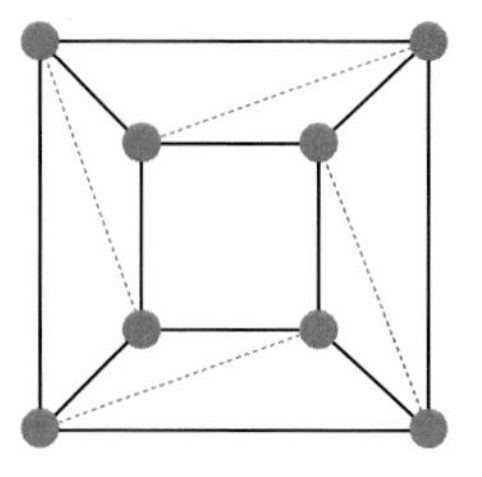

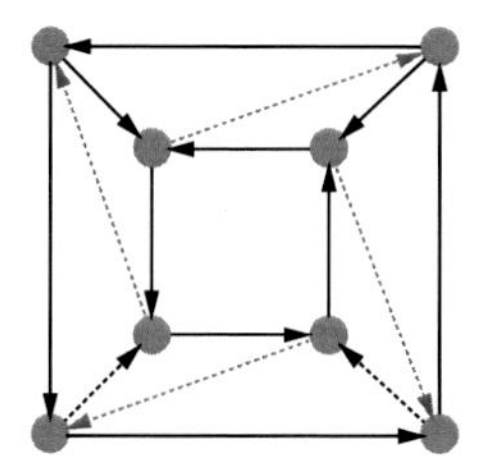

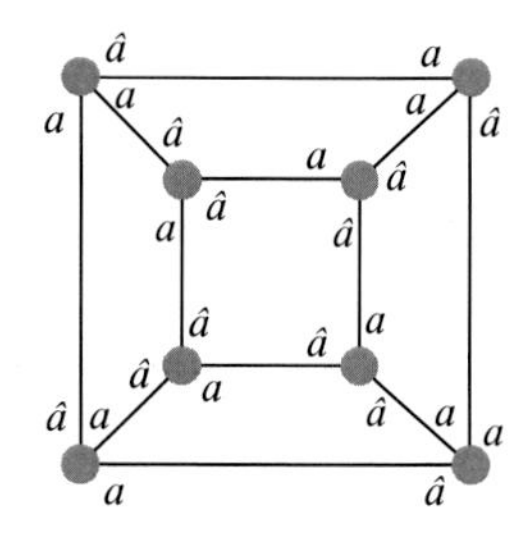

FIG. 2.6 Illustration of Algorithm 2.1: assembly design and a pot realizing a cube from its augmented Eulerian graph.

complete complex must have an equal number of hatted and un-hatted cohesive-ends. We conclude that $T^u(Cube) = 2$ and $B^u(Cube) = 1$.

For special classes of graphs, such as trees, we can improve on the upper bound of $T^u(G)$ given by Algorithm 2.1 for general graphs. In particular, a given tree may be realized by a pot with at most one more tile-type than the number of distinct vertex degrees in the tree. This is achieved by rooting the tree and creating an oriented graph with all edges directed away from the root. Note that every vertex other than the root will have exactly one incoming edge. If the root vertex has a degree different from all other vertices, the resulting number of tiles equals the number of different vertex degrees in the tree, otherwise the assembly design gives a pot with exactly one more tile type than this lower bound.

EXERCISES

[2.1] Use Algorithm 2.1 to determine an optimal pot for the 3×3 grid graph on nine vertices in the unconstrained setting.

[2.2] Find an example of a tree where the number of tiles needed is one more than the number of different vertex degrees. Find an example of a tree where the number of tiles needed is exactly the number of different vertex degrees.

[2.3] Recall that a graph H is a *covering* of a graph G if there is a surjection f: $V(H) \to V(G)$, such that for each $v \in H$, restricting f to the neighborhood of v gives a bijection onto the neighborhood of $f(v)$ in G. Demonstrate that the tile set $P = \{t_1 = \{a_1, a_2, a_3\},\ t_2 = \{a_1, a_2, \hat{a_3}\},\ t_3 = \{\hat{a_1}, \hat{a_2}^2\},\ t_4 = \{\hat{a_1}^3\}\}$ can be used to create both the complete graph K_4 and a cube (which is a covering of this graph).

[2.4] Prove that if a pot P realizes a graph G, then it also realizes any covering graph H of G.

[2.5] Open Problem: For the unconstrained case, characterize the trees for which the upper bound is tight.

2.3.3 Flexible Tiles, Constrained Case

Yields of experimental assemblies may improve if the pot is constrained so that the target graph is the unique graph with the smallest number of vertices realizable by the pot. This situation is the most commonly used experimentally, so we take that $T(G)$ and $B(G)$ are invariants for G in the constrained case. If the target graph does not contain any loops, this constraint implies that no tile may have hatted and un-hatted cohesive-ends with the same type: such a tile would mean that the pot could also realize a complex of the same or smaller size as the target, but containing a loop, thus violating the uniqueness or the size constraint. The constraint also implies that in an assembly design for the target graph G, adjacent vertices cannot have the same tile type, as shown in Fig. 2.7, for example. This is because, for a tile to attach to a copy of itself, it must have at least one pair of arms with complementary cohesive-ends. Since the arms are

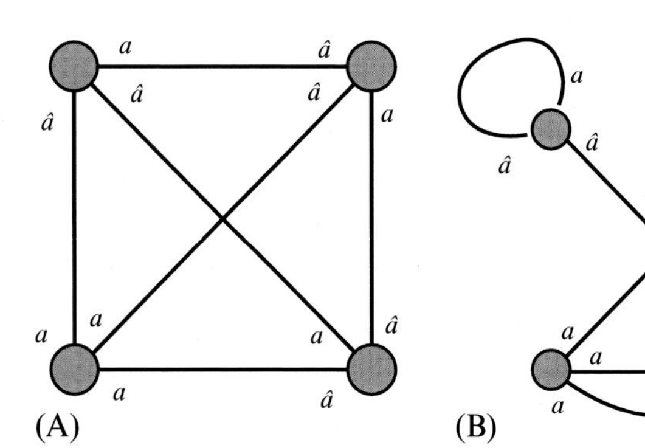

FIG. 2.7 (A) A target graph realized by a pot P in Fig. 2.5. (B) Another undesired graph of the same size realized by the same pot.

flexible, these two can then form an undesired loop (or nonloop edge, if the loop itself were desired). Because adjacent tiles cannot have the same type, the chromatic number is a lower bound on the number of tile types needed to build any loop-free target graph. In particular, K_n, the complete graph on n vertices, would require n tile types to assemble in this setting. Therefore, $T(K_n) = n$.

Unfortunately, the chromatic number is not a good lower bound, as it can be arbitrarily far from the actual bound. For example, a cycle C_n has chromatic number 2 if n is even and 3 if n is odd, but it was shown in [37] that $T(C_n) = \lceil \frac{n}{2} \rceil + 1$. This result follows from considering the bond-edges. If the assembly design of a cycle has three or more bond-edges of the same type, then smaller complete complexes can form, as shown in Fig. 2.8. The figure also illustrates that one has to be also careful when there are two edges with the same label. Combining the fact that each bond-edge type can be used at most twice with directing the edges from hatted to un-hatted, it can be shown

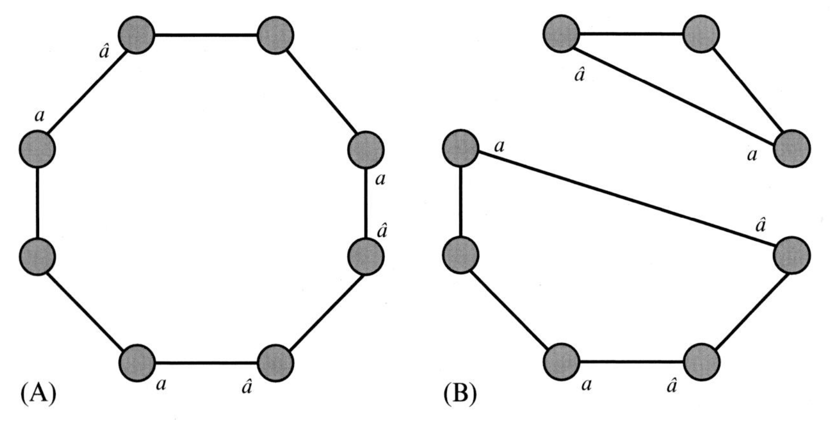

FIG. 2.8 (A) An assembly design for a cycle with repeated bond-edge type. (B) Two smaller cycles obtained from the same pot.

that $T(C_n) = \lceil \frac{n}{2} \rceil + 1$ (see the Exercises). Similarly, also from [37], a complete bipartite graph $K_{m,n}$ has $T(K_{m,n}) = m + 1$ where $m \geq n$.

A table of $T(G)$ and $B(G)$ in various laboratory settings for some common classes of graphs may be found in [37].

EXERCISES

[2.1] Prove that $T(C_n) = \lceil \frac{n}{2} \rceil + 1$ in the constrained setting.

[2.2] Open Problem: In the constrained setting, it is not known whether a pot always exists that simultaneously minimizes the number of tile types and bond-edge types to realize a graph G. Determine the relationship between $T(G)$ and $B(G)$ for flexible tiles in the constrained setting.

2.3.4 The Matrix of a Pot

A useful tool for modular assembly is the matrix of a pot, introduced in [38], and further developed in [37]. This matrix encodes the combinatorial structure of the tiles in a pot, so that linear algebra can be used to analyze the graphs realized by the pot.

Let $P = \{t_1, \ldots, t_p\}$, be a pot containing p different tile types, and define $A_{i,j}$ to be the number of cohesive-ends of type a_i on tile t_j and $\hat{A}_{i,j}$ to be the number of cohesive-ends of type $\hat{a}_i$ on the same tile t_j.

Suppose a target graph G with n vertices is realized by P, using R_j tiles of type j for $j = 1, \ldots, p$. Then the total number of tiles used must be n, and the number of un-hatted and hatted cohesive-ends used is the same for each cohesive-end type. This means that:

$$\sum_j R_j = n, \quad \text{and} \quad \sum_j R_j(A_{i,j} - \hat{A}_{i,j}) = 0 \quad \text{for all } i. \tag{2.1}$$

If we write $z_{i,j} = A_{i,j} - \hat{A}_{i,j}$, and let r_i be the proportion of tile type t_i used in realizing a graph G with n vertices so that $r_i = R_i/n$, then Eq. (2.1) becomes

$$\sum_j r_j = 1, \quad \text{and} \quad \sum_j r_j z_{i,j} = 0 \quad \text{for all } i. \tag{2.2}$$

It is often expedient to encode this system of equations in an augmented matrix, $M(P)$, called the *construction matrix of the pot*:

$$M(P) = \left[\begin{array}{cccc|c} z_{1,1} & z_{1,2} & \cdots & z_{1,p} & 0 \\ \vdots & \vdots & \ddots & \vdots & \vdots \\ z_{m,1} & z_{m,2} & \cdots & z_{m,p} & 0 \\ 1 & 1 & \cdots & 1 & 1 \end{array}\right]. \tag{2.3}$$

The solution space of the construction matrix of a pot P is called the *spectrum* of P, and is denoted $\mathcal{S}(P)$. The spectrum can be used to characterize P as well as the output $\mathcal{O}(P)$.

It is known that there is a graph realized by a pot P if and only if $\mathcal{S}(P) \neq 0$ [38]. One can also observe that the output of a pot contains a graph with an assembly design using all of the tile types in the pot if and only if there is a solution for $M(P)$ with all nonzero entries. The spectrum of a pot P can be described geometrically as a linear combination of its extreme solutions corresponding to vertices of a convex hull [38].

The design strategies for obtaining pots with a minimal number of tile types realizing a given graph asks for a specific type of solutions in the spectrum of a pot.

If a graph G of size n is realized by the pot P using R_j tiles of type t_j, then the vector $(1/n)\langle R_1, \ldots, R_p\rangle$ is a solution to the system associated with the construction matrix $M(P)$. Conversely, it was shown in [37, 38] that if $\langle r_1, \ldots, r_p\rangle$ is a solution, and there is a positive integer n such that $nr_j \in \mathbb{Z}_{\geq 0}$ for all j, then there is a graph of size n realized by P using nr_j tiles of type t_j. In the following two examples, we show how the spectrum of a pot can provide information about the output of the pot.

Example 2.2. Consider the pot $P_1 = \{t_1 = \{a_1, \hat{a}_2\}, t_2 = \{\hat{a}_1, a_2\}\}$ and the corresponding construction matrix:

$$M(P_1) = \left[\begin{array}{rr|r} 1 & -1 & 0 \\ -1 & 1 & 0 \\ 1 & 1 & 1 \end{array}\right]. \tag{2.4}$$

The spectrum of P_1 is $\mathcal{S}(P_1) = \{\, \frac{1}{2r}\langle r, r\rangle \mid r \in \mathbb{Z}^+ \,\}$. A cycle of length 8 can be constructed from this pot using four tiles of each type (see Fig. 2.9), but it is not the smallest target graph that can be constructed. The solution $(1/4)\langle 2, 2\rangle$ corresponds to a cycle of length 4 and $(1/2)\langle 1, 1\rangle$ to a digon, a multiedge on two vertices. In fact, any graph with an even number of vertices, all of degree 2, can be realized by P_1, which is a collection of even-length cycles. This means that $\mathcal{O}(P) = \{\, G \mid \#V(G) \text{ is even and } \forall v \in V(G), \deg(v) = 2 \,\}$.

Example 2.3. Consider the pot $P_2 = \{t_1 = \{a_1^2\}, t_2 = \{\hat{a}_1, a_2\}, t_3 = \{\hat{a}_2, a_3\}, t_4 = \{\hat{a}_3, a_4\}, t_5 = \{\hat{a}_4^2\}\}$ and its corresponding construction matrix:

$$M(P_2) = \left[\begin{array}{rrrrr|r} 2 & -1 & 0 & 0 & 0 & 0 \\ 0 & 1 & -1 & 0 & 0 & 0 \\ 0 & 0 & 1 & -1 & 0 & 0 \\ 0 & 0 & 0 & 1 & -2 & 0 \\ 1 & 1 & 1 & 1 & 1 & 1 \end{array}\right]. \tag{2.5}$$

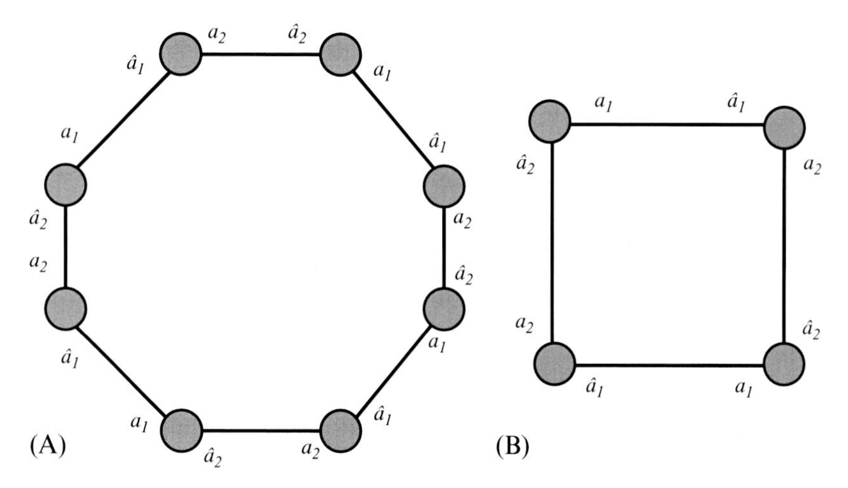

FIG. 2.9 Cycles of length 8 and 4 realized by P_1.

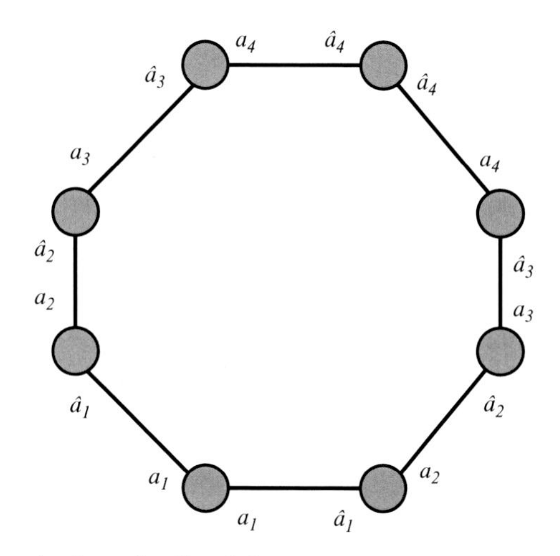

FIG. 2.10 Construction for cycle of length 8.

The solution $(1/8)\langle 1, 2, 2, 2, 1\rangle$ corresponds to a cycle of length 8 (see Fig. 2.10); however, in this case the pot does not realize any smaller complete complex. Moreover, the spectrum of P_2 consists of multiples of this vector, $S(P) = \{ \frac{1}{8r}\langle r, 2r, 2r, 2r, r\rangle \mid r \in \mathbb{Z}^+ \}$. Hence only cycles of length $8r$ for some r can be realized (see Exercise 2.3 in Section 2.3.2).

EXERCISES

[2.1] Recall from Exercise 2.3 in Section 2.3.2 that the tile set $P = \{t_1 = \{a_1, a_2, a_3\},\ t_2 = \{a_1, a_2, \hat{a}_3\},\ t_3 = \{\hat{a}_1, \hat{a}_2^{\,2}\},\ t_4 = \{\hat{a}_1^{\,3}\}\}$ can be used to create both the complete graph K_4 and a cube (which is a covering of this graph). Prove that no complete complex smaller than K_4 can be built with this pot in two ways, one using the matrix of the pot, and one using a geometric argument. Determine the spectrum of P and describe the output of P.

[2.2] Find the smallest graph that can be realized by the pot $P = \{t_1 = \{a_1, a_2, a_3, a_4\}, t_2 = \{\hat{a}_1^{\,3}\}, t_3 = \{\hat{a}_2^{\,3}\}, t_4 = \{\hat{a}_3^{\,3}\}, t_5 = \{\hat{a}_4^{\,3}\}\}$. Determine the spectrum $\mathcal{S}(P)$ and describe $\mathcal{O}(P)$, the output of P.

2.4 RIGID TILES

Molecular building blocks can be produced with additional supports along the edges that fix the inter-arm angles and make the arms fairly rigid, as for example in [21, 23, 39]. Thus, assembly design strategies that realize the potentials of these building blocks are also of interest. One significant challenge in this setting is mathematical representation of the rigid tiles themselves. Because the arms of these building blocks are rigid, each cohesive-end has a specific geometric orientation with respect to the tile, so that the attachment of two arms fixes the relative positions of the two tiles involved. To capture this geometric constraint, [40] adapted ideas from woodworking techniques to model rigid tiles. In particular, we can use *half-lap splice joints*, or more simply *half-laps*. This allows for a concrete mathematical formalism and general scheme for developing design strategies for rigid tiling. The model from [40] is given below, following that exposition closely.

We restrict our attention to simple graphs, without loops or multiedges, and do not allow palindromic cohesive-ends. For rigid tiles, we assume that: the arms are straight, rigid, and of unit length; the arms are fixed at specified angles with respect to each other about the vertex; the arms do not twist, compress, or elongate; a tile has at least two arms; and two rigid tiles connect only at the cohesive-ends, forming a straight edge when they connect.

We model rigid tiles as assemblages of wooden beams, where each arm is given by a rectangular prism with a *lap* at the cohesive-end. The thickness of the prism is not taken into consideration at this time. The lap is made up of the *cheek*, the surface parallel to the face that we call the *lap plane*, and the *shoulder*, the surface perpendicular to the face and cheek, as shown in Fig. 2.11.

Fig. 2.12 shows a pot with two rigid tile types and the resulting unique complex (geometric graph), a cube, that can be realized by this pot.

A critical aspect of this model is a basic mechanism for specifying the angle of the lap plane relative to the tile. This mechanism uses the plane containing a line through the core of the arm under consideration and a canonically

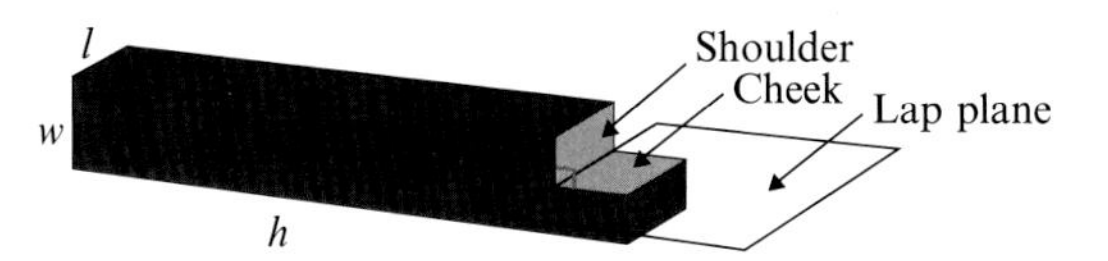

FIG. 2.11 A half-edge using the half-lap model, showing the shoulder perpendicular to the cheek. *(Reprinted with permission from M.M. Ferrari, A. Cook, A. Houlihan, R. Rouleau, N.C. Seeman, G. Pangborn, J. Ellis-Monaghan, Design formalism for DNA self-assembly of polyhedral skeletons using rigid tiles, J. Math. Chem. 56 (5) (2018) 1365–1392.)*

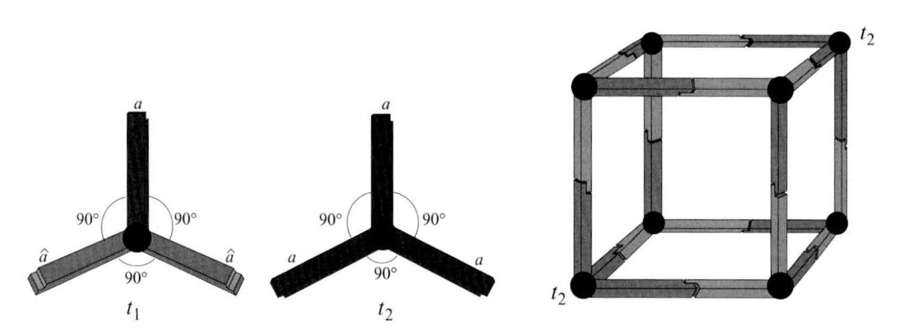

FIG. 2.12 Rigid cube. *(Reprinted with permission from M.M. Ferrari, A. Cook, A. Houlihan, R. Rouleau,N.C. Seeman, G. Pangborn, J. Ellis-Monaghan, Design formalism for DNA self-assembly of polyhedral skeletons using rigid tiles, J. Math. Chem. 56 (5) (2018) 1365–1392.)*

determined second line through the vertex of the tile and the midpoint of the arm. We then fix the angle of the lap plane with respect to the plane formed by these two lines.

When two rigid tiles bond along an edge, their positions with respect to each other are fixed. Thus, a second critical attribute of this model is a means of determining the relative angle between two tiles. The following definition of rigid tiles facilitates computing the various angles and geometric positions. It uses the fact that if we think of the vertex being positioned at the origin, the arm directions can be given by vectors in $\mathbb{R}^3$. Once an arm direction is fixed, its lap plane can be specified by the normal vector to the lap plane.

Definition 2.4. A *rigid tile with n arms* is a set of triples $t = \{(\vec{v}_i, \vec{n}_i, a_i)\}_{i=1,\ldots,n}$, where $\vec{v}_i$ and $\vec{n}_i$ are vectors in $\mathbb{R}^3$ such that $\vec{v}_i$'s are all distinct, $\vec{n}_i$'s are unit, and the a_i's are cohesive-end labels in $\Sigma \cup \hat{\Sigma}$. We consider rigid tiles modulo proper rigid transformations in space, that is, rotation and translation, but without reflection, so that $t_1 = \{(\vec{v}_i, \vec{n}_i, a_i)\}_{i=1,\ldots,n}$ is *equivalent to* $t_2 = \{(\vec{u}_i, \vec{m}_i, a_i)\}_{i=1,\ldots,n}$, if there is a roto-translation matrix S such that the set $\{(\vec{v}_i, \vec{n}_i, a_i)\}_{i=1,\ldots,n}$ equals the set $\{(S\vec{u}_i, S\vec{m}_i, a_i)\}_{i=1,\ldots,n}$.

A rigid tile $t = \{(\vec{v}_i, \vec{n}_i, a_i)\}_{i=1,\ldots,n}$ differs from the flexible tile in precise specification of the arm/lap directions. That is, each cohesive-end a_i is now part of a triple $(\vec{v}_i, \vec{n}_i, a_i)$ where $\vec{v}_i$ specifies the arm direction and length while $\vec{n}_i$ specifies the lap plane normal vector.

The definition of rigid tile requires additional constraints on the definition of assembly design. In particular, when two tiles attach to form an edge e with endpoints $\{u, v\}$, they are positioned relative to each other such that the vector at (u, e) is parallel but oppositely directed to the vector at (v, e), and the normals to the lap planes are likewise parallel and oppositely directed. The cohesive-ends must be the same symbol, but one hatted and the other un-hatted. More precisely, we have the following definition.

Definition 2.5. A *rigid graph* G is a straight edge embedding of a graph G in $\mathbb{R}^3$, that is, the vertices of G are points in $\mathbb{R}^3$ and the edges of G are nonintersecting line segments between the corresponding end vertices.

For example, the cube in Fig 2.12 is a rigid graph that corresponds, as a straight edge embedding, to the cube to the right in Fig. 2.6.

If $u, v \in \mathbb{R}^3$ are endpoints of an edge e for a rigid graph, then vectors $\vec{e} = u - v$ and $-\vec{e} = v - u$ are associated with a line segment between u and v. A rigid graph G also has its half-edges $H(G)$ defined as the collection of half-line segments obtained from the edges with corresponding endpoints. In particular, $H(G) = \{\pm\frac{1}{2}\vec{e} \mid e \in E(G)\}$.

For an assembly design λ of G we set the half-edge (v, e) with vector $\vec{w}_{(v,e)} = \frac{1}{2}\vec{e}$ if $\lambda(v, e)$ is un-hatted and $\vec{w}_{(v,e)} = -\frac{1}{2}\vec{e}$ if $\lambda(v, e)$ is hatted.

A set of *tiles associated with an assembly design* λ of a rigid graph G is a set of rigid tiles $P_\lambda(G) = \{\, t_v \mid v \in V(G)\,\}$ where $t_v = \{(\vec{w}_{(v,e)}, \vec{n}_{e,v}, \lambda(v, e)) \mid e \in E(G), e \in \mu(v)\,\}$ such that for each edge $e \in E(G)$ with $\mu(e) = \{v, u\}$ we have $\vec{n}_{e,v} = -\vec{n}_{e,u}$.

Consider the rigid cube in Fig. 2.12 and the assembly strategy indicated with the labeling of the half-edges as depicted in the figure. The set of tiles assembling the cube are $t_1 = \{(\vec{e}_1, -\vec{n}_1, \hat{a}), (\vec{e}_2, -\vec{n}_2, \hat{a}), (\vec{e}_3, -\vec{n}_3, a)\}$ and $t_2 = \{(\vec{e}_1, \vec{n}_1, a), (\vec{e}_2, \vec{n}_2, a), (\vec{e}_3, \vec{n}_3, a)\}$ where the e_i are the standard unit vectors, and the normals are $\vec{n}_1 = \langle 0, \sqrt{2}/2, \sqrt{2}/2\rangle$, $\vec{n}_2 = \langle\sqrt{2}/2, 0, \sqrt{2}/2\rangle$, and $\vec{n}_3 = \langle\sqrt{2}/2, \sqrt{2}/2, 0\rangle$. If we assume that the vertex of the bottom-left corner in the cube labeled t_2 is placed at the origin, then the other vertex labeled t_2 is positioned at $(2, 2, 2)$, diagonally opposite from the origin. In this case the tile t_2 is translated and rotated. Similarly, one can deduce the transformations that yield the locations of the tiles t_2 in this assembly design.

We end this section with a discussion about a possible way to determine the relative positions of two rigid tiles that have bonded along an edge. The following definitions from [40] provide a framework for convex targets.

Definition 2.6. We say that a rigid tile is *convex* if its vertex and all of its arm-ends are on the boundary of its convex hull.

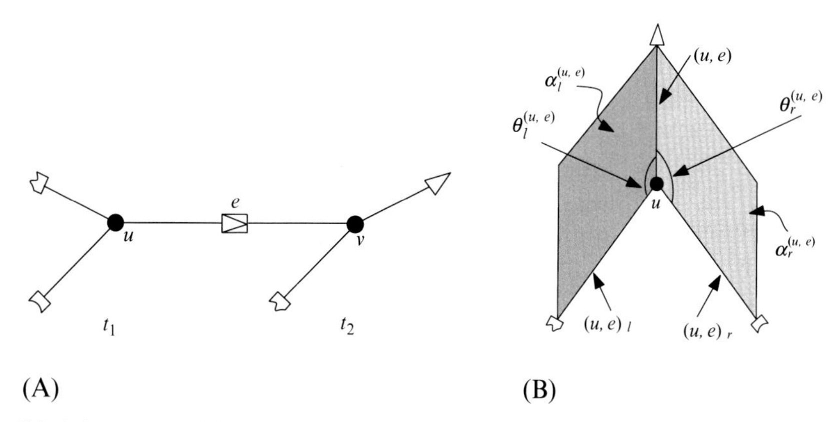

FIG. 2.13 (A) Two rigid tiles attached together to form the edge $e = uv$, (B) notations of the half-edges incident with u with respect to the half-edge (u, e). *(Reprinted with permission from M.M. Ferrari, A. Cook, A. Houlihan, R. Rouleau, N.C. Seeman, G. Pangborn, J. Ellis-Monaghan, Design formalism for DNA self-assembly of polyhedral skeletons using rigid tiles, J. Math. Chem. 56 (5) (2018) 1365–1392.)*

Let t_1 and t_2 be two convex rigid tiles whose central vertices are, respectively, u and v, and assume t_1 and t_2 bond together using two complementary arms to form the edge e with $\mu(e) = \{u, v\}$. Since t_1 is convex, we can fix an orientation of its convex hull by viewing the tile as resting on its arm ends with the vertex rising above, and ordering the edges counterclockwise.

To define the bond-angle, we first need to identify the relevant arms in a consistent way. Thus, we focus on the half-edge (u, e) (see Fig. 2.13A). Moving counterclockwise, the next half-edge is the *left arm relative to* (u, e), denoted $(u, e)_l$. The remaining arm is the *right arm relative to* (u, e), denoted $(u, e)_r$. Let $\alpha_l^{(u,e)}$ be the plane defined by the point u and the (nonparallel) directions given by the arms (u, e) and $(u, e)_l$. Likewise, let $\alpha_r^{(u,e)}$ be the plane defined by u and the vectors parallel to (u, e) and $(u, e)_r$.

Definition 2.7. The *left interarm angle* is the angle between (u, e) and $(u, e)_l$ in the plane $\alpha_l^{(u,e)}$, denoted $\theta_l^{(u,e)}$.

Definition 2.8. The *right interarm angle* is the angle between (u, e) and $(u, e)_r$ in $\alpha_r^{(u,e)}$, denoted $\theta_r^{(u,e)}$ (see Fig. 2.13B).

When we consider t_2 and the half-edge (v, e), in a similar way we obtain the planes $\alpha_l^{(v,e)}$ and $\alpha_r^{(v,e)}$, as well as the left and right interarm angles $\theta_l^{(v,e)}$ and $\theta_r^{(v,e)}$, as in Fig. 2.14. These definitions enable us to define the bond angle, the essential measure of how two tiles are oriented with respect to each other when they bond.

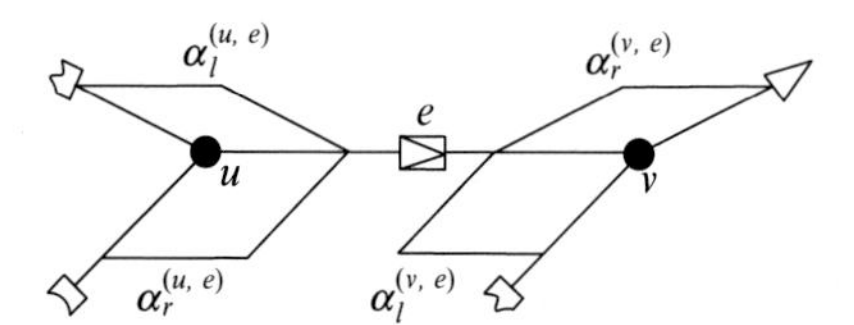

FIG. 2.14 The bond angle of t_1 and t_2 along e. *(Reprinted with permission from M.M. Ferrari, A. Cook, A. Houlihan,R. Rouleau, N.C. Seeman, G. Pangborn, J. Ellis-Monaghan, Design formalism for DNA self-assembly of polyhedral skeletons using rigid tiles, J. Math. Chem. 56 (5) (2018) 1365–1392.)*

Definition 2.9. The *bond angle* of t_1 and t_2 along e is then the angle between the planes $\alpha_l^{(u,e)}$ and $\alpha_l^{(v,e)}$.

Using this half-lap framework, the minimum number of tile types for three-regular Platonic and Archimedean polyhedral skeletons is two, as shown in [40], with one bond-edge type needed for the three-regular Platonic solids. However, design strategies for many other convex polyhedra remain to be determined.

EXERCISES

[2.1] Determine a pot for assembling a tetrahedron from rigid tiles, including determining the lap planes and intertile angles.

[2.2] Open Problem: Determine pots for rigid tile assembly of various polyhedra beyond the Platonic and Archimedean solid skeletons given in [40].

2.5 COMPUTATION BY SELF-ASSEMBLY

In this section, we observe how problems in graph theory can be solved by self-assembly. For example, one can design a pot such that the output of a pot is of certain size if and only if the output represents a solution to a given graph theoretical problem. We illustrate this with the well-known three-colorability problem in graph theory.

Let G be a graph with vertices V and edges E. The graph G is *three-colorable* if there is a surjective (onto) function $f\colon V \to \{r, g, b\}$ such that if there is an edge e with $\mu(e) = \{v, w\}$ then $f(v) \neq f(w)$. In other words, G is three-colorable if it is possible to assign one color to each vertex such that no two adjacent vertices are colored with the same color. The n-colorability of a graph is defined similarly.

Note that every two-colorable graph with more than two vertices is also three-colorable (by choosing one of the vertices to be colored with the third color). The converse, however, does not hold. An example of a three-colorable graph is presented in Fig. 2.15, left.

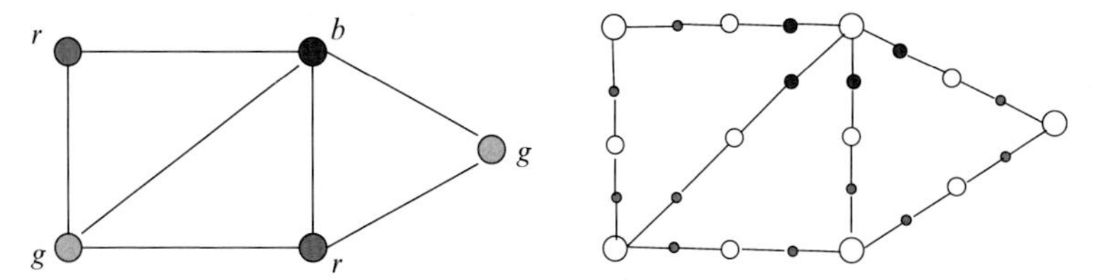

FIG. 2.15 A three-colorable graph (*left*). The corresponding subdivided graph and bond-edge types indicated with *colors* (*right*).

We design a pot P such that the smallest graph realized by P has $\#V + \#E$ tiles if and only if G is three-colorable. Otherwise, if G is not three-colorable, then P realizes only larger graphs. The design is such that the graph realized by P is a subdivision G' of G where each edge of G is split into two edges by inserting a vertex in the middle. An example of a subdivided graph with bond-edge types indicated with colors on the new edges is depicted in Fig. 2.15, right.

We indicate the three colors red, green, and blue with the set $C = \{r, g, b\}$. The set of labels is

$$\Sigma = \{(u, c, v) \mid \exists e \in E(G), \mu(e) = \{u, v\}, c \in C\}.$$

For each $v \in V(G)$, we include three tile types in P. If the vertices $u_1, \ldots, u_k$ are adjacent to v, then the three tiles are $t_v^c = \{ (v, c, u_i) \mid i = 1, \ldots, k\}$ for $c \in C$. For each edge $e \in E(G)$ with end vertices $\mu(e) = \{u, v\}$, we add six 2-arm tiles of the form $t_e^{c,c'} = \{ \widehat{(u, c, v)}, \widehat{(v, c', u)} \}$ for $c \neq c'$, see Fig. 2.16.

The solution of the three-colorability problem for a graph G then can be determined with the following two steps.

Algorithm 2.2.

1. Combine all tile types of pot P and allow the complementary cohesive-ends to hybridize.
2. Determine whether a DNA complex of size $\#V + \#E$ has formed.
   a. Remove partially formed 3D DNA structures with open ends that have not been matched.
   b. From the graphs realized in the above steps, remove by gel electrophoresis the graphs that are larger than G (possible dimers or trimers, see Exercise 2.3 in Section 2.3.2).
   c. If there are structures remaining realizing a graph, then we conclude that the graph is three-colorable.

Observe that $\mathcal{O}(P)$ contains a graph G with $\#V + \#E$ tiles if and only if G is three-colorable. Moreover, one can show that this is the smallest output of

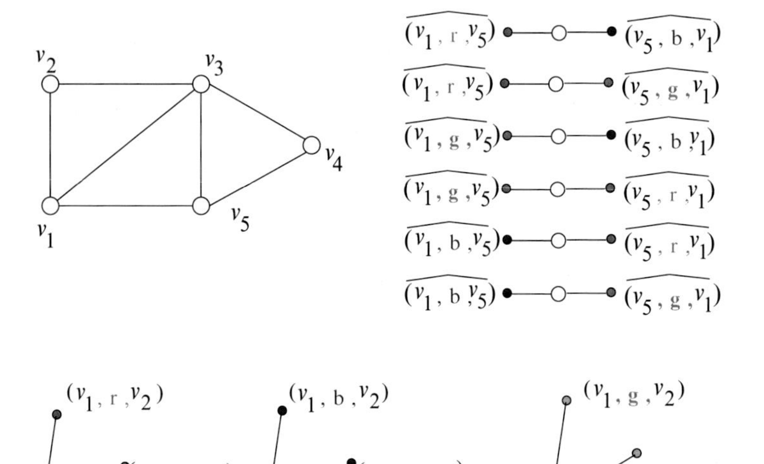

FIG. 2.16 A graph G (*top left*), a set of tile types for edge incident to v_1 and v_5 (*top right*), and at the *bottom*, three tile types representing vertex v_1.

P [41]. We note that the number of laboratory steps in this procedure does not depend on the number of vertices (or edges) in the graph.

Other well-known computational problems such as the Hamiltonian path problem, 3SAT, and others can be solved by pots that realize graph structures [41]. A solution of the three-colorability problem with design of a molecular complex representing a graph (as described earlier) was experimentally obtained in [27].

EXERCISES

[2.1] Prove that for a graph G, Algorithm 2.2 produces a pot P where the smallest graph realized by P represents subdivided G if and only if G is three-colorable.

[2.2] Given integer $k > 0$ and a graph G, design a pot P that will realize all cliques in G of size k. Determine a procedure to compute the maximal clique in G.

[2.3] Design a pot P for a graph G that solves the Hamiltonian cycle problem. That is, identify a graph $\hat{G}$ that can be realized by a pot if and only if G has a Hamiltonian cycle.

2.6 CONCLUSION

The renowned 20th century statistician George Box once said: "All models are wrong. Some models are useful." This chapter presents only one way to describe self-assembly of graphs and polyhedra. Even with these models there are several attributes that warrant further consideration, and improving the models is always a new direction for future research.

A very simple, but subtle, consideration in the tile-based assembly models is the best definition of "size." In this chapter we use the number of vertices to describe the size of a graph-like complex. However, the molecular weight of a DNA construct is more directly related to the number of edges, and even more, the length of the edges. How might results discussed here be adapted to a setting where size is given by the number of edges? How can we characterize the output of the pot according to the number of edges in the graphs realized by the pot? This can be done by adjusting the augmented matrix of a pot, and in this case, discussion about the spectrum of a pot may need to be developed as in [38].

In the flexible tile models, we make the simplifying assumption that the strands are completely flexible and arbitrarily long. In fact, the stacking of DNA strands about the center of a branched junction molecule is not well-understood, and this can determine the interarm angles to some degree. Furthermore, the strands are not in general arbitrarily long, and their lengths correlate to the amount they can bend. How might fixed arm lengths and/or limited flexibility be incorporated into an assembly model?

There are very limited results about pots that consist of rigid tiles. In particular, there are no convenient ways of describing an output of a pot in this setting and any general analysis in this direction will improve our understanding of the self-assembly process. For the 2D case, that is, for 2D rigid tiles assembling on the plane, an initial comparison with the flexible tile model is done in [30]. No such comparison has been done for general 3D rigid tiles. Some advances have been made for rigid tiles that assemble in periodic graphs [42].

There are numerous other examples where self-assembly has led to interesting mathematical questions, and conversely, where mathematics has informed the design of self-assembly questions. For example, Chen and Seeman were able to construct the first polyhedral object by DNA self-assembly, a cube, in 1991 [19] by first creating circular strands that traced the faces of the cubes, which then self-assembled to form the cube. This technique leads to the following mathematical question: what are the minimum and maximum numbers of circular single strands of DNA needed to construct the graph with a detachment-free bidirectional double covering of the edges. This question was addressed in [43–45].

2.7 RESOURCE MATERIALS

Below is a list of additional resources for this chapter, including background material, as well as methods and applications from the literature.

1. Some basic graph theory background is needed in this area, including degree sequences, Euler circuits, Hamilton cycles, directed graphs, and some basic algorithms. Any introductory graph theory book will have this material, for example, the first three chapters of [46].
2. The following are some articles by the authors of this chapter that give further details on the methods and applications described here: [36–38, 41, 47, 48].
3. Here are also some survey articles and books on self-assembly, particularly DNA-based self-assembly: [49–54].
4. For those who may wish to use this chapter as a resource for undergraduate research experiences, Ref. [55] describes the structure and practices of Ellis-Monaghan and Pangborn's research group with undergraduates.

ACKNOWLEDGMENTS

Anna Cook provided many of the figures and exercise solutions, and the authors also thank her for feedback as this chapter developed. Joanna Ellis-Monaghan and Greta Pangborn have been in part supported by NSF grant EFRI-1332411. Nataša Jonoska has been supported in part by NSF grant CCF-1526485 and NIH grant GM109459.

REFERENCES

[1] N.C. Seeman, DNA junctions and lattices, J. Theor. Biol. 99 (1982) 237–247.
[2] L.M. Adleman, Molecular computation of solutions to combinatorial problems, Science 266 (5187) (1994) 1021–1024.
[3] H. Yan, S. Park, G. Finkelstein, J. Reif, T. LaBean, DNA-templated self-assembly of protein arrays and highly conductive nanowires, Science 301 (2003) 1882–1884.
[4] T.H. LaBean, H. Li, Constructing novel materials with DNA, Nano Today 2 (2) (2007) 26–35.
[5] N.C. Seeman, An overview of structural DNA nanotechnology, Mol. Biotechnol. 37 (3) (2007) 246–257.
[6] P. Yin, H. Yan, X.G. Daniell, A.J. Turberfield, J.H. Reif, A unidirectional DNA walker that moves autonomously along a track, Angew. Chem. Int. Ed. Engl. 43 (37) (2004) 4906–4911.
[7] H. Gu, J. Chao, S.-J. Xiao, N.C. Seeman, A proximity-based programmable DNA nanoscale assembly line, Nature 465 (7295) (2010) 202–205.
[8] T. Omabegho, R. Sha, N.C. Seeman, A bipedal DNA Brownian motor with coordinated legs, Science 324 (5923) (2009) 67.
[9] K. Lund, A.J. Manzo, N. Dabby, N. Michelotti, A. Johnson-Buck, J. Nangreave, S. Taylor, R. Pei, M.N. Stojanovic, N.G. Walter, E. Winfree, H. Yan, Molecular robots guided by prescriptive landscapes, Nature 465 (2010) 206–210.
[10] A.K. Geim, K.S. Novoselov, The rise of graphene, Nat. Mater. 6 (3) (2007) 183–191.
[11] H. Kim, S. Yang, S.R. Rao, S. Narayanan, E.A. Kapustin, H. Furukawa, A.S. Umans, O.M. Yaghi, E.N. Wang, Water harvesting from air with metal-organic frameworks powered by natural sunlight, Science (2017) https://doi.org/10.1126/science.aam8743.
[12] N.R. Kallenbach, R.I. Ma, N.C. Seeman, An immobile nucleic acid junction constructed from oligonucleotides, Nature 305 (1983) 829–831.

[13] Y. Wang, J.E. Mueller, B. Kemper, N.C. Seeman, The assembly and characterization of 5-arm and 6-arm DNA junctions, Biochemistry 30 (1991) 5667–5674.
[14] E. Winfree, Algorithmic Self-assembly of DNA, PhD Thesis, California Institute of Technology, Pasadena, CA, 1998.
[15] E. Winfree, F. Liu, L.A. Wenzler, N.C. Seeman, Design and self-assembly of two-dimensional DNA crystals, Nature 394 (6693) (1998) 539–544.
[16] H. Wang, R.J. Di Gate, N.C. Seeman, An RNA topoisomerase, Proc. Natl Acad. Sci. USA 93 (18) (1996) 9477–9482.
[17] D. Liu, G. Chen, U. Akhter, T.M. Cronin, Y. Weizmann, Creating complex molecular topologies by configuring DNA four-way junctions, Nat. Chem. 8 (10) (2016) 907–914.
[18] D. Liu, Y. Shao, G. Chen, Y.-C. Tse-Dinh, J.A. Piccirilli, Y. Weizmann, Synthesizing topological structures containing RNA, Nat. Commun. 8 (2017), https://doi.org/10.1038/ncomms14936.
[19] J. Chen, N.C. Seeman, Synthesis from DNA of a molecule with the connectivity of a cube, Nature 350 (6319) (1991) 631–633.
[20] Y. Zhang, N.C. Seeman, Construction of a DNA-truncated octahedron, J. Am. Chem. Soc. 116 (5) (1994) 1661–1669.
[21] Y. He, T. Ye, M. Su, C. Zhang, A.E. Ribbe, W. Jiang, C. Mao, Hierarchical self-assembly of DNA into symmetric supramolecular polyhedra, Nature 452 (7184) (2008) 198–201.
[22] W. Sun, E. Boulais, Y. Hakobyan, W.L. Wang, A. Guan, M. Bathe, P. Yin, Casting inorganic structures with DNA molds, Science 346 (6210) (2014) 1258361.
[23] R. Iinuma, Y. Ke, R. Jungmann, T. Schlichthaerle, J.B. Woehrstein, P. Yin, Polyhedra self-assembled from DNA tripods and characterized with 3D DNA-PAINT, Science 344 (6179) (2014) 65–69.
[24] J. Zheng, J.J. Birktoft, Y. Chen, T. Wang, R. Sha, P.E. Constantinou, S.L. Ginell, C. Mao, N.C. Seeman, From molecular to macroscopic via the rational design of a self-assembled 3D DNA crystal, Nature 461 (7260) (2009) 74–77.
[25] N. Jonoska, P. Sa-Ardyen, N.C. Seeman, Computation by self-assembly of DNA graphs, Genet. Program Evolvable Mach. 4 (2003) 123–137.
[26] P. Sa-Ardyen, N. Jonoska, N. Seeman, Self-assembling DNA graphs, Nat. Comput. 4 (2003) 427–438.
[27] G. Wu, N. Jonoska, N.C. Seeman, Construction of a {DNA} nano-object directly demonstrates computation, Biosystems 98 (2) (2009) 80–84.
[28] N.C. Seeman, Nucleic acid junctions and lattices, J. Theor. Biol. 99 (2) (1982) 237–247.
[29] N. Jonoska, G.L. McColm, A Computational Model for Self-assembling Flexible Tiles, vol. 3699, Springer, Berlin, Heidelberg, 2005, pp. 142–156.
[30] N. Jonoska, G.L. McColm, A. Staninska, Expectation and variance of self-assembled graph structures, in: DNA Computing, Springer, 2006, pp. 144–157.
[31] N.C. Seeman, Macromolecular design, nucleic acid junctions and crystal formation, J. Biomol. Struct. Dyn. 3 (1) (1985) 11–34.
[32] N. Jonoska, G.L. McColm, A. Staninska, Spectrum of a Pot for DNA Complexes, vol. 4287, Springer, Berlin, Heidelberg, 2006, pp. 83–94.
[33] A. Staninska, The graph of a pot with DNA molecules, in: Proceedings of the 3rd Annual Conference on Foundations of Nanoscience (FNANO'06), 2006, pp. 222–226.
[34] C. Tian, X. Li, Z. Liu, W. Jiang, G. Wang, C. Mao, Directed self-assembly of DNA tiles into complex nanocages, Angew. Chem. Int. Ed. 126 (31) (2014) 8179–8182.
[35] P. Sa-Ardyen, N. Jonoska, N.C. Seeman, Self-assembling DNA graphs, in: DNA Computing, Springer, New York, NY, 2003, pp. 1–9.

[36] N. Jonoska, G.L. McColm, Complexity classes for self-assembling flexible tiles, Theor. Comput. Sci. 410 (4) (2009) 332–346.

[37] J. Ellis-Monaghan, G. Pangborn, L. Beaudin, D. Miller, N. Bruno, A. Hashimoto, Minimal Tile and Bond-Edge Types for Self-Assembling DNA Graphs, Springer, Berlin, Heidelberg, 2014, pp. 241–270.

[38] N. Jonoska, G.L. McColm, A. Staninska, On stoichiometry for the assembly of flexible tile DNA complexes, Nat. Comput. 10 (3) (2011) 1121–1141.

[39] W.M. Shih, J.D. Quispe, G.F. Joyce, A 1.7-kilobase single-stranded DNA that folds into a nanoscale octahedron, Nature 427 (6975) (2004) 618–621.

[40] M.M. Ferrari, A. Cook, A. Houlihan, R. Rouleau, N.C. Seeman, G. Pangborn, J. Ellis-Monaghan, Design formalism for DNA self-assembly of polyhedral skeletons using rigid tiles, J. Math. Chem. 56 (5) (2018) 1365–1392.

[41] N. Jonoska, S.A. Karl, M. Saito, Three dimensional DNA structures in computing, BioSystems 52 (1) (1999) 143–153.

[42] N. Jonoska, M. Krajcevski, G. McColm, Counter machines and crystallographic structures, Nat. Comput. 15 (2015) 97–113.

[43] N. Jonoska, M. Saito, Boundary Components of Thickened Graphs, Revised Papers of 7th International Meeting on DNA Based Computers, vol. 2340, Springer LNCS, 2002, pp. 70–81.

[44] J. Ellis-Monaghan, Transition polynomials, double covers, and biomolecular computing, Congressus Numerantium 166 (2004) 181–192.

[45] N. Jonoska, N.C. Seeman, G. Wu, On existence of reporter strands in DNA-based graph structures, Theor. Comput. Sci. 410 (15) (2009) 1448–1460.

[46] A. Tucker, Applied Combinatorics, sixth ed., Wiley, London, 2012.

[47] J. Ellis-Monaghan, G. Pangborn, Using DNA self-assembly design strategies to motivate graph theory concepts, Math. Model. Nat. Phenom. 6 (6) (2011) 96–107.

[48] J. Ellis-Monaghan, G. Pangborn, N.C. Seeman, S. Blakeley, C. Disher, M. Falcigno, B. Healy, A. Morse, B. Singh, M. Westland, Design tools for reporter strands and DNA origami scaffold strands, Theor. Comput. Sci. 671 (6) (2017) 69–78.

[49] Y. Ke, Designer three-dimensional DNA architectures, Curr. Opin. Struct. Biol. 27 (2014) 122–128.

[50] D. Luo, The road from biology to materials, Mater. Today 6 (11) (2003) 38–43.

[51] J. Nangreave, D. Han, Y. Liu, H. Yan, DNA origami: a history and current perspective, Curr. Opin. Chem. Biol. 14 (5) (2010) 608–615.

[52] W.-Y. Qiu, Z. Wang, G. Hu., Chemistry & Mathematics of DNA Polyhedra (DNA: Properties and Modifications, Functions and Interactions, Recombination and Applications), Nova Science Publishers, UK, 2010.

[53] J. Pelesko, Self Assembly: The Science of Things That Put Themselves Together, Taylor & Francis Group, LLC, Boca Raton, Florida, 2007.

[54] N.C. Seeman, Structural DNA Nanotechnology, Cambridge University Press, Cambridge, MA, 2016.

[55] J. Ellis-Monaghan, G. Pangborn, An example of practical organization for undergraduate research experiences, PRIMUS 23 (9) (2013) 805–814.

FURTHER READING

[56] M.P. Ball, Available from: https://commons.wikimedia.org/wiki/File:Ligation.svg. (accessed 07.07.18).

Chapter 3

Graphs Associated With DNA Rearrangements and Their Polynomials

Robert Brijder*, Hendrik Jan Hoogeboom†, Nataša Jonoska‡ and Masahico Saito‡

**Department WET-INF, Hasselt University, Diepenbeek, Belgium, †Department of Computer Science (LIACS), Leiden University, Leiden, The Netherlands, ‡Department of Mathematics and Statistics, University of South Florida, Tampa, FL, United States*

3.1 INTRODUCTION

In the last couple of decades, methods from graph theory have been more prominently used in descriptions of a variety of biological processes, as well as in the development of algorithms for genetic sequence analysis. For example, some of the main expansion in algorithms used for DNA sequencing and recent shotgun approaches are based on Eulerian cycles and paths within so-called de Bruijn graphs [1]. This chapter concentrates on DNA recombination events modeled by directed graphs and graph polynomials that capture some aspects of these processes. It has been known for over 60 years that chromosomal DNA rearrangements on an evolutionary scale can lead to species-specific differences through gene recombination events [2, 3]. The studies of rearrangements in mammalian genomes have observed that these events are more common within the same chromosome (intrachromosomal) rather than across different chromosomes [4]. Evolutionary recombination events have also been modeled with a double-cut-and-join operation, representing molecules as cycles and paths and the recombination through cuts and joins of these cycles and paths. The number of such operations can be used as a measure of genetic distance for evolutionary analysis in the design of evolutionary trees [1, 5, 6]. On a developmental scale, DNA rearrangements can specify gene expression, most commonly involving DNA deletion [7–10], and these events can be found even in humans [11]. Recently it has been observed that recombination events also occur in mitochondrial DNA across a variety of species [12, 13].

Algebraic and Combinatorial Computational Biology. https://doi.org/10.1016/B978-0-12-814066-6.00003-9

DNA recombination has been theoretically studied with models for DNA sequence reorganization (e.g., [14–17]), topological processes of knotting (e.g., [18–21]), as well as studies of RNA-guided recombination [22, 23].

In this chapter we concentrate on a theoretical model that describes DNA recombination processes in certain species of ciliates [9, 14, 24, 25]. In this process, called *gene assembly*, it has been observed that an additional molecule (called a template) takes part in the recombination process [26]. Based on these observations, a theoretical model with four-regular graphs (possibly with two degree-1 vertices) representing the molecule(s) at the time of recombination was introduced in [14, 23], called an *assembly graph*. This model describes the precursor molecule as an Eulerian circuit in the graph, while each degree-4 vertex represents the location of the homologous recombination. The product molecule after recombination is modeled as a special path in the graph visiting all vertices. We use aspects of the general theory of Eulerian circuits in four-regular graphs which was initiated in a seminal paper by Kotzig [27] and was further developed by Bouchet, relating it to delta-matroids [28], isotropic systems [29], and multimatroids [30].

In a four-regular graph, the sequence of vertices listed in the order visited by an Eulerian circuit forms a (double occurrence) word. In [14], three types of rewriting rules on double occurrence words (DOWs) are proposed and studied that reflect feasible DNA recombination processes. The recombination events can be modeled from a graph theory perspective by deviating from the Eulerian circuit at the degree-4 vertices, which is described by changing transitions at each vertex (see Section 3.3).

Motivated by polynomial invariants in graph and knot theory (see, e.g., [31–33]), a polynomial related to DNA rearrangements, called the *assembly polynomial*, has been defined and studied in [34, 35]. The number of connected components after change of transitions at every vertex in an assembly graph corresponds to the number of molecules after all recombination events [35]. Experimentally, circular DNA molecules excised from the precursor DNA have been observed after the rearrangements [36]. Such molecules correspond to the cyclic connected components of the assembly graph after change of transitions at all vertices. The assembly polynomial is a formal description of the possible number of connected components (molecules) *after* recombination. The recombination process, on the other hand, occurs in cascades as a sequence of events, rather than simultaneously. In terms of graphs, this is modeled by performing changes of transitions only at subsets of vertices, and the changed transitions describe partially rearranged molecules. The *rearrangement polynomial* introduced in Section 3.5.4 is a formal description of the possible number of connected components (molecules) *during* recombination.

In the next section we give the biological background for the DNA rearrangement processes in the model species of ciliates *Oxytricha* and *Stylonychia*. Section 3.2.1 introduces the biological notions, while Section 3.2.2 uses one example to introduce the description of the process with four-regular graphs.

This example, used throughout the chapter to illustrate all notions and computations, comes from an observed situation in a biological data. Section 3.3 gives the mathematical preliminaries. The graphs are defined through half-edges, and the notion is further used to define transitions at vertices and circuits. The mathematical model for DNA rearrangement is introduced in Section 3.4 where we illustrate the recombination process through the same example from Section 3.2.2. Section 3.5 defines graph polynomials and introduces both the assembly polynomial and the rearrangement polynomial. We give a recursive algorithm for their computation and illustrate this algorithm with examples.

3.2 GENE ASSEMBLY IN CILIATES

3.2.1 Biological Background

Ciliates are single-cell organisms that possess hair-like organelles called cilia. They are ubiquitous in every habitat of our planet and are genetically and evolutionary most diverse organisms. Even some closely related species can evolutionary be as distant as a cucumber and an elephant. Frequently, species of ciliates are used as model organisms for studies of some biological processes. The mathematical models that we present here use as model organisms several genera of ciliates, such as *Oxytricha* and *Stylonychia*. These species undergo massive DNA rearrangement during their sexual reproduction, in a process called *gene assembly*, and in this chapter we present models that describe this process. First, we provide a brief description of the biological phenomenon and introduce the biological notions that are modeled in our mathematical descriptions. We refer the reader to the reviews [37, 38] and references therein for more details. All ciliate species are unicellular Eukaryote organisms and have two types of nuclei, a germline *micronucleus* (MIC) and a transcriptionally active somatic *macronucleus* (MAC). The macronuclear DNA of *Oxytricha trifallax* consists of about 16,000 chromosomes each appearing in multiple copies. Each of these chromosomes is short (compared with chromosomes in other species) and most encode one to two genes, occasionally more, but never more than eight. During sexual reproduction, new micronuclei from the two parental cells are formed, the old MAC from the two mating cells disintegrate, and new MAC form from one of the new MIC. Within this process, the MAC chromosomes assemble through extensive DNA processing events which we describe with our models below.

The DNA processing includes deletion of all so-called “junk” DNA by eliminating 90%–98% of the MIC DNA. Because the genes in the MIC are interrupted by intervening DNA segments, called *internal eliminated sequences* (IESs), the deletion process eliminates all these segments. The IESs interrupt micronuclear coding regions, so each MAC gene may appear as several segments, called *macronuclear destined sequences* (MDSs) in the MIC. Moreover, the order of these MDS segments can be permuted or inverted (i.e., rotated by

180 degrees) in the MIC relative to the order of these segments in the MAC. As noted in [39], about 3000 genes are scrambled in the MIC of *O. trifallax* with the number of MDSs varying from 2 to over 100. During the MAC development, besides the deletion of all of the IESs, the MDSs are rearranged (translated, inverted, etc.) to form the gene-sized chromosomes. Therefore, assembly of the new macronuclear chromosomes may require any combination of the following three processes: descrambling (permuting) of MDSs, inversion of MDSs, and deletion of IESs.

There are several theoretical models that have attempted to describe these rearrangement processes [14, 23, 41–43]. In [26], it was experimentally demonstrated that an additional molecule, a template RNA (or DNA) guides the recombination process. This model was theoretically proposed in [23, 43].

From early studies of the recombination process in these and related species of ciliates it was observed that there are repeated DNA segments of 2–30 nucleotides at the end of the ith MDS and the beginning of the $(i+1)$-st MDS. These repeats appear as a single copy in the rearranged MAC gene. It is hypothesized that during the rearrangement, the template uses these repeats to align and facilitate joining of the consecutive MDSs (where by "consecutive" we mean consecutive in the MAC, so, e.g., M_i and M_{i+1} are consecutive MDSs). Therefore, the repeated segments, called *pointers*, are used in almost all theoretical models of this process. The pointers can be labeled with symbols (letters form an alphabet) and the arrangement of MDSs in the MIC can be described with a list of pointer labels. Because every pointer in the MIC appears in two copies, the list is a word where every symbol appears twice, called a DOW. In this chapter, the rearrangement process is described through graphs and DOWs.

3.2.2 Motivational Example

Fig. 3.1 shows an example of a genetic rearrangement in *O. trifallax*. The precursor MIC sequence in this educational example can be described as $I_0\,\overline{M}_2\,I_1\,M_1\,I_2\,\overline{M}_3\,I_3\,\overline{M}_4\,I_4$ where M_i denotes the ith MDS that occurs in the ith place in the rearranged MAC gene, and the I_i's, for $i \in \{1,\ldots,5\}$, denote the IESs which are excised sequences. The bars, such as in $\overline{M}_2$ of MDS M_2, indicate inverted MDSs in the MIC relative to their orientation in the MAC.

In the MAC sequence, the consecutive M_is are joined together on their overlapping pointers. Since these pointers are both a biological phenomenon and characteristic for the rearrangement (describing that, e.g., the end of M_2 must be glued to the start of M_3), we write them explicitly as $p_{i-1}\,M_i\,p_i$ to denote the ith MDS with its pointers. When the segment M_i is inverted, its pointers also become inverted and we write $\overline{p}_i\,\overline{M}_i\,\overline{p}_{i-1}$. Now the extended notation for the MIC is $I_0\,\overline{p}_2\,\overline{M}_2\,\overline{p}_1\,I_1\,M_1\,p_1\,I_2\,\overline{p}_3\,\overline{M}_3\,\overline{p}_2\,I_3\,\overline{M}_4\,\overline{p}_3\,I_4$ (see Fig. 3.2). We can further abstract this notation to write $\overline{p}_2\,\overline{p}_1\,p_1\,\overline{p}_3\,\overline{p}_2\,\overline{p}_3$ or simply $\overline{2}\,\overline{1}\,1\,\overline{3}\,\overline{2}\,\overline{3}$ or even 2 1 1 3 2 3. This representation is in the form of a DOW, signed in the former case

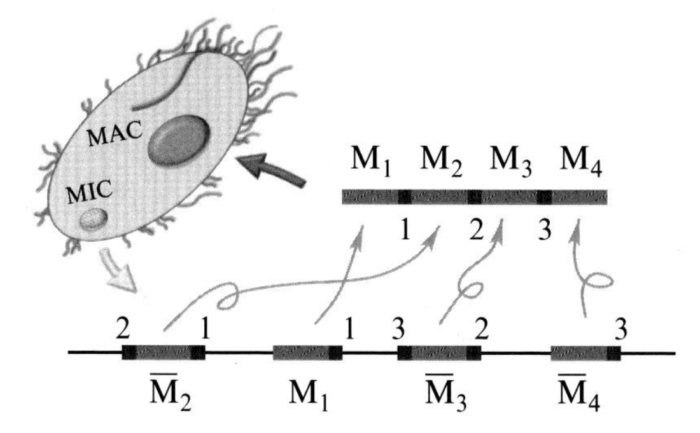

FIG. 3.1 Schematics of the gene segments rearrangement in *Oxytricha trifallax*. The precursor MIC DNA sequence has gene segments MDSs (red segments including their blue ends) interrupted with IESs (black lines). MDSs may be inverted (as in $\overline{M}_2$) or appear in a scrambled order (such as $\overline{M}_2$ being before M_1). Each MDS is flanked by pointers (blue ends). This is a simplified depiction of a rearrangement in actual MIC and MAC sequences, numbered OXYTRI_MIC_87901 and OXYTRI_MAC_19159, respectively [40].

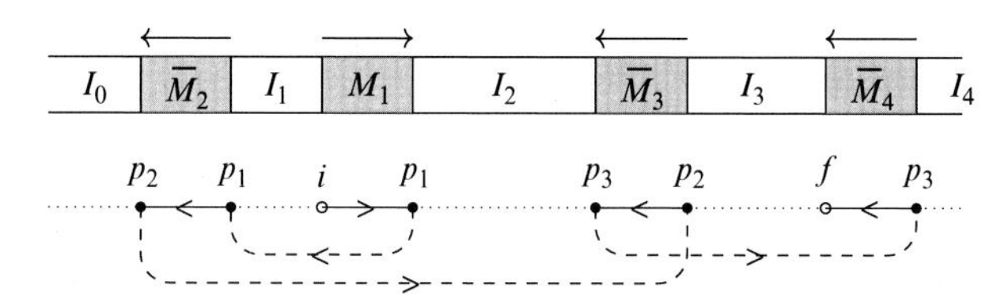

FIG. 3.2 Scrambled gene sequence with IESs I_i numbered in their order on the MIC and MDSs M_j oriented and ordered following the MAC. The MAC can be read from the MIC by starting in M_1 and jumping each time from a pointer at the end of an MDS to its copy at the beginning of the consecutive MDS.

and unsigned in the latter case. Observe that the first and the last MDS have only one pointer as they are recombined with only a single other MDS. Likewise, the rearranged MAC segment can be denoted as $M_1\, p_1\, M_2\, p_2\, M_3\, p_3\, M_4$. In the MAC, each of the pointers occurs only once.

Both the MIC and MAC sequences can be represented in a single graph, as we illustrate using the example of Fig. 3.2. Start by introducing vertices p_1, p_2, p_3 corresponding to the pointers of the MIC sequence. Add edges that follow the segments of the MIC between the pointers. We have, for instance, an edge between p_2 and p_1 with label $\overline{M}_2$, a loop at p_1 with label $I_1\, M_1$, an edge between p_1 and p_3 with label I_2, etc. Finally, we have an edge between the last pointer of the MIC sequence and the first pointer with label $I_4\, I_0$ that represents both the right-most segment I_4 and the left-most segment I_0 of the MIC. By construction, the graph has an Eulerian cycle (see Section 3.3) tracing the MIC sequence $I_0\, \overline{M}_2\, I_1\, M_1\, I_2\, \overline{M}_3\, I_3\, \overline{M}_4\, I_4$. If we traverse the edges at each of the pointers in a different way, as indicated with the double (red) line, we can trace the MAC sequence $M_1\, M_2\, M_3\, M_4$. Note, however, that we

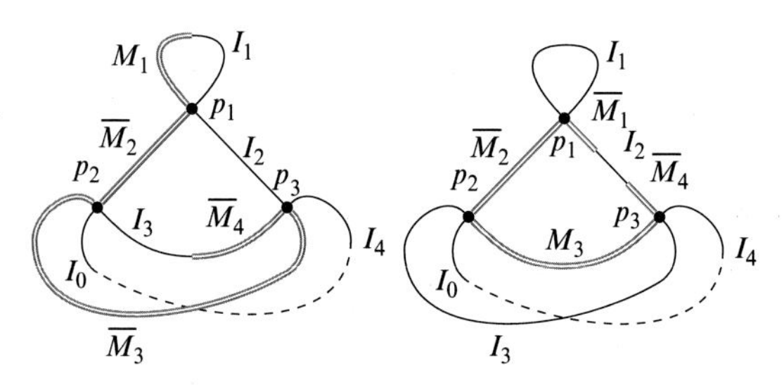

FIG. 3.3 Two isomorphic graphs representing DNA assembly of two different MIC sequences. The black line corresponds to the MIC segment, dotted line is added to connect the end-points (see Section 3.3). The MDS layout within the MIC is annotated by double lines. The MDS sequence in MIC read $\overline{M}_2M_1\overline{M}_3\overline{M}_4$ (*left*) and $\overline{M}_2\overline{M}_1\overline{M}_4M_3$ (*right*), respectively.

have to read the edges with inverted MDSs against their original direction, hence $\overline{M}_2$ in opposite direction becomes M_2. In fact, if we recombine not only the consecutive MDSs, but also the remaining edges at the vertices, we see that after recombination the MAC actually consists of a long-strand reading $I_1\,M_1,\ldots,M_4\,\overline{I}_3\,\overline{I}_0\,\overline{I}_4\,\overline{I}_2$, indicating the location and arrangement of the IES segments after the rearrangement.

Consider another example, where the MIC is described by $I_0\,\overline{M}_2\,I_1\,\overline{M}_1\,I_2\,\overline{M}_4\,I_3\,M_3\,I_4$. Applying the above graph construction, we obtain a graph that is structurally equal, that is, isomorphic (both examples spell out the same unsigned pointer sequence $p_2\,p_1\,p_1\,p_3\,p_2\,p_3$), but the MDSs lie on different edges; see Fig. 3.3 (right). If we trace the MAC in the resulting graph, we find a circular molecule $M_1\cdots M_4\,\overline{I}_2$, and observe that after recombination two (circular) molecules have been excised: $I_0\,\overline{I}_3\,I_4$ and I_1.

3.3 MATHEMATICAL PRELIMINARIES

To properly describe the DNA rearrangement process, we use undirected multigraphs where loops are allowed. Moreover, since we consider circuits in G where we distinguish between the two ways in which we can traverse a loop, we need to distinguish the two ends of a loop. This leads to a notion of multigraph that is defined using the concept of half edges. For notational convenience, we call this type of multigraph simply a graph.

Definition 3.1. A (undirected) *graph* G is a four-tuple (V,H,E,ϵ), where V and H are finite sets, E is a partition of H in (unordered) pairs, and $\epsilon\colon H\to V$ is a function. The elements of V, H, and E are called *vertices*, *half edges*, and *edges* of G, respectively.

A vertex v and half edge h are called *incident* if $\epsilon(h)=v$, we also say that v is the *endpoint* of h. A vertex v and an edge e are called *incident* if v is incident to an $h\in e$. An edge e is a *loop* if it is incident to exactly one vertex. The *degree* of a vertex is the number of half edges incident to it.

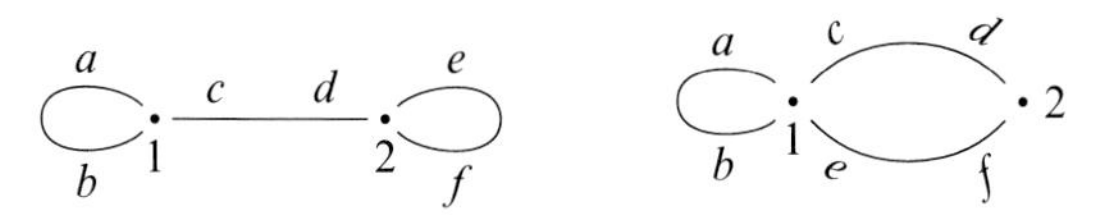

FIG. 3.4 Graphs G_1 (*left*) and G_2 (*right*) of Example 3.1.

Example 3.1. Fig. 3.4 shows two graphs G_1 and G_2, each with the same sets $V = \{1, 2\}$ of vertices, $H = \{a, b, c, d, e, f\}$ of half-edges, and $E = \{\{a, b\}, \{c, d\}, \{e, f\}\}$ of edges. For G_1, the endpoint function ϵ is defined by $\epsilon(h) = 1$ for $h \in \{a, b, c\}$ and $\epsilon(h) = 2$ for $h \in \{d, e, f\}$. Thus G_1 has two loops, $\{a, b\}$ and $\{e, f\}$, and one edge incident to vertices 1 and 2. For G_2, the endpoint function ϵ is defined by $\epsilon(h) = 1$ for $h \in \{a, b, c, e\}$ and $\epsilon(h) = 2$ for $h \in \{d, f\}$. Thus G_2 has one pair of mutually parallel edges and one loop.

A *single transition of G at vertex v* is an (unordered) pair of half edges, each incident to v. For example, the pair $\{a, c\}$ is a single transition at vertex 1 for both graphs at Fig. 3.4. We use single transitions to indicate a way to pass through a vertex with a path or a circuit in a graph. In this chapter, by a "circuit" we mean an unoriented closed path without a designated starting vertex. Before we define a circuit in this setting, we recall the more familiar notion of a path and an *oriented* closed path *with* a designated starting vertex. A *path* is a sequence $\pi = (h_0, \dots, h_{2k+1})$ of distinct half edges of G with $k \geq 0$ such that for all $i \in \{0, \dots, k\}$, $\{h_{2i}, h_{2i+1}\}$ is an edge of G, and for $i \in \{1, \dots, k\}$, $\{h_{2i-1}, h_{2i}\}$ is a single transition of G at vertex $\epsilon(h_{2i-1}) = \epsilon(h_{2i})$. The *starting vertex* of π is $\epsilon(h_0)$ and the *ending vertex* of π is $\epsilon(h_{2k+1})$. The *length* of π is $k + 1$. If a path $\pi = (h_0, \dots, h_{2k+1})$ is such that $\epsilon(h_0) = \epsilon(h_{2k+1})$, implying that $\{h_0, h_{2k+1}\}$ is a single transition at vertex $\epsilon(h_0) = \epsilon(h_{2k+1})$, then π is called an *oriented closed path* with starting vertex $\epsilon(h_0)$. The idea here is that we start this closed path in vertex $\epsilon(h_0)$ (the starting vertex) and proceed in the direction of h_0. The other half edge of the edge containing h_0 is h_1. So after traversing h_0, we traverse h_1 after which we visit vertex $\epsilon(h_1) = \epsilon(h_2)$ to proceed further in the direction of h_2, and so on. When $k = 0$, the path becomes (h_0, h_1) which can be seen as a directed version of a single edge. If this path is moreover closed, then it can be seen as a directed version of a loop.

We now define the notion of a circuit as a specific set of single transitions.

Definition 3.2. A *circuit* C of a graph G is a nonempty set of mutually disjoint single transitions of G such that the half edges in $\cup C$ can be ordered $(h_0, \dots, h_{k-1})$ to represent an oriented closed path.

Example 3.2. Consider again the graph G_2 from Example 3.1 depicted on the right-hand side of Fig. 3.4. The nonempty set $C = \{\{b, c\}, \{d, f\}, \{a, e\}\}$ of mutually disjoint single transitions is a circuit of G_2. Indeed, for example, $\pi = (c, d, f, e, a, b)$ is an oriented closed path of G_2. This circuit traverses all edges of G_2. Informally, we can represent π as $1\overset{c\,d}{—}2\overset{f\,e}{—}1\overset{a\,b}{—}1$ to make the relation between vertices and (half) edges explicit.

By "swapping" the single transitions at vertex 1 from $\{a,e\},\{b,c\}$ to $\{a,c\},\{b,e\}$ we obtain another circuit $C' = \{\{a,c\},\{d,f\},\{b,e\}\}$, for which a corresponding oriented closed path can be represented by $1\overset{c\,d}{—}2\overset{f\,e}{—}1\overset{b\,a}{—}1$. Note that C' also traverses all edges of G_2: the only difference with the C is that the loop at vertex 1 is traversed in the opposite direction.

A *circuit partition* of G is a set of circuits P partitioning the set of half edges. Note that not all graphs have circuit partitions. For example, the graph to the left-hand side of Fig. 3.4 has no circuit partition (see Exercise 3.1). Denote the set of circuit partitions of G by $\mathcal{P}(G)$. A circuit C is called *Eulerian* in G if $\cup C = H$, in other words, if $\{C\}$ is a circuit partition of G. An *Euler system* in G is a circuit partition P of G such that every circuit of P is Eulerian for some connected component of G. A graph is called *Eulerian* if it has an Euler system (in the literature, the notion of Eulerian is sometimes defined in a stronger sense where it is additionally assumed that the graph is connected). It is well known that a graph is Eulerian if and only if it has a circuit partition and this holds if and only if each vertex has even degree.

We now consider an alternative way to look at circuit partitions. A *transition* at a vertex v of G is a partition in (unordered) pairs of the set of half edges incident to v (consequently, the elements of a transition are single transitions). Observe that only a vertex of even degree has a transition, and a vertex has a unique transition if its degree is two. Vertices with higher even degrees have more than one transition. A *transition system* of G is a set that contains exactly one transition for each vertex. Denote the set of transition systems of G by $\mathcal{T}(G)$. Each transition system T of G uniquely determines the circuit partition P_T of G consisting of those circuits C where $C \subseteq \cup T$. Conversely, each circuit partition P of G uniquely determines a transition system T_P of G consisting of those transitions t that contain single transitions of circuits of P.

Example 3.3. Consider again the graph G_2 from Example 3.1. There is only one single transition at vertex 2, namely $\{d,f\}$. Thus $t_2 = \{\{d,f\}\}$ is the unique transition at 2. Note that $t_1 = \{\{a,b\},\{c,e\}\}$ is a transition at 1. Hence $T = \{t_1,t_2\}$ is a transition system of G_2. The circuit partition P_T corresponding to T consists of two circuits, $\{\{a,b\}\}$ and $\{\{d,f\},\{c,e\}\}$.

Note that there are precisely three transitions at a vertex v of degree 4. Each of the transitions can be uniquely described in terms of a fixed Euler system P of G (assuming such P exists, i.e., G is Eulerian). Indeed, since P is an Euler system, there is a circuit $C = \{h_0,\ldots,h_{k-1}\}$ of P with $\epsilon(h_i) = v = \epsilon(h_{i+1 \bmod k})$ and $\epsilon(h_j) = v = \epsilon(h_{j+1 \bmod k})$ for some distinct and odd $i,j \in \{1,\ldots,k-1\}$. We say that at vertex v, the transition

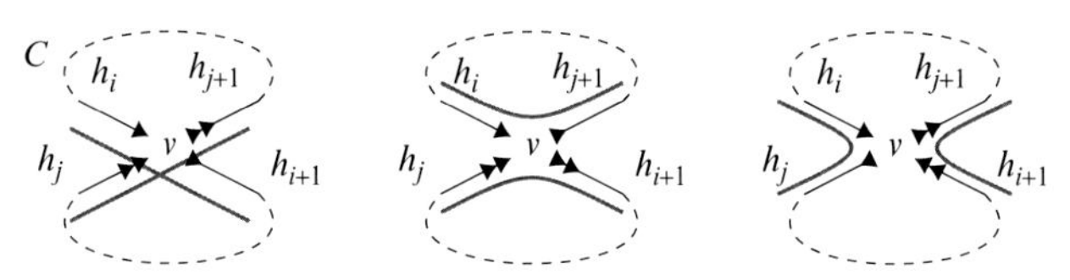

FIG. 3.5 Three possible transitions at a degree-4 vertex v with respect to a fixed circuit $C = \ldots \underline{h_i}\, v\, \underline{h_{i+1}} \ldots \underline{h_j}\, v\, \underline{h_{j+1}} \ldots$. The half edges indicated with $h_i, h_{i+1}, h_j, h_{j+1}$ are incident to v. Half edges forming a single transition have identical arrow heads, and we use this notation in subsequent figures to indicate transitions at vertices.

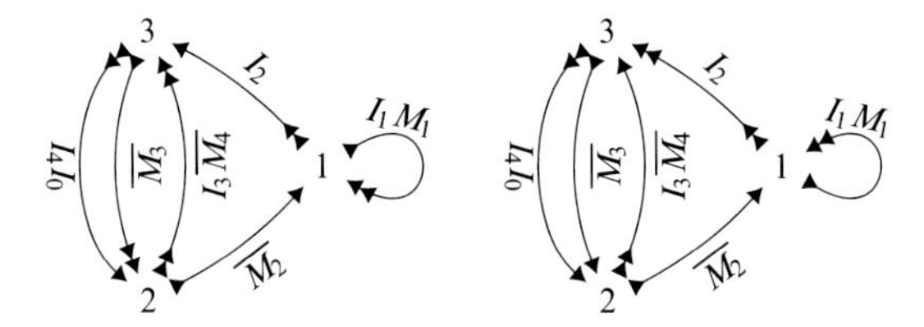

FIG. 3.6 Depiction of two Eulerian circuits of the four-regular graph corresponding to the left-hand side of Fig. 3.3. The Eulerian circuit on the *left-hand side* represents the MIC sequence and the circuit on the *right-hand side* represents the MAC sequence. See Example 3.4.

- $\{\{h_i, h_{i+1 \mod k}\}, \{h_j, h_{j+1 \mod k}\}\}$ *follows* C (Fig. 3.5, left)
- $\{\{h_i, h_{j+1 \mod k}\}, \{h_i, h_{j+1 \mod k}\}\}$ is *orientation-consistent* with C (Fig. 3.5, middle)
- $\{\{h_i, h_j\}, \{h_{i+1 \mod k}, h_{j+1 \mod k}\}\}$ is *orientation-inconsistent* with C (Fig. 3.5, right)

Note that these definitions are well-defined, that is, they are independent of the chosen circuit orientation or starting vertex.

A graph G is called *k-regular* if every vertex of G has degree k. In this chapter we are mostly interested in four-regular graphs.

Example 3.4. Fig. 3.6 depicts two Eulerian circuits of the four-regular graph corresponding to the left-hand side of Fig. 3.3. Each Eulerian circuit is indicated by marking, for each single transition of the corresponding transition system, its half edges by identical arrow heads. If one reads the labels of the edges following the Eulerian circuit C_1 on the left-hand side of Fig. 3.6 it reads the MIC sequence: $I_0\, \overline{M}_2\, I_1\, M_1\, I_2\, \overline{M}_3\, I_3\, \overline{M}_4\, I_4$. The Eulerian circuit C_2 on the right-hand side reads the MAC sequence (in reverse): $I_3\, \overline{M}_4\, \overline{M}_3\, \overline{M}_2\, \overline{M}_1\, \overline{I}_1\, I_2$. Here, when we traverse an edge in the opposite direction from that in the MIC, we invert its label (which happens only at the loop).

At each vertex, the transition taken for C_1 (MIC sequence) differs from the transition taken for C_2 (MAC sequence). Note that each transition of C_2 is either (i) orientation-consistent with C_1 whenever the two corresponding pointers in the MIC have the same orientation, or (ii) orientation-inconsistent whenever they are in the opposite orientation (i.e., pointer p_1).

EXERCISES

[3.1] Prove that a graph has a circuit partition if and only if every vertex has even degree.

[3.2] Explicitly determine a four-tuple (V, H, E, ϵ) that defines the four-regular graph shown in Fig. 3.6.

[3.3] **(a)** Determine the transition systems T and T' that correspond to the left and right circuit partitions P_T and $P_{T'}$, respectively, of the graph in Fig. 3.6.

(b) Find another transition system T'' for the graph in Fig. 3.6 that is different than T and T'. Determine the circuit partition $P_{T''}$ and check whether it is Eulerian. At every vertex, indicate whether the transitions in T, T', and T'' are orientation-consistent with circuits C_T and/or $C_{T'}$.

3.4 MATHEMATICAL MODELS FOR GENE REARRANGEMENT

3.4.1 Graphs Obtained From Double Occurrence Words

In this section we formalize the gene rearrangement process that occurs during gene assembly in some species of ciliates as detailed in Section 3.2.1. The mathematical model is based on four-regular graphs and Eulerian circuits of these graphs which can be represented by words where every symbol occurs twice.

Recall that a DOW is a word w such that each letter of w occurs exactly twice in w. For any set Σ, denote by $\overline{\Sigma} = \{\overline{s} \mid s \in \Sigma\}$ a disjoint copy of Σ. We take that $\overline{\overline{s}} = s$. A *signed* DOW over $\Sigma \cup \overline{\Sigma}$ is a word $\hat{w}$ over $\Sigma \cup \overline{\Sigma}$ such that by removing all the bars from $\hat{w}$ we obtain a DOW. For a signed DOW $\hat{w}$, a letter $p \in \Sigma \cup \overline{\Sigma}$ such that p or $\bar{p}$ occurs in $\hat{w}$ is called *odd* if p occurs exactly once in $\hat{w}$, and *even* otherwise.

Each DOW w describes an Eulerian circuit $C(w)$ of a four-regular graph G_w: indeed, the letters of w describe the vertices that are traversed when tracing $C(w)$.

Definition 3.3. Let $w = p_0 p_1 \cdots p_{2k-1}$ be an (unsigned) DOW. Then the four-regular graph G_w is defined to have the symbols p occurring in w as vertices, half edges $0, \ldots, 4k-1$ and edges $\{2i, 2i+1\}$ for $i \in \{0, \ldots, 2k-1\}$ with $\epsilon(2i) = p_i$ and $\epsilon(2i+1) = p_{i+1 \bmod 2k}$. The corresponding Eulerian circuit $C(w)$ equals $\{\{1, 2\}, \{3, 4\}, \ldots, \{4k-1, 0\}\}$.

By construction, $C(w)$ spells out w: $p_0 \overset{0\,1}{—} p_1 \overset{2\,3}{—} p_2 \overset{4}{—} \cdots \overset{4k-3}{—} p_{2k-1} \overset{4k-2\,4k-1}{—} p_0$. Note that each pair $\{2i-1, 2i\}$ is one of the two single transitions of the transition at vertex p_i corresponding to $C(w)$.

One naturally associates a *signed* DOW $\hat{w}$ to an MIC sequence. Indeed, $\hat{w}$ can be obtained by simply recording the pointers we encounter when reading the MIC sequence from "left" to "right." A pointer p that is inverted in the MIC sequence appears as $\overline{p}$ in $\hat{w}$.

Example 3.5. Consider again the MIC sequence from Fig. 3.1. When recording the successive pointers (taking into account inversion), we obtain the signed DOW $\hat{w} = \overline{2}\,\overline{1}\,1\,\overline{3}\,\overline{2}\,\overline{3}$, cf. Fig. 3.2 and Section 3.2.2.

Consider a signed DOW $\hat{w}$. Let us define the four-regular graph $G_{\hat{w}}$ and the Eulerian circuit $C(\hat{w})$ to be equal to G_w and $C(w)$, respectively, where w is the unbarred version of $\hat{w}$.

Note that the Eulerian circuit $C(\hat{w})$ represents the MIC sequence. The MAC sequence corresponds to a different circuit partition, which we denote by $P_{\text{MAC}}(\hat{w})$. Consider the case where pointer p occurs twice unbarred in $\hat{w}$. This happens when both MDSs M_{p-1} and M_p occur in the forward direction in the MIC: $M_{p-1}\,p\,I_i$ and $I_j\,p\,M_p$. In the MAC sequence, the two consecutive MDSs appear merged at the pointer p and the corresponding sequence reads $M_{p-1}\,p\,M_p$ with a complementary sequence $I_j\,p\,I_i$. This means that the transition following the MAC sequence at p is orientation-consistent with $C(w)$. The situation is similar when both pointers p occur barred in $\hat{w}$.

However, when one copy of p occurs as p and the other copy is its barred version $\overline{p}$ in $\hat{w}$, that is, p is odd in $\hat{w}$, one of the MDSs is in inverted orientation in the MIC sequence with respect to the corresponding sequential MDS in the MAC. Then the situation in the MIC sequence is, for example, $M_{p-1}\,p\,I_i$ and $\overline{M}_p\,\overline{p}\,I_j$. Now the two consecutive MDSs can be recombined at the pointer p by traversing M_p in $P_{\text{MAC}}(\hat{w})$ in the reverse orientation, that is, by exiting p against its original direction. In this case, the transition following the MAC rearrangement at p is orientation-inconsistent with $C(w)$. The similar, symmetric, situation arises when M_{p-1} is in a reversed orientation with respect to M_p.

Example 3.6. Let $w = 2\,1\,1\,3\,2\,3$. The graph G_w and the Eulerian circuit $C(w)$ are depicted on the left-hand side of Fig. 3.7. The Eulerian circuit $C(w)$ goes "straight through" the vertices, as indicated by the red crosses inside the vertices.

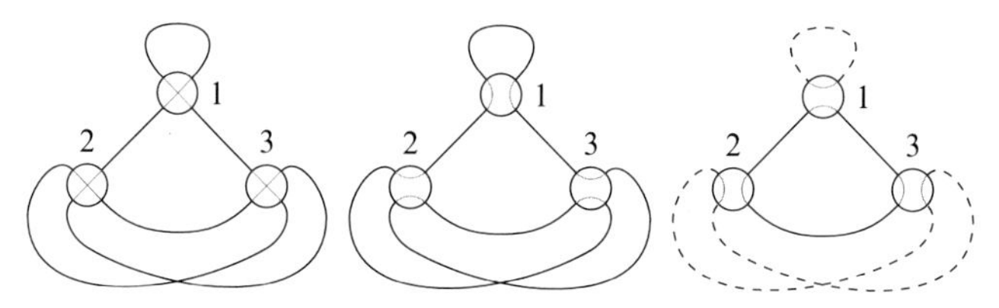

FIG. 3.7 (*Left diagram*) Graph G_w with circuit $C(w)$ for DOW $w = 2\,1\,1\,3\,2\,3$. (*Middle and right diagrams*) Signed versions $\hat{w} = \overline{2}\,\overline{1}\,1\,\overline{3}\,\overline{2}\,\overline{3}$ and $\hat{w}' = \overline{2}\,\overline{1}\,\overline{1}\,\overline{3}\,2\,3$ induce different circuit partitions corresponding to the red portion in Fig. 3.2, associated with MAC sequences. See Example 3.6.

Consider the signed DOW $\hat{w} = \overline{2}\,\overline{1}\,1\,\overline{3}\,\overline{2}\,\overline{3}$ from Example 3.5. Compare with the left-hand side of Fig. 3.3. Its even pointers are 2 and 3, and pointer 1 is odd. The circuit partition $\mathcal{P}_{\text{MAC}}(\hat{w})$ is obtained from $C(\hat{w})$ by changing the transitions at 2 and 3 so they are orientation-consistent with $C(w)$ and the one at 1 so it is orientation-inconsistent with $C(w)$; see Fig. 3.7 (middle). The resulting circuit partition consists of a single Eulerian circuit.

Now consider $\hat{w}' = \overline{2}\,\overline{1}\,\overline{1}\,\overline{3}\,2\,3$, which is another signed version of w; this time with 1 even and 2 and 3 odd, and compare it with the right-hand side of Fig. 3.3. The resulting circuit partition $\mathcal{P}_{\text{MAC}}(\hat{w}')$ contains three circuits, see the right-hand side of Fig. 3.7.

Remark 3.1. Both $\hat{w} = \overline{2}\,\overline{1}\,1\,\overline{3}\,\overline{2}\,\overline{3}$ and $\hat{w}' = \overline{2}\,\overline{1}\,\overline{1}\,\overline{3}\,2\,3$ are so-called "realistic" signed DOWs, meaning we can find MDSs matching the DOWs: $\overline{M}_2 M_1 \overline{M}_3 \overline{M}_4$ and $\overline{M}_2 \overline{M}_1 \overline{M}_4 M_3$, respectively. This implies that we can indicate which edges represent MDSs, as shown in Fig. 3.3. However, not every DOW is a representation of an MDS sequence, for example, the DOW 1 1 3 2 2 3 is not because the second MDS has pointers 1 and 2, but there is no subsequence 1 2 or 2 1 in this DOW. We note that this sequence defines a graph that is isomorphic to the graph of the sequence 1 1 2 3 3 2 whose signed version $1\,1\,2\,\overline{3}\,\overline{3}\,\overline{2}$ corresponds to an MDS sequence $M_1 M_2 \overline{M}_4 \overline{M}_3$ and appears in several MIC segments of the *O. trifallax* genome [40].

3.4.2 Double Occurrence Words Corresponding to Graphs

We now describe an alternative approach that uses rigid vertex graphs. A *rigid vertex* is a vertex with a preassigned cyclic order of its incident half-edges. Diagrammatically, a rigid vertex v can be considered as a small disk, as we have depicted in some of our figures, such that incident half-edges touch the boundary of the disk at a point. If one traces the boundary of the disk clockwise, half-edges incident to v are encountered in the order of the preassigned cyclic order or its reverse. A rigid vertex is sometimes called a fat vertex.

A DNA rearrangement model based on spatial graphs with rigid vertices was introduced in [23]. This model describes an MIC sequence as a spatial graph, called an assembly graph, where vertices represent the homologous recombination sites (pointers), while an MAC sequence is represented as a particular path in the graph. Examples of such graphs are depicted in Fig. 3.3. An *assembly graph* is a connected graph, where all vertices are rigid vertices of degrees (valency) 1 or 4. A vertex of degree 1 is called an *end-point*. As depicted in Fig. 3.3, the end-points I_0 and I_4 can be connected (see dotted lines) to obtain a four-regular rigid vertex graph.

A path that represents an MIC sequence is realized in an assembly graph as the *transverse* path that starts from one end-point and travels along the graph going straight through every vertex visited, to another end-point. For convenience, by "assembly graph" we mean from now on an assembly graph that has an Eulerian transverse path (i.e., represents at most a single MIC sequence). Such graphs have two end-points. In this case, similarly as above, if the two edges incident to end-points are joined to form a single edge, then the transverse path corresponds to the circuit partition $\{C\}$ and C is Eulerian that represents an MIC sequence. In the assembly graph model, the single transition corresponding to the transverse path is fixed at every vertex as in the left part of Fig. 3.5, $\{\{h_i, h_{i+1}\}, \{h_j, h_{j+1}\}\}$. A single transition that corresponds to a rearrangement at a vertex v is a pair of half-edges incident to v that are nonconsecutive with respect to the cyclic order specified at the rigid vertex. A transverse path is seen in Fig. 3.7 (left) as a path that goes straight through every vertex.

Let Γ be an assembly graph with the set of degree-4 vertices $V = \{v_1, \ldots, v_n\}$. The DOW corresponding to Γ is obtained in a way that is opposite to the construction in Section 3.4.1. Starting from an end-point (call it initial), write down the sequence of vertices in the order they are encountered along the transverse path going straight through the vertices up to the terminal end-point. This process yields an unsigned DOW over alphabet V. Conversely, as described in Section 3.4.1, given a DOW w, an assembly graph can be constructed that corresponds to it.

A path that represents an MAC gene is realized as a *polygonal path* that makes a ("90-degree") turn at every vertex. This corresponds to the change of the single transition at every vertex visited by the path. If the graph represents a rearrangement of a single MAC nanochromosome, the polygonal path must visit every vertex (i.e., the recombination must appear at every pointer sequence) and such polygonal path is called a *Hamiltonian polygonal path.* In Fig. 3.3, the Hamiltonian polygonal paths are represented by a double line. It is not the case that every assembly graph contains a Hamiltonian polygonal path. Thus there are DOWs that do not represent MIC sequences with a single MAC sequence. For example, the graph with end-points represented by the DOW 1 1 2 3 3 2 4 4 has no single Hamiltonian polygonal path [22] (see Exercise 3.4). This is another situation, besides those in Remark 3.1, where a DOW may not correspond to a "realistic" MDS sequence. In this case, however, there is no possible relabeling of the pointers that can correspond to an MDS list.

EXERCISES

[3.1] Sketch all assembly graphs for DOWs with $\Sigma = \{1, 2, 3\}$.

[3.2] Find all Hamiltonian polygonal paths in the assembly graph 1 2 1 3 2 4 3 4.

[3.3] **(a)** Determine a graph that corresponds to $M_1 \overline{M_2}\, \overline{M_3}\, \overline{M_4}$ and $M_1\, M_3\, M_5\, \overline{M_4}\, \overline{M_2}$.

(b) Indicate the transition system that corresponds to the MIC sequence and the one that corresponds to the MAC sequence.

[3.4] Prove that the graph G_w with $w = 1\,1\,2\,3\,3\,2\,4\,4$ has no Hamiltonian polygonal path.

[3.5] A DOW is called *reducible* if it is the concatenation of two nonempty DOWs. For example, the DOW 11233244 from Exercise 3.4 is reducible since it is the concatenation of 11 with 233244. Find an irreducible DOW whose corresponding graph has no Hamiltonian polygonal path.

3.5 GRAPH POLYNOMIALS

In this section, we describe a way to capture different situations during rearrangement. Since a graph can represent many rearrangement combinations, the assembly polynomial describes the number of molecules obtained after the recombination. By contrast, the rearrangement polynomial that we introduce here describes the number of molecules during the rearrangement. Both polynomials are derived as specializations of the transition polynomial which is our starting point.

3.5.1 Transition Polynomial

We denote the number of connected components of a graph G by $c(G)$. Recall that a transition at a vertex v is a partition of incident half-edges into pairs, and a transition system is a collection of fixed transitions at every vertex.

We now recall the definition of the (weighted) transition polynomial [31], which is directly inspired by the Martin polynomial [32].

Definition 3.4. Let G be an Eulerian graph without isolated vertices and let W be a function which assigns a variable or constant, called a weight, to every transition at a vertex of G. The (weighted) *transition polynomial* of G is

$$q(G, W; x) = \sum_{T \in \mathcal{T}(G)} W(T)\, x^{|P_T| - c(G)},$$

where we define $W(T) = \prod_{t \in T} W(t)$. Recall that P_T is the circuit partition corresponding to a transition system T.

In the literature, the term "multivariate" is sometimes used as a synonym for "weighted." As we shall see, the weights can be used to introduce extra variables to the polynomial. Notice that the exponents of $q(G, W; x)$ are nonnegative since $|P| - c(G) \geq 0$ for every circuit partition P of a graph G without isolated vertices. If G is connected, then $c(G) = 1$.

We point out that the *Martin polynomial* [32] is equal to $q(G, W; x - 1)$ with all weights equal to 1.

Remark 3.2. Note that when $G = G_1 \cup G_2$ is the disjoint union of two graphs, one has $q(G, W; x) = q(G_1, W; x) \cdot q(G_2, W; x)$. Indeed, any contribution $W(T)\, x^{|P_T| - c(G)}$ to $q(G, W; x)$ splits into contributions $W(T_1)\, x^{|P_{T_1}| - c(G_1)}$ and

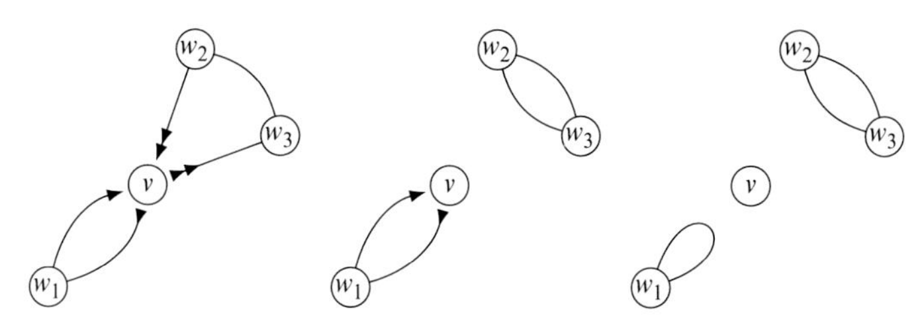

FIG. 3.8 Vertex reduction, see Example 3.7.

$W(T_2)\, x^{|P_{T_2}|-c(G_2)}$ to $q(G_1, W; x)$ and $q(G_2, W; x)$, where T is the disjoint union of T_1 and T_2.

As we will see in Theorem 3.1, the transition polynomial satisfies a recursive relation from which it can be computed. We first need some auxiliary notions (which can be defined for arbitrary Eulerian graphs). We consider only the case of four-regular graphs.

Let G be an Eulerian graph and t be a transition at a vertex v of G that does not contain a loop of G—that is, no single transition of t consists of both half edges of a loop. A *vertex reduction* of G at t, denoted by G/t, is obtained from G by: (1) replacing, for each single transition $s = \{h_1, h_2\} \in t$, the two edges $\{h_1, h_1'\}$ and $\{h_2, h_2'\}$ with half-edges in s by a single edge $\{h_1', h_2'\}$, and (2) deleting the now isolated vertex v. Note that vertex reduction is not defined on transitions that contain a loop. Indeed, otherwise v would not be isolated but incident to that loop, and so deletion of v cannot happen while simultaneously retaining the loop.

Example 3.7. Fig. 3.8 illustrates a vertex reduction at v. The transition t of v is indicated by the arrows in Fig. 3.8 (left). For (1), in two successive steps, edges are merged according to single transitions of t, as shown in Fig. 3.8 (middle and right). For (2) (not depicted), the isolated vertex is removed. As shown, a vertex reduction may introduce parallel edges as well as loops.

A transition t of G is called *singular* if for all transition systems T of G with $t \in T$, $|P_T| > c(G)$. A *cut vertex* is a vertex v that has a singular transition. Thus, intuitively, a singular transition at v "cuts" the connected component to which v belongs in two pieces. Since a transition that contains a loop is singular, every looped vertex is a cut vertex. For example, vertex v in Fig. 3.8 is a cut vertex as the connected component to which v belongs is split using a particular transition. Note that each vertex of degree 4 has at most one singular transition (see Fig. 3.5).

We are now ready to state the recursive relation for the transition polynomial in the case when G is four-regular.

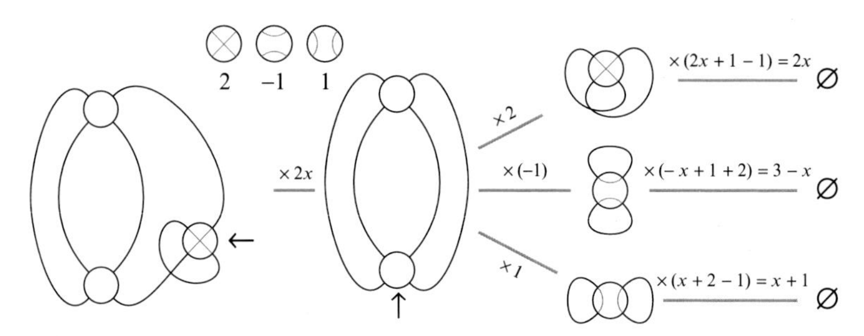

FIG. 3.9 Recursive computation of the weighted polynomial of the four-regular graph to the *left*, see Example 3.8.

Theorem 3.1 (Jaeger [31]). *Let G be a four-regular graph and let W be a weight function of G. For vertices v of G, denote by T_v the set of transitions of v.*

- *If G is the empty graph, then $q(G, W; x) = 1$.*
- *If v is a cut vertex of G, then for every $t' \in T_v$ that does not contain a loop at v,*

$$q(G, W; x) = \left(W(t_s)\, x + \sum_{t \in T_v \setminus \{t_s\}} W(t) \right) q(G/t', W; x),$$

  where t_s is the singular transition at v.
- *If v is not a cut vertex of G, then*

$$q(G, W; x) = \sum_{t \in T_v} W(t)\, q(G/t, W; x).$$

In particular, if v is a cut vertex, then all three polynomials $q(G/t, W; x)$, for $t \in T_v$, are equal, except when t contains a loop, in which case G/t is not defined.

Example 3.8. We compute the weighted transition polynomial of the four-regular graph depicted on the left-hand side of Fig. 3.9, by applying Theorem 3.1 recursively. The weights for the transitions at the vertices are also given in Fig. 3.9.

Let us first apply Theorem 3.1 to the cut vertex indicated by an arrow, where the singular transition is indicated at the vertex (inside the circle). Vertex reduction can be applied on either one of the other two transitions of this vertex. In both cases, when we apply vertex reduction, we obtain the same four-regular graph, see Fig. 3.9 (middle). The multiplicative factor from Theorem 3.1 equals $2x - 1 + 1 = 2x$.

In the second step, we apply vertex reduction at the vertex indicated by an arrow, which is not a cut vertex. Applying vertex reductions on the three transitions at that vertex, we obtain the three single-vertex graphs with two loops as indicated in Fig. 3.9 (right). The multiplicative factors are the three respective transition weights 2, -1, and 1. In a third and final step, the three four-regular graphs can be reduced to the empty graph, which has polynomial 1, using the indicated singular transitions. The respective multiplicative factors are $2x - 1 + 1$, $2 - x + 1$, and $2 - 1 + x$.

The resulting polynomial is therefore

$$2x \cdot [\ (2 \cdot 2x) + (-1 \cdot (3 - x)) + (1 \cdot (x + 1))\ ] = 2x\,(6x - 2). \tag{3.1}$$

Remark 3.3. The Tutte polynomial $T(G)$ of a graph G is recursively defined using the operations of deletion and contraction. These operations correspond to two vertex reductions when we consider the four-regular directed medial graph $\vec{G}_m$ of G. This ties the Tutte polynomial to the Martin polynomial [32].

3.5.2 Assembly Polynomial

In this section we consider assembly graphs where the end-points with their incident edges have been merged to obtain a four-regular graph (see Fig. 3.3). Furthermore, we consider assembly graphs that are defined by a DOW, as in Section 3.4.

Definition 3.5. Let w be a DOW. The *assembly polynomial* of w is the two-variable polynomial

$$S(w; p, x) = q(G_w, W; x),$$

where for all transitions t of G_w,

$$W(t) = \begin{cases} 0 & \text{if } t \text{ follows } C(w), \\ p & \text{if } t \text{ is orientation-consistent with } C(w), \\ 1 & \text{if } t \text{ is orientation-inconsistent with } C(w). \end{cases}$$

The weight p assigned to orientation-consistent transitions acts as a second variable for the polynomial. We often abbreviate $S(w; p, x)$ by $S(w)$.

We call a transition system $T \in \mathcal{T}(G_w)$, such that $T \cap T_{\{C(w)\}} = \emptyset$, a *smoothing* of $C(w)$. The only nonzero terms in the assembly polynomial are obtained through a smoothing transition system. Thus the assembly polynomial of the DOW w can be obtained by enumerating the smoothings T of $C(w)$ while counting the number of resulting circuits $|P_T|$ and the number of orientation-consistent transitions $\pi(T)$.

Explicitly, the assembly polynomial of w is

$$S(w) = \sum_{\substack{T \in \mathcal{T}(G_w), \\ T \cap T_{\{C(w)\}} = \emptyset}} p^{\pi(T)} x^{|P_T| - 1}.$$

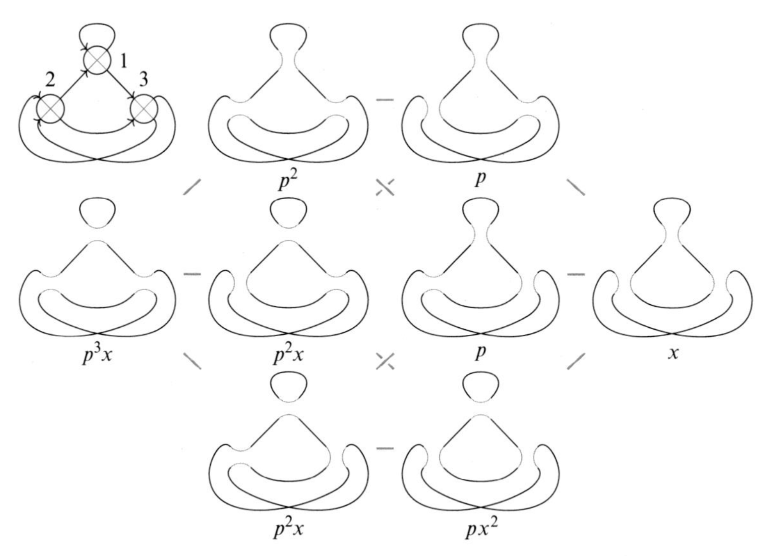

FIG. 3.10 (*Top left*) For $w = 113{,}232$ the graph G_w with circuit $C(w)$. (*Diagrams left to right, top to bottom*) Eight smoothings, all three, two, one, and zero orientation-consistent transitions with contributions p^3, p^2, p, and p^0, respectively. By taking the connected component counts, we obtain the assembly polynomial $S(w) = p^3x^1 + p^2(x^0 + x^1 + x^1) + p(x^0 + x^0 + x^2) + p^0x^1$.

Clearly, the assembly polynomial of a DOW does not change if it is cyclically permuted, $S(w) = S(vu) = S(uv)$, or reversed, $S(w) = S(w^R)$, because G_w and $C(w)$ are invariant under these operations [22].

Example 3.9. For $w = 1\,1\,3\,2\,3\,2$ the assembly polynomial is

$$S(w) = p^3x + 2p^2x + p^2 + px^2 + 2p + x = (px + 1)(p^2 + 2p + x);$$

see Fig. 3.10.

Since there are precisely $\binom{n}{k}$ smoothings with k orientation-consistent transitions on a four-regular graph with n vertices, we obtain the following observation.

Lemma 3.1. *Let w be a DOW of length $2n$. Then $S(w; p, 1) = (p + 1)^n$.*

If a DOW is the concatenation of two nonempty DOWs, then the corresponding assembly polynomial is the product of the corresponding pairs of polynomials. This is from [35, Lemma 6.5], but the proof here follows from Remark 3.2.

We say that DOWs v and w are *disjoint* if vw is a DOW.

Lemma 3.2. *If DOWs v and w are disjoint, then $S(vw) = S(v) \cdot S(w)$.*

Proof. Consider DOW $z = 0v0w$, where 0 is a new symbol. Note that 0 is a cut vertex for G_z and the orientation-consistent transition at 0 is singular. Then $G_{vw} = G_z/t$ and $G_v \cup G_w = G_z/t'$, where t and t' are the transitions of G_z at 0 that follow $C(z)$ and are orientation-consistent with $C(z)$, respectively. The circuits $C(vw)$ and $C(v), C(w)$ define transition weights W compatible to those of $C(z)$.

Hence, by Remark 3.2 and the comment following Theorem 3.1, we obtain

$$\begin{aligned} S(vw) &= q(G_{vw}, W; x) = q(G_z/t, W; x) = q(G_z/t', W; x) \\ &= q(G_v \cup G_w, W; x) = q(G_v, W; x)\, q(G_w, W; x) = S(v)S(w). \end{aligned}$$

□

Since cyclic permutations of a DOW do not change the assembly polynomial, we have, by Lemma 3.2, that if w contains two consecutive copies of the same letter, say 1, we have $S(w) = S(11u) = S(11) \cdot S(u) = (px+1) \cdot S(u)$. Furthermore, if uvw and v are DOWs, then we have $S(uvw) = S(vwu)$ where wu is a DOW, hence $S(uvw) = S(v)S(wu)$. Iterating this property, assembly polynomials for some DOWs can be obtained easily. For example, $S(112332) = S(11) \cdot S(2332) = S(11) \cdot S(2233) = S(11) \cdot S(22) \cdot S(33) = (px+1)^3 = S(112233)$. For more complicated DOWs we need other techniques.

3.5.3 Reduction Rules for the Assembly Polynomial

We now derive a set of recursive rules to compute the assembly polynomial of a DOW by reducing the symbols in the DOW. One such recursive rule is given by Lemma 3.2, but that rule is not useful for DOWs that are not the concatenation of two disjoint nonempty DOWs, such as $S(1212)$.

As the assembly polynomial is a specialization of the transition polynomial, we can always apply Theorem 3.1 to recursively compute it. Thus $S(w) = q(G_w, W; x)$, where W is the weight function that assigns the transitions according to their orientation with respect to $C(w)$. Assuming that v is not a cut vertex of G_w, then $q(G_w, W; x) = p \cdot q(G_w/t, W; x) + q(G_w/t', W; x)$, where t and t' are orientation-consistent, and orientation-inconsistent, respectively.

To truly formulate a recursive relation for $S(w)$, we have to translate the resulting terms like $q(G_w/t, W; x)$ back into a polynomial of the form $S(w')$, where w' represents G_w/t. There are some issues that complicate this approach, cf. Fig. 3.5:

- The DOW w can be split into two parts that may have symbols in common. This means that there is no obvious Eulerian circuit $C(w')$ for G_w/t.
- Part of w' becomes reversed. As a consequence, some vertices in $C(w')$ are traversed differently, and hence the transitions may change their orientation with respect to the circuit. This means that some orientation-inconsistent

transitions for $C(w)$ become orientation-consistent for $C(w')$ and those transitions have weights p although not so originally.

The first issue can be solved by carefully choosing a set of transitions that together do not break the original circuit. To overcome the second issue, we generalize the assembly polynomial to *signed* DOWs. The bars allow one to record the changes relative to circuit orientations of parts of $C(w)$, while the weight p in the polynomial can be assigned to appropriate transitions as they switch from orientation-inconsistent to orientation-consistent transitions (and conversely).

Definition 3.6. The *assembly polynomial* of a *signed* DOW $\hat{w}$ is the two-variable polynomial $S(\hat{w}; p, x) = q(G_{\hat{w}}, W; x)$, where for all transitions t of $G_{\hat{w}}$,

$$W(t) = \begin{cases} 0 & \text{if } t \text{ follows } C(\hat{w}), \\ p & \text{if } t \text{ is orientation-consistent (orientation-inconsistent, resp.) with} \\ & C(\hat{w}) \text{ and vertex } z \text{ belonging to } t \text{ is even (odd, resp.) in } \hat{w}, \\ 1 & \text{otherwise.} \end{cases}$$

Note that the above-defined assembly polynomial for a signed DOW $\hat{w}$ coincides with the assembly polynomial for a DOW when $\hat{w}$ is a DOW (i.e., when $\hat{w}$ does not have bars).

Example 3.10. Let $\hat{w} = \bar{1}\,\bar{1}\,3\,\bar{2}\,3\,2$. We can reuse Fig. 3.10 for $w = 1\,1\,3\,2\,3\,2$ as $G_{\hat{w}} = G_w$ and $C(\hat{w}) = C(w)$, but we have to "flip" the weights at vertex 2. The assembly polynomial is therefore equal to

$$p{\cdot}1{\cdot}p{\cdot}x^1 + 1{\cdot}1{\cdot}p{\cdot}x^0 + p{\cdot}p{\cdot}p{\cdot}x^1 + p{\cdot}1{\cdot}1{\cdot}x^1 + 1{\cdot}p{\cdot}p{\cdot}x^0 + 1{\cdot}1{\cdot}1{\cdot}x^0 + p{\cdot}p{\cdot}p{\cdot}x^2 + 1{\cdot}p{\cdot}1{\cdot}x^1,$$

which is

$$p^3x + p^2x^2 + p^2x + p^2 + 2px + p + 1 = (px + 1)(p^2 + px + p + 1).$$

Lemmas 3.1 and 3.2 are still valid for signed DOWs.

Example 3.11. One computes

$$S(\varepsilon) = 1, \quad S(11) = px + 1, \quad \text{and} \quad S(1\bar{1}) = p + x,$$

as well as

$$S(1\,2\,1\,2) = p^2 + 2p + x, \quad S(1\,2\,1\,\bar{2}) = p^2 + px + p + 1, \quad \text{and}$$
$$S(1\,2\,\bar{1}\,\bar{2}) = p^2x + 2p + 1.$$

For a signed DOW $\hat{w}$ we use $\hat{w}^{RC}$ to denote its *reverse complement* which is obtained from $\hat{w}$ by a string reversal and (un)barring all its letters:

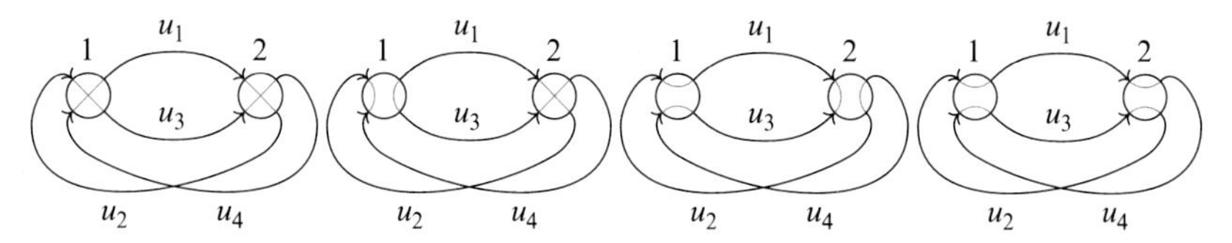

FIG. 3.11 Illustration of the proof of Lemma 3.3.

if $\hat{w} = p_1 p_2 \cdots p_n$ then $\hat{w}^{\mathrm{RC}} = \overline{p}_n \cdots \overline{p}_2 \overline{p}_1$. Reverse complement turns out to keep the assembly polynomial invariant: $S(\hat{w}) = S(\hat{w}^{\mathrm{RC}})$ (see Exercise 3.4).

We present the basic reduction rule for signed DOWs for which Lemma 3.2 is not useful.

Lemma 3.3. *Let* $\hat{w} = 1\, u_1\, 2\, u_2\, 1\, u_3\, 2\, u_4$ *be a signed DOW. Then*

$$S(\hat{w}) = S(u_1\, 2\, u_2\, [\, u_3\, 2\, u_4\,]^{\mathrm{RC}}) + p \cdot S(u_1\, [u_4\, u_3]^{\mathrm{RC}}\, u_2) + p^2 \cdot S(u_1\, u_4\, u_3\, u_2).$$

Proof. The Eulerian circuit $C(\hat{w})$ is abstractly depicted in the first diagram of Fig. 3.11. The smoothings of $C(\hat{w})$ can be divided into those for which we consider a transition at 1 that is orientation-inconsistent (second diagram) and those for which the transition at 1 is orientation-consistent. For the orientation-consistent transition at 1 we further distinguish the orientation-inconsistent transition at 2 (third diagram) and the orientation-consistent transition at 2 (fourth diagram). In all three cases, the proposed string traces the new circuit after the vertex reduction for which the transition was fixed.

The reversal of a segment v in w swaps consistent and inconsistent orientations at vertices that occur only a single time in v, but this is nullified by adding a single bar to those vertices. □

Example 3.12. We compute $S(1\,2\,3\,1\,2\,3)$ by applying Lemma 3.3, taking $u_1 = u_3 = \varepsilon$ and $u_2 = u_4 = 3$. Thus $S(1\,2\,3\,1\,2\,3) = S(2\,3\,\overline{3}\,\overline{2}) + p \cdot S(\overline{3}\,3) + p^2 \cdot S(3\,3) = (p + x)^2 + p(p + x) + p^2(px + 1) = p^3 x + 3p^2 + 3px + x^2$.

When we start the reduction on one or two odd pairs of pointers instead of only even pointer pairs, we simply adjust the weights accordingly.

Corollary 3.1. *We have*

$$S(1\, u_1\, 2\, u_2\, 1\, u_3\, \overline{2}\, u_4) = S(u_1\, 2\, u_2\, [\, u_3\, \overline{2}\, u_4\,]^{\mathrm{RC}}) + p^2 \cdot S(u_1\, [u_4\, u_3]^{\mathrm{RC}}\, u_2) + p \cdot S(u_1\, u_4\, u_3\, u_2),$$

$$S(1\, u_1\, 2\, u_2\, \overline{1}\, u_3\, \overline{2}\, u_4) = p \cdot S(u_1\, 2\, u_2\, [\, u_3\, \overline{2}\, u_4\,]^{\mathrm{RC}}) + p^2 \cdot S(u_1\, [u_4\, u_3]^{\mathrm{RC}}\, u_2) + S(u_1\, u_4\, u_3\, u_2).$$

The assembly polynomial is computed recursively using the above formulas as follows. If a DOW is of the form stated in Lemma 3.3 or Corollary 3.1, then the polynomial reduces to those of shorter strings. Eventually, we reach those DOWs that do not have "interleaved" symbols in them, that is, a DOW not of the form $\cdots 1 \cdots 2 \cdots 1 \cdots 2 \cdots$. Such DOWs are equal (up to cyclic permutation) to the concatenation of two nonempty DOWs, and then Lemma 3.2 applies.

If we consider DOWs up to cyclic permutation and renaming of symbols as equivalent, for example, 122133, 221331, and 112332 are all equivalent, then there are five distinct DOWs of three symbols. We list these, and their polynomials, in the following table.

DOW w	$S(w)$	See
112233	$1+3px+3p^2x^2+p^3x^3=(px+1)^3$	Below Lemma 3.2
112323	$x+2p+p^2+px^2+2xp^2+xp^3$	Example 3.9
112332	$1+3px+3p^2x^2+p^3x^3$	
121323	$1+2p+px+2p^2+xp^2+xp^3$	
123123	$3px+x^2+3p^2+xp^3$	Example 3.12

3.5.4 Rearrangement Polynomial

We now describe a graph polynomial that considers all "partial" rearrangements of MIC sequences into MAC sequences, where the MIC sequence is specified as a signed DOW $\hat{w}$. It considers all transition systems where transitions at each vertex follow either the circuit representing the MIC sequence or the circuit representing the MAC sequence. Thus, at vertex v, a transition with nonzero weight can: (a) follow $C(w)$, or (b) be either orientation-consistent when v is even in $\hat{w}$, or be orientation-inconsistent when v is odd in $\hat{w}$, but not both. Like the assembly polynomial, it is a specialization of the general transition polynomial.

Definition 3.7. Let $\hat{w}$ be a signed DOW. The *rearrangement polynomial* of $\hat{w}$ is

$$R(\hat{w}) = R(\hat{w}; x) = q(G_w, W; x),$$

where for all transitions t of G_w, $W(t) = 1$ if t follows $C(w)$ or $P_{\text{MAC}}(w)$; and $W(t) = 0$, otherwise.

Explicitly, the rearrangement polynomial of u is

$$R(\hat{w}; x) = \sum_{T \in \mathcal{T}(G_w),\ T \subseteq T_{\{C(w)\}} \cup T_{P_{\text{MAC}}(w)}} x^{|P_T|-1}.$$

Note that for an *unsigned* DOW w, the definition of the rearrangement polynomial reduces to a summation over all transition systems where each

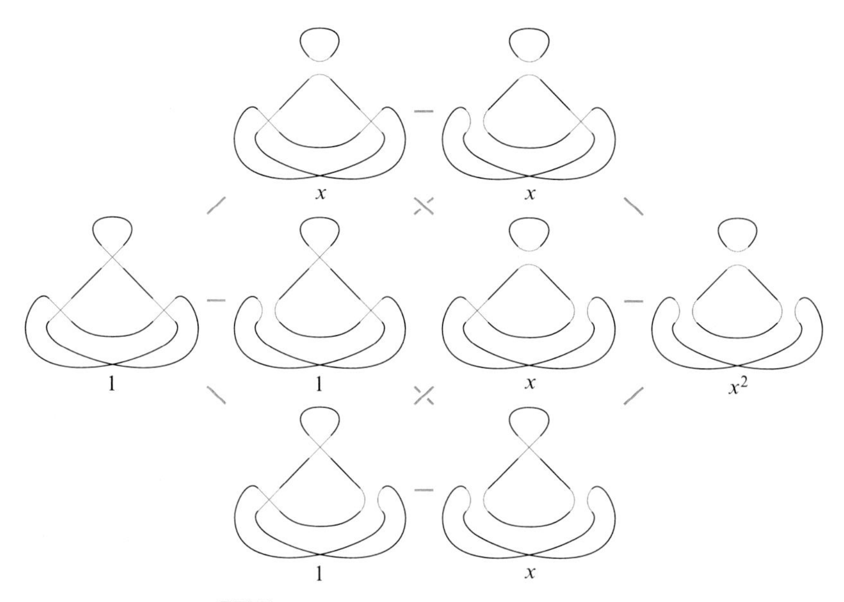

FIG. 3.12 For $\hat{w} = \overline{2}\,\overline{1}\,\overline{1}\,\overline{3}\,2\,3$, the eight transition systems that are considered by the rearrangement polynomial. At every vertex, the transitions with nonzero weight either follow $C(w)$ or $P_{\mathrm{MAC}}(w)$. The *leftmost diagram* represents the initial circuit $C(w)$ (i.e., the MIC sequence), while the *rightmost diagram* represents the *full* rearrangement (i.e., the MAC sequence), see Fig. 3.3 (*right*). The terms contributing to $R(\hat{w})$ are indicated under the each diagram.

transition either follows $C(w)$ or is orientation-consistent with $C(w)$. This matches the Martin polynomial for directed four-regular graphs [32].

Example 3.13. For $\hat{w} = \overline{2}\,\overline{1}\,\overline{1}\,\overline{3}\,2\,3$ the rearrangement polynomial equals $R(\hat{w}) = x^2 + 4x + 3$; see Fig. 3.12.

Lemmas 3.1 and 3.2 (and their proofs) can be easily transferred from the assembly polynomial to the rearrangement polynomial.

Lemma 3.4. *Let $\hat{w}$ be a signed DOW of length $2n$. Then $R(\hat{w}; 1) = 2^n$.*

Lemma 3.5. *If signed DOWs $\hat{v}$ and $\hat{w}$ are disjoint, then $R(\hat{v}\hat{w}) = R(\hat{v}) \cdot R(\hat{w})$.*

EXERCISES

[3.1] Show that $q(G, W; -2) = 0$ when G is four-regular and nonempty (i.e., contains at least one vertex) and $W(t) = 1$ for all transitions t of G. *Hint*: Use Theorem 3.1.

[3.2] Compute the rearrangement polynomial for the four-regular graph on the left-hand side of Fig. 3.3.

[3.3] Use recursion to compute the assembly polynomial for signed DOWs $1\,2\,3\,4\,1\,2\,3\,4$ and $1\,2\,\bar{1}\,\bar{3}\,\bar{2}\,\bar{4}\,\bar{3}\,4$.

[3.4] Prove that for a signed DOW $\hat{w}$, $S(\hat{w}) = S(\hat{w}^{\mathrm{RC}})$.

[3.5] (Project) Find a set of recursive relations characterizing the rearrangement polynomial in a similar way as was done for the assembly polynomial in Section 3.5.3.

[3.6] (Open problem, project) Determine the signed DOWs $\hat{w}$ such that $S(\hat{w}; 1, x) = R(\hat{w}; x)$.

3.6 GENERALIZATIONS

This chapter introduced and studied basic properties of two graph polynomials motivated by DNA rearrangements, the assembly and the rearrangement polynomial, which are both specializations of the (weighted) transition polynomial.

DOWs have appeared often in graph theory. We call a pair of symbols p, q *interlaced* in a DOW w, if w is of the form $w = \cdots p \cdots q \cdots p \cdots q \cdots$. In this way, one naturally associates a simple graph, called a *circle graph*, to a DOW: the vertices are the symbols of w and there is an edge between two symbols if and only if these symbols are interlaced. Note that distinct DOWs may have a common circle graph, for example, the circle graphs of DOWs $1\,2\,3\,4\,4\,3\,2\,1$ and $1\,1\,2\,2\,3\,3\,4\,4$ both consist of four isolated vertices because they have no interlaced symbols. It is well known that if two DOWs have a common circle graph, then the four-regular graphs corresponding to these two DOWs have the same transition polynomial [44, 45]. This is observed in [46] for the special case of the assembly polynomial. Therefore, the assembly and rearrangement polynomials of this chapter can be defined directly for circle graphs instead of signed DOWs. This leads to the notions of the *interlace polynomial* [47] (which was originally motivated by DNA sequencing by hybridization), the *bracket polynomial for graphs* [48], and the *Tutte-Martin polynomial* [44, 45]. These graph polynomials are defined for arbitrary simple graphs instead of circle graphs, and can thus be seen as generalizations of the assembly and rearrangement polynomials that we considered in this chapter. In particular, the assembly polynomial can be seen as a special case of the bracket polynomial [34]. Such more general graph polynomials have been studied in even larger generality using the notions of delta-matroids and multimatroids in [49]. A direct link between DNA recombination in ciliates and multimatroids, that is, without using the notions of DOWs and circle graphs, is provided in [50].

ACKNOWLEDGMENTS

Robert Brijder is a postdoctoral fellow of the Research Foundation, Flanders (FWO). This work has been supported in part by the NSF grant CCF-1526485 and the NIH grant R01 GM109459.

REFERENCES

[1] P.A. Pevzner, H. Tang, M.S. Waterman, An Eulerian path approach to DNA fragment assembly, Proc. Natl Acad. Sci. 98 (17) (2001) 9748–9753.

[2] T. Dobzhansky, On the sterility of the interracial hybrids in *Drosophila pseudoobscura*, Proc. Natl Acad. Sci. 19 (4) (1933) 397–403.

[3] H.A. Orr, Dobzhansky, Bateson, and the genetics of speciation, Genetics 144 (4) (1996) 1331–1335.

[4] P. Pevzner, G. Tesler, Genome rearrangements in mammalian evolution: lessons from human and mouse genomes, Genome Res. 13 (1) (2003) 37–45.

[5] P. Biller, L. Guéguen, C. Knibbe, E. Tannier, Breaking good: accounting for fragility of genomic regions in rearrangement distance estimation, Genome Biol. Evol. 8 (5) (2016) 1427–1439.

[6] S. Yancopoulos, O. Attie, R. Friedberg, Efficient sorting of genomic permutations by translocation, inversion and block interchange, Bioinformatics 21 (16) (2005) 3340–3346.

[7] S. Beermann, The diminution of heterochromatic chromosomal segments in Cyclops (Crustacea, Copepoda), Chromosoma 60 (4) (1977) 297–344.

[8] S.A. Gerbi, Unusual chromosome movements in SCIARID flies, Results Probl. Cell Differ. 13 (1986) 71–104.

[9] D.M. Prescott, The DNA of ciliated protozoa, Microbiol. Rev. 58 (2) (1994) 233–267.

[10] J.J. Smith, C. Baker, E.E. Eichler, C.T. Amemiya, Genetic consequences of programmed genome rearrangement, Curr. Biol. 22 (16) (2012) 1524–1529.

[11] P.J. Stephens, C.D. Greenman, B. Fu, F. Yang, G.R. Bignell, L.J. Mudie, E.D. Pleasance, K.W. Lau, D. Beare, L.A. Stebbings, S. McLaren, M.-L. Lin, D.J. McBride, I. Varela, S. Nik-Zainal, C. Leroy, M. Jia, A. Menzies, A.P. Butler, J.W. Teague, M.A. Quail, J. Burton, H. Swerdlow, N.P. Carter, L.A. Morsberger, C. Iacobuzio-Donahue, G.A. Follows, A.R. Green, A.M. Flanagan, M.R. Stratton, P.A. Futreal, P.J. Campbell, Massive genomic rearrangement acquired in a single catastrophic event during cancer development, Cell 144 (1) (2011) 27–40.

[12] G. Aguileta, D.M. de Vienne, O.N. Ross, M.E. Hood, T. Giraud, E. Petit, T. Gabaldón, High variability of mitochondrial gene order among fungi, Genome Biol. Evol. 6 (2) (2014) 451–465.

[13] B.F. Lang, M. Jakubkova, E. Hegedusova, R. Daoud, L. Forget, B. Brejova, T. Vinar, P. Kosa, D. Fricova, M. Nebohacova, P. Griac, L. Tomaska, G. Burger, J. Nosek, Massive programmed translational jumping in mitochondria, Proc. Natl Acad. Sci. 111 (16) (2014) 5926–5931.

[14] A. Ehrenfeucht, T. Harju, I. Petre, D. Prescott, G. Rozenberg, Computation in Living Cells—Gene Assembly in Ciliates, Springer Verlag, Berlin, 2004.

[15] G. Fertin, A. Labarre, I. Rusu, E. Tannier, S. Vialette, Combinatorics of Genome Rearrangements, MIT Press, Cambridge, MA, 2009.

[16] T. Head, Formal language theory and DNA: an analysis of the generative capacity of specific recombinant behaviors, Bull. Math. Biol. 49 (1987) 737–759.

[17] D. Sankoff, M. Blanchette, Phylogenetic invariants for genome rearrangements, J. Comput. Biol. 6 (3–4) (2004) 431–445.

[18] D. Buck, DNA topology. Applications of knot theory, Proc. Symp. Appl. Math. 66 (2009) 47–79.

[19] I.K. Darcy, Modeling protein-DNA complexes with tangles, Comput. Math. Appl. 55 (5) (2008) 924–937.

[20] K. Shimokawa, M. Vazquez, DNA and the knot theory, Sugaku (Japanese) 63 (2) (2011) 237–242.

[21] D.W. Sumners, Lifting the curtain: using topology to probe the hidden action of enzymes, Not. AMS 42 (1995) 528–537.

[22] A. Angeleska, N. Jonoska, M. Saito, DNA recombination through assembly graphs, Discret. Appl. Math. 157 (14) (2009) 3020–3037.

[23] A. Angeleska, N. Jonoska, M. Saito, L.F. Landweber, RNA-guided DNA assembly, J. Theor. Biol. 248 (4) (2007) 706–720.

[24] R. Brijder, M. Daley, T. Harju, N. Jonoska, I. Petre, G. Rozenberg, Computational nature of gene assembly in ciliates, in: G. Rozenberg, T.H.W. Bäck, J.N. Kok (Eds.), Handbook of Natural Computing, vol. 3, chap. 37, Springer, New York, NY, 2012, pp. 1233–1280.

[25] R. Brijder, H.J. Hoogeboom, The algebra of gene assembly in ciliates, in: N. Jonoska, M. Saito (Eds.), Discrete and Topological Models in Molecular Biology, Natural Computing Series, Springer, Berlin, Heidelberg, ISBN 978-3-642-40192-3, 2014, pp. 289–307.

[26] M. Nowacki, V. Vijayan, Y. Zhou, K. Schotanus, T.G. Doak, L.F. Landweber, RNA-mediated epigenetic programming of a genome-rearrangement pathway, Nature 451 (7175) (2008) 153–158.

[27] A. Kotzig, Eulerian lines in finite 4-valent graphs and their transformations, in: Theory of Graphs, Proceedings of the Colloquium, Tihany, Hungary, 1966, Academic Press, New York, NY, 1968, pp. 219–230.

[28] A. Bouchet, Greedy algorithm and symmetric matroids, Math. Program. 38 (2) (1987) 147–159.

[29] A. Bouchet, Isotropic systems, Eur. J. Comb. 8 (3) (1987) 231–244.

[30] A. Bouchet, Multimatroids I. Coverings by independent sets, SIAM J. Discret. Math. 10 (4) (1997) 626–646.

[31] F. Jaeger, On transition polynomials of 4-regular graphs, in: G. Hahn, G. Sabidussi, R.E. Woodrow (Eds.), Cycles and Rays, NATO ASI Series, vol. 301, Kluwer, 1990, pp. 123–150, https://doi.org/10.1007/978-94-009-0517-7_12.

[32] P. Martin, Enumérations eulériennes dans les multigraphes et invariants de Tutte-Grothendieck (Ph.D. thesis), Institut d'Informatique et de Mathématiques Appliquées de Grenoble (IMAG), 1977, Available from: http://tel.archives-ouvertes.fr/tel-00287330_v1/.

[33] W.T. Tutte, A contribution to the theory of chromatic polynomials, Can. J. Math. 6 (1954) 80–91.

[34] R. Brijder, H.J. Hoogeboom, Graph polynomials motivated by gene rearrangements in ciliates, in: A. Beckmann, E. Csuhaj-Varjú, K. Meer (Eds.), 10th Conference on Computability in Europe (CiE 2014), vol. 8493 of Lecture Notes in Computer Science, Springer, 2014, pp. 63–72.

[35] J. Burns, E. Dolzhenko, N. Jonoska, T. Muche, M. Saito, Four-regular graphs with rigid vertices associated to DNA recombination, Discret. Appl. Math. 161 (10–11) (2013) 1378–1394.

[36] S.L. Tausta, L.A. Klobutcher, Detection of circular forms of eliminated DNA during macronuclear development in *E. crassus*, Cell 59 (6) (2017) 1019–1026.

[37] J.R. Bracht, W. Fang, A.D. Goldman, E. Dolzhenko, E.M. Stein, L.F. Landweber, Genomes on the edge: programmed genome instability in ciliates, Cell 152 (3) (2013) 406–416.

[38] V.T. Yerlici, L.F. Landweber, Programmed genome rearrangements in the *Ciliate Oxytricha*, Microbiol. Spectr. 2 (6) (2014) MDNA3-0025-2014, https://doi.org/10.1128/microbiolspec.MDNA3-0025-2014.

[39] X. Chen, J.R. Bracht, A.D. Goldman, E. Dolzhenko, D.M. Clay, E.C. Swart, D.H. Perlman, T.G. Doak, A. Stuart, C.T. Amemiya, R.P. Sebra, L.F. Landweber, The architecture of a scrambled genome reveals massive levels of genomic rearrangement during development, Cell 158 (5) (2014) 1187–1198.

[40] J. Burns, D. Kukushkin, X. Chen, L.F. Landweber, M. Saito, N. Jonoska, Recurring patterns among scrambled genes in the encrypted genome of the ciliate *Oxytricha trifallax*, J. Theor. Biol. 410 (2016) 171–180.

[41] A. Ehrenfeucht, T. Harju, G. Rozenberg, Gene assembly through cyclic graph decomposition, Theor. Comput. Sci. 281 (1–2) (2002) 325–349.

[42] L. Kari, L.F. Landweber, Computational power of gene rearrangement, in: E. Winfree, D.K. Gifford (Eds.), DNA Based Computers V, vol. 54 of DIMACS Series in Discrete Mathematics and Theoretical Computer Science, 1999, pp. 207–216.

[43] D.M. Prescott, A. Ehrenfeucht, G. Rozenberg, Template-guided recombination for IES elimination and unscrambling of genes in stichotrichous ciliates, J. Theor. Biol. 222 (3) (2003) 323–330.

[44] A. Bouchet, Tutte-Martin polynomials and orienting vectors of isotropic systems, Graphs Comb. 7 (3) (1991) 235–252.

[45] A. Bouchet, Graph polynomials derived from Tutte-Martin polynomials, Discret. Math. 302 (1–3) (2005) 32–38.

[46] E. Dolzhenko, K. Valencia, Invariants of graphs modeling nucleotide rearrangements, in: N. Jonoska, M. Saito (Eds.), Discrete and Topological Models in Molecular Biology, Natural Computing Series, Springer, Berlin, Heidelberg, ISBN 978-3-642-40192-3, 2014, pp. 309–323.

[47] R. Arratia, B. Bollobás, G.B. Sorkin, The interlace polynomial of a graph, J. Comb. Theory Ser. B 92 (2) (2004) 199–233.

[48] L. Traldi, L. Zulli, A bracket polynomial for graphs, I, J. Knot Theory Its Ramifications 18 (12) (2009) 1681–1709.

[49] R. Brijder, H.J. Hoogeboom, Interlace polynomials for multimatroids and delta-matroids, Eur. J. Comb. 40 (2014) 142–167.

[50] R. Brijder, Recombination faults in gene assembly in ciliates modeled using multimatroids, Theor. Comput. Sci. 608 (2015) 27–35.

Chapter 4

The Regulation of Gene Expression by Operons and the Local Modeling Framework

Matthew Macauley*, Andy Jenkins† and Robin Davies‡
*School of Mathematical and Statistical Sciences, Clemson University, Clemson, SC, United States, †Department of Mathematics, University of Georgia, Athens, GA, United States, ‡Biomedical Sciences, Jefferson College of Health Sciences, Roanoke, VA, United States

4.1 BASIC BIOLOGY INTRODUCTION

Gene expression must be carefully regulated in order for a cell to maintain homeostasis. In prokaryotes, genes are often clustered together and cotranscribed, meaning that small groups of genes governing a specific cellular function will be either all "ON" or all "OFF." Such a cluster of genes is called an *operon*, and these come in four basic types. Operons can be inducible or repressible, depending on whether they are ON or OFF by default. Additionally, the method of regulation can be positive or negative, depending on whether a repressor protein prevents transcription or activator protein stimulates it. Classically, mathematical biologists have modeled operons using differential equations derived from the laws of mass-action kinetics [1]. More recently, they have been modeled using discrete approaches such as Boolean and logical networks [2, 3]. In this chapter, we introduce the concept of a *local model* over a finite set, which is basically a collection of functions without specifying anything about dynamics or how they are composed. This unifies both Boolean and logical networks into a single framework. Given a local model, the functions can be updated *synchronously*, defining a finite dynamical system (FDS), or *asynchronously*, as is often done with logical models. Naturally, there are other update schemes as well, but we will just consider these two basic ones in this chapter. After comparing the so-called synchronous and asynchronous "state spaces" of local models, we will prove bijections between local models, synchronous state spaces, and asynchronous state spaces. Also in this chapter, we will introduce two basic operons in the model organism *Escherichia*

Algebraic and Combinatorial Computational Biology. https://doi.org/10.1016/B978-0-12-814066-6.00004-0

coli: the lactose (*lac*) and arabinose (*ara*) operons, and we will model these gene regulatory networks as local models. Finally, we will show how to use computational algebra and freely available software tools for their analysis. Specifically, we will use the Macaulay2 computational algebra software [4], the Gene Interaction Network simulation (GINsim) [5], and a new online crowd-sourced platform called TURING: *Algorithms for Computation with Finite Dynamical Systems* [6].

This chapter is organized as follows. In the remainder of Section 4.1, we introduce the *biological background* relevant to this chapter. After starting with the central dogma of molecular biology, we discuss gene regulation by operons before ending with details of the lactose (*lac*) and arabinose (*ara*) operons in *E. coli*. In Section 4.2, we introduce the *mathematical background* relevant to this chapter. We begin with differential equation models of the *lac* and *ara* operons, how they both exhibit the phenomenon of bistability, before finishing with a brief discussion of existing discrete modeling frameworks such as Boolean networks. Section 4.3 begins with a quick review of only the basics about rings, ideals, and finite fields that we will use. Next, we introduce the local modeling framework. In Section 4.4, we introduce published local models of the *lac* and *ara* operons. In Section 4.5, we analyze the fixed points of these models using computational algebra and the Macaulay2 software. In Section 4.6, we briefly discuss a number of freely available software packages for the simulation and analysis of local models, before focusing in on two: GINsim and TURING. Finally, we conclude in Section 4.7 with a brief summary of more advanced and specialized topics involving local models.

4.1.1 The Central Dogma and Gene Regulation

In all living cells, genetic information is stored as deoxyribonucleic acid (DNA). A section of DNA which encodes a protein is called a gene. The instructions for producing the protein are encoded in the sequence of the bases adenine, cytosine, guanine, and thymine (A, C, G, and T, respectively) which make up the DNA. When a cell needs to make a particular protein, it copies the DNA sequence in the gene into another molecule, ribonucleic acid (RNA). This RNA copy, called a messenger RNA (mRNA), contains the sequence information necessary to produce the protein. The ribosome, the cell's protein-synthesizing machinery, binds the mRNA and, using the instructions in the mRNA, links together the correct amino acids in the correct order to make the protein.

This order of information transfer, from DNA to RNA (transcription) and from RNA to protein (translation), was described by Francis Crick as the "Central Dogma" of molecular biology [7]. Both transcription and translation are energy-intensive processes. Efficient use of cellular resources requires that proteins are made only when needed. Efficient species are successful species, and successful organisms can regulate the transcription of their genes such that only the necessary genes are transcribed and only the necessary proteins are produced. This control of transcription is known as gene regulation, and modeling of gene regulation is the topic of this chapter.

4.1.2 Types of Operons

Escherichia coli is a rod-shaped bacterium that lives in the guts of mammals and birds. It is the most widely studied prokaryotic model organism due to its simple genome that has been fully sequenced, and its ease of cultivation and reproduction in a laboratory. Since it lives in the lower intestines, its diet is largely shaped by that of its host. Bacteria such as *E. coli* use *cellular respiration* to convert biochemical energy from nutrients into ATP, a form of energy they can utilize for cellular processes. Common nutrients include sugars, amino acids, and fatty acids. The preferred sugar for *E. coli* is glucose because it is a fairly simple molecule and does not need to be broken down further to be metabolized in the process of glycolysis, the first step of cellular respiration. If glucose is not available, then the bacteria can process other sugars, such lactose, sucrose, or arabinose, just to name a few. These sugars are more complex or, in the case of arabinose, have structures sufficiently different from glucose that they cannot be used in glycolysis. For example, lactose (milk sugar) contains a glucose molecule bound to a galactose molecule, and sucrose is glucose bound to fructose. In order for *E. coli* to metabolize a complex sugar such as lactose or sucrose, it needs to be able to both transport these sugars into the cell, and then break them down into glucose and its companion sugar which can both be used in glycolysis. Transport is done by a *transporter protein*, which can be thought of as a "cellular doggy door." Once inside the cell, these sugar molecules are broken apart by an enzyme, which can be thought of as "molecular scissors." However, these doggy door proteins or molecular scissors are not universal—different proteins are required to transport lactose and arabinose into the cell, and different enzymes are needed to modify them for metabolism.

The proteins and enzymes needed for the transport and metabolism of sugars such as lactose and arabinose are *gene products*, and it takes considerable cellular energy to produce them through transcription and translation. As such, this is only done when they are needed to carry out their functions. For example, if these sugars are not present, the genes will not be transcribed. Alternatively, if the sugars are present but so is glucose, the preferred energy source, then the genes should still not be transcribed. Since *E. coli* cells obviously do not have a central computer or brain that governs transcription and translation, this functionality must be automatic. In other words, somehow, the genes that code for these transport proteins and enzymes must be automatically disabled or activated, depending on which sugars are present.

Some genes, such as those that produce gene products for critical cellular functions, are always transcribed. Others are only transcribed if needed. The complex process of how these genes are turned ON or OFF is called gene regulation. A *gene regulatory network* is a collection of gene products and other biomolecules that regulate each other to serve a specific purpose for the cell. There are many ways in which this is done, and some of these types are more prevalent in different types of organisms, for example, prokaryotes versus eukaryotes, or plants versus mammals. A common method of gene regulation in prokaryotes is through the use of operons. An *operon* is a cluster of genes that are required for a particular cellular task, and are transcribed together. When the genes are being transcribed, we say that the operon is ON. When transcription is blocked, we say the operon is OFF. The rate at which the genes are being transcribed determines their *expression levels*. Both theoretically (by the laws of mass-action kinetics; beyond the scope of this article) and in practice, these expression levels are either very low (called "*basal*"), or thousands of times higher.

The genes required for the transport and metabolism of lactose and arabinose are located in operons called the *lactose (lac) operon* and the *arabinose (ara) operon*, respectively. An *operon* is a region of the genome, usually consisting of several thousand base pairs. The beginning of an operon is known as the *promoter region*, and is where transcription is initiated. An enzyme called *RNA polymerase* is bound to the promoter region, and when the operon turns ON, this enzyme unzips the DNA and reads it. Downstream from the promoter region is the *operator region*. This is where transcription factors bind to regulate transcription. For example, a repressor protein can block transcription by binding to a site in the operator region. Conversely, an activator protein can increase transcription levels by binding to a site near the promoter region (which we will call a *promoter-associated region*) and facilitating RNA polymerase binding at the promoter. Finally, the genes in the operon, called the *structural genes*, are located downstream of the operator region. It should be noted that operons can have several promoter-associated and operator regions, allowing binding of several different activator and repressor proteins. Fig. 4.1 shows the distinct promoter and operator regions and structural genes of the *lac* operon.

During the process of transcription, the RNA polymerase unzips the DNA and reads it, creating a complementary copy of the DNA in the form of messenger RNA (mRNA). This is a single-stranded nucleic acid with four nitrogenous bases: guanine, adenine, cytosine, but uracil instead of thymine. This newly created mRNA strand binds to RNA-protein complexes called *ribosomes*, analogous to how a paper Scantron multiple choice test is fed into a machine that reads it. Each triplet of bases that is read codes for 1 of 20 amino acids. The end result is a long chain of amino acids called a *protein*, which is the *gene product*. This process of creating a gene product from an mRNA strand is called *translation*. Recall that the expression level of a gene is the *rate* at which it is being transcribed, and this is measured by the concentration of mRNA in the cell. The expression level will generally be proportional to the concentration of the gene products as well. That said, mRNA and proteins are organic molecules and they have a finite shelf-life. Once they are created, they start to break down. The half-lives of these molecules are generally on the scale of minutes, though of course it varies widely depending on the gene product. Accurate mathematical models need to take this degradation into consideration.

In addition to being either inducible or repressible, an operon's mechanism of regulation can be either *positive* or *negative*. Loosely speaking, an operon with positive regulation means that a "molecular key" is needed to initiate transcription. By contrast, negative regulation simply means that the operon is turned OFF by a mechanism that blocks transcription. Let us pause to summarize all four types of operons, and assign a nonbiological analogy to each.

In a *negative inducible operon*, transcription is normally blocked. This is most commonly done by a *repressor protein* that binds to the operator region of the operon, blocking the RNA polymerase from attaching to the DNA strand. To turn the operon on, an *inducer* binds to the repressor protein and changes its molecular shape, causing it to fall off the DNA strand so transcription can begin. As a real-world analogy, think of the repressor protein as a U-lock for a bike. The inducer represents the key that is needed to remove the lock so the bike can move.

In a *negative repressible operon*, the genes are normally transcribed. Repressor proteins can turn the operon OFF, but they are inactivated by default. A molecule called a *corepressor* is needed to bind to the repressor protein to change its shape so it can properly bind to the DNA and block transcription. As a real-world analogy, think of a repressible operon as a long-distance train that can travel for several days without stopping. The repressor protein can be thought of as the brakes that are just inches from the wheels, but do not function unless the engineer intervenes by pulling a lever. The human hand that pulls the lever to activate the brakes and stop the train is analogous to adding a corepressor to activate the repressor protein so it can stop transcription.

A *positive inducible operon* is normally OFF. *Activator proteins* required for transcription to begin are normally unable to bind to the DNA strand. An inducer is needed to bind to them to change their shape so they can bind to the operon and initiate transcription. As a real-life analogy, consider a car that is outfitted with a breathalyser test required to start the engine. The key represents the activator protein, but more is needed: a physical breath of (alcohol-free) air represents the inducer that activates the key that starts the engine.

A *positive repressible operon* would have activator proteins normally bound to the DNA, allowing transcription to take place by default. Turning OFF the operon would require a repressor protein to bind to these activator proteins, thereby changing their molecular shape and causing them to fall off the DNA. There are no known examples of positive repressible operons. Interestingly, it is not easy to come up with a mechanical real-world analogy of this either. The practical ways to disengage a vehicle or engine—applying brakes or a shut-down valve—are examples of negative repressible regulation. A positive repressible mechanism would be something that causes the starter or engine *key* to stop functioning. It is possible to imagine such an engineered scenario, for example, consider a drone that is designed so that if it gets too close to an aircraft, the hand-held controller automatically disengages and lands the drone. That said, such examples are less natural, both in real-world mechanical devices as well as in biological networks.

The following table summarizes the four types of regulation found in operons, and gives an example of each type.

	Inducible	**Repressible**
Negative	*lac*	*trp*
Positive	*ara*	Does it exist?

4.1.3 Two Well-Known Operons in *E. coli*

4.1.3.1 The Lactose Operon

The *lac* operon regulates the transport and metabolism of lactose in *E. coli*. Since it is an inducible operon, transcription is blocked by a repressor protein unless the operon's gene products are needed. This happens precisely when there is lactose but no glucose present, and we need to explain exactly how the operon achieves this.

When a car is illegally parked, police may attach a *wheel clamp* or *parking boot* and charge a high fee to remove it. In order for the car's owner to drive it away, two things are needed: (i) a police key to unlock and remove the clamp and (ii) the driver's key to turn ON the engine. If either of these steps is skipped, the car will be stuck in the parking lot. The steps needed to turn ON the *lac* operon are somewhat analogous. The repressor protein can be thought of as the wheel clamp, the key that unlocks it is called allolactose, and a protein complex called *cAMP-CAP* can be thought of as the "engine key."

Allolactose is an *isomer* of lactose, which means that they have the same chemical formulas, with the only difference being the geometric arrangement of the molecules. When lactose is present, some of it will be converted to allolactose. When allolactose is present, it binds to the repressor protein and changes its shape so it can no longer attach onto the DNA strand and block transcription. Since allolactose is only present when lactose is, the "wheel clamp" of the *lac* operon is removed only when lactose is present.

Even without the repressor protein attached to the DNA strand, one more condition is needed for transcription to begin. Specifically, the cAMP-CAP protein complex must be bound at a site adjacent to the promoter region of the operon. This "molecular key" is comprised of two molecules bound together: the *catabolite activator protein* (CAP) and *cyclic adenosine monophosphate* (cAMP). Glucose decreases the synthesis of cAMP, thereby reducing the production of cAMP-CAP complexes. In other words, when present, glucose results in the warping of the molecular key need to start the engine (initiate

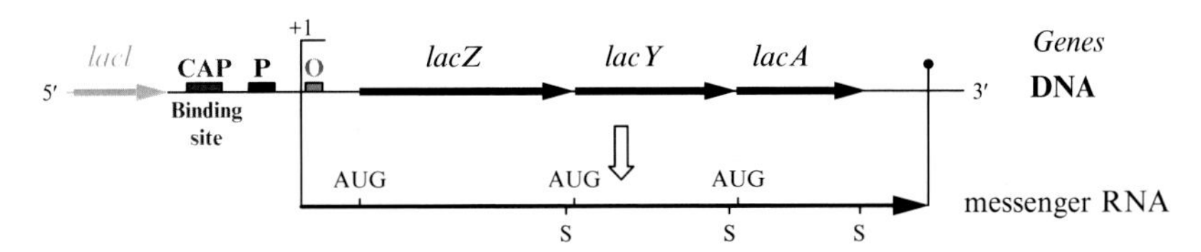

FIG. 4.1 The *lac* operon. The P denotes the promoter region, where transcription is initiated. The O denotes the operator region, where the lac repressor binds to block transcription. The *lacZ*, *lacY*, and *lacA* regions are the structural genes. *(From Wikipedia Commons, by G3pro/CC BY.)*

transcription), rendering it useless. Additionally, glucose inhibits transcription of the *lac* operon in another more basic way: it results in the production of another protein which binds to the transporter protein and prevents it from bringing lactose into the cell. This repression mechanism by glucose is called *inducer exclusion*, and it can be thought of as "jamming the doggy door." Interestingly, glucose does not use this repression method in the *ara* operon.

The *lac* operon consists of three structural genes, *lacZ*, *lacY*, and *lacA*, which are shown in Fig. 4.1. Each of these codes for a distinct gene product and these are denoted with uppercase letters: LacZ, LacY, and LacA. The LacA protein is not involved in the gene regulatory network, but the other two gene products are. These gene products are better known by their common names: β-galactosidase (LacZ) and *lac* permease (LacY), and their functions are listed follows.

- *β-galactosidase* is an enzyme, which can be thought of as "molecular scissors," as it cleaves lactose into glucose and galactose. It also facilitates the conversion of lactose into its isomer allolactose.
- *lac permease* is the transporter protein, which can be thought of as the "doggy door" that lets lactose come into the cell.

It is important to remember that even when the operon is OFF, there are trace amounts of β-galactosidase and *lac* permease in the cell. These occur because every once in a while when the stars align, the repressor protein may briefly fall off of the DNA strand, and a few rogue mRNA strands will be transcribed before another repressor protein molecule reattaches. Thus, one can think of the genes in the operon as being like a "leaky faucet." However, these imperfections are crucial for the operon's capability to turn itself ON when the first few molecules of lactose appear.

For example, suppose the operon is OFF and lactose arrives outside the cell. A tiny bit will seep into the cell through the few molecules of *lac* permease in the membrane. A few of these molecules will get converted into allolactose by the trace amounts of β-galactosidase. This will cause the repressor protein that is bound to the DNA to fall off. If glucose is not present, then transcription of the *lac* operon will begin, until another repressor protein molecule reattaches. As a result, the concentration of β-galactosidase and *lac* permease increases.

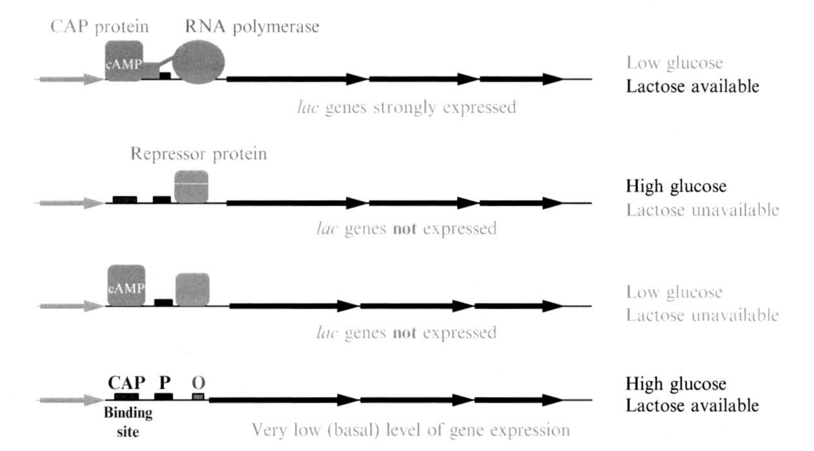

FIG. 4.2 The *lac* operon depending on whether lactose and glucose are present. The operon is ON in the *upper panel*. In the *middle two panels*, only trace amounts of transcription occur. In the *bottom panel*, a low level of transcription occurs, because the CAP protein is not bound and therefore the RNA polymerase does not bind efficiently at the promoter. *(From Wikipedia Commons, by G3pro/CC BY.)*

There are now more doggy door proteins to bring in lactose and more molecular scissors to convert lactose into allolactose, which will in turn ensure that fewer of the wheel clamps can bind (i.e., most of the *lac* repressor proteins are inactivated). This has a snowball effect, and soon the flood gates open and the operon is ON. This situation is shown in the upper panel of Fig. 4.2.

While the genes in the *lac* operon are being transcribed, lactose is being broken down into glucose and galactose. When all the lactose is used up, there will be no more allolactose present to bind to the repressor protein molecules, so they will re-attach to the operator region and block transcription. This is how the operon automatically turns itself OFF when lactose is depleted.

4.1.3.2 The Arabinose Operon

Like the *lac* operon, the *ara* operon is inducible, so its default state is off. Unlike the *lac* operon, which uses a repressor protein to block transcription, in the *ara* operon, the DNA strand physically loops, preventing the RNA polymerase from binding to, unzipping, and reading the DNA. This loop is formed by an *AraC protein* dimer binding to the operator site, as shown in the left-hand side of Fig. 4.3. If arabinose is present, then it binds to the AraC protein, altering the shape of the dimer and preventing it from forming the DNA loop, as shown in the right-hand side of Fig. 4.3. In this arabinose-bound form, the AraC protein acts as an activator and initiates transcription. Just like in the *lac* operon, the *ara* operon has a CAP binding site, and the cAMP-CAP complex serves as an "engine key" to start transcription. Since glucose decreases cAMP levels,

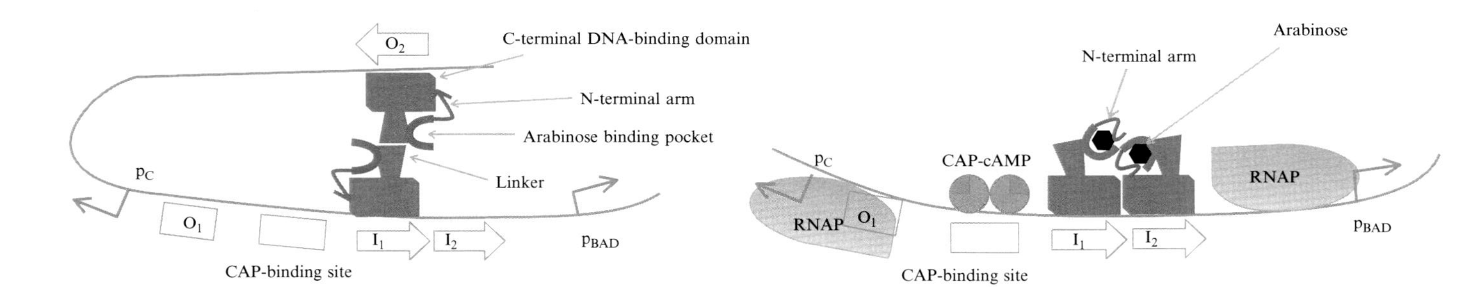

FIG. 4.3 When arabinose is not present, the AraC protein forms a dimer which binds to the DNA strand in two places, forming a DNA loop and blocking transcription, as shown in the *left panel*. When arabinose is present, it binds to the AraC protein, breaking the DNA loop, so transcription can begin. This is shown in the image on the right. *(From Wikipedia Commons, by Marckemi/CC BY-SA 3.0.)*

and hence the formation of the cAMP-CAP complex, the operon will be off whenever glucose is present, regardless of whether or not there is arabinose.

The structural genes of the *ara* operon are called *araB*, *araA*, and *araD* or together, *araBAD*. They code for enzymes needed to metabolize arabinose via the pentose phosphate pathway, which is a sequence of chemical reactions that converts arabinose to xylulose 5-phosphate. Also in the operon is the *araC* gene which codes for AraC, though this is transcribed in the opposite direction. Transport of arabinose into the cell is achieved through genes *araC* and the genes *araE* and *araFGH* that are all located upstream of the operon. The arabinose-bound AraC protein acts as an activator and induces transcription of these "transporter genes," resulting in production of the transport proteins that they code for. The *araE* and *araFGH* genes are controlled by two promoters, and as such, they produce two distinct strands of RNA: one for *araE* and one for *araFGH*. Technically, these are two transport systems: the *araE* gene product controls the low-affinity high-capacity transport system, and the *araFGH* gene products control the high-affinity low-capacity system, though these details are unimportant for the scope of this chapter. Unlike the lactose transport proteins, glucose does not repress the arabinose transport proteins by inducer exclusion. We will see later how that affects the dynamics of the respective Boolean models. Even though the AraC protein represses transcription when not bound to arabinose, the operon is said to be positive inducible because the arabinose-bound AraC protein acts as an activator by inducing binding of RNA polymerase to the promoter regions of the *araBAD*, *araE*, *araFGH*, and *araC* genes.

4.2 CONTINUOUS AND DISCRETE MODELS OF BIOLOGICAL NETWORKS

4.2.1 Differential Equation Models

Originally, mathematical biologists primarily used continuous methods such as systems of ordinary differential equations (ODEs) to model molecular networks. In this setting, each variable $x_i(t)$ represents the concentration of a particular biomolecule such as a gene product. The end result is a system of nonlinear differential equations with many rate constants that cannot be solved explicitly, and require numerical methods to analyze.

This is best seen through examples, though it is not necessary to explain every detail to come away with the main ideas. Both the *lac* and *ara* operons have been modeled with systems of ODEs. First, consider the *lac* operon, and let $M(t)$, $B(t)$, $A(t)$, $P(t)$, and $L(t)$ represent the concentrations of mRNA, β-galactosidase, allolactose, *lac* permease, and intracellular lactose, respectively, as functions of the elapsed time t. All of these are functions of time because they can change quite rapidly; on the order of seconds or minutes. By contrast, let L_e be the concentration of extracellular lactose. While this does not stay constant forever, it changes much slower than the concentrations of the intracellular biomolecules do. For example, if the host organism consumes milk, then *E. coli* in its gut will be exposed to high levels of lactose for several hours. We say that M, B, A, P, and L are *variables* and the constant L_e is a *parameter*. The following model was proposed by Yildirim [8]:

$$\begin{aligned}
M'(t) &= \alpha_M \tfrac{1+K_1[e^{-\mu\tau_M}A(t-\tau_M)]^n}{K+K_1[e^{-\mu\tau_M}A(t-\tau_M)]^n} + \Gamma_0 - (\gamma_M + \mu)M(t), \\
B'(t) &= \alpha_B e^{-\mu\tau_B} M(t-\tau_B) - (\gamma_B + \mu)B(t), \\
A'(t) &= \alpha_A B(t)\tfrac{L(t)}{K_L+L(t)} - \beta_A B(t)\tfrac{A(t)}{K_A+A(t)} - (\gamma_A + \mu)A(t), \\
P'(t) &= \alpha_P e^{-\mu(\tau_B+\tau_P)} M(t-(\tau_B+\tau_P)) - (\gamma_P + \mu)P(t), \\
L'(t) &= \alpha_L P(t)\tfrac{L_e}{K_{L_e}+L_e} - \beta_{L_e} P(t)\tfrac{L(t)}{K_{L_e}+L(t)} - \alpha_A B(t)\tfrac{L(t)}{K_L+L(t)} - (\gamma_L+\mu)L(t).
\end{aligned}$$

One of the most striking features of this model, in addition to how incredibly complicated and intimidating it looks, is how many constants are included. Without going into too many details, the five constants $\alpha_M, \dots, \alpha_L$ and both β_A, β_{L_e} are *rate constants* from Michaelis-Menden equations that model biochemical reactions.[1] The constants τ_M, τ_B, and τ_A are time-delays from transcription and/or translation, because these processes are not instantaneous. The constants $\gamma_M, \dots, \gamma_L$ describe the degradation (i.e., decay) of the corresponding organic molecules. Since concentration is mass over volume, and the cells are rapidly growing, a dilution constant μ is needed to account for this growth rate. The constant Γ_0 describes the basal (or trace) concentration of mRNA in the cell, even if the genes are not being expressed. (Recall the "leaky faucet" analogy.) Finally, the constants such as K, K_1, K_L, K_A, and K_{L_e} all arise from Hill functions in Michaelis-Menten kinetics. In all, there are 22 constants in this model which are either experimentally determined or estimated by fitting to data; see [8] for details. What makes matters worse is that for many of these constants, estimates from the published literature can differ by orders of magnitude; see [9, Section 2.6.1] and the references therein. This is more the case for the rate constants than it is for the time delays, because scientists can accurately measure the length that certain biochemical processes take.

It should be no surprise that systems of ODEs such as this cannot be solved explicitly. In principle, one primary advantage of ODEs is that they are quantitative models, amenable to tools of calculus and analysis. However, any model with 22 unknown constants is in practice, qualitative at best. Moreover, it is fair to ask whether such a model can even have practical value. The authors of [8] provided strong evidence that it does, and they do this by analyzing the fixed points. Though the precise numerical values may not provide information that is completely reliable, their qualitative characteristics are still useful. For example, it was shown in [10] that this model exhibits *bistability* for medium levels of

1. Chapter 9 is on biochemical reaction networks.

lactose, which is an important phenomenon that we will discuss soon. Our discrete models also exhibit bistability, which is an important part of their model validation.

A strong argument for the validity of ODE models with many unknown rate constants is the robustness that biological networks must possess. The survival of a complex organism requires its myriad of biological networks, such as operons, gene regulatory, metabolic, and signaling networks, to be able to function under a wide range of changes to its environment and across different genotypes. For example, different individual organisms might have vastly different diets and metabolisms, or they could be living in environments with a wide range of temperatures or pH levels. For these organisms to survive and evolve as a species for millions of years, such networks that control basic biological functions need to be very robust to such changes. Thus, it is reasonable to expect that even if the rate constants in an ODE model are not exactly what they are in real life, or even if some differ by orders of magnitude, the basic qualitative features of the model, such as the number or type (e.g., stable vs. unstable) of fixed points, should still be preserved.

Some years after publishing the aforementioned ODE model of the *lac* operon, Yildirim modeled the *ara* operon with differential equations [11]. In this model, $A(t)$, $E(t)$, and $F(t)$ represent concentrations of intracellular arabinose, *araE* mRNA, and *araFGH* mRNA, respectively. The following three differential equations were proposed, where the parameter A_e represents the concentration of extracellular arabinose:

$$\begin{aligned}
A'(t) &= \frac{A_e V_E E(t)}{K_E + A_e} + \frac{A_F V_F F(t)}{K_F + A_e} - (\mu + \gamma_A)A(t) \\
E'(t) &= \alpha_E + \frac{V_{m_E}(A(t))^n}{K_{m_E}^n + (A(t))^n} - (\mu + \gamma_E)E(t) \\
F'(t) &= \alpha_F + \frac{V_{m_F}(A(t))^n}{K_{m_F}^n + (A(t))^n} - (\mu + \gamma_F)F(t).
\end{aligned}$$

As in the prior *lac* operon model, the constants γ_A, γ_E, and γ_F represent degradation rates, and μ describes loss of concentration due to cell growth. The constants V_E, V_F, $V_{mE}V_{mF}$, K_E, K_F, K_{mE}, and K_{mF} arise from Michaelis-Menten functions because these transport systems can be effectively modeled using basic mass-action kinetics.

Though we will not analyze the aforementioned ODE models in detail, we will discuss a few qualitative aspects of them. To begin, we can associate each system with a directed multigraph called a *wiring diagram* that describes whether variables have a positive, negative, or zero influence on each other. Consider a biomolecule A in a molecular network with concentration $A(t)$. Another biomolecule B is said to *activate* A if increasing $B(t)$ causes $A(t)$ to increase. In contrast, if increasing B it causes $A(t)$ to decrease, then B *represses*

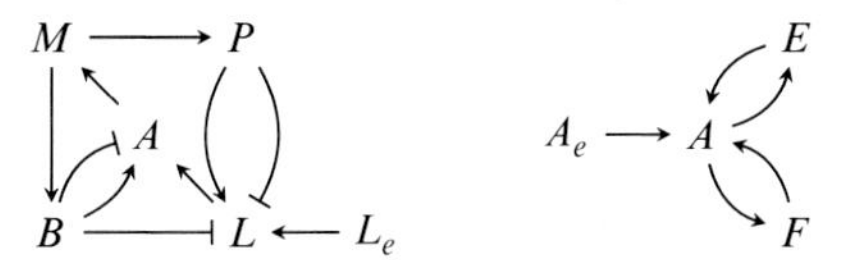

FIG. 4.4 The wiring diagrams of the differential equation models of the *lac* operon (*left*) and the *ara* operon (*right*). Notice how the parameters are characterize by being sinks.

or *inhibits* or even *degrades*[2] A. In an ODE model, if B activates (respectively, represses or inhibits) A, then $B(t)$ appears in the function for $A'(t)$ with a positive (respectively, negative) sign. We include an edge $A \longrightarrow B$ if A activates B, and an edge $A \dashv B$ if A represses or inhibits B. Sometimes we use $A \mathrel{-\!\!-\!\!\bullet} B$ for repression/inhibition because it can stand out more visually, especially if the nodes are circled, as in Fig. 4.9. Of course, it is possible for A to both activate and repress/inhibit B, in which case we include both edges. For example, in the ODE model of the *lac* operon above, β-galactosidase activates allolactose, because it converts lactose into allolactose. However, it also acts as an inhibitor because it cleaves allolactose into glucose and galactose. The wiring diagrams of the ODE models of the *lac* and *ara* operons are shown in Fig. 4.4.

Note that in these wiring diagrams, we omit all self-loops, because it is not biologically accurate to say that each biomolecule inhibits itself. The terms like $-(\gamma_M + \mu)M(t)$ that appear in $M'(t)$ simply describe the fact that the concentration is decreasing due to degradation and cell growth.

In any model that involves an underlying network or graph, there is always considerable interest in understanding how the underlying structure affects the global dynamics, and this is not just limited to models of molecular networks that are the focus of this chapter. This is at the heart of several other chapters in this volume alone. For example, the focus on Chapter 8 is how the structure of a neural network affects the emergent dynamics of a system of threshold linear ODEs arising as simple models in neuroscience. Chapter 9 discusses how the structure of a chemical reaction network affects the global dynamics of a corresponding system of ODEs that model the reactions under the laws of mass action. In all of these examples, including the ones here, the focus is on the *long-term* behavior of the dynamics, especially fixed points.

One way to characterize the structure of an underlying network is by the presence or absence of *motifs*, which are small often repeating subnetworks [12]. One common example are feedback loops, and these come in several types. For example, the directed cycle $A \longrightarrow M \longrightarrow B \dashv A$ of length 3 in Fig. 4.4

2. There is a difference between repressing, inhibiting, and degrading. Biologically speaking, repressors interfere with gene expression (decreasing the production of a protein), and inhibitors interfere with the action of enzymes (decreasing their activity). Degradation means that the molecule in question is being chemically broken down. In the wiring diagram, though, all of these processes are indicated by the same negative edge symbol.

is called a *negative feedback loop*. The length 2 cycle $A \longrightarrow E \longrightarrow A$ is a *positive feedback loop*. It should be apparent how this generalizes: a feedback loop is any directed cycle, and it is positive if it contains an even number of negative edges. In 1981, Thomas made a general conjecture [13] that:

- positive feedback loops are necessary for *multistationarity*, and
- negative feedback loops are necessary for *cyclic attractors*.

In this context, multistationarity simply means having more than one stable fixed point. We will not define cyclic attractors in an ODE context, but it roughly means that there is periodicity that is not a fixed point. We will revisit both of these themes when we discuss discrete models later in this chapter. Thomas' conjectures have been proven in various frameworks, from differential equations [14] to Boolean networks [15, 16] to chemical reaction networks (Chapter 9).

4.2.2 Bistability in Biological Systems

The most common special case of multistationarity in biological networks is that of bistability. This means that a system has two stable steady states, and in classical differential equation models, these are usually separated by an unstable steady state. A simple example of bistability occurs in ecology in the *threshold equation*,

$$y' = -ry(y - T)(y - M), \quad 0 < T < M. \tag{4.1}$$

Here, $y(t)$ represent the population of a species. The steady states, or fixed points, are found by setting $y' = 0$ and solving for y. Clearly, there are three of them: the stable fixed point $y = 0$ represents extinction, the unstable fixed point $y = T$ is the "extinction threshold"; a population for which $y < T$ is said to be endangered, and the stable fixed point $y = M$ is the natural "carrying capacity." Fig. 4.5 shows the phase portrait of this differential equation for $r = 0.2$, as well as several solution curves. The stability of a fixed point can be determined by linearizing near it and seeing whether small perturbations shrink or grow. But that is not the focus of this chapter.

In a biological system like the *lac* (respectively, *ara*) operon, lactose (respectively, arabinose) is the *inducer*,[3] because as its concentration increases, the expression levels of the genes increase as well. However, this change is highly nonlinear. In fact, there is usually a sharp threshold between when expression levels are near zero ("basal") and when they are very high (the gene is being expressed). In many biological networks, it has been observed that when there are medium levels of this inducer, the target gene can be expressed in

3. Technically, the allolactose is the inducer of the *lac* operon because it directly bonds with the repressor protein. However, lactose indirectly induces the operon, and it will be easier and harmless to consider it the inducer for this exercise.

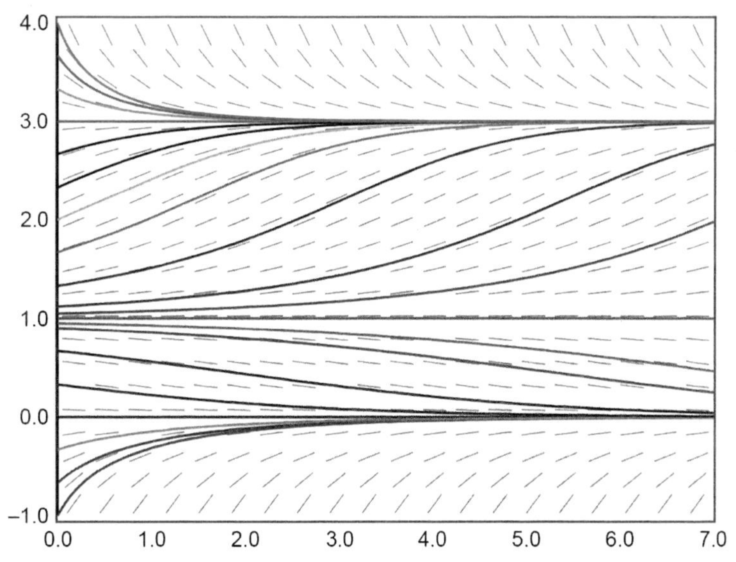

FIG. 4.5 The phase plot and several solutions of the threshold ODE. The carrying capacity is $M = 3$ and the extinction threshold is $T = 1$.

some cells and unexpressed in other cells. This is the phenomenon of *bistability*, which has been shown to exist in a wide range of biological systems, including both the *lac* and *ara* operons.

As an explicit example, let L_e be the concentration of extracellular lactose, and suppose that $L_e \approx 0$. The operon is OFF, and so the steady-state concentrations of the biomolecules such as mRNA, allolactose, and the gene products are all approximately zero. As L_e increases, the steady-state concentration X^* of a generic product X (e.g., $X = M, B, A, P$, or L in the ODE model from [8]) increases marginally, though it is still considered basal. However, when L_e is sufficiently high, say $L_e > \tau^{\uparrow}$ these steady-state concentrations jump to thousands of times higher, a hallmark of the operon being ON. Once the operon is ON, external lactose levels need to drop well below the "up-threshold" $\tau^{\uparrow}$ before the fixed points drop down to basal levels. That is, not until L_e drops below some "down-threshold" $\tau^{\downarrow} < \tau^{\uparrow}$ does the operon switch OFF. This means that when L_e is in this "middle range," that is, $\tau^{\downarrow} < L_e < \tau^{\uparrow}$, the operon could be either ON or OFF, and this depends on the recent history. Fig. 4.6 shows this bistability. The parameter L_e is on the horizontal axis, and for any fixed value of L_e, the steady state(s) of the generic product X are given by the intersection of the curve with a vertical line at that value of L_e. The bistable region can be identified by where a vertical line intersects the curve in three places. The top and the bottom point of intersection are the stable fixed points, and the middle point is an unstable fixed point that is never observed in practice. It separates the two stable fixed points much like how the extinction threshold $y = T$ separated the stable fixed points of $y = 0$ and $y = M$ in the threshold ODE in Eq. (4.1).

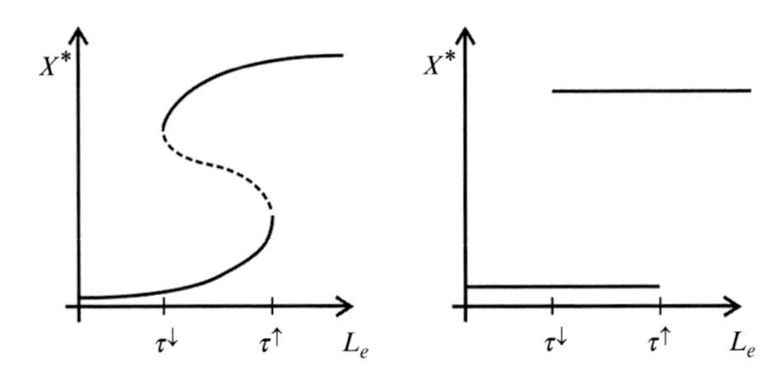

FIG. 4.6 At *left* is a curve showing how the steady states of a gene product X is a function of the parameter L_e. On the *right* is the Boolean version of this, where the only steady-state values of X^* are 0 (basal) or 1 (high).

As an analogy of this, suppose that one keeps their house heated to 65 degrees in winter and cooled to 75 degrees in summer. If it is 65 or 75 degrees inside, then you know what season it is. However, if it is 70 degrees inside, you cannot deduce from that alone whether it is spring or fall. If you look at the recent history, and saw that during the previous month, it was 73 degrees, then you can be fairly confident that it is probably fall. However, if last month it was 67 degrees inside, then it is probably spring. Similarly, if there are medium levels of lactose, then the operon should be ON in cells in an environment where lactose levels are decreasing from high concentration. The operon should be OFF if lactose levels are increasing from a lower concentration. In other words, cells that were raised in a lactose-rich environment will be expressing the *lac* genes when lactose levels decrease down to the middle range, but cells that were raised in a lactose-starved environment will not express these genes when lactose concentration increases up to the same level.

4.2.3 Discrete Models of Biological Networks

Recall that due to the highly nonlinear relationship between the expression levels of genes and their inducers and repressors, the concentration of a given gene product is usually either almost zero (basal) or thousands of times higher. This can be approximated by considering those expression levels as being OFF or ON. In 1969, the theoretical biologist Stuart Kauffman proposed modeling gene regulatory networks using Boolean functions [17]. Soon after, biologist René Thomas published a discrete formalization of a discrete modeling framework that he called *logical models* [18]; others call these *automata networks* [19]. In all of these settings, there is a finite set $[n] = \{1, \ldots, n\}$ of nodes, and each one has a state x_i from a finite set $S_i = \{0, \ldots, r_i - 1\}$. Obviously, the Boolean case is when each $r_i = 2$, but in many settings, there are more than two states, such as if one wants to distinguish between low, medium, and high levels.

Once the individual functions are chosen in a model, the next step is to decide how to compose them to generate the dynamics. One of the most common ways

$$\begin{cases} f_1(x_1, x_2) = \overline{x_2} \\ f_2(x_1, x_2) = \overline{x_1} \end{cases}$$

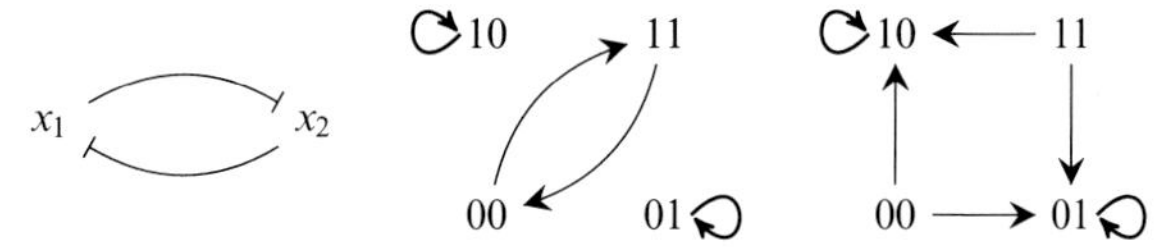

FIG. 4.7 From *left to right*: a simple Boolean network; its two-node wiring diagram; its phase space under a synchronous update; its phase space under general asynchronous update.

is to just update them synchronously. If there are n nodes and the functions are Boolean, then this defines a map

$$f \colon \mathbb{F}_2^n \longrightarrow \mathbb{F}_2^n, \quad f \colon (x_1, \ldots, x_n) \longmapsto (f_1(x_1, \ldots, x_n), \ldots, f_n(x_1, \ldots, x_n)).$$

Of course, molecular networks have no central clock, so this assumption of synchronicity is not biologically sound. Nevertheless, it is often used out of convenience, with the argument made that since biological systems are so robust, this artificial relic does not make much of a difference. Also, fixed points do not depend on the update order, and these are arguably the most important features of a model. Other models update the functions asynchronously, in some fixed order [20] or randomly [19]. The latter is arguably more natural, because it allows nodes to update at any time, and it does not use an artificial permutation update order. Important features include the strongly connected components (SCCs), especially the terminal ones (definitions to be given later). Of course, there are many other variants such as block sequential update [21] or some sort of stochastic scheme; see Chapter 5 for more details. Fig. 4.7 shows a simple example of a two-node Boolean model and the four-node phase space under both a synchronous and general asynchronous update. The details of both of these will be formalized in the next section; the goal here is to just give the intuition. Informally, the synchronous phase space (second from right) shows the result of updating f_1 and f_2 at the same time to the states $(x_1, x_2) \in \mathbb{F}_2^2$ of the system, and the asynchronous phase space (far right) shows the result of updating only one f_i at time. We will revisit this example later.

In the last few decades, scientists have proposed numerous Boolean and/or discrete models of all sorts of biological systems. Boolean models, often just called "Boolean networks," are easily the most common. Not surprisingly, there is no general consensus about what a "discrete model" or "Boolean network" actually means. Some authors speak of a "Boolean network" as a collection of functions $(f_1, \ldots, f_n)$, where $f_i \colon \mathbb{F}_2^n \to \mathbb{F}_2$. Others speak of it as any function $\mathbb{F}_2^n \to \mathbb{F}_2^n$; that is, the function defined by updating each f_i synchronously. It is not clear right away whether such a function necessarily arises from synchronously updating individual "local functions" $f_1, \ldots, f_n$ (it does; stay tuned). Other authors define a Boolean network as, for example, a triple consisting of a finite set V of vertices, a collection of Boolean functions, and an update mechanism, such as synchronous or asynchronous. Some include the underlying graph in the definition, whereas others say the graph is a consequence of the functions. There are many other variants in the literature, as there is no agreed upon standard definition for what a Boolean network, or even a generic model, should consist of.

Our next task will be to define and study a natural modeling framework that unifies many of these scattered definitions. *One of our driving philosophies is that a discrete model of a molecular network should consist of the local functions, and nothing else.* Whether the modeler chooses to update them synchronously, asynchronously, stochastically, or some other way, should be seen as part of *model validation*.

4.3 LOCAL MODELS

4.3.1 Polynomial Rings and Ideals for the Nonexpert

The algebraic objects that are central to this chapter and Chapters 5 and 6 are polynomial rings and ideals. These are advanced topics that can get quite abstract, and thus are usually relegated to a senior-level "modern algebra" course for math majors. However, the special cases of rings and ideals that will arise in this chapter are much more concrete, and are things that people who have never even taken calculus are probably unknowingly familiar with. We will spend the remainder of this short section giving fancy names to these fairly pedestrian objects, without actually defining them in formal mathematical terms. *Readers who know what rings and ideals are may safely skip the remainder of this short section.* Proofs for the propositions, theorems, etc., that are only stated in this section can be found in any introductory abstract algebra book, for example, [22].

A *field* is a set of numbers where one can add, subtract, multiply, and divide (except by 0), with the result remaining in the set. Common examples include the rational numbers (fractions) $\mathbb{Q}$, the real numbers $\mathbb{R}$, and the complex numbers $\mathbb{C}$. Nonexamples include the nonnegative real numbers $\mathbb{R}_{\geq 0}$ (cannot subtract) and the integers $\mathbb{Z}$ (cannot divide). The most common fields that we will see in this chapter are *finite fields*, such as $\mathbb{F}_2 = \{0, 1\}$, where arithmetic is done modulo 2 (i.e., 0 = even and 1 = odd), and $\mathbb{F}_p = \{0, 1, \ldots, p-1\}$, where arithmetic is done modulo p (i.e., divide by p and take the remainder). In this setting, the concept of "division" just means "what you multiply by to get back to 1." Though we never need to write negative signs when working with these fields, sometimes we do for convenience or clarity. For example, $2 = -1$ in $\mathbb{F}_3$, but if we write

$$f(x) = (x-1)(x^2+x+2) = (x+2)(x^2+x+2),$$

it is more immediately obvious that 1 is a root of f from the first expression than it is from the second.

A *ring* is an algebraic object where one can add, subtract, and multiply, but not necessarily divide. The most familiar example is the set of integers, $\mathbb{Z}$. In this chapter, the most common examples will be *polynomial rings*, with coefficients coming from a ring like $\mathbb{Z}$, or a field $\mathbb{F}$ (we say the ring is "over $\mathbb{Z}$" or "over $\mathbb{F}$"). When allowing just a single variable x, we denote the rings of all polynomials as $\mathbb{F}[x]$, and when using multiple variables (e.g., x and y), we will write $\mathbb{F}[x, y]$. Note that this is a ring because the sum, difference, or product of any two polynomials is still a polynomial. However, it is not a field because the quotient of two polynomials is generally not a polynomial. A ring is *commutative* if the order of multiplication does not matter. All rings that we will encounter in this chapter will be commutative. A common nonexample is the ring of $n \times n$ matrices with coefficients from a field $\mathbb{F}$; note that AB generally does not equal BA.

The last advanced concept in this section is that of an *ideal* of a ring. This is a set I that is invariant under multiplication. For example, the even integers $2\mathbb{Z}$ is an ideal of $\mathbb{Z}$, because multiplying an even integer by anything in $\mathbb{Z}$ remains even. The polynomials that have zero as a root, that is, those that can be written as $f(x) = x \cdot g(x)$ for an arbitrary polynomial $g(x) \in \mathbb{F}[x]$, are an ideal of $\mathbb{F}[x]$. Every ideal has a (nonunique) set of generators. For example, in the polynomial ring $\mathbb{Z}[x]$, the following set I is an ideal, generated by the two polynomials x and 2:

$$I = \langle x, 2 \rangle = \big\{ x \cdot h_1(x) + 2 \cdot h_2(x) \mid h_1, h_2 \in \mathbb{Z}[x] \big\}.$$

The reader should take a few moments to verify that (i) this is precisely the set of polynomials with integer coefficients whose constant term is even, (ii) this is indeed an ideal, because multiplying any such polynomial by *anything* in $\mathbb{Z}[x]$ still yields a polynomial with an even constant term, and (iii) there is no way to generate this ideal with only one polynomial.

4.3.2 Finite Fields

A commutative ring with $1 \neq 0$ is an *integral domain* if the product of nonzero elements is always nonzero. Fields must be integral domains because if $ab = 0$, for $a, b \neq 0$, then a cannot have a multiplicative inverse. However, integral domains need not be fields; the integer $\mathbb{Z}$ is a basic example of this. Perhaps surprisingly, such examples only exist for infinite rings.

Proposition 4.1. *Every finite integral domain is a field.*

A simple example of a finite ring that is not integral is $\mathbb{Z}_4 = \{0, 1, 2, 3\}$, because $2 \cdot 2 = 0$. It is similarly easy to see that $\mathbb{Z}_n$ is nonintegral for any composite n. To summarize, we know that $\mathbb{F}_p$ is a finite field for every prime p. Such a field is said to be *prime*. We need a little bit more theory before we can properly describe nonprime finite fields.

Throughout this section, $\mathbb{F}$ will be a finite field. By assumption, $\mathbb{F}$ contains the multiplicative identity $1 \neq 0$.

Definition 4.1. The *characteristic* of $\mathbb{F}$, denoted $\mathrm{Char}(\mathbb{F})$, is the smallest positive integer n for which $n \cdot 1 := \underbrace{1 + \cdots + 1}_{n \text{ times}} = 0$.

It is easy to show that Char($\mathbb{F}$) must be prime. Therefore, $\mathbb{F}$ contains $\mathbb{F}_p = \{0, 1, \ldots, p-1\}$ as a subfield. Since $\mathbb{F}$ is closed under addition and scalar multiplication, it is a *vector space* over $\mathbb{F}_p$. This means that it has some *basis* $x_1, \ldots, x_k$, and every element can be written uniquely as $x = a_1x_1 + \cdots + a_kx_k$ for $a_i \in \mathbb{F}$. As a simple consequence of this, the *order* of $\mathbb{F}$ (the number of elements it contains) is $|\mathbb{F}| = p^k$. Thus, the order of any finite field must be a prime power. We can use this vector space technique to say more.

Proposition 4.2. *If K and L are finite fields with $K \subseteq L$ and $|K| = p^m$ and $|L| = p^n$, then m divides n.*

Proof. We have $\mathbb{F}_p \subseteq K \subseteq L$. Then L is not only a $\mathbb{F}_p$-vector space, but also a K-vector space.

Let $x_1, \ldots, x_k$ be a basis for L over K. Every $x \in L$ can be written uniquely as $x = a_1x_1 + \cdots + a_kx_k$, where $a_k \in K$. Clearly, there are $|K| = p^m$ choices for each a_k, and thus $p^n = |L| = (p^m)^k$. □

Thus far, we know that every finite field $\mathbb{F}$ must have order $|\mathbb{F}| = p^k$ for some $k \in \mathbb{N}$. We are of course very familiar with the case when $k = 1$. To show that $k > 1$ is possible, we will explicitly construct a nonprime field $\mathbb{F}_4$ of order 4; recall that $\mathbb{Z}_4 = \{0, 1, 2, 3\}$ is *not* a field. First, note that the polynomial $f(x) = x^2 + x + 1$ is *irreducible* over $\mathbb{F}_2$, that is, it does not factor as a product $f(x) = g(x)h(x)$ of lower degree polynomials (if it did, it would have a root, but $f(0) = f(1) = 1 \neq 0$). Consider the ideal generated by this polynomial,

$$I = \langle x^2 + x + 1 \rangle = \big\{(x^2 + x + 1)h(x) \mid h \in \mathbb{F}_2[x]\big\}.$$

In the quotient ring R/I, we have $x^2+x+1 = 0$, or equivalently, $x^2 = -x-1 = x+1$. Thus, R/I only has four elements:

$$0 + I, \quad 1 + I, \quad x + I, \quad (x+1) + I.$$

As with the quotient group (or ring) $\mathbb{Z}/n\mathbb{Z}$, we usually drop the "I," and just write

$$R/I = \mathbb{F}_2[x]/\langle x^2 + x + 1 \rangle \cong \{0, \ 1, \ x, \ x+1\}.$$

It is easy to check that this is a field.

Exercise 4.1. Construct addition and multiplication tables for the finite field $\mathbb{F}_4 = \{0, 1, x, x+1\}$.

This method works to construct any finite field of order $q = p^k$.

Theorem 4.1. *There exists a finite field $\mathbb{F}_q$ of order q, which is unique up to isomorphism, if and only if $q = p^k$ for some prime p. If $k > 1$, then this field is* *isomorphic to the quotient ring*

$$\mathbb{F}_p[x]/\langle f \rangle,$$

where f is any irreducible polynomial of degree k.

Nonprime finite fields arise in a number of applications, from biology to information theory. For example, much of the error correcting techniques in coding theory are built using mathematics over $\mathbb{F}_{2^8} = \mathbb{F}_{256}$. This is what allows a CD or DVD to play despite scratches [23].

Exercise 4.2. Construct the finite field of order 8, and include the addition and multiplication tables.

4.3.3 Functions Over Finite Fields

Though they are often (incorrectly) used interchangeably, there is an important distinction between *polynomials* over a finite field $\mathbb{F}$ and *functions* over $\mathbb{F}$. To see why these are different, note that there are infinitely many polynomials over $\mathbb{F}$ but only finitely many functions $\mathbb{F} \to \mathbb{F}$. For example, over $\mathbb{F}_2$, both $f(x) = x^2$ and $g(x) = x$ are different polynomials (i.e., different elements in $\mathbb{F}_2[x]$) but they define the same function because $f(a) = g(a)$ for all $a \in \mathbb{F}_2$. In the remainder of this section, we will understand precisely when this happens.

Throughout, let p be prime and $\mathbb{F} = \mathbb{F}_q$ be a field of order $q = p^k$. Every $f \in \mathbb{F}[x]$ canonically defines a function $\mathbb{F} \to \mathbb{F}$, by $c \mapsto f(c)$. The set $\mathbb{F}[x]$ of polynomials over $\mathbb{F}$ is infinite, but there are only q^q functions $\mathbb{F} \to \mathbb{F}$. The following basic result from number theory gives explicit conditions for when two polynomials determine the same function.

Theorem 4.2 (Fermat's Little Theorem)**.** *If p is prime, then $a^p \equiv a \pmod p$.*

Fermat's little theorem implies that x^p and x define the same function over the prime field $\mathbb{F}_p$. Therefore, elements in the quotient ring $\mathbb{F}_p[x]/I$, where $I = \langle x^p - x \rangle$, have the form

$$(a_{p-1}x^{p-1} + \cdots + a_1x + a_0) + I, \quad a_i \in \mathbb{F}, \tag{4.2}$$

because every instance of x^p can be replaced with x. Since there are p choices for each a_i in Eq. (4.2), there are p^p elements in $\mathbb{F}_p[x]/I$. The following result says that though polynomials should be thought of as elements in $\mathbb{F}_p[x]$, functions can be identified with elements in the quotient ring $\mathbb{F}_p[x]/I$.

Proposition 4.3. *For a prime p, there is a bijection between functions $\mathbb{F}_p \to \mathbb{F}_p$ and elements in the quotient ring $\mathbb{F}_p[x]/\langle x^p - x \rangle$.*

Proof. Clearly, every element f of $\mathbb{F}_p[x]/\langle x^p - x\rangle$, which is of the form in Eq. (4.2), describes a function $\mathbb{F}_p \to \mathbb{F}_p$, by $c \mapsto f(c)$. However, these two sets have the same size, whence the bijection. □

Fermat's little theorem is actually a special case of a more general result about finite fields $\mathbb{F} = \mathbb{F}_q$.

Proposition 4.4. *For any finite field $\mathbb{F}$, the multiplicative group $\mathbb{F}^* := \mathbb{F}\backslash\{0\}$ is cyclic. Since $|\mathbb{F}^*| = q - 1$, this means that*

$$a^q = a \quad \textit{for any } a \in \mathbb{F}_q.$$

The proof of Proposition 4.4 carries through almost verbatim upon replacing each p with q in Eq. (4.2) and Proposition 4.3.

Corollary 4.1. *There is a bijection between functions $\mathbb{F}_q \to \mathbb{F}_q$ and elements in the quotient ring $\mathbb{F}_q[x]/\langle x^q - x\rangle$.*

Corollary 4.1 tells us that every single variable function from a finite field to itself can be written uniquely as a polynomial with maximum degree $q - 1$, that is, an element in the quotient ring $\mathbb{F}_q[x]/\langle x^q - x\rangle$.

This carries over quite easily to multivariate functions, and these are the objects that arise in local models. Every $f \in \mathbb{F}[x_1, \ldots, x_n]$ canonically defines a function

$$\mathbb{F}^n \longrightarrow \mathbb{F}, \quad (c_1, \ldots, c_n) \longmapsto f(c_1, \ldots, c_n).$$

Note that the set $R = \mathbb{F}[x_1, \ldots, x_n]$ of polynomials is infinite, but there are only $q^{(q^n)}$ functions $\mathbb{F}^n \to \mathbb{F}$.

Elements in the quotient ring R/I, where $I = \langle x_1^q - x_1, \ldots, x_n^q - x_n\rangle$, are sums of monomials with each exponent from $0, \ldots, q - 1$:

$$f = \sum c_\alpha x^\alpha, \quad x^\alpha := x_1^{\alpha_1} x_2^{\alpha_2} \cdots x_n^{\alpha_n}, \quad \alpha = (\alpha_1, \ldots, \alpha_n) \in \mathbb{Z}_q^n,$$

where the sum is taken over all q^n monomials, and $c_\alpha \in \mathbb{F}$. This is called the *algebraic normal form* of $f \in R/I$. Since there are q choices for each coefficient c_α in $\mathbb{F}$, and q^n monomials x^α, there are $q^{(q^n)}$ elements in $\mathbb{F}[x_1, \ldots, x_n]/\langle x_1^p - x_1, \ldots, x_n^p - x_n\rangle$, which we will still refer to as *polynomials*. Since each such polynomial canonically defines a function $\mathbb{F}^n \to \mathbb{F}$, and there are also $q^{(q^n)}$ such functions, the following result is immediate.

Proposition 4.5. *There is a bijection between functions $\mathbb{F}_q^n \to \mathbb{F}_q$ and polynomials in the quotient ring*

$$R/I = \mathbb{F}[x_1, \ldots, x_n]/\langle x_j^q - x_j\rangle := \mathbb{F}[x_1, \ldots, x_n]/\langle x_1^q - x_1, \ldots, x_n^q - x_n\rangle.$$

Proof. Each polynomial in R/I canonically defines a function $f\colon \mathbb{F}_q^n \to \mathbb{F}_q$. Since both of these sets have cardinality $q^{(q^n)}$, this correspondence is bijective. □

We will finish this section by returning to the Boolean setting for some examples to motivate the theory we just developed. A *Boolean function* is a function $f\colon \mathbb{F}_2^n \to \mathbb{F}_2$, and there are three primary ways to represent it:

1. using *Boolean logic*, for example, AND, OR, NOT ($\wedge$, $\vee$, $\overline{}$),
2. as a square-free polynomial, that is, as an element of $\mathbb{F}_2[x_1, \dots, x_n]/\langle x_j^2 - x_j\rangle$, and
3. as a *truth table*.

By "truth table," we simply mean a table that has explicitly shows all 2^n inputs and the corresponding output. As an example, consider the function that outputs 0 if and only if $x = y = z = 1$. The three ways to express it are below:

- $f(x, y, z) = \overline{x \wedge y \wedge z} = \overline{x} \vee \overline{y} \vee \overline{z}$
- $f(x, y, z) = 1 + xyz$
-

x	1	1	1	1	0	0	0	0
y	1	1	0	0	1	1	0	0
z	1	0	1	0	1	0	1	0
$f(x, y, z)$	0	1	1	1	1	1	1	1

Note that the Boolean logical expressions need not be unique, for example, due to De Morgan's laws: $\overline{x \wedge y} = \overline{x} \vee \overline{y}$, and $\overline{x \vee y} = \overline{x} \wedge \overline{y}$. However, it is straightforward to go between a Boolean logical expression and a square-free polynomial using the rules in the following table along with De Morgan's laws.

Boolean Operation	Logical Expression	Polynomial
AND	$z = x \wedge y$	$z = xy$
OR	$z = x \vee y$	$z = x + y + xy$
NOT	$z = \overline{x}$	$z = 1 + x$

Exercise 4.3. Convert the following Boolean expressions into square-free polynomials over $\mathbb{F}_2$, and construct their truth tables.

(a) $f(x, y, z) = x \wedge (\overline{y} \vee z)$
(b) $f(x, y, z) = x \wedge \overline{y} \wedge z$
(c) $f(x, y, z) = \overline{x \vee y \vee z}$

Exercise 4.4. Convert the following square-free polynomials over $\mathbb{F}_2$ into Boolean logical expressions, and construct their truth tables.

(a) $f(x, y, z) = x + y + z$
(b) $f(x, y, z) = xy + z$
(c) $f(x, y, z) = x + y + xyz + 1$

Exercise 4.5. For the Boolean function given by the following truth table, write it in Boolean logical form and as a square-free polynomial over $\mathbb{F}_2$.

x	1	1	1	1	0	0	0	0
y	1	1	0	0	1	1	0	0
z	1	0	1	0	1	0	1	0
$f(x, y, z)$	0	1	1	0	1	0	0	0

4.3.4 Boolean Networks and Local Models

Our definition of a local model over a finite set will naturally give rise to a synchronous phase space and an asynchronous phase space. We will then show that every possible function that "could be" a synchronous or asynchronous phase space uniquely determines a local model. We will start with the Boolean case for motivation, and then see why most of the theory carries over by simplifying replacing 2 with q.

Classically, a *Boolean network* (BN) is an n-tuple $f = (f_1, \ldots, f_n)$ of Boolean functions, where $f_i \colon \mathbb{F}_2^n \to \mathbb{F}_2$. This defines an *FDS map*

$$f \colon \mathbb{F}_2^n \longrightarrow \mathbb{F}_2^n, \quad x = (x_1, \ldots, x_n) \longmapsto \big(f_1(x), \ldots, f_n(x)\big).$$

Any function from a finite set to itself can be described by a directed graph with every node having out-degree 1. For a BN, this is called the *phase space* or *state space*. To distinguish this from the asynchronous phase space (which is technically not even a dynamical system phase space), we will include the word "synchronous" in the definition.

Definition 4.2. The *synchronous phase space* of a BN is the digraph with vertex set $\mathbb{F}_2^n$ and edges $\{(x, f(x)) \mid x \in \mathbb{F}_2^n\}$.

Running Example 4.1. Consider the Boolean network $(f_1, f_2, f_3) = (\overline{x_2}, x_1 \wedge x_3, \overline{x_2})$. The synchronous phase space is shown in Fig. 4.8.

As we have already said, every Boolean network $f = (f_1, \ldots, f_n)$ defines a synchronous phase space $f \colon \mathbb{F}_2^n \to \mathbb{F}_2^n$. Perhaps surprisingly, the converse holds as well.

Proposition 4.6. *Every function $f \colon \mathbb{F}_2^n \to \mathbb{F}_2^n$ is the synchronous phase space of a Boolean network $f = (f_1, \ldots, f_n)$.*

Proof. Clearly, every BN canonically defines a function $\mathbb{F}_2^n \to \mathbb{F}_2^n$. To show the converse, it suffices to show that these sets have the same cardinality.

To count functions $\mathbb{F}_2^n \to \mathbb{F}_2^n$, we count synchronous phase spaces. Each of the 2^n nodes has 1 outgoing edge, and 2^n destinations. Thus, there are $(2^n)^{2^n} = 2^{(n2^n)}$ synchronous phase spaces.

To count BNs: there are $2^{(2^n)}$ choices for each f_i, and so $(2^{(2^n)})^n = 2^{(n2^n)}$ possible BNs. □

Since each function $f\colon \mathbb{F}_2^n \to \mathbb{F}_2^n$ can be expressed as the synchronous phase space of a Boolean network $f = (f_1, \dots, f_n)$, and each $f_i\colon \mathbb{F}_2^n \to \mathbb{F}_2$ uniquely defines a polynomial f in $\mathbb{F}_2[x_1, \dots, x_n]/\langle x_j^2 - x_j \rangle$, the following result is immediate.

Corollary 4.2. *Every function $f = \mathbb{F}_2^n \to \mathbb{F}_2^n$ can be written as an n-tuple of square-free polynomials over $\mathbb{F}_2$. That is,*

$$f = (f_1, \dots, f_n), \quad f_i \in \mathbb{F}_2[x_1, \dots, x_n]/\langle x_j^2 - x_j \rangle.$$

These all carry over to generic finite fields, but we will carefully re-define things first.

Definition 4.3. Let X_i be a finite set for $i = 1, \dots, n$. A *local model* is a collection $f = (f_1, \dots, f_n)$ of *local functions*, where each $f_i\colon X_1 \times \cdots \times X_n \to X_i$.

If $X_1 = \cdots = X_n = \mathbb{F}$, as will usually be the case, then we say that f is a *local model over* $\mathbb{F}$. A Boolean network is simply a local model over $\mathbb{F}_2$.

Throughout this chapter, we will only consider local models over a finite field $\mathbb{F}$. This is not always the case. For example, many logical models have some nodes with Boolean states and others that are ternary or more; see [19] for some examples.

Our definition of local model captures the idea that the model should just consist of the local entities and their interactions. The update mechanism, whether synchronous, asynchronous, stochastic, or something else, is usually artificial, and is imposed by the modeler for sake of *model validation*. Definition 4.3 also is general enough that it encompasses frameworks such as (finite) cellular automata, Boolean networks, logical models, graph dynamical systems, automata networks, and many others, as special cases. Every local model automatically comes with a dynamical system map if one chooses to update the functions synchronously.

Definition 4.4. Every local model $f = (f_1, \dots, f_n)$ over $\mathbb{F}$ defines an FDS, by iterating the map

$$f\colon \mathbb{F}^n \longrightarrow \mathbb{F}^n, \qquad x = (x_1, \dots, x_n) \longmapsto \big(f_1(x), \dots, f_n(x)\big).$$

The synchronous *phase space* of f is the representation of the FDS map as a directed graph: the with vertex set is $\mathbb{F}^n$ and there are directed edges of the form $(x, f(x))$ for each $x \in \mathbb{F}^n$.

Definition 4.4 simply says that every local model $f = (f_1, \ldots, f_n)$ over $\mathbb{F}$ defines an FDS map $f\colon \mathbb{F}^n \to \mathbb{F}^n$, or phase space. However, it is not clear a priori whether a generic function $f\colon \mathbb{F}^n \to \mathbb{F}^n$ necessarily arises as the FDS map of a local model over $\mathbb{F}$. The next result says that it does.

Proposition 4.7. *Let $\mathbb{F} = \mathbb{F}_2$ and $R/I = \mathbb{F}[x_1, \ldots, x_n]/\langle x_j^q - x_j \rangle$. There is a bijection between the following sets:*

1. *local models $f = (f_1, \ldots, f_n)$ over $\mathbb{F}$,*
2. *functions $f\colon \mathbb{F}^n \to \mathbb{F}^n$, and*
3. *n-tuples of polynomials $(f_1, \ldots, f_n)$ in $(R/I)^n$.*

Proof. Every local model defines a unique function $f\colon \mathbb{F}^n \to \mathbb{F}^n$ and a unique n-tuple of polynomials $(f_1, \ldots, f_n)$ in the quotient ring $(R/I)^n$. Thus, it suffices to show that there are the same number of each of these three objects.

Since there are $q^{(q^n)}$ functions $f_i\colon \mathbb{F}^n \to \mathbb{F}$, there are clearly

$$\left(q^{(q^n)}\right)^n = q^{nq^n}$$

local models $f = (f_1, \ldots, f_n)$ over $\mathbb{F}$. The functions from $\mathbb{F}^n$ to itself can be enumerated by counting synchronous phase spaces: each one has q^n vertices, and exactly one outgoing edge from each. Clearly, there are q^n possible destinations for each edge, which means that there are

$$(q^n)^{q^n} = q^{nq^n}$$

synchronous phase spaces, or equivalently, functions $f\colon \mathbb{F}^n \to \mathbb{F}^n$. Finally, since $|R/I| = q^{q^n}$, there are clearly $(q^{q^n})^n = q^{nq^n}$ elements in $(R/I)^n$. □

The equivalence of (1) and (2) in Proposition 4.7 can be thought of as saying that every graph that is a potential candidate for being the synchronous phase space of a local model (i.e., it has vertex set $\mathbb{F}^n$ and uniform out-degree 1), is one.

Another important feature of a local model is its wiring diagram. Just like how we defined this concept for ODE models in Section 4.2.1, the *wiring diagram* of a local model $f = (f_1, \ldots, f_n)$ is a directed graph that encodes which functions depend on which variables. The vertex set is $V = \{1, \ldots, n\}$ and we include a directed edge from i to j if f_j depends on the variable x_i. As we did for ODEs, such an edge is often *signed* depending on whether the interaction is positive or negative. In the Boolean case, a positive interaction basically means that x_i appears in f_j, and a negative interaction means that $\overline{x_i}$ appears in f_j. The former is usually denoted by $i \longrightarrow j$ and the latter by $i \dashv j$ or sometimes $i \longrightarrow\bullet\, j$. We are intentionally not providing details on wiring diagrams and how to formally define positive and negative interactions in $\mathbb{F}_2$, let alone in other finite fields, because that it is not needed for the remainder of this chapter, and because wiring diagrams are one of the central objects in Chapter 5. Wiring diagrams will be formally defined there. However, we should mention two important points. First of all, interactions can be neither positive nor negative (often, we say such are *both* positive and negative, e.g., the logical XOR function). Second, the idea of a signed (positive vs. negative) interaction does not even make sense in a nonprime field $\mathbb{F} = \mathbb{F}_{p^k}$ $(k > 1)$, because there is not a natural ordering of the elements of $\mathbb{F}_{p^k}$ like there is in $\mathbb{F}_p$.

4.3.5 Asynchronous Boolean Networks and Local Models

The synchronous phase space of a local model is the directed graph on vertex set $\mathbb{F}^n$ generated by composing the local functions synchronously. A different type of directed graph results if the local functions are applied individually and asynchronously. Of course, one cannot compose f_i with f_j because the domains and codomains are different. However, by expanding the codomain, this can be done rather easily. Consider a local model $f = (f_1, \ldots, f_n)$. For each local function $f_i\colon \mathbb{F}^n \to \mathbb{F}$, the function

$$F_i\colon \mathbb{F}^n \longrightarrow \mathbb{F}^n, \qquad x = (x_1, \ldots, x_i, \ldots, x_n) \longmapsto (x_1, \ldots, f_i(x), \ldots, x_n)$$

updates only the ith node. Now, updating the ith node followed by the jth node is simply the composition $F_j \circ F_i$.

Definition 4.5. The *asynchronous phase space* of $(f_1, \ldots, f_n)$ is the directed multigraph with vertex set $\mathbb{F}^n$ and edge set $\{(x, F_i(x)) \mid i = 1, \ldots, n;\ x \in \mathbb{F}^n\}$.

By construction, each of the q^n nodes (elements of $\mathbb{F}^n$) has n outgoing edges; one corresponding to the application of each function $F_1, \ldots, F_n$. Thus, the entire asynchronous phase space has nq^n edges. However, many of these edges are self-loops, and these are usually omitted for clarity. Each nonloop edge of the asynchronous phase space connects two vertices that differ in exactly one bit. All edges are of the form $(x, x + ke_i)$, where e_i is the ith standard unit basis vector and $k \in \mathbb{F}$. The loops are those for which $k = 0$.

If we assume that time is discrete, and that at any time t, exactly one node is updated, say F_i with probability p_i so that $p_1 + \cdots + p_n = 1$, then the asynchronous phase space becomes a discrete time Markov chain. If time is assumed to be continuous, then transition rates can be assigned to define a continuous time Markov chain [24]. One can think of the asynchronous phase space as the Markov chain graph without the probabilities. Unlike the synchronous phase space, which is the actual phase space of a discrete dynamical

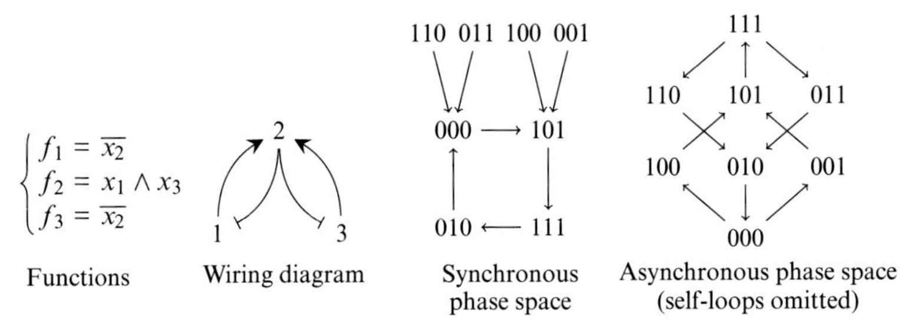

FIG. 4.8 The functions, wiring diagram, and phase spaces of the Boolean network from Running Example 4.1.

system—iterations of the map $f\colon \mathbb{F}^n \to \mathbb{F}^n$, the asynchronous phase space is not the actual phase space of any dynamical system map. *Therefore, unless we specify otherwise, the term "phase space" will refer to the "synchronous phase space."*

Running Example 4.1 (continued). Consider the Boolean network $(f_1,f_2,f_3) = (\overline{x_2}, x_1 \wedge x_3, \overline{x_2})$. The wiring diagram, synchronous phase space, and asynchronous phase space are shown in Fig. 4.8.

Recall how Proposition 4.7 says that every graph that potentially "could be" the synchronous phase space of a local model, is one. Similarly, the next result says that every multigraph that "could be" the asynchronous phase space of a local model, is one.

Theorem 4.3. *Let $G = (\mathbb{F}^n, E)$ be a directed multigraph with the following* "local property" *(definition):*

> *For every $x \in \mathbb{F}^n$: E contains exactly n edges – one each of the form $(x, x + k_i e_i)$, where $k_i \in \mathbb{F}$ (repeats of self-loops allowed).*

Then G is the asynchronous phase space of some local model $(f_1, \ldots, f_n)$ over $\mathbb{F}$.

Proof. There are $q^{(nq^n)}$ local models, and each one canonically determines a unique asynchronous phase space, that is, a digraph $G = (\mathbb{F}^n, E)$ with the "local property." Thus, it suffices to show there are exactly $q^{(nq^n)}$ such digraphs.

Each of the q^n nodes $x \in \mathbb{F}^n$ has n outgoing edges (including loops). Each edge has q possible destinations: $x + k_i e_i$ for $k_i \in \mathbb{F}$. This gives q^n choices for how to wire the edges from each node, for all q^n nodes. Thus, there are $(q^n)^{q^n} = q^{(nq^n)}$ digraphs with the "local property." □

To summarize, there are $q^{(nq^n)}$ local models $(f_1, \ldots, f_n)$ over $\mathbb{F}$. Each one gives rise to:

- a *synchronous phase space*: the FDS map $\mathbb{F}^n \to \mathbb{F}^n$ and
- an *asynchronous phase space*: a directed multigraph $G = (\mathbb{F}^n, E)$ with the "local property."

Moreover, there are exactly $q^{(nq^n)}$ maps $\mathbb{F}^n \to \mathbb{F}^n$ and $q^{(nq^n)}$ graphs with the local property. Therefore, these correspondences are bijective.

Exercise 4.6. Consider the local model $(f_1,f_2,f_3) = (x_1 \vee \overline{x_2}, x_1, \overline{x_1} \wedge x_3)$. Draw the wiring diagram, synchronous phase space, and asynchronous phase space.

Exercise 4.7. How many local models over $\mathbb{F}_3$ are there on n nodes, for $n = 2, 3, 4, 5$?

Exercise 4.8. How many local models over $\mathbb{F}_2$ are there on n nodes, for $n = 2, 3, 4, 5$?

Exercise 4.9. How may local models over $\mathbb{F}_2$ have the 4-cycle $000 \to 101 \to 111 \to 010 \to 000$ in their phase space, like the one in Fig. 4.8? Use a counting argument as in the proof of Proposition 4.7.

4.3.6 Phase Space Structure

The two most common ways to describe the dynamics of a local model are by its synchronous and asynchronous phase spaces. The structure of a synchronous phase space is less complicated; points lie either on a directed cycle or on a path that leads into a cycle.

Definition 4.6. Let $f = (f_1, \ldots, f_n)$ be a local model over $\mathbb{F}$. A point $x \in \mathbb{F}^n$ in the synchronous phase space is:

- a *transient point* if $f^k(x) \neq x$ for all $k \geq 1$; and
- a *periodic point* if $f^k(x) = x$ for some $k \geq 1$.

Periodic points on a cycle of length $k = 1$ are called *fixed points.*

The asynchronous phase space can be more complicated. For $x, y \in \mathbb{F}^n$, we write $x \sim y$ when there is a directed path from x to y and a directed path from y to x. The resulting equivalence classes are the SCCs of the phase space. An SCC is *terminal* if it has no outgoing edges from it.

Definition 4.7. Let $f = (f_1, \ldots, f_n)$ be a local model over $\mathbb{F}$. A point $x \in \mathbb{F}^n$ in the asynchronous phase space:

- is *transient* if it is not in a terminal SCC;
- lies on a *cyclic attractor* if its terminal SCC is a chordless k-cycle; and
- lies on a *complex attractor* otherwise.

A cyclic attractor of size $k = 1$ is called a *fixed point.*

Proposition 4.8. *The fixed points of a local model are the same under synchronous and asynchronous update.*

Exercise 4.10. Prove Proposition 4.8.

Exercise 4.11. For the local model from Running Example 4.1 (see Fig. 4.8), determine the transient points, periodic point, and fixed points of the synchronous phase space. Then find the SCCs of the asynchronous phase space. Characterize the transient points, the cyclic attractors, and the complex attractors.

Exercise 4.12. Recall the local model $f = (f_1, f_2, f_3) = (x_1 \vee \overline{x_2}, x_1, \overline{x_1} \wedge x_3)$ from Exercise 4.6. In the asynchronous phase space of f, classify the fixed, transient, and periodic points, as well as the cyclic and complex attractors.

4.4 LOCAL MODELS OF OPERONS

In this section, we will describe published local models of the *lac* and *ara* operons. Both were the first published Boolean models of these respective operons. The first is a Boolean model of the *lac* operon, published by Veliz-Cuba and Stigler in 2011 [3]. The second is a Boolean model of the *ara* operon, published by Jenkins and Macauley in 2017 [2]. After introducing these models in this section, we will learn how to find their fixed points using computational algebra in Section 4.5. In Section 4.6, we will see how to use freely available software packages to analyze and visualize the dynamics, under both a synchronous and asynchronous update.

4.4.1 A Boolean Model of the *lac* Operon

In order for a model to capture bistability, it must be able to distinguish between (at least) three levels of inducer concentrations: high, medium, and low (basal). One way to do that is to increase the states from two to three, that is, use $\mathbb{F}_3$ instead of $\mathbb{F}_2$. However, this is not as natural for the other variables, and it adds needless complexity to the model. As an alternative, we can introduce new variables to denote "at least medium levels." Specifically, denote these new variables with the subscript m, which stands for "at least medium." For example, the *lac* operon Boolean model has the following *parameters*.

- L_e = high levels of extracellular lactose
- L_{em} = (at least) medium levels of extracellular lactose
- G_e = extracellular glucose

Throughout, the subscript e denotes "extracellular" or "external." Low or "basal" levels of extracellular lactose are denoted by $(L_e, L_{em}) = (0, 0)$, medium levels by $(L_e, L_{em}) = (0, 1)$, and high levels by $(L_e, L_{em}) = (1, 1)$. The fourth case, when $(L_e, L_{em}) = (1, 0)$ is simply ignored because it is meaningless. These quantities are chosen for parameters because they change over much larger time-scales than the concentrations of the biomolecules that represent the variables do. For example, if the host consumes lactose or glucose, then that sugar will be available outside the cell for several hours before it degrades below meaningful levels. By contrast, concentrations of cell and gene products can change in a matter of minutes or even seconds. The following are the *variables*, which all represent concentrations of biomolecules that are *inside* the cell.

- M = mRNA
- P = *lac* permease transporter protein
- B = β-galactosidase
- C = cAMP-CAP protein complex
- R = *lac* repressor protein
- A = allolactose
- L = lactose

To capture bistability, we also need to be able to distinguish between high, medium, and low concentrations of some of these biomolecules as well. As before, we do this by adding an additional variable with an m subscript for "at least medium." The choice of which biomolecules to allow medium concentrations is up to the modeler. But it should include allolactose, because that is the direct inducer of the operon. If the model allows for medium concentration of allolactose, then it is reasonable to do this for lactose as well. The authors of [3], the first published *lac* operon model, also did this for the repressor protein. However, this is not necessary; the Boolean model for the *lac* operon in [9, Chapter 1] did not include this.

Next, we will propose and justify the Boolean functions for each of the variables. In all of these, we use *will be* to refer to the state of a variable at the next time-step, $t + 1$. For example *X will be present if Y and Z are present* is expressed as $X(t + 1) = Y(t) \wedge Z(t)$. We say that the state of X is a function of the states of Y and Z, and write this as $f_X = Y \wedge Z$.

- mRNA will be transcribed if the repressor protein is not present (i.e., it is bound to allolactose) and the cAMP-CAP protein complex is present. Thus, $f_M = C \wedge \overline{R}$.
- The gene products *lac* permease and β-galactosidase will be present if mRNA is present. Thus, $f_P = M$ and $f_B = M$.
- The cAMP-CAP protein complex will be present if there is no glucose. Thus, $f_C = \overline{G_e}$.

- High levels of the repressor protein will be present if there is no allolactose. Thus, $f_R = \overline{A} \wedge \overline{A_m}$.
- At least medium levels of the repressor protein will be present if there are already high levels of the repressor protein (in which case it will not degrade to basal levels in the next time-step), or there is no allolactose. Thus, $f_{R_m} = R \vee (\overline{A} \wedge \overline{A_m})$.
- High levels of allolactose will be present if there are high levels of lactose and β-galactosidase available. Thus, $f_A = B \wedge L$.
- There will be (at least) medium levels of allolactose available if there are at least medium levels of lactose. In this case, even basal levels of β-galactosidase are sufficient to transform lactose into allolactose. Thus, $f_{A_m} = L \vee L_m$.
- To have high levels of lactose, we need high levels of extracellular lactose, no glucose, and *lac* permease available. Thus, $f_L = P \wedge L_e \wedge \overline{G_e}$.
- To have at least medium levels of lactose, we need no extracellular glucose, and one of the two following to be present: (i) high levels of extracellular lactose, in which case some will enter the cell through basel permease transport, or (ii) medium levels of extracellular lactose and the transporter protein. Thus, $f_{L_m} = \overline{G_e} \wedge ((P \wedge L_{em}) \vee L_e)$.

Let us pause to point out several implicit assumptions. First, we are assuming that the biomolecules represented by the variables degrade in one time-step, with the exception of those that exist in high levels, which degrade in two time-steps. For example, if β-galactosidase is present but mRNA is not, then in the next time-step, the β-galactosidase concentration will have degraded to basal levels. However, if there are high levels of allolactose but no lactose, then at the next time-step, it will decay to medium levels, and then to trace levels in the following time-step. Next, the functions above are certainly not the only ones that we could have proposed. For example, consider the statement *high levels of the repressor protein will be present if there is no allolactose*, which we modeled by $f_R = \overline{A} \wedge \overline{A_m}$. We could have instead chosen $f_R = \overline{A_m}$, because if there are not even medium levels of lactose present, there cannot be high levels.

At this point, we have a local model. The local functions and wiring diagram are shown in Fig. 4.9. To analyze the model, it will be convenient to rename the variables by using numbers instead of letters, as follows:

$$(M, P, B, C, R, R_m, A, A_m, L, L_m) = (x_1, x_2, x_3, x_4, x_5, x_6, x_7, x_8, x_9, x_{10}). \quad (4.3)$$

Though this is not necessary, a good rule of thumb is that *humans prefer letters, computers prefer numbers*.

Next, we convert the update functions from Boolean expressions into square-free polynomials over $\mathbb{F}_2$. Though it is not difficult to do this by hand, it is much easier with a computer algebra package. We will soon see how to do this in Section 4.4.2 for our *ara* operon model. The polynomial form of the *lac* operon functions is follows.

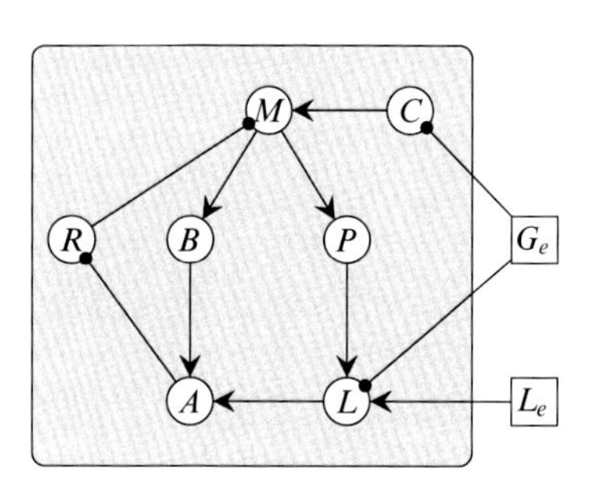

$$\begin{cases} f_M = \overline{R} \wedge \overline{R_m} \wedge C \\ f_P = M \\ f_B = M \\ f_C = \overline{G_e} \\ f_R = \overline{A} \wedge \overline{A_m} \\ f_{R_m} = (\overline{A} \wedge \overline{A_m}) \vee R \\ f_A = L \wedge B \\ f_{A_m} = L \vee L_m \\ f_L = \overline{G_e} \wedge P \wedge L_e \\ f_{L_m} = \overline{G_e} \wedge ((P \wedge L_{em}) \vee L_e) \end{cases}$$

FIG. 4.9 The 10-variable Boolean model of the *lac* operon and its wiring diagram. The *shaded region* represents the cell. Parameters are *boxed* and variables are *circled*. The pairs of variables representing high and medium levels (e.g., A and A_m, R and R_m, etc.) have been collapsed into one node each. Notice how the parameters are characterized by being sources (no incoming edges) in the wiring diagram.

$$\begin{cases} f_1 = x_4(x_5+1)(x_6+1) \\ f_2 = x_1 \\ f_3 = x_1 \\ f_4 = G_e + 1 \\ f_5 = (x_7+1)(x_8+1) \\ f_6 = x_5 + (x_7+1)(x_8+1) + x_5(x_7+1)(x_8+1) \\ f_7 = x_3 x_9 \\ f_8 = x_9 + x_{10} + x_9 x_{10} \\ f_9 = x_2(G_e+1)L_e \\ f_{10} = (x_2 L_{em} + L_e + x_2 L_{em} L_e)(G_e+1). \end{cases} \tag{4.4}$$

At this point, let us pause to point out what we have so far. We have proposed a model of 10 Boolean functions that involve 10 Boolean variables and 3 Boolean parameters, G_e, L_e, and L_{em}, which remain constant. Of course, we could make these variables, by letting $x_{11} = G_e$, $x_{12} = L_e$, and $x_{13} = L_{em}$, and with functions $f_{G_e} = G_e$, $f_{L_e} = L_e$, and $f_{L_{em}} = L_{em}$ (i.e., $f_i = x_i$ for $i = 11, 12, 13$), so these values do not change. This would result in having one local model $(f_1, \ldots, f_{13})$ of 13 functions.

There are eight choices for the parameter vector $(G_e, L_e, L_{em}) \in \mathbb{F}_2^3$. However, the two where $L_e = 1$ and $L_{em} = 0$ are meaningless, and so will be ignored. Each of the remaining six gives us a distinct local model $f = (f_1, \ldots, f_{10})$. Even though our original formal definition of a local model does not allow for parameters, we can safely relax this requirement and consider

Eq. (4.4) to be a local model with 10 variables and 3 parameters. Alternatively, we could use the aforementioned 13-variable local model.

Exercise 4.13. Find and characterize all feedback loops in the wiring diagram in Fig. 4.9.

Exercise 4.14. Note that in Fig. 4.9, the "medium concentration" nodes A_m, L_{em}, L_m, and R_m have been collapsed into the corresponding "high concentration" nodes. Draw the wiring diagram of the local model without this reduction; it should have 13 nodes. Find and characterize all feedback loops in this wiring diagram.

4.4.2 A Boolean Model of the *ara* Operon

We will construct our model for the *ara* operon much like we did for the *lac* operon. This time, there will be four parameters:

- A_e = high levels of extracellular arabinose
- A_{em} = (at least) medium levels of extracellular arabinose
- Ara_- = AraC protein, unbound to arabinose
- G_e = extracellular glucose

As before, the subscript e stands for "extracellular" and the subscript m stands for "at least medium." Low or "basal" levels of extracellular arabinose are given by $(A_e, A_{em}) = (0,0)$, medium levels by $(A_e, A_{em}) = (0,1)$, and high levels by $(A_e, A_{em}) = (1,1)$. The fourth case, when $(A_e, A_{em}) = (1,0)$, is simply ignored. The unbound AraC protein is chosen as a parameter because the cell generally keeps a small amount ($\approx$20 molecules) of the AraC protein available. However, if this protein is depleted, either by chance or by a mutation, then the operon cannot carry out its functionality. Our model needs to be able to handle this. Our choices of Boolean variables are listed in following.

- M_S = mRNA of the structural genes (ara_{BAD})
- M_T = mRNA of the transport genes (ara_{EFGH})
- E = enzymes AraA, AraB, and AraD, coded for by ara_{BAD}
- T = transporter proteins, coded for by ara_{EFGH}
- A = intracellular arabinose
- C = cAMP-CAP protein complex
- L = DNA loop
- Ara_+ = arabinose-bound AraC protein

Though it is not on the above list, we will also introduce a variable A_m so we can distinguish between high, medium, and low levels of intracellular arabinose, the inducer of the operon. The variable L is 1 if the DNA is looped, and 0 if it is not looped. Finally, note that the difference between the arabinose-bound AraC protein Ara_+, which we represent by a variable, and the parameter Ara_- which represents the unbound version. The former acts as an activator and the latter as a repressor. Our justification of the Boolean rules in the model follows.

- The mRNA of the structural genes will be transcribed if the cAMP-CAP protein complex and the arabinose-bound AraC protein are present. The DNA also needs to be unlooped. Thus, the Boolean function is $f_{M_S} = C \wedge Ara_+ \wedge \overline{L}$.
- The mRNA of the transport genes will be transcribed if the cAMP-CAP protein complex and the arabinose-bound AraC protein are present. The DNA loop does not block transcription of these genes. Thus, the Boolean function is $f_{M_T} = C \wedge Ara_+$.
- The enzymes that metabolize arabinose will be present if the mRNA of the operon's structural genes is transcribed. Thus, the Boolean function is $f_E = M_S$.
- The transport proteins will be present if the mRNA of the transport genes is present. Thus, the Boolean function is $f_T = M_T$.
- There will be high levels of arabinose in the cell if there are high levels of extracellular arabinose and the transporter proteins are present. Thus, the Boolean function is $f_A = A_e \wedge T$.
- There are two ways that there will be at least medium levels of intracellular arabinose: either there are medium levels of extracellular arabinose and the transport proteins are present, or there are high levels outside cell, in which case some will seep inside through basal levels of transport protein. Thus, the Boolean function is $f_{A_m} = (A_{em} \wedge T) \vee A_e$.
- The cAMP-CAP protein complex will be present if there is no glucose. Thus, the local function is $f_C = \overline{G_e}$.
- The DNA will be looped if the AraC protein is present, but not bound to arabinose. Thus, the local function is $f_L = Ara_- \wedge \overline{Ara_+}$.
- The AraC protein will be in its arabinose-bound form if the unbound form is present as well as at least medium levels of arabinose. Thus, the local function is $f_{Ara_+} = Ara_- \wedge (A \vee A_m)$.

The wiring diagram and functions of this local model are shown in Fig. 4.10. Like we did with the *lac* operon, we will number our variables according to the following order:

$$(A, A_m, Ara_+, C, E, L, M_S, M_T, T) = (x_1, x_2, x_3, x_4, x_5, x_6, x_7, x_8, x_9). \tag{4.5}$$

Our next step for both the *lac* and *ara* models is to convert the functions into square-free polynomials over $\mathbb{F}_2$. Instead of doing this by hand, we will learn how to do this using computational algebra software.

Exercise 4.15. Find and characterize all feedback loops in the wiring diagram in Fig. 4.10.

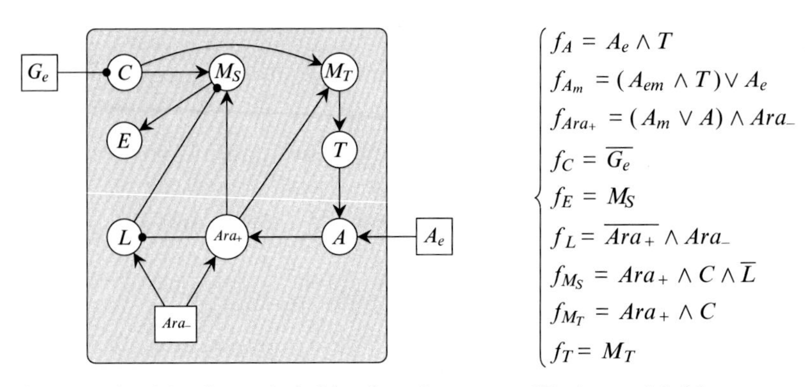

FIG. 4.10 The wiring diagram logical functions of our proposed Boolean model of the *ara* operon. *Circles* represent variables and *rectangles* represent parameters. The two nodes for intracellular arabinose (A and A_m) are collapsed into one, as are the two nodes for extracellular arabinose (A_e and A_{em}).

Exercise 4.16. Note that in Fig. 4.10, the "medium concentration" nodes A_{em} and A_m have been collapsed into the corresponding "high concentration" nodes. Draw the wiring diagram of the local model without this reduction; it should have 13 nodes. Find and characterize all feedback loops in this wiring diagram.

4.5 ANALYZING LOCAL MODELS WITH COMPUTATIONAL ALGEBRA

There are a number of computer algebra software packages that can be used to analyze local models. Examples include Macaulay2 [4], Sage [25], and Singular [26]. We will use Macaulay2, which like Singular, is a free open-source program specifically designed for computational algebra. It can be downloaded, or it can be run on a web server, or it can be run from within Sage. To do the latter, simply type the following command from Sage:

```
%default_mode macaulay2
```

In this section, we will analyze our Boolean model of the *ara* operon and then leave the similar analysis of the *lac* operon to the reader, as a series of exercises. We will only introduce the most basic Macaulay2 commands to get our desired results, so that even the reader with no programming experience can follow. We encourage the interested reader to explore more efficient ways to get the same results, using, for example, loops and functions. The book [27] is a good way to learn more about the language.

4.5.1 Computing the Fixed Points

As we have previously mentioned, one of the most important parts of validating a model, especially a qualitative one such as a Boolean network, is to verify that the fixed points are what should be expected biologically. Doing this with a local model just amounts to solving a particular system of polynomial equations, which is an easy task for a computational algebra software package such as Macaulay2.

The current URL for the Macaulay2 web server is http://web.macaulay2.com/, though there is no guarantee that it will always be at that address. To begin, we type the following command into Macaulay2 to define the quotient ring Q of nine-variable Boolean functions:

```
Q = ZZ/2[x1,x2,x3,x4,x5,x6,x7,x8,x9] /
    ideal(x1^2-x1, x2^2-x2,
    x3^2-x3, x4^2-x4, x5^2-x5, x6^2-x6, x7^2-x7,
    x8^2-x8, x9^2-x9);
```

For convenience, we will define shortcuts for the AND and OR functions, $a \mid b := a + b + ab$ and $a\&b := ab$, by typing the following commands:

```
RingElement | RingElement :=(x,y)->x+y+x*y;
RingElement & RingElement :=(x,y)->x*y;
```

This allows us to enter the Boolean expressions into Macaulay2 directly, rather than first convert them into polynomials over $\mathbb{F}_2$ by hand. For example, if we enter

```
(x1 | x2) & x3
```

which is short for $(x_1 \vee x_2) \wedge x_3$, the output will be

```
x1*x2*x3 + x1*x3 + x2*x3
```

A semicolon at the end of a line of Macaulay2 code is optional, and it suppresses the output. Next, we need to define the parameters, which we will call `Ae`, `Aem`, `Ara`, and `Ge`. To begin, we will set the parameters to $(A_e, A_{em}, Ara_-, G_e) = (1, 1, 1, 0)$. Note that the following does *not* work for setting $A_e = 1$:

```
Ae = 1;
f1 = Ae | x9;
```

The error returned is that the binary operator `&` cannot be applied to objects of different types, and `Ae` is of class `ZZ` (an integer) but `x1` is of class `Q` (the quotient ring defined earlier). There is a simple "hack" to get around this, though. Since $1 + 1 = 0$ in $\mathbb{F}_2$, entering the following

```
Ae = x1+x1+1
```

ensures that $A_e = x_1 + x_1 + 1 = 2x_1 + 1 = 1$, the multiplicative identity in Q. Similarly, to set $G_e = 0$, we can type `Ge = x1+x1` to get the zero element of Q. The following four lines set our parameters to what we want:

```
Ae = x1+x1+1;
Aem = x1+x1+1;
Ara = x1+x1+1;
Ge = x1+x1;
```

Now, we type in the *ara* operon model. The following lines can be entered separately or all at once. To enter them separately, hit "enter" after each semicolon. Alternatively, if you hit "shift + enter," you can start typing the next line, and then you can hit "enter" once, at the very end.

```
f1 = Ae & x9;
f2 = (Aem & x9) | Ae;
f3 = (x1 | x2) & Ara;
f4 = 1+Ge;
f5 = x7;
f6 = (1+x3) & Ara;
f7 = x3 & x4 & (1+x6);
f8 = x3 & x4;
f9 = x8;
```

If we type the previous commands without the semicolons, then we would see the output of each line. Alternatively, to see the polynomial form of these functions with the parameter vector $(A_e, A_{em}, Ara_-, G_e) = (1, 1, 1, 0)$, type in the following:

```
(f1, f2, f3, f4, f5, f6, f7, f8, f9)
```

Since we did not use a semicolon, the output will be

```
(x9, 1, x1*x2 + x1 + x2, 1, x7, x3 + 1, x3*x4*x6
   + x3*x4, x3*x4, x8)
```

Next, we want to find the fixed points of this local model. This amounts to solving the nine-variable nonlinear system

$$\{f_i = x_i \mid i = 1, \ldots, 9\}. \tag{4.6}$$

Of course, over $\mathbb{F}_2$, the equation $f_i = x_i$ is the same as $f_i + x_i = 0$. Notice that for any solution to this system and any Boolean polynomials $h_1, \ldots, h_9$,

$$(f_1 + x_1)h_1 + \cdots + (f_9 + x_9)h_9 = 0.$$

Thus, the solutions to the system in Eq. (4.6) are precisely the polynomials that are in the *ideal* $I = \langle f_i + x_i \mid i = 1, \ldots, 9\rangle$. We can compute this abstract set of polynomials in Macaulay2 with the following command:

```
I = ideal(f1+x1, f2+x2, f3+x3, f4+x4, f5+x5, f6+x6,
          f7+x7, f8+x8, f9+x9);
```

Once again, this ideal I is a set of polynomials whose common zeros (or "solutions") are the fixed points of the local model. In algebraic geometry, this set is called the *variety* of I. To concretely describe the ideal I, we need a generating set. There are many ways to do this, but one of the best ways is to compute a *Gröbner basis*. A Gröbner basis of an ideal is a particularly "nice" generating set, in that it admits a number of desirable theoretical andcomputational properties. There are entire books written on these objects, and at this point, the mathematically inclined reader who does not know about *Gröbner bases* is strongly encouraged to explore the fundamentals; see [28] for a good introduction. For the purposes of this chapter, the following theorem encapsulates all that we need to know about Gröbner bases.

Theorem 4.4. *Consider an ideal* $I = \langle p_1, \ldots, p_k \rangle$ *of a polynomial ring* $\mathbb{F}[x_1, \ldots, x_n]$. *If* $\mathcal{G} = \{g_1, \ldots, g_m\}$ *is a Gröbner basis of I, then the following two systems of equations have the same solutions:*

$$\begin{cases} p_1 = 0 \\ \vdots \\ p_k = 0 \end{cases} \qquad \begin{cases} g_1 = 0 \\ \vdots \\ g_m = 0. \end{cases}$$

To compute a Gröbner basis for our ideal I in Macaulay2, we type the following:

```
G = gens gb I
```

The output is:

```
| x9+1 x8+1 x7+1 x6 x5+1 x4+1 x3+1 x2+1 x1+1 |
```

This means that the set

$$\mathcal{G} = \{x_1 + 1,\ x_2 + 1,\ x_3 + 1,\ x_4 + 1,\ x_5 + 1,\ x_6,\ x_7 + 1,\ x_8 + 1,\ x_9 + 1\}$$

is a Gröbner basis for I. This describes a system of nine equations that is trivial to solve by inspection! Clearly, $x_6 = 0$ and all other $x_i = 1$. Thus, there is a unique solution to our original nonlinear system from Eq. (4.6), which is

$$(x_1, x_2, x_3, x_4, x_5, x_6, x_7, x_8, x_9) = (1, 1, 1, 1, 1, 0, 1, 1, 1).$$

This is the unique fixed point of our local model of the *ara* operon with parameter vector $(A_e, A_{em}, Ara_-, G_e) = (1, 1, 1, 0)$. It is important to make sure that this fixed point makes biological sense, and fortunately it does. This particular choice of parameters describes the situation where there are high levels of extracellular arabinose, the unbound AraC protein is available, and there is no glucose. Under these circumstances, the operon should be ON, and indeed it is. The fixed point corresponds to where all the biomolecules and gene products are present, but the DNA is not looped (because $x_6 = L = 0$).

In order to validate this model, it is necessary to check the fixed points under the other 11 parameter vectors, $(A_e, A_{em}, Ara_-, G_e)$. Recall that we ignore the four vectors for which $A_e = 1$ and $A_{em} = 0$ because they are meaningless. We will skip most of the computations and summarize the results. For 10 of the remaining parameter vectors, the model has a unique fixed point which is shown in Table 4.1. These are not all the same, but it is easy to check that they all accurately describe the biology. For example, in some cases the DNA is unlooped or the cAMP-CAP protein complex is present, but the operon is still OFF.

TABLE 4.1 The Fixed Points of the *ara* Operon Model for Each Choice of Parameters

Initial Condition(s) $x = (A_e, A_{em}, Ara_-, G_e)$	Fixed Point(s) $(A, A_m, Ara_+, C, E, L, M_S, M_T, T)$	Operon State
(0, 0, 0, 0) (0, 1, 0, 0)	(0, 0, 0, 1, 0, 0, 0, 0, 0)	OFF
(1, 1, 0, 0)	(0, 1, 0, 1, 0, 0, 0, 0, 0)	OFF
(0, 0, 0, 1) (0, 1, 0, 1)	(0, 0, 0, 0, 0, 0, 0, 0, 0)	OFF
(1, 1, 0, 1)	(0, 1, 0, 0, 0, 0, 0, 0, 0)	OFF
(0, 0, 1, 0)	(0, 0, 0, 1, 0, 1, 0, 0, 0)	OFF
(0, 0, 1, 1) (0, 1, 1, 1)	(0, 0, 0, 0, 0, 1, 0, 0, 0)	OFF
(1, 1, 1, 1)	(0, 1, 1, 0, 0, 0, 0, 0, 0)	OFF
(1, 1, 1, 0)	(1, 1, 1, 1, 1, 0, 1, 1, 1)	ON
(0, 1, 1, 0)	(0, 0, 0, 1, 0, 1, 0, 0, 0) (0, 1, 1, 1, 1, 0, 1, 1, 1)	OFF ON

The exceptional case occurs with the parameter vector $(A_e, A_{em}, Ara_-, G_e) = (0, 1, 1, 0)$, which we can input to Macaulay2 as:

```
Ae = x1+x1;
Aem = x1+x1+1;
Ara = x1+x1+1;
Ge = x1+x1;
```

This time, the output of the command `G = gens gb I` is

```
| x8+x9 x7+x9 x6+x9+1 x5+x9 x4+1 x3+x9 x2+x9 x1 |
```

It is not quite as visually obvious what the solution to this system is. However, it is also not difficult. Right away, we can see that $x_1 = 0$ and $x_4 = 1$. Next, note that over $\mathbb{F}_2$, the equation $x_2 + x_9 = 0$ means that $x_2 = x_9$. Similarly, we deduce that $x_2 = x_3 = x_7 = x_8 = x_9$. The equation $x_6 + x_9 + 1 = 0$ means that $x_6 = x_9 + 1$. Putting this together, we have two solutions:

$$(x_1, x_2, x_3, x_4, x_5, x_6, x_7, x_8, x_9) = (0, c, c, 1, c, 1 + c, c, c, c), \quad c \in \mathbb{F}_2.$$

The solution with $c = 0$ corresponds to the case when none of the biomolecules and gene products are present, except the cAMP-CAP protein complex ($C = x_4 = 1$), and the DNA is looped ($L = x_6 = 1$). This describes the case when the operon is OFF and there is no glucose. The solution when $c = 1$ corresponds to the case when all of the biomolecules are present, except high levels of intracellular arabinose ($A = x_1 = 0$) and the DNA is not looped ($L = x_6 = 0$). This accurately describes the case when the operon under medium levels of arabinose concentration. Together, these two fixed points predict the bistability of the *ara* operon under medium levels of arabinose.

Exercise 4.17. For each choice of parameter vector $(A_e, A_{em}, Ara_-, G_e)$ shown in Table 4.1, explain in one to two sentences why the fixed point reached makes biological sense.

Exercise 4.18. Repeat the computation of the fixed points for the parameter vector $(A_e, A_{em}, Ara_-, G_e) = (0, 1, 1, 0)$ in Macaulay2, but this time, use the ring of Boolean polynomials

```
R = ZZ/2[x1,x2,x3,x4,x5,x6,x7,x8,x9];
```

instead of the quotient ring Boolean functions R/I, where $I = \langle x_j^2 - x_j \rangle$. The resulting Gröbner basis might have x_j^2. However, to find the fixed points, you can manually replace that by x_j. Check that you still get the same two fixed points.

Exercise 4.19. In this section, we computed the fixed points of the 9-variable *ara* operon model for all 12 choices of the parameter vector $(A_e, A_{em}, Ara_-, G_e)$.

Repeat the steps taken in this section but for the 10-variable *lac* operon model, and for all 6 choices of the parameter vector (L_e, L_{em}, G_e). Ignore the two cases where $L_e = 0$ and $L_{em} = 1$, as they are meaningless. Specifically, use Macaulay2 to compute the fixed points and interpret each biologically. Does each fixed point accurately describe the biology? Does this model exhibit bistability?

Exercise 4.20. As an alternative to a 9-variable, 4-parameter local model of the *ara* operon, we could instead create a 13-variable local model over the quotient ring

```
Q = ZZ/2[x1,x2,x3,x4,x5,x6,x7,x8,x9,Ae,Aem,Ara,Ge] /
    ideal(x1^2-x1, x2^2-x2, x3^2-x3, x4^2-x4, x5^2-x5,
    x6^2-x6, x7^2-x7, x8^2-x8, x9^2-x9, Ae^2-Ae,
    Aem^2-Aem, Ara^2-Ara, Ge^2-Ge);
```

Define the functions $f_1, \ldots, f_9$ as before, and then define the remaining functions to be constants, for example,

```
f_Ae = 1;
f_Aem = 1;
f_Ara = 1;
f_Ge = 0;
```

Carry out the steps performed in this chapter for at least the parameter vectors $(A_e, A_{em}, Ara_-, G_e) = (1, 1, 1, 0)$ and $(0, 1, 1, 0)$. Do you get the same number of fixed points, and why or why not? Do these make biological sense?

Exercise 4.21. (Project.) Repeat the previous exercise, but define the last four functions as

```
f_Ae = Ae;
f_Aem = Aem;
f_Ara = Ara;
f_Ge = Ge;
```

What are the similarities and differences from this to the previous problem? Do the results still make sense biologically?

4.5.2 Longer Limit Cycles

Thus far, our *ara* operon model validation has not hit any snags. In all 12 cases of the parameter vector $(A_e, A_{em}, Ara_-, G_e)$, the fixed point(s) accurately predict the biology. Additionally, the model correctly predicts the bistability of the system when there is no glucose and medium levels of arabinose. The probability of this happening by chance in all 12 cases is astronomically small. While that is a promising sign, it is important to be aware that simply finding the fixed points is not sufficient to fully validate a local model. We have only found the fixed points, which fortunately, are always independent of the update order. However, we have not eliminated the possibility that there are periodic points that are not fixed points—for example, longer limit cycles in the synchronous phase space, or cyclic or complex attractors in the asynchronous phase space.

Unfortunately, identifying longer limit cycles is not easy to do algebraically like finding fixed points is, where we just had to solve a system of equations. Instead, we often resort to using a software package specifically designed for analyzing local models. In Section 4.6, we will learn about two software packages that can do this, called GINsim and TURING, respectively. But first, we will finish this section with the results obtained using that software to analyze the *ara* operon model.

For each of the 11 parameter vectors $(A_e, A_{em}, Ara_-, G_e)$ that have a unique fixed point, the synchronous phase space indeed has a unique connected component. In other words, regardless of initial condition, the state vector $(x_1, \ldots, x_9)$ will end up in the (unique) fixed point that accurately describes the biology. However, the bistable case is different. Instead of having two connected components like we want, the phase space actually consists of six components: two with 24 nodes each that lead into the fixed points, but also one with 80 nodes that leads into a two-cycle, $(0,0,1,1,0,1,0,0,1) \longrightarrow (0,1,0,1,0,0,0,1,0)$, and three with 128 nodes each, that lead into the following four cycles:

$$(0,1,0,1,0,1,0,0,0) \longrightarrow (0,0,1,1,0,1,0,0,0) \longrightarrow (0,0,0,1,0,0,0,1,0) \longrightarrow (0,0,0,1,0,1,0,0,1),$$

$$(0,1,0,1,1,1,0,0,1) \longrightarrow (0,1,1,1,0,1,0,0,0) \longrightarrow (0,0,1,1,0,0,0,1,0) \longrightarrow (0,0,0,1,0,0,1,1,1),$$

$$(0,0,1,1,0,0,1,1,1) \longrightarrow (0,1,0,1,1,0,1,1,1) \longrightarrow (0,1,1,1,1,1,0,0,1) \longrightarrow (0,1,1,1,0,0,0,1,0).$$

In other words, for most of the initial conditions, the state vector $(x_1,\ldots,x_9)$ does *not* actually reach one of the fixed points that it should.

It now becomes crucial to ask if this is due to a fundamental problem with the model, and whether it can be resolved. It turns out that in this case, the issue is due to the artificial synchronous update. To motivate this, let us revisit the local model $(f_1,f_2) = (\overline{x_2},\overline{x_1})$ from Fig. 4.7. The synchronous phase space has two fixed points, 01 and 10, and a two-cycle, $00 \leftrightarrow 11$. However, in the asynchronous phase space, the two fixed points are the only two terminal SCCs. In other words, if one starts at 00 or 11 and updates the nodes perfectly in sync (which never happens in a real molecular network), then the global state will oscillate between 00 and 11. However, then moment that a single node is updated individually, the system gets stuck in one of the two fixed points. In other words, we should never really expect to see this "artificial" two-cycle in practice—we say that it is a *artifact of synchrony*. It can be verified with, for example, GINsim or TURING, that this is what happens in the *ara* operon model as well. In the bistable case $(A_e, A_{em}, Ara_-, G_e) = (0,1,1,0)$, six connected components in the synchronous phase space "merge," and the only nontransient points (those that lie on terminal SCCs) are the two fixed points.

It is important to mention that in some models, longer limit cycles are expected. For example, there are local models of *cell cycles*, which are the sequential events that make up cell division [29]. In [30], a local model was developed for the two ways that the lambda phage virus can replicate: via a *lytic* life cycle or a *lysogenic* cycle. In these examples and others, the limiting behavior should not be fixed points, but rather either periodic cycles (if synchronous) or cyclic and complex attractors (if asynchronous).

4.6 SOFTWARE FOR LOCAL MODELS

In the previous section, we used a software package specifically designed for computational algebra and algebraic geometry to find the fixed points of a local model. There are also a number of freely available software tools specifically designed to simulate and analyze local models. Perhaps the most well known is the GINsim [5], which was developed by Denis Thieffry's research group in France in the early 2000s. Also around that time, another tool was developed and hosted at the Virginia Bioinformatics Institute. It was originally called the *Discrete Visualizer of Dynamics* (DVD), but it was replaced by a newer version called *Analysis of Dynamic Algebraic Models* (ADAM). More recently, that has been overhauled and replaced with a new crowd-sourced version called *TURING: Algorithms for Computation with Finite Dynamical Systems* [6]. There is a popular R package named BoolNet that contains tools for the analysis of synchronous, asynchronous, and probabilistic Boolean networks [31]. The similarly named BooleanNet is an open source software toolbox written in Python for the analysis of Boolean networks [32]. Other tools include the *Discrete Dynamics Lab* [33], the GenYsis Toolbox [34], the web-based Boolean network simulator BooleSim [35], and the web-based GDSCalc: *Graph Dynamical Systems Calculator* [36].

These software packages become available, change format, and move their web locations with great rapidity. As such, any book that includes detailed how-to instructions on one of these is just taking a snapshot in time and will likely become quickly out-of-date. With this in mind, our goal for this chapter will be to briefly introduce two of these software packages and their capabilities, and let the reader further explore if desired. Our choice of software is influenced by several reasons. GINsim is a good choice because it is well established and widely used, especially by those in the logical modeling community (local models under an asynchronous update order). We also choose TURING because of the crowd-sourcing feature. Though it is very new and bare-bones, anybody can write and submit algorithms to the project, even with very limited computer programming knowledge.

4.6.1 GINsim

GINsim is a freely available software package that can be run on Windows, Mac, or Linux/UNIX provided that a recent version of Java is installed. The website http://ginsim.org/ contains documentation, a tutorial, a repository of several dozen models along with references to the original papers in which they were published, and several companion tools written by third parties. A more detailed description of GINsim can be found in the book chapter [37] that should be freely available online. This 17-page chapter uses an example of a local model of the network that governs the lysogenic and lytic reproductive cycles network in the lambda phage (virus). These type of viruses have two very different ways to reproduce, and like the operons in the chapter, the decision is governed by a self-regulatory molecular network [38].

Once on the GINsim website, the user has the option to download the program with or without dependencies. Unless one is proficient with compiling software, it is probably easiest to download and run the version with dependencies. GINsim is run from the command-line—a shell terminal in Linux/UNIX, or the Terminal program in MacOS, or the Command prompt in Windows. The user should download the GINsim.jar file and then navigate to the file's path on the command line. As of the writing of this book, the following command launches the most recent version of GINsim, provided the following file is downloaded in the current directory:

```
java -jar GINsim-2.9.4-with-deps.jar
```

This opens a new window. Begin by clicking the button: "New model." We will use GINsim to create a toy three-node local model from Running Example 4.1,

$$f = (f_1, f_2, f_3) = (\overline{x_2}, x_1 \wedge x_3, \overline{x_2}) = (1 + x_2, x_1x_3, 1 + x_2). \tag{4.7}$$

At the top of the new window are the following buttons.

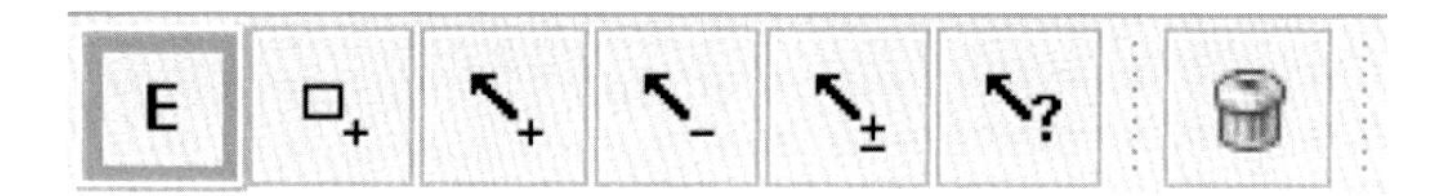

Double-click the second button, which is "Add components." Clicking anywhere in the top white space creates a square note labeled "G0." Click the "Id box" and change G0 to x1. Repeat these steps to create nodes x2 and x3. At this point, you should see three square nodes.

Next, we want to add edges. Double-click the fourth button; the negative arrow. Next, click node x2 and drag the cursor to x1 and then release it on that node. This will create a red[4] arrow x2 ⊣ x1. Next, create a negative edge from x2 to x1, and positive edges from x1 to x2 and from x3 to x2. Hover the cursor over the first button, double-click E (for "Edit"), and a box saying "Select/Move" will appear. Double-click on that button and then on the node x2, and you should see the following:

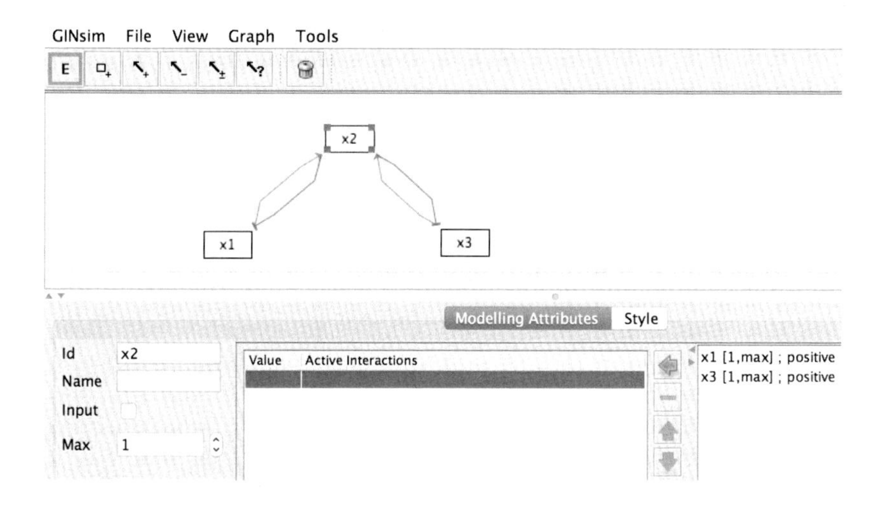

The empty middle box above is where the truth table belongs. To create it, click the x1 [1,max] row from the rightmost column and then click the green left arrow to move it to the middle column. Repeat with the x3 [1,max] row. Next, hold "Ctrl" (or "Cmd" if on an Mac) and select both rows, and click the green left arrow. Finally, hold "Ctrl" to unselect both rows, and click the green left arrow. At this point, the middle column should have four rows, as shown in the following image on the left:

4. The pdf version of this chapter has full colors. It is available online with a ScienceDirect subscription, which many academic institutions have.

Value	Active Interactions
0	x1
0	x3
1	x1 x3
0	(basal value)

Value	Active Interactions
1	x1
1	x3
1	x1 x3
0	(basal value)

This is how to describe the Boolean function $f(x_1, x_3) = x_1 \wedge x_3$. Each row corresponds to an entry in the truth table. For example, the first row says that when x1 is the only active interaction (i.e., $x_1 = 1$ but $x_3 = 0$), then the output of the function is 0; that is, $f(1, 0) = 0$. The next three rows say that $f(0, 1) = 0$, $f(1, 1) = 1$, and $f(0, 0) = 0$, respectively.

You can change the entries in the first ("Value") column by double-clicking and then typing the new value. For example, doing this for the first two rows in the above changes the Boolean function to $f(x_1, x_3) = x_1 \vee x_3$, which is shown on the image above on the right. Next, repeat this process for nodes x1 and x3. In both cases, add the active interaction x2 with Value 0, and then deselect x2 [1,max] and click the green arrow to add "(basal value)" to the Active interactions with Value 1. This sets the local functions to $f_1 = \overline{x_2}$ and $f_3 = \overline{x_2}$. At any point, you can click *File* and *Save* to save the current model as a file which is given a .zginml extension.

To compute the state space, click "Tools" from the top menu and then select "Run simulation." The default option is asynchronous, but it should be clear how to change that to "synchronous." Clicking "Run" will open a new window with the desired state space. It will likely not be laid out in a visually pleasing manner, but the user can drag around the nodes, as desired. The following images are two examples from the previous local model: the synchronous state space on the left, and the asynchronous state space on the right.

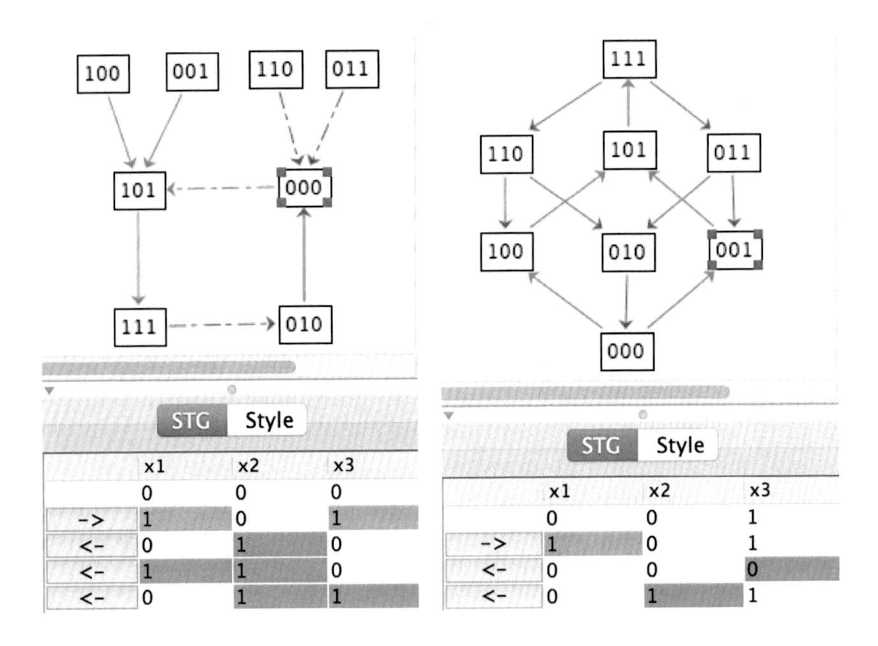

Solid green (respectively, red) arrows denote a state transition where the number of 1s increases (respectively, decreases) by one. Dashed arrows indicate a change of more than one bit. Obviously, dashed arrows cannot exist in the asynchronous state space. If you click on a node in the state space, then below it shows all of the edges to and from that node. This can be particularly helpful when the state space is large. In the earlier examples, the transitions are shown for the node $x = (0, 0, 0) \in \mathbb{F}_2^3$ at left and $(0, 0, 1) \in \mathbb{F}_2^3$ at right.

We invite the reader to explore other GINsim features on their own, such as the options under the *Graph* and *Tools* menu. For example, under *Graph* is the option *Color SCC*, and under *Tools* are options such as *Compute stable states* (i.e., fixed points) and *Construct SCC graph*. This is how one can verify what was mentioned in the previous section about the *ara* operon model: that with the parameter vector $(A_m, A_{em}, Ara_-, G_e) = (0, 1, 1, 0)$, the synchronous phase space has six connected components (two fixed points, one 2-cycle, and three 4-cycles), and the asynchronous phase spaces have only two fixed points as terminal SCCs.

4.6.2 TURING: Algorithms for Computation With FDSs

TURING is the newest version of a software package developed by Reinhard Laubenbacher's research group. The earliest version was originally called *Discrete Visualizer of Dynamics* (DVD) and that was replaced by *Analysis of Dynamic Algebraic Models* (ADAM). The previous versions actually had more hard-coded features, but people could not contribute. By contrast, TURING is a crowd-sourcing software that has only the bare-bone capabilities built in, but a framework for people to develop algorithms that anyone can use without prior coding knowledge. TURING debuted online in 2017 and is still in its infancy. In this section, we will show how to use TURING to visualize the phase space of a Boolean network, summarize existing algorithms, and discuss how to contribute new features with the hope the interested reader will get involved.

Much like GINsim, TURING has the capability to plot the (synchronous) phase space of a local model. It can be accessed at http://www.discretedynamics.org/. On the website, one should see the following links:

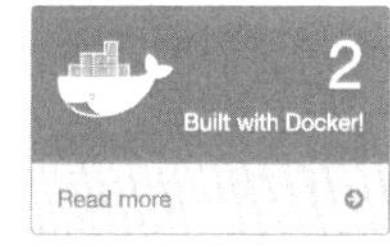

Clicking "Join Us" allows one to create an account, though this is only necessary for those who want to contribute to TURING. The individual who only wants to run the existing algorithms does not gain anything from having an account. There are two ways to contribute to TURING:

(i) adding new algorithms; and
(ii) adding new workflows.

The former should be self-explanatory. The latter can be thought of as utilize multiple existing algorithms in sequence, for example, run *Algorithm A*, and then feed its output into *Algorithm B*, and so on. As of the writing of this book (early 2018), there are no existing Workflows in TURING, but they are coming.

To get a feel for what it like to run an algorithm in TURING, click either the Algorithms tab or the "New Algorithms" button. That will take you to a page with a list of algorithms and descriptions. Click on the one called Cyclone, which has description *Calculate Dynamics of a discrete dynamical system using exhaustive search.* All of these algorithms have a "Load sample data" button, and modifying the sample data is a good way to learn how it works.

We will do an explicit example now using the three-node local model from Running Example 4.1. Click "Load sample data" and now modify the code so it describes the model in Eq. (4.7). See the left half of Fig. 4.11 for what this should look like; it should be fairly self-explanatory. Click *RUN COMPUTATION*, and the synchronous phase space should appear in plain text, as shown on the right half of Fig. 4.11. Click the *visualization* tab to replace this right window with a .png image of this phase space, which is shown on the left in Fig. 4.12. This image will always be sized to fit the window, so for larger examples, it is best to view this by right-clicking and choosing "Save image as..." and then opening the file directly.

Exercise 4.22. Use TURING to visualize the synchronous phase space of our *ara* operon local model from Section 4.4.2. Use $(A_e, A_{em}, Ara_-, G_e) = (1,1,1,0)$ for the parameter vector, and then try $(A_e, A_{em}, Ara_-, G_e) =$

FIG. 4.11 Simulating our three-node running example $(\overline{x_2}, x_1 \wedge x_3, \overline{x_2})$ using the free crowd-sourced software *TURING: Algorithms for Computation with Finite Dynamical Systems.*

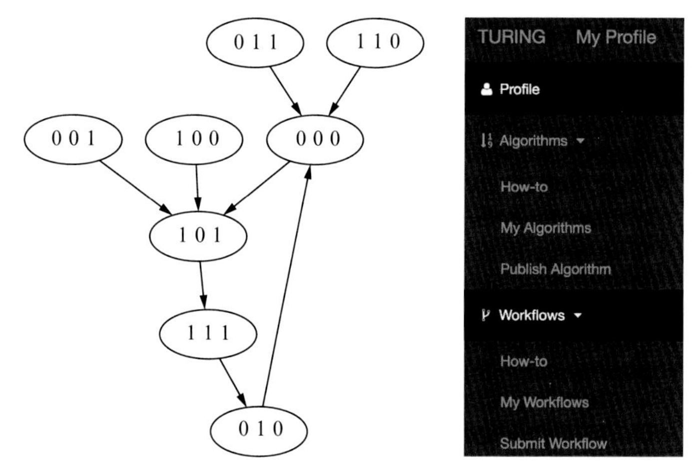

FIG. 4.12 *Left*: The synchronous phase space of the local model from Running Example 4.1 and Eq. (4.7) rendered with TURING. *Right*: The menu in TURING that one sees upon creating an account and logging in.

$(0, 1, 1, 0)$. It will be helpful to use Macaulay2 to convert the Boolean expressions into polynomials over $\mathbb{F}_2$, like we did in Section 4.5.

Let us pause to point out several features of the input for the Cyclone algorithm. Note that in line 4 of Fig. 4.11, we called the variables `x1 x2 x3`, but this is not necessary. We could have instead named them with letters, like we did when we proposed our *lac* and *ara* operon models. Also note that the algorithm allows for the variables to have a different number of states (line 5). For example, we could give one variable three states ($F_3 = \{0, 1, 2\}$) but make the rest of them binary. Finally, in line 6 (SPEED OF VARIABLES) the algorithm allows the variables to be updated at different rates.

As of the writing of this book in early 2018, TURING has seven algorithms on local models published by contributors and available on the website. The following is a list of them with descriptions, not including Cyclone, which was already discussed.

- SDDS. This algorithm analyzes a nondeterministic version of local models called *stochastic discrete dynamical systems*, introduced in [39]. These are basically local models along with probabilities of the functions being applied at each update step. For more information, read on to Chapter 5.
- SDDS Control. This implements control theory algorithms for the aforementioned SDDS that were introduced in [40]. These are also covered in Section 5.5.
- Discretize. Given time-series data, such as expression levels or concentrations of gene products, this algorithm discretizes it, so it can fit into the framework of a local model, using a method from [41]. The states can be binary or more refined. An example of where this was used is discussed

later in this book (Section 6.5), where time-series data from a gene network in *Caenorhabditis elegans* was discretized into states from $\mathbb{F}_7 = \{0, \ldots, 6\}$.

- BasicRevEng. A way to build, or "reverse-engineer" a local model from only knowing part of its state space. This was developed in [42]; see Chapter 3 for a nice survey and tutorial. This is briefly in Section 6.6.
- BN Reduction. This algorithm takes a local model whose phase space is too large to compute, and reduces it by eliminating variables while preserving key features such as its fixed points. This was developed in [43]; see [44, Chapter 6] for a nice survey and tutorial.
- Gfan. This algorithm takes in discretized time-series data and reverse-engineers local models using an advanced algebraic object called a *Gröbner fan*, which in some sense describes all possible Gröbner bases that can arise from the data by varying the term order. This was developed in [45]; see also [9, Section 3.5] for more information.

Users can contribute new algorithms in TURING using a program called AlgoRun. This program was designed to package algorithms in bioinformatics and computational science using so-called virtual containers called *dockers*. This technology was created by the company Docker, Inc., in 2013 [46], and it eliminates the need for the user to write code, thereby also doing away with common problems such as broken libraries or missing software dependencies. Dockers are portable across platforms and can be an attractive to using virtual machine(s).

Adding an *algorithm* to TURING requires one to visit https://hub.docker.com/ to download and install Docker on their local machine. The AlgoRun website http://algorun.org/ contains both documentation and examples of how to use AlgoRun to create algorithms for TURING, and then how to publish them to the AlgoRun website.

Another way to contribute to TURING is to add a *workflow*, or chains of AlgoRun algorithms. This is done with a drag-n-drop visual program called AlgoPiper, where the workflows are called *pipelines*. AlgoPiper is run on the cloud; see http://algopiper.org/ for documentation and example pipelines. Since AlgoPiper is very new, the only existing pipelines involve algorithms for nucleic acid databases (e.g., DNA and RNA). However, seeing how these were built with AlgoRun algorithms should be useful for those looking to design similar pipelines for local models in TURING. Fig. 4.13 shows what the AlgoPiper interface looks like.

We will conclude this section by describing a theoretical basic algorithm that does not yet exist in TURING, that should be fairly straight-forward to implement.[5] Recall from Section 4.5 that when we analyze our local models of the *lac* and *ara* operons, we had to consider all possible parameter vectors (6

---

5. Of course, there is a decent chance that such an algorithm will exist by the publication date of this book, or soon after.

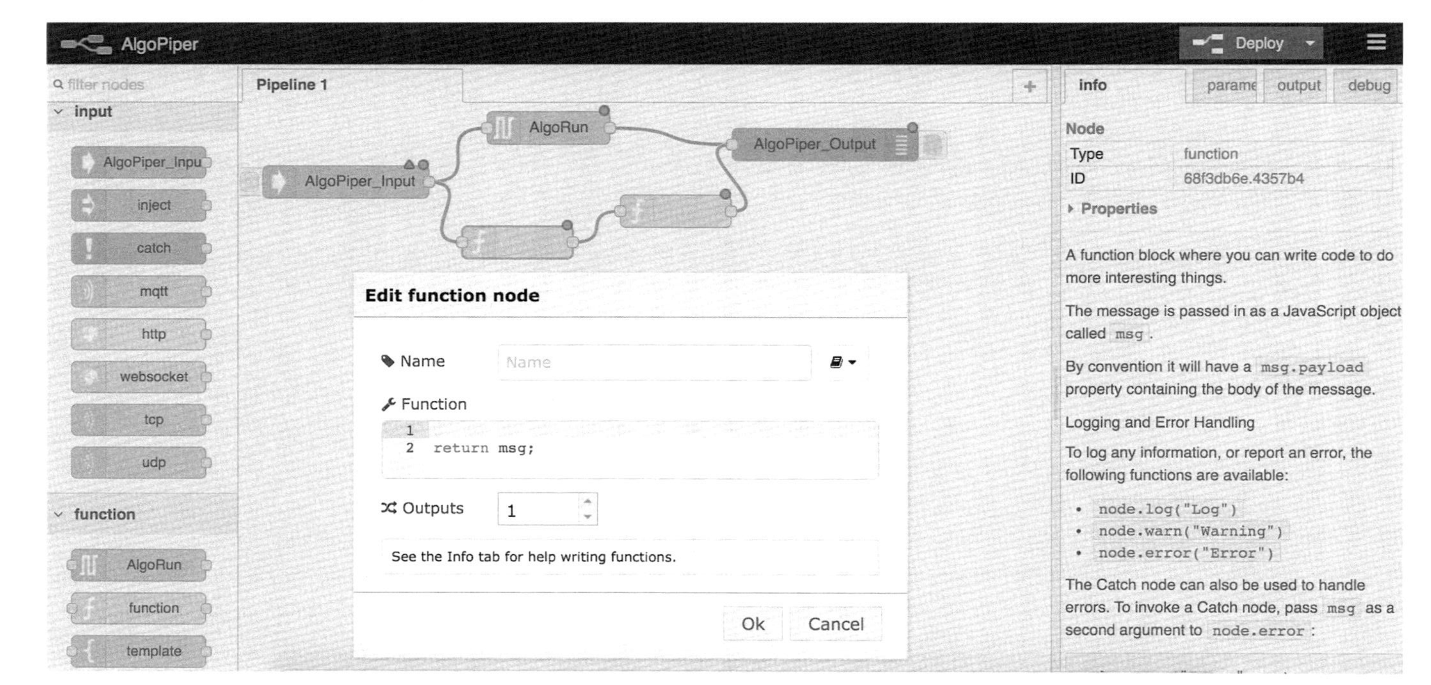

FIG. 4.13 AlgoPiper creates workflows, or pipelines, by stringing together AlgoRun algorithms.

for the *lac* and 12 for the *ara* operon) and compute the fixed points separately. Similarly, if we want to visual the phase space using GINsim or TURING, we would have to enter these individually. Using TURING, we could create a workflow that allows the capability of the user specifying parameters as inputs, for example, `Ae=0; Aem=1; Ara=1; Ge=0;` that would get fed into Cyclone before the phase space was created. The functions in Cyclone would be defined using these parameters, and so to create the phase space with a different parameter vector would only require the user to change the input parameters, rather than change the actual functions in plain text, as in the left half of Fig. 4.11. This feature was actually built into ADAM, the prior version of TURING. However, TURING has only the bare-bones features hard-coded, and the crowd-sourcing mission leaves the design of these type of algorithms up to the user to contribute.

4.7 CONCLUDING REMARKS

In this chapter, we introduced the concept of a *local model*. These have been around for decades, though under various definitions and using various names—Boolean networks, logical models, graph dynamical systems, automata networks, generalized cellular automata, and more. One goal of our simple definition was to unify all of these into a single framework that was both simple but also flexible. We emphasize that the *model* should consist of the individual functions *only*, whereas the update scheme is part of model validation. We saw this explicitly with the *ara* operon model, where the functions were sound, but updating the functions together introduced "artifacts of synchrony," which went away under an asynchronous update.

We will conclude this chapter with a brief summary of more advanced and specialized topics involving local models, as well as current and future research. Chapter 5 describes a stochastic variant, called *stochastic discrete dynamical systems* (SDDS), where probabilities are assigned to the functions. The focus of the chapter after that Chapter 6 is how to reconstruct the wiring diagram of a local model given only partial information of the phase space. This can be done using mathematical tools from the field of *combinatorial commutative algebra*. Specifically, one can encode the coordinates which could have caused differences in function outputs with square-free monomials. These monomials generate an ideal, and its *primary decomposition* gives all sets of minimal dependencies, or minimal wiring diagrams. One can take this question further, and ask: given partial information of the phase space, what are the *functions* that fit the data? It turns out that this set has an "affine-like" coset structure, of the form $(f_1 + I, \ldots, f_n + I)$, where $(f_1, \ldots, f_n)$ is any one particular model that fits the data, and I is the set (an ideal) of all functions that vanish on the data. The former can be found via, for example, Lagrange interpolation, or using the Chinese remainder theorem. The latter can be found computationally using (e.g., Macaulay2). See [9, Chapter 3] for a nice survey of this.

A common challenge with local models is their sheer size. Even just a Boolean model has a phase space with 2^n nodes, and many models have dozens of nodes, which make simulating or visualizing their dynamics prohibitive, if not impossible. Well-known examples of this include a 60-node Boolean model of the segment polarity gene network in the fruit fly [47], and a 23-node network on T-helper cell differentiation [48]. With examples such as these, it is desirable to "reduce" the model down to a smaller network by solving for and eliminating variables in a way that preserves key features in the phase space such as fixed points. See [44, Chapter 6] for a survey of how this is done using a synchronous update order, and [49] for an asynchronous update. A recent paper [50] took this further and showed how one can preserve not only the fixed points, but also features such as certain complex attractors.

A common theme when studying any class of models defined over a graph is *how the network structure affects the dynamics*. This arose in Section 4.2, when we discussed R. Thomas' conjectures about positive feedback loops being necessary for multistability [15] and negative feedback loops being necessary for cyclic attractors [16]. In [51], the authors studied other effects of negative feedback loops on the dynamics. Another classic result along these lines is that if the wiring diagram has no directed cycles, then there is a unique fixed point, though there could be other periodic cycles or complex attractors [52]. This has since been generalized by Shih and Dong by restricting these directed cycles to subgraphs of the wiring diagram [53]. In [54], Richard extended this by proving several fixed points theorems in terms of "forbidden subnetworks" in the wiring diagram.

Another approach to understanding how graph structure affects dynamics is analyzing whether small errors, or *perturbations*, tend to die out or propagate throughout the network. As an analogy, if we revisit the threshold equation from Eq. (4.1) and Fig. 4.1, $y = M$ is a stable fixed point, because if the population deviates slightly, it will self-correct back to $y = M$. On the other hand, $y = T$ is an unstable fixed point, because if the population deviates slightly, then it will diverge away from $y = T$, and end up approaching either $y = 0$ (extinct) or $y = M$ (carrying capacity). Now, consider the synchronous phase space of a local model, and take a random state $x = (x_1, \ldots, x_n) \in \mathbb{F}^n$. If we introduce a small perturbation, for example, change 1 or 2 bits, to get a new state $x' \in \mathbb{F}^n$, we can ask if this error will self-correct so the trajectories of x and x' will converge, or if it the error will snowball, causing the trajectories of x and x' to diverge and up in different limit cycles. Of course, questions like this depend on the state chosen and the particular bit(s) that are flipped, so they are generally investigated by sampling thousands of times and averaging the results. Kauffman proposed studying large *random Boolean networks* (RBNs), where each node is randomly wired to K other nodes, and randomly assigned a K-variable Boolean function. A random state is chosen and perturbed, and it was determined whether such perturbations tend to die out or propagate [17]. It turns out that for $K = 1$, perturbations tend to die out; such networks are called *stable*, or *frozen*. The (synchronous) phase spaces of such local models are often characterized by having lots of fixed points. On the other hand, for $K \geq 3$, perturbations tend to propagate, and such phase spaces often have large limit cycles. These networks are said to be *chaotic*. The case when $K = 2$ was the so-called "critical threshold" between these two dynamical regimes, and RBNs with $K = 2$ display characteristics of both stable and chaotic systems. There is evidence to suggest that many "real-world" networks lie in this critical threshold, as they must be robust enough to survive, but flexible enough to adapt and evolve [55]. This type of large-scale random Boolean network approach arises from statistical physics; see [56] for a nice survey article.

A line of questions similar to the aforementioned random Boolean network approach of how does the network topology affect dynamics, is *how do the functions in a local model affect dynamics*? Certain types of functions tend to arise in local models of biological networks more than others. For examples, most gene products or enzymes are either activators or inhibitors, but not both. Such Boolean functions with this property are said to be *unate*; see Chapter 6 for a more detailed discussion on this. Another class of functions that tend to appear in biological models are *canalizing* Boolean functions. These are characterized by having a variable that can determine the output no matter the values of the other variables; like a shut-down valve or a gene knock-out. They were inspired by the concept of canalization, which was originated in the 1940s by the geneticist Waddington [57]. These have been generalized to *nested canalizing functions* [58], and further generalized to *k-canalizing functions* [59]. See [44, Chapter 5] for a survey on this, as well as the differences between stable, chaotic, and critical networks in this setting. People have studied the dynamics of local models built with other functions, for example, Post classes [60], monomials (AND functions) [61], OR functions [62], AND-NOT functions [63], just to name a few. For all of these, people want to understand how the properties of the functions affect the dynamics of the local models built with them. Sometimes, the goal is to prove a precise theorem about the phase space, for example, the presence or absence of fixed points, or the sizes of the limit cycles. Other times, one just wants to prove whether random networks tend to be more stable or chaotic. For example, it has been shown that Boolean networks built with canalizing functions are more stable (i.e., less chaotic) than those with random functions [64].

Current research involving local models falls into several different categories, both biological and mathematical, and there is plenty of room for future research. One direction is the development of new biological models, and there are no shortage of potential systems. For example, *E. coli* alone has around 700 operons. There are more complicated systems such as regulons and modulons, and other types of biochemical networks—signaling, metabolic, protein-protein interaction, and so on. In some cases, the algebraic tools can predict interactions and dependencies that have not been observed biologically; see Section 6.5 for more on this. Many of these are "low-hanging fruit," accessible to students, and can be especially productive when worked on with a collaborative team of both mathematicians and biologists. Ongoing research on the mathematical side includes further exploring the relationship between the structure of the network and functions, to the dynamics of the system. More advanced topics include how to utilize theorems and algorithms from computational algebraic geometry to answer questions about local models; see Chapters 3 and 6 for examples of this. Many current products involve both mathematics and biology, such as the development and application of stochastic variants of local models; read on to Chapter 5 for more details on this. Of course, more immediate and accessible future work includes contributing to crowd-sourced software such as TURING. There are very simple algorithms that could be added without much trouble, and these can be done by undergraduate students. For example, it should not be difficult to add the capabilities to have parameters in a local model, or to count the number of fixed points or connected components. We hope that the interested reader will create an account and consider getting involved in this ongoing work.

REFERENCES

[1] M.C. Mackey, M. Santillán, N. Yildirim, Modeling operon dynamics: the tryptophan and lactose operons as paradigms, C. R. Biol. 327 (3) (2004) 211–224.

[2] A. Jenkins, M. Macauley, Bistability and asynchrony in a Boolean model of the L-arabinose operon in *Escherichia coli*, Bull. Math. Biol. 79 (8) (2017) 1778–1795.

[3] A. Veliz-Cuba, B. Stigler, Boolean models can explain bistability in the lac operon, J. Comp. Biol. 18 (6) (2011) 783–794.

[4] D.R. Grayson, M.E. Stillman, Macaulay2, a software system for research in algebraic geometry, Available from: https://faculty.math.illinois.edu/Macaulay2/. (Accessed 25 July 2018).

[5] A.G. Gonzalez, A. Naldi, L. Sanchez, D. Thieffry, C. Chaouiya, GINsim: a software suite for the qualitative modelling, simulation and analysis of regulatory networks, Biosystems 84 (2) (2006) 91–100.

[6] A. Hosny, R. Laubenbacher, TURING: algorithms for computation with finite dynamical systems, 2018, Available from: http://www.discretedynamics.org/. (Accessed 25 July 2018).

[7] M. Cobb, 60 years ago, Francis Crick changed the logic of biology, PLoS Biol. 15 (9) (2017) e2003243.

[8] N. Yildirim, M.C. Mackey, Feedback regulation in the lactose operon: a mathematical modeling study and comparison with experimental data, Biophys. J. 84 (5) (2003) 2841–2851.

[9] R. Robeva, T. Hodge, Mathematical Concepts and Methods in Modern Biology: Using Modern Discrete Models, Academic Press, London, 2013.

[10] N. Yildirim, M. Santillan, D. Horike, M.C. Mackey, Dynamics and bistability in a reduced model of the lac operon, Chaos 14 (2) (2004) 279–292.

[11] N. Yildirim, Mathematical modeling of the low and high affinity arabinose transport systems in *Escherichia coli*, Mol. BioSyst. 8 (4) (2012) 1319–1324.

[12] U. Alon, An Introduction to Systems Biology: Design Principles of Biological Circuits, CRC Press, Boca Raton, FL, 2006.

[13] R. Thomas, On the relation between the logical structure of systems and their ability to generate multiple steady states or sustained oscillations, in: Numerical Methods in the Study of Critical Phenomena, Springer, 1981, pp. 180–193.
[14] E.H. Snoussi, Necessary conditions for multistationarity and stable periodicity, J. Biol. Syst. 6 (1) (1998) 3–9.
[15] É. Remy, P. Ruet, D. Thieffry, Graphic requirements for multistability and attractive cycles in a Boolean dynamical framework, Adv. Appl. Math. 41 (3) (2008) 335–350.
[16] A. Richard, Negative circuits and sustained oscillations in asynchronous automata networks, Adv. Appl. Math. 44 (4) (2010) 378–392.
[17] S.A. Kauffman, Metabolic stability and epigenesis in randomly constructed genetic nets, J. Theor. Biol. 22 (3) (1969) 437–467.
[18] R. Thomas, Boolean formalization of genetic control circuits, J. Theor. Biol. 42 (3) (1973) 563–585.
[19] E. Goles, S. Martínez, Neural and Automata Networks: Dynamical Behavior and Applications, vol. 58, Springer Science & Business Media, New York, NY, 2013.
[20] H.S. Mortveit, C.M. Reidys, An Introduction to Sequential Dynamical Systems (Universitext), Springer Verlag, Berlin, 2007.
[21] E. Goles, M. Noual, Block-sequential update schedules and Boolean automata circuits, in: Automata 2010: 16th Intl. Workshop on CA and DCS, Discrete Math. Theor. Comput. Sci., 2010, pp. 41–50.
[22] D.S. Dummit, R.M. Foote, Abstract Algebra, vol. 3, Wiley, Hoboken, NJ, 2004.
[23] W.C. Huffman, V. Pless, Fundamentals of Error-Correcting Codes, Cambridge University Press, Cambridge, MA, 2010.
[24] G. Stoll, E. Viara, E. Barillot, L. Calzone, Continuous time Boolean modeling for biological signaling: application of Gillespie algorithm, BMC Syst. Biol. 6 (1) (2012) 116.
[25] SageMath Inc., CoCalc Collaborative Computation Online, 2018, Available from: https://cocalc.com/. (Accessed 25 July 2018).
[26] W. Decker, G.M. Greuel, G. Pfister, H. Schönemann, SINGULAR 4-1-0: a computer algebra system for polynomial computations, 2017, Available from: http://www.singular.uni-kl.de. (Accessed 25 July 2018).
[27] D. Eisenbud, D.R. Grayson, M. Stillman, B. Sturmfels, Computations in Algebraic Geometry With Macaulay 2, vol. 8, Springer, New York, NY, 2001.
[28] D.A. Cox, J. Little, D. O'Shea, Ideals, Varieties, and Algorithms: An Introduction to Computational Algebraic Geometry and Commutative Algebra, Springer, 2015.
[29] A. Fauré, A. Naldi, C. Chaouiya, D. Thieffry, Dynamical analysis of a generic Boolean model for the control of the mammalian cell cycle, Bioinformatics 22 (14) (2006) e124–e131.
[30] D. Thieffry, R. Thomas, Dynamical behaviour of biological regulatory networks-II. Immunity control in bacteriophage lambda, Bull. Math. Biol. 57 (2) (1995) 277–297.
[31] C. Müssel, M. Hopfensitz, H.A. Kestler, BoolNet—an R package for generation, reconstruction and analysis of Boolean networks, Bioinformatics 26 (10) (2010) 1378–1380.
[32] I. Albert, J. Thakar, S. Li, R. Zhang, R. Albert, Boolean network simulations for life scientists, Source Code. Biol. Med. 3 (1) (2008) 1–8.
[33] A. Wuensche, Discrete Dynamics Lab, in: Artificial Life Models in Software, Springer, 2009, pp. 215–258.
[34] A. Garg, A. Di Cara, I. Xenarios, L. Mendoza, G. De Micheli, Synchronous versus asynchronous modeling of gene regulatory networks, Bioinformatics 24 (17) (2008) 1917–1925.
[35] M. Bock, T. Scharp, C. Talnikar, E. Klipp, BooleSim: an interactive Boolean network simulator, Bioinformatics 30 (1) (2014) 131–132.

[36] S.H.E. Abdelhamid, C.J. Kuhlman, M.V. Marathe, H.S. Mortveit, S.S. Ravi, GDSCalc: a web-based application for evaluating discrete graph dynamical systems, PLoS ONE 10 (8) (2015) e0133660.
[37] C. Chaouiya, A. Naldi, D. Thieffry, Logical Modelling of Gene Regulatory Networks With GINsim, in: Bacterial Molecular Networks: Methods and Protocols, Springer, New York, NY, 2012, pp. 463–479.
[38] M. Ptashne, A Genetic Switch: Phage Lambda Revisited, vol. 3, Cold Spring Harbor Laboratory Press, New York, NY, 2004.
[39] D. Murrugarra, A. Veliz-Cuba, B. Aguilar, S. Arat, R. Laubenbacher, Modeling stochasticity and variability in gene regulatory networks, EURASIP J. Bioinform. Syst. Biol. 2012 (1) (2012) 1–11.
[40] D. Murrugarra, A. Veliz-Cuba, B. Aguilar, R. Laubenbacher, Identification of control targets in Boolean molecular network models via computational algebra, BMC Syst. Biol. 10 (1) (2016) 94.
[41] E.S. Dimitrova, L.D. García-Puente, F. Hinkelmann, A.S. Jarrah, R. Laubenbacher, B. Stigler, M. Stillman, P. Vera-Licona, Parameter estimation for Boolean models of biological networks, Theor. Comput. Sci. 412 (26) (2011) 2816–2826.
[42] A.S. Jarrah, R. Laubenbacher, B. Stigler, M. Stillman, Reverse-engineering of polynomial dynamical systems, Adv. Appl. Math. 39 (4) (2007) 477–489.
[43] A. Veliz-Cuba, Reduction of Boolean network models, J. Theor. Biol. 289 (2011) 167–172.
[44] R. Robeva, Algebraic and Discrete Mathematical Methods for Modern Biology, Elsevier, Amsterdam, 2015.
[45] E.S. Dimitrova, A.S. Jarrah, R. Laubenbacher, B. Stigler, A Gröbner fan method for biochemical network modeling, in: Proc. Int. Symposium Symb. Algebraic Comput., ACM, 2007, pp. 122–126.
[46] C. Anderson, Docker [software engineering], IEEE Software 32 (3) (2015) 102-c3.
[47] R. Albert, H.G. Othmer, The topology of the regulatory interactions predicts the expression pattern of the segment polarity genes in *Drosophila melanogaster*, J. Theor. Biol. 223 (1) (2003) 1–18.
[48] L. Mendoza, A network model for the control of the differentiation process in Th cells, Biosystems 84 (2) (2006) 101–114.
[49] A. Naldi, E. Remy, D. Thieffry, C. Chaouiya, Dynamically consistent reduction of logical regulatory graphs, Theor. Comput. Sci. 412 (21) (2011) 2207–2218.
[50] J.G.T. Zañudo, R. Albert, An effective network reduction approach to find the dynamical repertoire of discrete dynamic networks, Chaos 23 (2) (2013) 025111.
[51] E. Sontag, A. Veliz-Cuba, R. Laubenbacher, A.S. Jarrah, The effect of negative feedback loops on the dynamics of Boolean networks, Biophys. J. 95 (2) (2008) 518–526.
[52] R. Robert, Iterations sur des ensembles finis et automates cellulaires contractants, Linear Algebra Appl. 29 (1980) 393–412.
[53] M.H. Shih, J.L. Dong, A combinatorial analogue of the Jacobian problem in automata networks, Adv. Appl. Math. 34 (1) (2005) 30–46.
[54] A. Richard, Fixed point theorems for Boolean networks expressed in terms of forbidden subnetworks, Theor. Comput. Sci. 583 (2015) 1–26.
[55] E. Balleza, E.R. Alvarez-Buylla, A. Chaos, S. Kauffman, I. Shmulevich, M. Aldana, Critical dynamics in genetic regulatory networks: examples from four kingdoms, PLoS ONE 3 (6) (2008) e2456.
[56] B. Drossel, Random Boolean Networks, Chap. 3, Wiley-VCH Verlag GmbH & Co., Weinheim, Germany, 2009, pp. 69–110.

[57] C.H. Waddington, Canalization of development and the inheritance of acquired characters, Nature 150 (3811) (1942) 563–565.

[58] S. Kauffman, C. Peterson, B. Samuelsson, C. Troein, Random Boolean network models and the yeast transcriptional network, Proc. Natl Acad. Sci. 100 (25) (2003) 14796–14799.

[59] Q. He, M. Macauley, Stratification and enumeration of Boolean functions by canalizing depth, physica D 314 (2016) 1–8.

[60] I. Shmulevich, H. Lähdesmäki, E.R. Dougherty, J. Astola, W. Zhang, The role of certain Post classes in Boolean network models of genetic networks, Proc. Natl Acad. Sci. 100 (19) (2003) 10734–10739.

[61] O. Colón-Reyes, R. Laubenbacher, B. Pareigis, Boolean monomial dynamical systems, Ann. Comb. 8 (4) (2005) 425–439.

[62] A.S. Jarrah, R. Laubenbacher, A. Veliz-Cuba, The dynamics of conjunctive and disjunctive Boolean network models, Bull. Math. Biol. 72 (6) (2010) 1425–1447.

[63] A. Veliz-Cuba, K. Buschur, R. Hamershock, A. Kniss, E. Wolff, R. Laubenbacher, AND-NOT logic framework for steady state analysis of Boolean network models, Appl. Math. Inf. Sci. 7 (4) (2013) 1263–1274.

[64] I. Shmulevich, S.A. Kauffman, Activities and sensitivities in Boolean network models, Phys. Rev. Lett. 93 (4) (2004) 048701.

Chapter 5

Modeling the Stochastic Nature of Gene Regulation With Boolean Networks

David Murrugarra* and Boris Aguilar†

**Department of Mathematics, University of Kentucky, Lexington, KY, United States, †Institute for Systems Biology, Seattle, WA, United States*

5.1 INTRODUCTION

The mathematical and computational methods described in this chapter will be applied for the modeling and simulation of gene regulatory networks. Gene Regulatory Networks (GRN) are representations of the intricate relationships among genes, proteins, and other substances that are responsible for the expression levels of mRNA and proteins. The amount of these gene products and their temporal patterns characterize specific cell states or phenotypes. Thus, GRN play a key role in the understanding of the various functions of cells and cellular components, and can aid in the design of intervention strategies for the control of biological systems.

Many dynamic modeling approaches have been used over the last six decades to develop computational tools for analyzing the dynamics of GRN. As a result, a large variety of models exists today [1–9]. Generally, dynamic models can be classified according to how the time and the population of gene products are treated. There exist methods based on continuous gene populations and continuous time such as systems of ordinary differential equations [1–4]; discrete populations and continuous time such as models based on the Gillespie formulation and their generalizations [5–9]; and discrete population and discrete time frameworks such as Boolean Networks (BN), logical models, local models, and their stochastic variants [10–14].

BN are a class of computational models in which genes can only be in one of two states: ON or OFF. They were introduced as models for GRN by Kauffmann [15] and Thomas [16]. More generally, multistate models allow

Algebraic and Combinatorial Computational Biology. https://doi.org/10.1016/B978-0-12-814066-6.00005-2

genes to take on more than two states, such as for considerations of three states for low, medium, and high. BN and multistate models have been efficiently used to model biological systems such as the yeast cell cycle network [17], Th regulatory network [18], the *lac* operon [19], the *p53-mdm2* complex [14, 20, 21], the flower development network in *A. thaliana* [22], and many others [23–28].

To incorporate the inherent stochasticity of the gene regulatory processes into the models, stochastic versions of BN have been developed. In the modeling context, the stochasticity often stems from the updating schedule. Standard updating schedules include synchronous, where all the nodes are updated at the same time, and asynchronous such as where a randomly selected node is updated at each time step. The synchronous update produces deterministic dynamics while the asynchronous update is often used to generate stochastic dynamics. In this chapter, we will consider a more general stochastic setup that considers propensity parameters for updating each node. The framework called Stochastic Discrete Dynamical Systems (SDDS) [14] considers two propensity parameters. One propensity is used when the update has a positive impact on the variable, that is, when the update makes the variable increase its value. The other propensity is used when the update has a negative impact, that is, when the update makes the variable decrease its value. Another class of discrete stochastic models are Probabilistic Boolean Networks (PBN), see [12, 29] for a comprehensive treatment. These put probability distributions on the individual update functions at each node.

BN and their stochastic extensions such as SDDS and PBN can be studied under the general framework of Markov chains, which gives access to a large variety of computational methods for model analysis. For instance, to study the long-term dynamics of SDDS and PBN, we can compute the stationary distribution. If it exits, this gives the probability that in the long-run, the system will end up at a given state. Theoretical results that guarantee the existence of a stationary distribution will be discussed in this chapter.

Stochastic extensions of BN, including SDDS and PBN, offer additional features for analysis and simulations but they also add the complexity of parameter estimation of the stochastic components. In this chapter, we will present a method for estimating the propensity parameters for SDDS. The method is based on adding noise to the system using the Google PageRank algorithm to make the system ergodic and thus guaranteeing the existence of a stationary distribution. Then with the use of a genetic algorithm, the propensity parameters can be estimated.

In this chapter, we will also discuss the problem of control of discrete stochastic models. Both SDDS and PBN along with their control parameters can be studied under the framework of Markov decision processes (MDP) [30]. For instance, in the MDP research literature, a standard method, which we will discuss in this chapter, is the infinite-horizon method with a discounting factor [30]. Here, we will describe the method adapting the terminology for SDDS. This was also described in the context of control of PBN [31].

This chapter is organized in the following way. In Section 5.1, we provide an introduction to stochastic modeling with BN. In Section 5.2, we will describe the stochastic framework to be used throughout the chapter. In Section 5.3, we will discuss techniques for studying the long-term dynamics with the SDDS framework. In Section 5.4, we will describe the PageRank algorithm which will be used for parameter estimation. In Section 5.5, we will discuss a method for parameter estimation of the SDDS propensity parameters. Finally, in Section 5.6, we will describe a method for optimal control for the SDDS framework.

5.2 STOCHASTIC DISCRETE DYNAMICAL SYSTEMS

We will start by describing the modeling framework that will be used in this chapter, SDDS, first introduced in [14]. This is similar in nature to the popular framework of PBN which will not be discussed here. We refer the reader to [12, 29] for a comprehensive description and analysis of PBN. Both of these frameworks are appropriate setups to model the effects of intrinsic noise on network dynamics.

An SDDS in the variables $x_1, \ldots, x_n$, which in this chapter represent genes, is defined as a collection of n triples

$$\hat{\boldsymbol{F}} = \left\{f_k, p_k^{\uparrow}, p_k^{\downarrow}\right\}_{k=1}^{n}, \tag{5.1}$$

where for $k = 1, \ldots, n$,

- f_k: $\{0, 1\}^n \rightarrow \{0, 1\}$ is the update function for x_k,
- $p_k^{\uparrow} \in [0, 1]$ is the activation propensity, and
- $p_k^{\downarrow} \in [0, 1]$ is the degradation propensity.

The stochasticity originates from the propensity parameters $p_k^{\uparrow}$ and $p_k^{\downarrow}$, which should be interpreted as follows: if there would be an activation of x_k at the next time step, that is, $x_k(t) = 0$, and $f_k(x_1(t), \ldots, x_n(t)) = 1$, then $x_k(t+1) = 1$ with probability $p_k^{\uparrow}$. The degradation probability $p_k^{\downarrow}$ is defined similarly.

An SDDS can be represented as a Markov chain by specifying its transition matrix which we describe next. The dynamics of the SDDS $\hat{\boldsymbol{F}} = \{f_k, p_k^{\uparrow}, p_k^{\downarrow}\}_{k=1}^{n}$, from a Markov chain point of view is defined by the transition probabilities between the states of the system. For an SDDS with n genes, there are 2^n possible vector states—elements of the set $S = \{0, 1\}^n$. For $x = (x_1, \ldots, x_n) \in S$ and $y = (y_1, \ldots, y_n) \in S$, the transition probability from x to y is given by

$$a_{x,y} = \prod_{i=1}^{n} Prob(x_i \rightarrow y_i), \tag{5.2}$$

where $Prob(x_i \to f_i(x))$ is the probability that x_i will change its value and is given by

$$Prob(x_i \to f_i(x)) = \begin{cases} p_i^{\uparrow}, & \text{if } x_i < f_i(x), \\ p_i^{\downarrow}, & \text{if } x_i > f_i(x), \\ 1, & \text{if } x_i = f_i(x). \end{cases} \quad (5.3)$$

The probability that x_i will maintain its current value is given by

$$Prob(x_i \to x_i) = \begin{cases} 1 - p_i^{\uparrow}, & \text{if } x_i < f_i(x), \\ 1 - p_i^{\downarrow}, & \text{if } x_i > f_i(x), \\ 1, & \text{if } x_i = f_i(x), \end{cases} \quad (5.4)$$

for all $i = 1, \ldots, n$. Notice that $Prob(x_i \to y_i) = 0$ for all $y_i \notin \{x_i, f_i(x)\}$.

The transition matrix is defined by

$$A = (a_{x,y})_{x,y \in S}. \quad (5.5)$$

In Markov chain notation, the transition probability $a_{x,y} = Prob(X_t = x \mid X_{t-1} = y)$ represents the probability of being in state x at time t given that system was in state y at time $t-1$.

For the type of stochastic system $\hat{F} = \{f_k, p_k^{\uparrow}, p_k^{\downarrow}\}_{k=1}^n$ described here, we can consider that there is an underlying deterministic system given by $F = (f_1, \ldots, f_n)$, where the dynamics is obtained by the synchronous update, that is, $y = F(x)$ for $x, y \in S$. It is straightforward to see that the stochastic and the deterministic systems have the same fixed points (see Exercise 5.1). However, for more complex attractors, such as periodic limit cycles or strongly connected components in the state space of SDDS, this is not true in general (see Exercise 5.5).

Exercise 5.1. Show that the deterministic system $F = (f_1, \ldots, f_n)$ and stochastic system $\hat{F} = \{f_k, p_k^{\uparrow}, p_k^{\downarrow}\}_{k=1}^n$ have the same set of fixed points.

Example 5.1. Cytotoxic T-cells are part of the immune system that fight against antigens by killing cancer cells and then going through controlled cell death (apoptosis) themselves. The T-cell large granular lymphocyte (T-LGL) leukemia is a disease where cytotoxic T-cells escape apoptosis and keep proliferating. A Boolean network model for this system was proposed in [26] and a reduced version was made for steady-state analysis in [25]. For this example, we will focus on the reduced model which is described next. The nodes of the model are

$$x_1 = S1P, \quad x_2 = FLIP,$$
$$x_3 = Fas, \quad x_4 = Ceramide,$$
$$x_5 = DISC, \quad x_6 = Apoptosis,$$

and the Boolean rules are

$$\begin{aligned} f_1 &= \overline{x_4} \wedge \overline{x_6}, \\ f_2 &= \overline{x_5} \wedge \overline{x_6}, \\ f_3 &= \overline{x_1} \wedge \overline{x_6}, \\ f_4 &= x_3 \wedge \overline{x_1} \wedge \overline{x_6}, \\ f_5 &= (x_4 \vee (x_3 \wedge \overline{x_2})) \wedge \overline{x_6}, \\ f_6 &= \overline{x_5} \vee \overline{x_6}. \end{aligned} \tag{5.6}$$

Fig. 5.1 shows the *wiring diagram* of this model, which is a signed graph showing the positive and negative influences of the nodes. This T-LGL system has two steady states: 000001 represents the normal state, where *Apoptosis* is ON, and 110000 represents the disease state, where *Apoptosis* is OFF. For simplicity, we will often continue to omit the spaces, commas, and parentheses when denoting vector states.

Exercise 5.2. Verify that 000001 and 110000 are indeed fixed points and that there are no other fixed points.

In Fig. 5.2, we show the state space of the *T-LGL* network with no stochasticity, that is, the transitions are solely determined by the regulatory functions given in Eq. (5.6). This state space graph has two components corresponding to the basins of attraction (the set of states leading to the attractor) of the two fixed points 000001 and 110000. From Fig. 5.2, one can see that one basin of attraction is much larger than the other.

In order to construct a stochastic model of the *T-LGL* network within the SDDS framework, we need to assign propensity values to each node of the model. As a starting point, we assign 0.9 to each one. In other words, we are

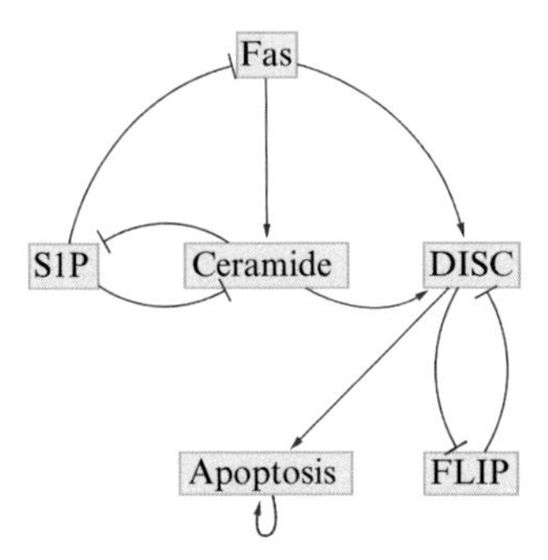

FIG. 5.1 The wiring diagram for the *T-LGL* network. *Arrowheads in blue* represent activations while *hammerheads in red* represent inhibitions. The inhibitory edges from Apoptosis to other nodes are not shown. *(Adapted from A. Saadatpour, R.-S. Wang, A. Liao, X. Liu, T.P. Loughran, I. Albert, R. Albert, Dynamical and structural analysis of a T cell survival network identifies novel candidate therapeutic targets for large granular lymphocyte leukemia, PLoS Comput. Biol. 7 (11) (2011) e1002267, https://doi.org/10.1371/journal.pcbi.1002267.)*

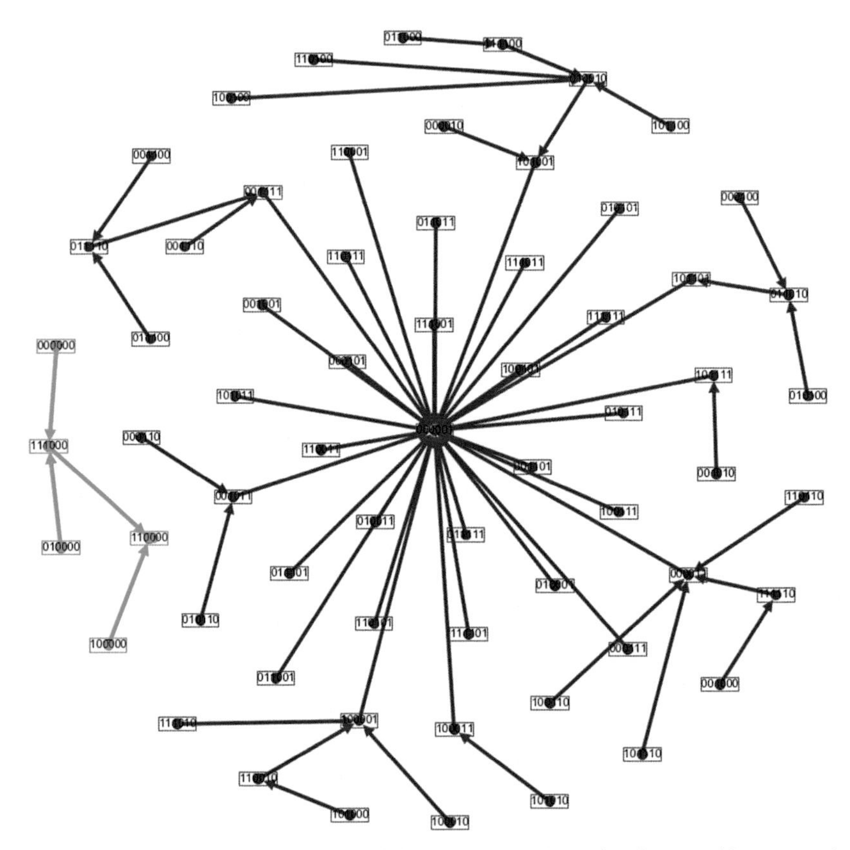

FIG. 5.2 The state space for the deterministic *T-LGL* network, that is, all propensities are equal to 1. The *green nodes* are in the basin of 110000 and the *red nodes* are in the basin of 000001. The edges are *colored* according to the target nodes.

giving a 90% chance of using the regulatory functions for each node and 10% chance of keeping the current value.

The transition matrix for this example is $A = (a_{i,j})$ for $i, j \in \{1, \ldots, 2^6\}$. That is, A is a 64×64 matrix where the binary states are ordered lexicographically. To calculate the entries of A, we use Eq. (5.2). Let us compute few entries, and leave the remaining entries to the reader. For instance, to compute $a_{1,1}$, we will need to compute the transition probability from 1 to 1. In the lexicographic order, 1 corresponds to 000000, thus we want to compute the transition probability from $x = 000000$ to $y = 000000$. We note that the synchronous update gives $F(x) = F(0, 0, 0, 0, 0, 0) = (1, 1, 1, 0, 0, 0)$, that is, with no stochasticity the system would transition from x to $z = (1, 1, 1, 0, 0, 0)$. However, under the SDDS framework, the system can transition from x to y with a probability that we will determine following. For x to transition into y, the update of the first three coordinates of x should not happen. The probability for updating each of the first three entries is 0.9. Thus, the probability that each of these updates will not happen is $1-0.9=0.1$. The last three entries do not change with the update functions, so they will remain the same with probability 1. Now if we multiply these probabilities we obtain the transition probability from x to y:

$$a_{1,1}=\prod_{i=1}^{n} Prob(x_i \rightarrow y_i)=0.1 \times 0.1 \times 0.1 \times 1 \times 1 \times 1=0.001, \quad (5.7)$$

where $Prob(x_i \rightarrow y_i)=1-0.9=0.1$ for $i=1,2,3$ and $Prob(x_i \rightarrow y_i)=1$ for $i=4,5,6$.

Next, we will calculate the transition probability from $x=000000$ to $y=111000$, which corresponds to the entry $(1,57)$. In this case, we need the update of the first three entries. Each of these updates happens with a 90% chance. Thus,

$$a_{1,57}=\prod_{i=1}^{n} Prob(x_i \rightarrow y_i)=0.9 \times 0.9 \times 0.9 \times 1 \times 1 \times 1=0.729, \quad (5.8)$$

where $Prob(x_i \rightarrow y_i)=0.9$ for $i=1,2,3$ and $Prob(x_i \rightarrow y_i)=1$ for $i=4,5,6$.

Finally, we will calculate the transition probability from $x=000000$ to $y=001000$, which corresponds to the entry $(1,9)$. In this case, we need the update of the third entry, which happen with a probability of 0.9. Thus,

$$a_{1,9}=\prod_{i=1}^{n} Prob(x_i \rightarrow y_i)=0.1 \times 0.1 \times 0.9 \times 1 \times 1 \times 1=0.009, \quad (5.9)$$

where $Prob(x_i \rightarrow y_i)=1-0.9=0.1$ for $i=1,2$, $Prob(x_i \rightarrow y_i)=0.9$ for $i=3$, and $Prob(x_i \rightarrow y_i)=1$ for $i=4,5,6$.

It can be shown that the only nonzero transitions from state 1 are to the states $1,9,17,25,33,41,49,57$. The synchronous transition from 1 to 57 is the most likely, occurring with a 72.9% chance, see Eq. (5.8). However, this is not always the case with different choices of propensity parameters.

Fig. 5.3 shows the state space of the *T-LGL* network under the SDDS framework, where all of the propensity parameters are equal to 0.9. Now, we will discuss some similarities and differences between the deterministic state spaces in Fig. 5.2 and the stochastic state space in Fig. 5.3. Note that in Fig. 5.2, each node has a unique next state where the transition happens with a 100% probability while in Fig. 5.3, from each node, the system can transition to several other states with probabilities given in the transition matrix. Both systems have two fixed points, 000001 and 110000. In Fig. 5.3, the most likely transitions correspond to the synchronous transition. However, in general, this need not happen; see Example 5.2. Finally, in both cases, all states of the system will eventually transition to the two fixed points. We will discuss techniques for studying the long-term behavior in the next section.

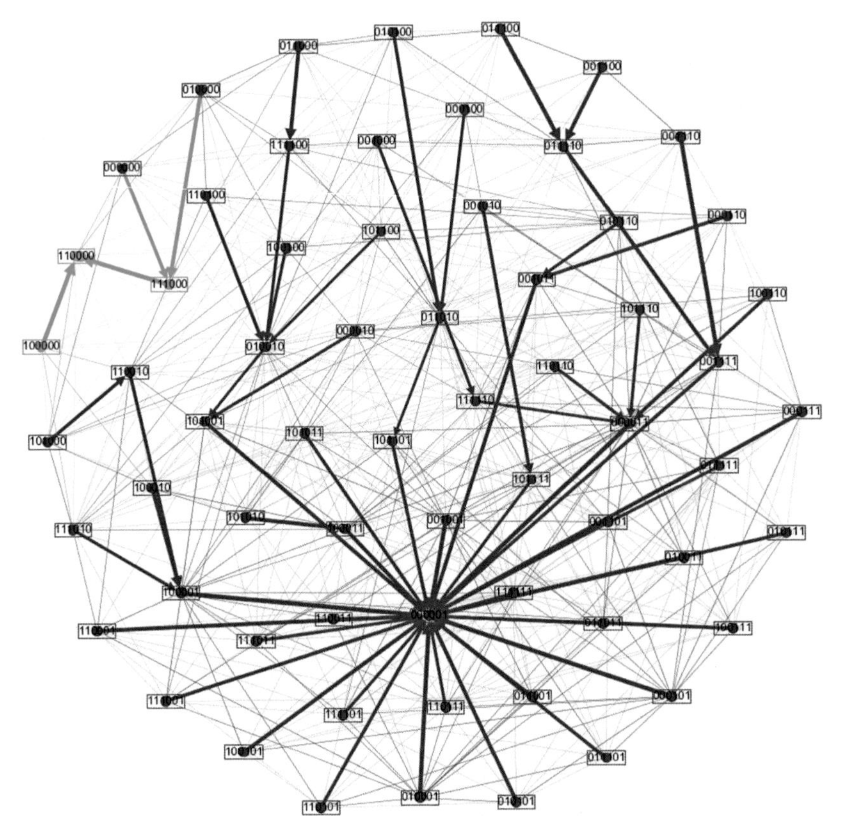

FIG. 5.3 The state space for the stochastic *T-LGL* network, where all propensities are equal to 0.9. *Green nodes* are in the basin of 110000, *red nodes* are in the basin of 000001, and *blue nodes* can transition to either fixed point. Edges are *colored* according to the target nodes. *Thick arrows* represent the most likely transitions and the remaining *arrows* represent transitions with smaller probabilities.

Exercise 5.3. Calculate the following transition probabilities of the *T-LGL* network where all propensities are 0.9:

1. From $x = 000000$ to $y = 010000$ (i.e., from state 1 to state 17).
2. From $x = 000000$ to $y = 011000$ (i.e., from state 1 to state 25).
3. From $x = 000000$ to $y = 100000$ (i.e., from state 1 to state 33).
4. From $x = 000000$ to $y = 101000$ (i.e., from state 1 to state 41).
5. From $x = 000000$ to $y = 110000$ (i.e., from state 1 to state 49).
6. From $x = 000000$ to $y = 111111$ (i.e., from state 1 to state 64).

Exercise 5.4. Use a software package to compute the transition matrix of the *T-LGL* network where all propensities are 0.9.

Exercise 5.5. Consider the following Boolean network, $F = (f_1, f_2, f_3)$: $\{0,1\}^3 \to \{0,1\}^3$, where $(f_1, f_2, f_3) = (x_2 \vee x_3, 0, x_1)$.

1. Show that the discrete system $F = (f_1, f_2, f_3)$ has two fixed points and a limit cycle of period 2.
2. Consider the associated SDDS $\hat{F} = \{f_k, p_k^{\uparrow}, p_k^{\downarrow}\}_{k=1}^{3}$ where each $p_k^{\uparrow} = p_k^{\downarrow} = 0.9$. Show that the limit cycle of the deterministic system is not preserved under the SDDS framework. That is, there is a probability of escaping from the states of the limit cycle into a different state.
3. Use a software package to compute the transition matrix of this network where all propensities are 0.9.
4. Use a software package to compute the transition matrix of this network where all propensities are 0.5.

Example 5.2. In this example, we will compare the dynamics of the *T-LGL* model discussed in Example 5.1 but with a different choice of propensity parameters. This time we will assign the value of 0.5 to all propensities. That means, for each node, we are assigning a 50% chance of using the regulatory function and a 50% chance of keeping its current value.

As we did in Example 5.1, let us analyze the transition probability from $x = 000000$ to $y = 000000$. We note that the synchronous update gives $F(x) = F(0,0,0,0,0,0) = (1,1,1,0,0,0)$, that is, with no stochasticity the system would transition from x to $z = (1,1,1,0,0,0)$. However, under the SDDS framework, the system can transition from x to y with a probability that we will determine next. For x to transition into y, the update of the first three coordinates of x should not happen. The probability for updating each of the first three entries is 0.5. Thus, the probability that each of these updates will not happen is $1 - 0.5 = 0.5$. The last three entries do not change with the update functions, so they will remain the same with probability 1. Now if we multiply these probabilities we obtain the transition probability from x to y:

$$a_{1,1} = \prod_{i=1}^{n} Prob(x_i \to y_i) = 0.5 \times 0.5 \times 0.5 \times 1 \times 1 \times 1 = 0.125, \quad (5.10)$$

where $Prob(x_i \to y_i) = 1 - 0.5 = 0.5$ for $i = 1,2,3$ and $Prob(x_i \to y_i) = 1$ for $i = 4,5,6$.

Next, we will calculate the transition probability from $x = 000000$ to $y = 111000$. This corresponds to the entry $(1,57)$. Thus,

$$a_{1,57} = \prod_{i=1}^{n} Prob(x_i \to y_i) = 0.5 \times 0.5 \times 0.5 \times 1 \times 1 \times 1 = 0.125, \quad (5.11)$$

where $Prob(x_i \to y_i) = 0.5$ for $i = 1,2,3$ and $Prob(x_i \to y_i) = 1$ for $i = 4,5,6$.

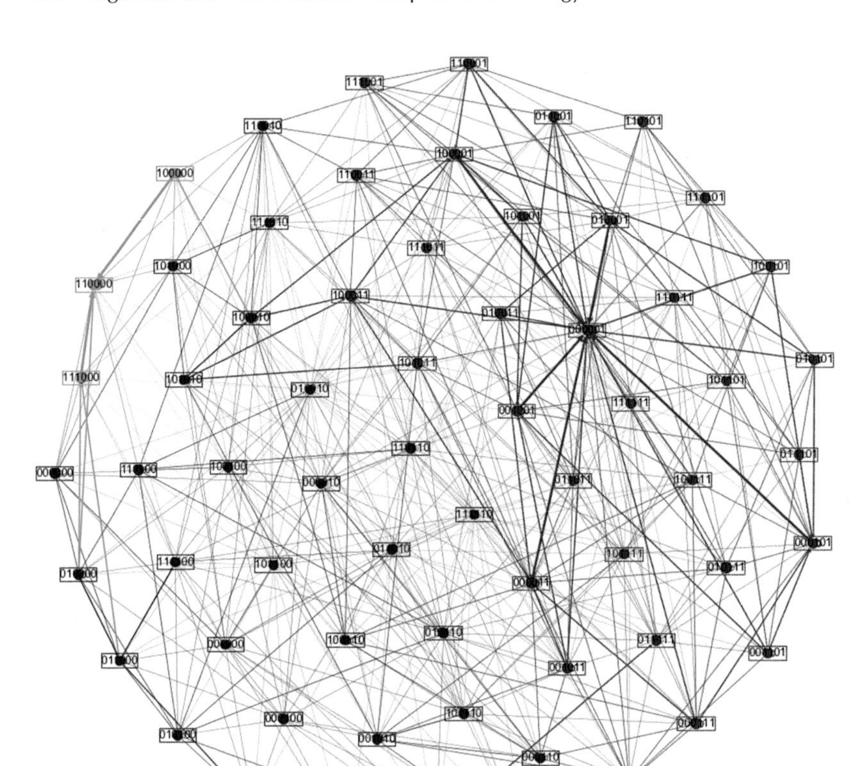

FIG. 5.4 State space for the stochastic *T-LGL* network with all propensities 0.5. *Green nodes* are in the basin of 110000, *red nodes* are in the basin of 000001, and *blue nodes* can transition to either fixed point. Edges are *colored* according to the target nodes. *Thick arrows* represent the most likely transitions and the remaining *arrows* represent transitions with smaller probabilities.

Finally, we will calculate the transition probability from $x = 000000$ to $y = 001000$. This corresponds to the entry $(1, 9)$. Thus,

$$a_{1,9} = \prod_{i=1}^{n} Prob(x_i \to y_i) = 0.5 \times 0.5 \times 0.5 \times 1 \times 1 \times 1 = 0.125, \quad (5.12)$$

where $Prob(x_i \to y_i) = 1 - 0.5 = 0.5$ for $i = 1, 2$, $Prob(x_i \to y_i) = 0.5$ for $i = 3$, and $Prob(x_i \to y_i) = 1$ for $i = 4, 5, 6$.

Notice that in this example, in contrast to Example 5.1, the synchronous transition from 1 to 57 is no longer the most likely transition; see Eq. (5.11).

Fig. 5.4 shows the state space of the *T-LGL* network under the SDDS framework with all propensity parameters 0.5. Here, we will compare the stochastic state spaces in Figs. 5.3 and 5.4. Both are similar but there are a few key differences. The main differences between these two state spaces are the weights of the edges that represent the transition probabilities. For instance, in Fig. 5.4, there are more transitions in green from the blue states, which means that, for this choice of parameters, more states will eventually transition to the fixed point 110000 than will using the parameters in Example 5.1. These differences will matter more when we study the long-term dynamics in Section 5.3.

Exercise 5.6. Calculate the following transition probabilities of the *T-LGL* network where all of the propensities are 0.5:

1. From $x = 000000$ to $y = 010000$ (i.e., from state 1 to state 17).
2. From $x = 000000$ to $y = 011000$ (i.e., from state 1 to state 25).
3. From $x = 000000$ to $y = 100000$ (i.e., from state 1 to state 33).
4. From $x = 000000$ to $y = 101000$ (i.e., from state 1 to state 41).
5. From $x = 000000$ to $y = 110000$ (i.e., from state 1 to state 49).
6. From $x = 000000$ to $y = 111111$ (i.e., from state 1 to state 64).

As in Example 5.1, we note that the only nonzero transitions from state 1 are to the states $1, 9, 17, 25, 33, 41, 49, 57$. As opposed to what happened in Example 5.1 where the transition corresponding to the synchronous transition was the most likely, in this example, all of these transitions have equal probability; see Exercise 5.6.

Exercise 5.7. Use a software package to compute the transition matrix of the *T-LGL* network where all the propensities are 0.5.

5.3 LONG-TERM DYNAMICS

In this section, we will discuss methods for studying the long-term dynamics in the SDDS framework. First, we will introduce the concept of stationary distributions. For each state of the system, the stationary distribution basically gives the probability that in the long-run the system will be in that state at a random "snapshot."

Remember that the transition probability $a_{x,y} = p(X_t = x \mid X_{t-1} = y)$ represents the probability of being in state x at time t given that system was in state y at time $t-1$. If $\pi_t = p(X_t = x)$ represents the probability of being in state x at time t, then we will assume that π is a row vector containing the probabilities of being in state x at time t for all $x \in S$. If π_0 is the initial distribution at time $t = 0$, then at time $t = 1$,

$$\pi_1 = \sum_{x \in S} \pi_0(x) a_{xy}. \tag{5.13}$$

If we iterate Eq. (5.13) and get to the point where

$$\pi = \sum_{x \in S} \pi(x) a_{xy} \tag{5.14}$$

then we say that the Markov chain has reached the stationary distribution π. The entries of π can be interpreted as occupation times for each state. They give the probability of being at a certain state in the long-run.

The following theorem ensures the existence of a unique stationary distribution π provided that the transition matrix is *regular*, that is, if some power $\mathbf{A}^k$ of it contains only strictly positive entries.

Theorem 5.1 (Perron-Frobenius). *If $\mathbf{A}$ is a regular $m \times m$ transition matrix with $m \geq 2$, then*

(a) *For any initial probability vector π_0, $\lim_{n\to\infty} \mathbf{A}^n \pi_0 = \pi$.*
(b) *The vector π is the unique probability vector which is an eigenvector of $\mathbf{A}$ associated with the eigenvalue $\lambda = 1$.*

A proof of Theorem 5.1 can be found in Chapter 10 of [32].

To study the long-term dynamics of the SDDS given in Eq. (5.5), we want to apply Theorem 5.1 to get the stationary distribution of its transition matrix $\mathbf{A}$. However, due to existence of fixed points, the transition matrix $\mathbf{A}$ of the SDDS might not be regular. To circumvent this issue, in Section 5.4, we will use a similar approach to Google's PageRank algorithm to add noise to the system to obtain a new regular transition matrix.

5.4 PAGERANK ALGORITHM

In this section, we will describe how to add noise to the system in order to obtain a regular transition matrix. We consider an SDDS, $\hat{F} = \{f_k, p_k^{\uparrow}, p_k^{\downarrow}\}_{k=1}^n$. To introduce noise into the system we consider the following matrix

$$\mathbf{G} = g\mathbf{A} + (1 - g)\mathbf{K}, \tag{5.15}$$

where $g \in [0, 1]$ is constant and $\mathbf{K}$ is a $2^n \times 2^n$ matrix all of whose entries are $1/2^n$. The matrix $\mathbf{G}$ in Eq. (5.15) is sometimes called the *Google Matrix* and it is regular. The Perron-Frobenius theorem guarantees a stationary distribution for $\mathbf{G}$,

$$\pi = \pi_{\mathbf{G}} = (\pi_1, \ldots, \pi_{2^n}). \tag{5.16}$$

This stationary distribution approximates the long-term dynamics of the SDDS. The importance of a state $x \in S$ can be measured by the size of the corresponding entry π_x in the stationary distribution. For instance, for ranking the importance of the states in a Markov chain, one can use the size of the corresponding entries in the stationary distribution. We will refer to this entry π_x as the *PageRank score* of x.

Example 5.3. Consider the *T-LGL* network discussed in Example 5.1, where all propensities are 0.9. Fig. 5.3 shows the state space of this network with this

TABLE 5.1 Top 10 PageRank Scores for the Stationary Distribution in Example 5.3

Rank	State ID	State	Score
1	2	000001	0.7401
2	49	110000	0.0843
3	34	100001	0.0108
4	4	000011	0.0100
5	16	001111	0.0075
6	42	101001	0.0070
7	19	010010	0.0066
8	10	001001	0.0059
9	57	111000	0.0049
10	31	011110	0.0048

Notes: The State ID is the decimal representation of the binary state in lexicographic order.

choice of parameters. We approximate the stationary distribution of this model using the PageRank approach described above. Let $\mathbf{A}$ be the transition matrix; see Exercise 5.4, and let $g = 0.9$. The Google matrix is thus

$$\mathbf{G} = g\mathbf{A} + (1 - g)\mathbf{K}, \tag{5.17}$$

where $\mathbf{K}$ is the 64×64 matrix where all entries are $1/64$.

From the Perron-Frobenius theorem, we can calculate the stationary distribution $\pi = \pi_{\mathbf{G}}$ by approximating the eigenvector of $\mathbf{G}$ associated with the eigenvalue $\lambda = 1$. There are many software packages for computing eigenvectors but in most of these cases it is enough to take a high power of the matrix $\mathbf{G}$ and then the columns will approximate the desired eigenvector.

We computed the stationary distribution and rank the states by the size of the corresponding entries in the stationary distribution. Table 5.1 shows the 10 largest entries. The top two correspond to the fixed points, 000001 and 110000. We conclude that in the long-term, 74% of the time the systems will be in 000001 (state 2 in the lexicographic order), which has the largest basin, and 8.4% of the time, the system will be in the second fixed point, 110000.

Exercise 5.8. Consider the *T-LGL* network where all propensities are 0.9.

1. Compute the stationary distribution using Eq. (5.16) with the following values for g:

TABLE 5.2 Top 10 PageRank Scores for the Stationary Distribution in Example 5.4

Rank	State ID	State	Score
1	2	000001	0.5786886
2	49	110000	0.1173941
3	4	000011	0.0223845
4	34	100001	0.0193110
5	10	001001	0.0178521
6	6	000101	0.0160818
7	18	010001	0.0137842
8	16	001111	0.0105385
9	15	001110	0.0097970
10	36	100011	0.0084878

Notes: The State ID is the decimal representation of the binary state in lexicographic order.

a. $g = 0.85$
b. $g = 0.80$
c. $g = 0.50$

2. Discuss the main differences between the stationary distributions from part 1.

Example 5.4. Consider the *T-LGL* network discussed in Example 5.2, where all propensities are 0.5. Fig. 5.4 shows the state space of this network with these parameters. We approximate the stationary distribution of this model using the PageRank approach. Let $\mathbf{A}$ be the transition matrix; see Exercise 5.7, and let $g = 0.9$. The Google matrix is

$$\mathbf{G} = 0.9\mathbf{A} + 0.1\mathbf{K}, \tag{5.18}$$

where $\mathbf{K}$ is the 64×64 matrix with all entries are $1/64$.

Table 5.2 shows the 10 largest entries of the stationary distribution. The top two correspond to the fixed points of the system: 000001 and 110000. We conclude that in the long-term, 57.8% of the time the systems will spend at 000001, and 11.7% of the time at 110000.

Exercise 5.9. Consider the *T-LGL* network where all propensities are 0.5.

1. Compute the stationary distribution using Eq. (5.16) with the following values for g:

   a. $g = 0.85$
   b. $g = 0.80$
   c. $g = 0.50$
2. Discuss the main differences between the stationary distributions from part 1.

5.5 PARAMETER ESTIMATION TECHNIQUES

In this section, we will describe a genetic algorithm for estimating the propensity parameters of an SDDS with the goal of having the stationary distribution approximate a desired stationary distribution. This method was introduced in [33] and uses the PageRank approach (see Section 5.4) for adding noise into the systems so that the Perron-Frobenius theorem can be used to estimate a stationary distribution.

Suppose that we have a desired stationary distribution $\pi^* = (\pi_1^*, \ldots, \pi_{2^n}^*)$. We can use a genetic algorithm to initialize a population of random propensity matrices and search for a propensity matrix $c = (p_i^{\uparrow}, p_i^{\downarrow})_{i=1}^n$ such that its stationary distribution $\pi(c) = (\pi_1, \ldots, \pi_{2^n})$ approaches π^*. That is, we search for propensity matrices such that the Euclidean distance between π and π^* is minimized, which is

$$c^* = \arg\min_c d(\pi(c), \pi^*). \tag{5.19}$$

The input of the algorithm is a Boolean network $F = (f_1, \ldots, f_n)$ and the output is an estimated propensity matrix c^*. Briefly, the genetic algorithm performs the following operations that imitate an evolutionary process.

1. Start with a population of propensity values
2. Do mutations
3. Do crossover
4. Fitness evaluation using the following fitness function

$$f(c) = \exp(-d(\pi(c), \pi^*)/\gamma), \text{ where } \gamma \text{ is a small constant.}$$

5. Do selection

The genetic algorithm in [33] starts with a population of random propensity matrices. Mutations are applied to each matrix by adding or subtracting small values to the entries. The crossover step swaps rows and columns. Finally, the algorithm applies selection by using the fitness function described earlier. Performing these operations to a population of matrices constitutes a generation, and this process is iterated for a number of generations. In the end, we select the fittest member as our estimated propensity matrix. The pseudocode of this genetic algorithm is given in [33]. It has been implemented in Octave/MATLAB and the code can be downloaded from http://www.ms.uky.edu/~dmu228/GeneticAlg/Code.html.

TABLE 5.3 Propensity Values Computed With the Genetic Algorithm

	x_1	x_2	x_3	x_4	x_5	x_6
$p_i^{\uparrow}$	0.4792	0.4826	0.7098	0.3911	0.0549	0.0314
$p_i^{\downarrow}$	0.0274	0.9602	0.5195	0.7758	0.8373	1.0000

TABLE 5.4 Ten Largest PageRank Scores in the Stationary Distribution From Example 5.5

Rank	State ID	State	Score
1	2	000001	0.2749
2	49	110000	0.2612
3	34	100001	0.1711
4	15	001110	0.0473
5	29	011100	0.0348
6	33	100000	0.0285
7	57	111000	0.0134
8	35	100010	0.0119
9	10	001001	0.0114
10	42	101001	0.0098

Notes: The State ID is the decimal representation of the binary state in lexicographic order.

Example 5.5. In this example, we will study the dynamics of the *T-LGL* model using the propensity parameters obtained with the genetic algorithm described previously.

Consider the *T-LGL* network and suppose that we want to use propensities such that the probability of reaching the fixed point with the smaller basin size is maximized, that is, we want to select propensity parameters that maximize the probability of ending at the fixed point 110000. To approximate these propensities we set the desired stationary distribution such that the entry for 110000 is 90% and the remaining 10% is distributed evenly among the other states.

The genetic algorithm gave the propensity parameters shown in Table 5.3. Again, we rank the states by the size of the corresponding entries in the stationary distribution. Table 5.4 shows the 10 largest entries. The top two still correspond to the fixed points, 000001 and 110000. We conclude that in the long-run, the system will spend 27.5% of the time at 000001 (state 2), the fixed point with the largest basin, and 26.1% of the time in the second fixed point, 110000.

As we did in Examples 5.1 and 5.2, we will compute the transition probability from $x = 000000$ to $y = 000000$. We note that the synchronous update gives $F(x) = F(0,0,0,0,0,0) = (1,1,1,0,0,0)$, that is, with no stochasticity the system would transition from x to $z = (1,1,1,0,0,0)$. However, under the SDDS framework, the system can transition from x to y with a probability that we will determine following. For x to transition into y, the update of the first three coordinates of x should not happen. The probability for updating each of the first three entries is $p_i^{\uparrow}$ for $i = 1,2,3$. Thus, the probability that each of these updates will not happen is $1 - p_i^{\uparrow}$. The last three entries do not change with the update functions, so they will remain the same with probability 1. Now if we multiply these probabilities we obtain the transition probability from x to y:

$$a_{1,1} = \prod_{i=1}^{n} Prob(x_i \to y_i) = 0.5208 \times 0.5174 \times 0.2902 \times 1 \times 1 \times 1 = 0.0782, \tag{5.20}$$

where $Prob(x_i \to y_i) = 1 - p_i^{\uparrow}$ for $i = 1,2,3$ and $Prob(x_i \to y_i) = 1$ for $i = 4,5,6$.

Next, the transition probability from $x = 000000$ to $y = 111000$ is

$$a_{1,57} = \prod_{i=1}^{n} Prob(x_i \to y_i) = 0.4792 \times 0.4826 \times 0.7098 \times 1 \times 1 \times 1 = 0.1642, \tag{5.21}$$

where $Prob(x_i \to y_i) = p_i^{\uparrow}$ for $i = 1,2,3$ and $Prob(x_i \to y_i) = 1$ for $i = 4,5,6$.

Finally, the transition probability from $x = 000000$ to $y = 001000$ is

$$a_{1,9} = \prod_{i=1}^{n} Prob(x_i \to y_i) = 0.5208 \times 0.5174 \times 0.7098 \times 1 \times 1 \times 1 = 0.1913, \tag{5.22}$$

where $Prob(x_i \to y_i) = 1 - p_i^{\uparrow}$ for $i = 1,2$, $Prob(x_i \to y_i) = p_i^{\uparrow}$ for $i = 3$, and $Prob(x_i \to y_i) = 1$ for $i = 4,5,6$.

Recall from Exercise 5.10 that the only nonzero transitions from state 1 are to the states 1, 9, 17, 25, 33, 41, 49, 57. The synchronous transition from 1 to 57 has a probability of 16.4%, and the transition from 1 to 9 has a probability of 19.13%, see Eqs. (5.21), (5.22). Thus, for the propensity parameters in Table 5.3, the transition corresponding to the synchronous update is not the most likely. Rather, the most-likely transition is from 1 to 9, see Eq. (5.22).

Fig. 5.5 shows the state space with the propensity parameters calculated by the genetic algorithm. As we did in Example 5.2, we will discuss some similarities and differences between the state spaces in Figs. 5.3–5.5. Although

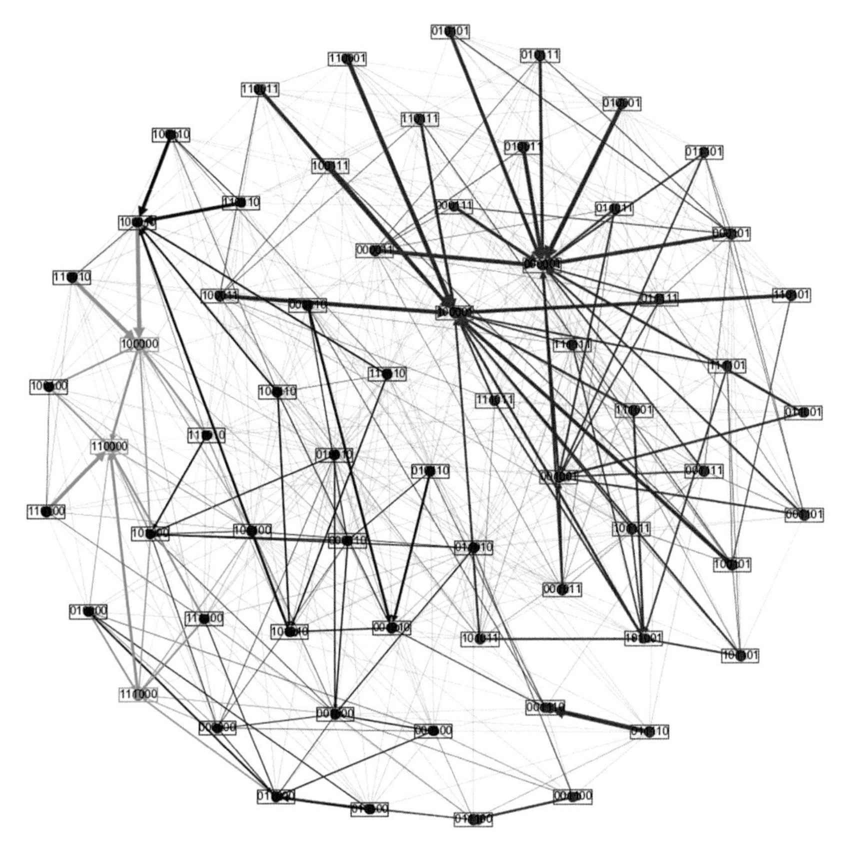

FIG. 5.5 State space for the stochastic *T-LGL* network, with the propensities given in Table 5.3. *Green nodes* are in the basin of 110000, *red nodes* are in the basin of 000001, and *blue nodes* can transition to either fixed point. Edges are *colored* according to the target nodes. *Thick arrows* represent the most likely transitions and the *remaining arrows* represent transitions with smaller probabilities.

they look similar, there are few key differences. The main differences are the edge weights that represent the transition probabilities. For instance, Fig. 5.5 has more green transitions from blue states, which means that more states will eventually transition to the fixed point 110000 than in Figs. 5.3 and 5.4. These small differences are reflected in the long-term dynamics. For instance, in Fig. 5.5, some transitions from the basin of the fixed point 000001 have a relatively high probability of reaching the other fixed point 110000. This is reflected in the increase of the PageRank score of the fixed point 110000, see Table 5.4.

Exercise 5.10. Calculate the following transition probabilities of the *T-LGL* network with the propensities in Table 5.3:

1. From $x = 000000$ to $y = 010000$ (i.e., from state 1 to state 17).
2. From $x = 000000$ to $y = 011000$ (i.e., from state 1 to state 25).
3. From $x = 000000$ to $y = 100000$ (i.e., from state 1 to state 33).
4. From $x = 000000$ to $y = 101000$ (i.e., from state 1 to state 41).
5. From $x = 000000$ to $y = 110000$ (i.e., from state 1 to state 49).
6. From $x = 000000$ to $y = 111111$ (i.e., from state 1 to state 64).

Exercise 5.11. Use a software package to compute the transition matrix of the *T-LGL* network with the propensities from Table 5.3.

Exercise 5.12. Consider the *T-LGL* network with the propensities from Table 5.3.

1. Compute the stationary distribution using Eq. (5.16) with the following values for g:
   a. $g = 0.85$
   b. $g = 0.80$
   c. $g = 0.50$
2. Discuss the main differences between the stationary distributions from part 1.

5.6 OPTIMAL CONTROL FOR SDDS

This section introduces a method for optimal control that can be used for general MDP. In the MDP research literature, this is usually called the infinite-horizon method with a discounting factor [30]. We will describe the method adapting the terminology for SDDS. This has also been applied to control of PBNs [31, 34]. In this section we will consider edge and node control in discrete stochastic networks. Details about this type of control in BN and a method for how to identify control targets can be found in [35].

Let $\hat{\boldsymbol{F}} = \{f_k, p_k^{\uparrow}, p_k^{\downarrow}\}_{k=1}^n$ be an SDDS and $\mathcal{G}$ be the *wiring diagram* associated with $\hat{\boldsymbol{F}}$. That is, $\mathcal{G}$ has nodes $x_1, \ldots, x_n$, and there is a directed edge from x_i to x_j if f_j depends on x_i. In the context of molecular network models, $\mathcal{G}$ represents the wiring diagram of the model. Next, consider the Boolean network $\mathbf{F} = (f_1, \ldots, f_n) : \mathbb{F}_2^n \to \mathbb{F}_2^n$ associated with $\hat{\boldsymbol{F}}$. We will show how to encode the edge and node controls by $\mathcal{F}$: $\mathbb{F}_2^n \times U \to \mathbb{F}_2^n$, such that $\mathcal{F}(x, 0) = \mathbf{F}(x)$. That is, the case of no control coincides with the original Boolean network.

Definition 5.1 (Edge Control)**.** Consider an edge $x_i \to x_j$ in the wiring diagram $\mathcal{G}$. The function $\mathcal{F}_j$: $\{0,1\}^n \times \{0,1\} \to \{0,1\}$,

$$\mathcal{F}_j(x, u_{i,j}) := f_j(x_1, \ldots, (u_{i,j} + 1)x_i, \ldots, x_n) \tag{5.23}$$

encodes the control of the edge $x_i \to x_j$. Specifically, each possible value of $u_{i,j} \in \{0, 1\}$ describes the following control settings:

- If $u_{i,j} = 0$, $\mathcal{F}_j(x, 0) = f_j(x_1, \ldots, x_i, \ldots, x_n)$. The control is inactive.
- If $u_{i,j} = 1$, $\mathcal{F}_j(x, 1) = f_j(x_1, \ldots, x_i = 0, \ldots, x_n)$. The control is active, and the action represents the removal of the edge $x_i \to x_j$.

Definition 5.2 (Node Control)**.** Consider the node x_i in the wiring diagram $\mathcal{G}$. The function $\mathcal{F}_i$: $\{0, 1\}^n \times \{0, 1\} \to \{0, 1\}$,

$$\mathcal{F}_i(x, u_i^-) := (u_i^- + 1)f_j(x) \tag{5.24}$$

encodes the control (knock-out) of the node x_i. Specifically, each $u_i^- \in \{0, 1\}$ describes the following control settings:

- For $u_i^- = 0$, $\mathcal{F}_i(x, 0) = f_i(x)$. The control is inactive.
- For $u_i^- = 1$, $\mathcal{F}_i(x, 1) = 0$. This represents the knock-out of the node x_i.

An SDDS with control is obtained by replacing the functions f_k for $\mathcal{F}_k$: $\{0, 1\}^n \times U \to \{0, 1\}$, where U is a set that denotes all possible control inputs.

5.6.1 Control Actions

Suppose that we identify a set E of control edges and a set V control nodes. Define a *control action a* as an array of binary elements of size $|U| = |E| + |V|$. The kth element of a corresponds to a control node u_l^- if $k < |V|$ and to a control edge $u_{i,j}$ if $|V| \leq k < |U|$. Thus, a value of 1 in a_k represents that the corresponding control intervention (node or edge) is being applied. Thus, an action array a is a combination of control edges and nodes that are being applied to the GRN simultaneously at a given time step. Let $A = \{0, 1\}^{|U|}$ be the set of all possible actions. Notice that the action $a = (0, \ldots, 0)$ represents the case where none of the control actions are applied.

5.6.2 Markov Decision Processes for SDDS

In this section, we define an MDP for the SDDS and the control actions defined in the previous sections. An MDP for the set of states S and the set of actions A consists of transition probabilities $P^a_{x,y}$ and associated costs $C(x, a, y)$, for each transition from state x to state y due to applying the action a.

5.6.2.1 *Transition Probabilities*

The application of an action a results in a new SDDS, $\hat{\boldsymbol{F}}'_{\boldsymbol{a}} = \{\mathcal{F}^a_k, p^{\uparrow}_k, p^{\downarrow}_k\}^n_{k=1}$. For each state-action pair (x, a), the transition probability $P^a_{x,y}$ from x to state y

upon execution of a is computed using Eq. (5.2) with f_k replaced by $\mathcal{F}_k^a$, that is,

$$P_{x,y}^a = \prod_{k=1}^{n} Prob_{\mathcal{F}_k^a}(x_k \to y_k),$$

where $Prob_{\mathcal{F}_k^a}(x_k \to y_k)$ is the probability that x_k will change its value under $\mathcal{F}_k^a$.

5.6.2.2 Cost Function

The cost $C(x, a, y)$ of going from state x to state y under action a can be written as a sum of two costs, one for actions and one for states:

$$C(x, a, y) = C_a + C_y. \tag{5.25}$$

The application of control edges and nodes has penalties, c_e and c_v, respectively, that represent expenses associated with the use of technologies and drugs required to silence nodes and edges. Thus, the action cost can be expressed as $C_a = c_v N_v + c_e N_e$, where N_v and N_e are the number of applied control nodes and edges in a given action a. The cost C_y of ending up in a state y is the weighted distance between y and a user-specified desirable state s^*. That is,

$$C_y = \sum_{k=1}^{N} w_k |y_k - s_k^*|,$$

where w_k are user-specified weights. Note that if all the weights are 1, then C_y is simply the *Hamming distance* between y and s^*, that is, the number of bits in which they differ.

5.6.2.3 Optimal Control Policies

A deterministic control policy π is defined as a set $\pi = \{\pi_0, \pi_1, \ldots\}$, where π_t: $S \to A$ associates a state $x(t)$ to an action a at time step t. We will formulate the optimal control problem for infinite horizon MDPs with a discounting factor as described in [31, 34]. Given a state $x \in S$, a control policy π, and a discounting factor $\gamma \in (0, 1)$, the cost function V^π for π is defined as

$$V^\pi(x) = \mathbf{E}\left[\sum_{t=0}^{\infty} \gamma^t C(x(t), a)\right], \tag{5.26}$$

where $C(x(t), a)$ represents the expected cost at step t for executing the policy π from state x, defined by

$$C(x(t), a) = \mathbf{E}[C(x, a, y)] = \sum_{y \in S} P_{x,y}^a C(x, a, y).$$

The expectation in Eq. (5.26) is over the random transition probabilities obtained under actions associated with π.

Our goal is to find the optimal policy $\pi^* = \{\pi_0^*, \pi_1^*, \ldots\}$, where $\pi_t^*\colon S \to A$, $t = 1, 2, \ldots$, that minimizes the function cost for all states. The cost function associated with π^* is

$$V^*(x) = \min_{\pi}\{V^{\pi}(x) \mid x \in S\}.$$

It was shown [34] that the optimal cost function V^* satisfies Bellman's principle,

$$V^*(x) = \min_{a \in A}\left\{C(x,a) + \gamma \mathbf{E}[V^*(y)]\right\}.$$

The optimal policy for the MDP defined for an SDDS is a stationary policy in which every state is associated with an action. We will determine π^* with the help of an iterative algorithm called *value iteration* [31, 34].

5.6.2.4 Value Iteration Algorithm

The optimal control policy will be calculated using the value iteration algorithm, which is described following. For any bounded cost function $V\colon S \to \mathbb{R}$, we define the function $TV\colon S \to \mathbb{R}$ by

$$TV(x) = \min_{a \in A}\left[C(x,a) + \gamma \sum_{y \in S} P_{x,y}^{a} V(x)\right].$$

Given a control policy $\pi\colon S \to A$, we define the function $T_\pi V\colon S \to \mathbb{R}$ by

$$T_\pi V(x) = \min_{a \in A}\left[C(x,a) + \gamma \sum_{y \in S} P_{x,y}^{\pi(x)} V(x)\right].$$

Next, we define the operators T and T_π recursively by

$$T^0 V(x) = V(x), \quad T^k V(x) = T(T^{k-1} V(x)), \quad k = 1, 2, \ldots$$

and

$$T_\pi^0 V(x) = V(x), \quad T_\pi^k V(x) = T_\pi(T_\pi^{k-1} V(x)), \quad k = 1, 2, \ldots$$

One can show (see [34]) that the optimal cost function V^* is the unique fixed point of the operator T. Thus

$$V^*(x) = \lim_{M \to \infty} T^M V(x),$$

and

$$V^*(x) = \min_{a \in A}\left[C(x,a) + \gamma \sum_{y \in S} P_{x,y}^{a} V^*(x)\right]. \tag{5.27}$$

Additionally, we have that

$$TV^* = T_\pi V^*. \tag{5.28}$$

Using Eq. (5.27), the cost function can be iteratively calculated by running the recursion

$$V_t(x) = \min_{a \in A} \left[C(x, a) + \gamma \sum_{y \in S} P^a_{x,y} V_{t-1}(x) \right] \tag{5.29}$$

for any initial bounded cost function V_0: $S \to \mathbb{R}$. Since this recursion will converge to V^* (from Eq. 5.28), we can run this iteration until some stopping criterion is met. This iteration is called the *value iteration* algorithm.

Example 5.6. We will apply the optimal control method just described to the *T-LGL* model from Example 5.1, where all propensities are 0.9.

We will consider the following two controllers:

1. the deletion of *FLIP* (*FLIP* = OFF)
2. the constant expression of *Fas* (*Fas* = ON)

Using the same labeling of the variables from Example 5.1, the controllers can be represented as

$$\begin{cases} 1. & x_2 = 0, \\ 2. & x_3 = 1. \end{cases} \tag{5.30}$$

These control nodes have been identified using the methods given in [35]. It can be shown that the simultaneous application of these controllers will result in a fixed point 001001 that is globally reachable; see Exercise 5.13.

Using these controllers, we can compute a control policy for this system. Since we have two controllers, there are four possible actions:

1. 00 (no intervention)
2. 01 (deletion of *FLIP*)
3. 10 (constant expression of *Fas*)
4. 11 both controllers

Fig. 5.6 shows the control policy where transitions are marked by colors:

- green arrows mean no control
- blue arrows represent the control of the node *FLIP* ($x_2 = 0$)
- orange arrows represent the control of the node *Fas* ($x_3 = 1$)
- red arrows represent the control of both nodes

Notice that in Fig. 5.6, only a few states require intervention. One such example is the disease state 110000 and the states in its basin of attraction. Also in Fig. 5.6, notice that the controllers are only needed transiently, that is,

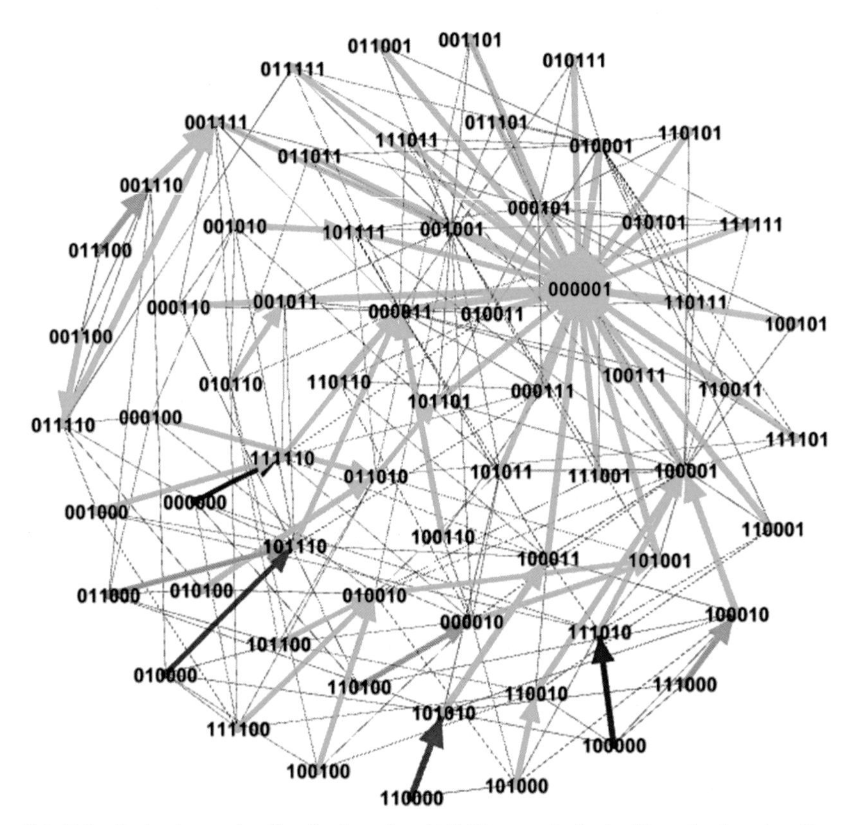

FIG. 5.6 Optimal control policy for the reduced *T-LGL* network obtained by value iteration. Two controls have been considered, *FLIP* = OFF ($x_2 = 0$) and *Fas* = ON ($x_3 = 1$). *Green arrows* represent no control, *blue arrows* represent the control of the node *FLIP* ($x_2 = 0$), *orange arrows* represent the control of the node *Fas* ($x_3 = 1$), and *red arrows* represent the control of both nodes. The *colored thick arrows* show the most likely transitions while *gray arrows* represent other possible transitions.

after one intervention, each state will transition into a state that does not require additional interventions to get to the desired fixed point.

Exercise 5.13. Compute the fixed points of the *T-LGL* network with the controllers given in Eq. (5.30).

5.7 DISCUSSION AND CONCLUSIONS

In this chapter, we presented a stochastic extension of deterministic BN, which we called SDDS. In this framework, each node is associated with a regulatory function and a pair of propensity parameters which describe the probabilities of using the regulatory functions. Importantly, the SDDS framework preserves some of the relevant features of the synchronous Boolean network, such as the set of fixed points. In Section 5.3, we described the long-term dynamics of SDDS using stationary distributions. In Section 5.4, we used the PageRank algorithm to add noise into the system to guarantee a unique stationary distribution. In Section 5.5, we described a method for estimating the SDDS propensity parameters using a genetic algorithm and the PageRank approach to search for a desired stationary distribution. Finally, in Section 5.6, we discussed an optimal control method based on techniques for MDP. The main limitation of this method is its scalability, as it requires one to calculate the transition matrices whose sizes grow exponentially with the size of the network.

REFERENCES

[1] J.J. Tyson, K. Chen, B. Novak, Network dynamics and cell physiology, Nat. Rev. Mol. Cell Biol. 2 (12) (2001) 908–916, https://doi.org/10.1038/35103078.

[2] J.J. Tyson, K.C. Chen, B. Novak, Sniffers, buzzers, toggles and blinkers: dynamics of regulatory and signaling pathways in the cell, Curr. Opin. Cell Biol. 15 (2) (2003) 221–231, https://doi.org/10.1016/S0955-0674(03)00017-6.

[3] C.P. Fall, A.S. Marland, J.M. Wagner, J.J. Tyson, Computational Cell Biology (Interdisciplinary Applied Mathematics), Springer, New York, NY, 2010.

[4] T. Toulouse, P. Ao, I. Shmulevich, S. Kauffman, Noise in a small genetic circuit that undergoes bifurcation, Complexity 11 (1) (2005) 45–51.

[5] D.T. Gillespie, Exact stochastic simulation of coupled chemical reactions, J. Phys. Chem. 81 (25) (1977) 2340–2361.

[6] D.T. Gillespie, Stochastic simulation of chemical kinetics, Annu. Rev. Phys. Chem. 58 (1) (2007) 35–55, https://doi.org/10.1146/annurev.physchem.58.032806.104637.

[7] D. Bratsun, D. Volfson, L.S. Tsimring, J. Hasty, Delay-induced stochastic oscillations in gene regulation, Proc. Natl Acad. Sci. USA 102 (41) (2005) 14593–14598, https://doi.org/10.1073/pnas.0503858102.

[8] A. Ribeiro, R. Zhu, S.A. Kauffman, A general modeling strategy for gene regulatory networks with stochastic dynamics, J. Comput. Biol. 13 (9) (2006) 1630–1639, https://doi.org/10.1089/cmb.2006.13.1630.

[9] A.S. Ribeiro, Stochastic and delayed stochastic models of gene expression and regulation, Math. Biosci. 223 (1) (2010) 1–11, https://doi.org/10.1016/j.mbs.2009.10.007.

[10] S. Kauffman, C. Peterson, B. Samuelsson, C. Troein, Random Boolean network models and the yeast transcriptional network, Proc. Natl Acad. Sci. 100 (25) (2003) 14796–14799.

[11] S. Kauffman, C. Peterson, B. Samuelsson, C. Troein, Genetic networks with canalyzing Boolean rules are always stable, Proc. Natl Acad. Sci. USA 101 (49) (2004) 17102–17107, https://doi.org/10.1073/pnas.0407783101.

[12] I. Shmulevich, E.R. Dougherty, S. Kim, W. Zhang, Probabilistic Boolean networks: a rule-based uncertainty model for gene regulatory networks, Bioinformatics 18 (2) (2002) 261–274.

[13] D. Murrugarra, R. Laubenbacher, Regulatory patterns in molecular interaction networks, J. Theor. Biol. 288 (2011) 66–72, https://doi.org/10.1016/j.jtbi.2011.08.015.

[14] D. Murrugarra, A. Veliz-Cuba, B. Aguilar, S. Arat, R. Laubenbacher, Modeling stochasticity and variability in gene regulatory networks, EURASIP J. Bioinform. Syst. Biol. 2012 (1) (2012) 5.

[15] S.A. Kauffman, Metabolic stability and epigenesis in randomly constructed genetic nets, J. Theor. Biol. 22 (3) (1969) 437–467.

[16] R. Thomas, R. D'Ari, Biological Feedback, CRC Press, Boca Raton, FL, 1990, ISBN 0849367662.

[17] F. Li, T. Long, Y. Lu, Q. Ouyang, C. Tang, The yeast cell-cycle network is robustly designed, Proc. Natl Acad. Sci. USA 101 (14) (2004) 4781–4786, https://doi.org/10.1073/pnas.0305937101.

[18] L. Mendoza, A network model for the control of the differentiation process in Th cells, Biosystems 84 (2) (2006) 101–114, https://doi.org/10.1016/j.biosystems.2005.10.004.

[19] A. Veliz-Cuba, B. Stigler, Boolean models can explain bistability in the *lac* operon, J. Comput. Biol. 18 (6) (2011) 783–794, https://doi.org/10.1089/cmb.2011.0031.

[20] M. Choi, J. Shi, S.H. Jung, X. Chen, K.-H. Cho, Attractor landscape analysis reveals feedback loops in the p53 network that control the cellular response to DNA damage, Sci. Signal. 5 (251) (2012) ra83.

[21] W. Abou-Jaoudé, D.A. Ouattara, M. Kaufman, From structure to dynamics: frequency tuning in the p53-Mdm2 network I. Logical approach, J. Theor. Biol. 258 (4) (2009) 561–577, https://doi.org/10.1016/j.jtbi.2009.02.005.

[22] E. Balleza, E.R. Alvarez-Buylla, A. Chaos, S. Kauffman, I. Shmulevich, M. Aldana, Critical dynamics in genetic regulatory networks: examples from four kingdoms, PLoS ONE 3 (6) (2008) e2456, https://doi.org/10.1371/journal.pone.0002456.

[23] M.I. Davidich, S. Bornholdt, Boolean network model predicts cell cycle sequence of fission yeast, PLoS ONE 3 (2) (2008) e1672, https://doi.org/10.1371/journal.pone.0001672.

[24] R. Albert, H.G. Othmer, The topology of the regulatory interactions predicts the expression pattern of the segment polarity genes in *Drosophila melanogaster*, J. Theor. Biol. 223 (1) (2003) 1–18.

[25] A. Saadatpour, R.-S. Wang, A. Liao, X. Liu, T.P. Loughran, I. Albert, R. Albert, Dynamical and structural analysis of a T cell survival network identifies novel candidate therapeutic targets for large granular lymphocyte leukemia, PLoS Comput. Biol. 7 (11) (2011) e1002267, https://doi.org/10.1371/journal.pcbi.1002267.

[26] R. Zhang, M.V. Shah, J. Yang, S.B. Nyland, X. Liu, J.K. Yun, R. Albert, T.P. Loughran, Jr, Network model of survival signaling in large granular lymphocyte leukemia, Proc. Natl Acad. Sci. USA 105 (42) (2008) 16308–16313, https://doi.org/10.1073/pnas.0806447105.

[27] T. Helikar, J. Konvalina, J. Heidel, J.A. Rogers, Emergent decision-making in biological signal transduction networks, Proc. Natl Acad. Sci. USA 105 (6) (2008) 1913–1918, https://doi.org/10.1073/pnas.0705088105.

[28] T. Helikar, N. Kochi, B. Kowal, M. Dimri, M. Naramura, S.M. Raja, V. Band, H. Band, J.A. Rogers, A comprehensive, multi-scale dynamical model of ErbB receptor signal transduction in human mammary epithelial cells, PLoS ONE 8 (4) (2013) e61757, https://doi.org/10.1371/journal.pone.0061757.

[29] I. Shmulevich, E.R. Dougherty, Probabilistic Boolean Networks—The Modeling and Control of Gene Regulatory Networks, SIAM, Philadelphia, PA, 2010, ISBN 978-0-89871-692-4.

[30] H.S. Chang, J. Hu, M.C. Fu, S.I. Marcus, Simulation-Based Algorithms for Markov Decision Processes, second ed., Springer, New York, NY, 2013.

[31] M.R. Yousefi, A. Datta, E.R. Dougherty, Optimal intervention strategies for therapeutic methods with fixed-length duration of drug effectiveness, IEEE Trans. Signal Process. 60 (9) (2012) 4930–4944.

[32] D.C. Lay, Linear Algebra and Its Applications, fourth ed., Pearson, Boston, MA, 2012.

[33] D. Murrugarra, J. Miller, A.N. Mueller, Estimating propensity parameters using Google PageRank and genetic algorithms, Front. Neurosci. 10 (2016) 513.

[34] R. Pal, A. Datta, E.R. Dougherty, Optimal infinite-horizon control for probabilistic Boolean networks, IEEE Trans. Signal Process. 54 (6-2) (2006) 2375–2387, https://doi.org/10.1109/TSP.2006.873740.

[35] D. Murrugarra, A. Veliz-Cuba, B. Aguilar, R. Laubenbacher, Identification of control targets in Boolean molecular network models via computational algebra, BMC Syst. Biol. 10 (1) (2016) 94.

Chapter 6

Inferring Interactions in Molecular Networks via Primary Decompositions of Monomial Ideals

Matthew Macauley* and Brandilyn Stigler†
*School of Mathematical and Statistical Sciences, Clemson University, Clemson, SC, United States,
†Department of Mathematics, Southern Methodist University, Dallas, TX, United States

6.1 INTRODUCTION

6.1.1 The Local Modeling Framework

Discrete models such as logical or Boolean networks (BNs) are popular choices for modeling biological systems, especially in molecular biology. Examples include gene regulatory networks, protein-protein interaction networks, signaling networks, and more. Chapter 4 introduced a general framework for these types of models called *local models*, and this includes logical and BNs as special cases. We encourage the reader to read Chapter 4 first. In the current chapter, we will study a particular mathematical problem that arises in modeling which we can tackle using computational algebra. Specifically, suppose we wish to understand a biological system from "partial data." That is, we know some input and output values, and the goal is simply to determine which update functions depend on which variables. Moreover, sometimes we wish to determine which nodes are activators and which nodes inhibitors. Such data might arise from gene silencing or gene knockout experiments.

Throughout this section, unless otherwise specified, $\mathbb{F}$ will denote a finite field, usually $\mathbb{F} = \mathbb{F}_p = \{0, 1, \ldots, p-1\}$, where the arithmetic is done modulo p. A common special case is the Boolean field $\mathbb{F}_2 = \{0, 1\}$. At times, we will work with $\mathbb{F}_3$, though it will often be particularly useful to use -1 instead of 2, so $\mathbb{F}_3 = \{0, 1, -1\}$. One can think of these three values as

Algebraic and Combinatorial Computational Biology. https://doi.org/10.1016/B978-0-12-814066-6.00006-4

"no interaction," "positive interaction," and "negative interaction," respectively. To make this chapter self-contained, we will begin with a few basic definitions, and then provide a motivating example. The following definitions are meant to be a quick summary of the concepts from Chapter 4 which are relevant here, and the exposition is more brief than it would be for a reader seeing this for the first time.

Definition 6.1. A *local model over* $\mathbb{F}$ is an n-tuple of "coordinate functions" $f = (f_1, \ldots, f_n)$, where $f_i \colon \mathbb{F}^n \to \mathbb{F}$. Each function f_i uniquely determines a function

$$F_i \colon \mathbb{F}^n \longrightarrow \mathbb{F}^n, \qquad F_i \colon (x_1, \ldots, x_n) \longmapsto (x_1, \ldots, \underbrace{f_i(x)}_{i\text{th coord.}}, \ldots, x_n), \tag{6.1}$$

where $x = (x_1, \ldots, x_n)$ is the sequence of variables. A BN is simply a local model over $\mathbb{F}_2$.

We want to emphasize that our definition of a local model does not include *how* the functions are updated, but rather how the nodes, which represent biological entities, interact with each other. The functions can be updated synchronously, asynchronously (i.e., sequentially), or some other way such as block-sequentially [1] or stochastically [2, 3]. For example, updating the functions synchronously, as

$$f \colon \mathbb{F}^n \longrightarrow \mathbb{F}^n, \quad f \colon (x_1, \ldots, x_n) \longmapsto (f_1(x), \ldots, f_n(x)),$$

defines a *finite dynamical system* (FDS) map. Depending on how it is done, updating the nodes asynchronously can lead to a directed multigraph; see Section 4.3.5 for more details, or a sequential dynamical system [4]. A stochastic update could lead to a Markov chain, or something variant (see, e.g., Chapter 5).

Regardless of the update scheme, every local model comes with a *wiring diagram* that describes the dependence of the coordinate functions on the variables. This can be defined in several ways, depending on how much and what type of information is desired. For example, sometimes, one wishes to keep track of whether the influence of a variable is positive, negative, both, or zero. Biologically, a positive interaction might mean that a variable is acting as an activator, like a transcription factor. A negative interaction might describe an inhibitor, like a repressor protein. In rarer cases, certain biomolecules can act as both an activator and inhibitor, depending on the circumstances. In some situations, this extra "signed" information is unnecessary or even unnatural. Perhaps the modeler is not concerned with this, or maybe the state space X (in this chapter, $X = \mathbb{F}$) does not have a natural order. Nonprime finite fields such as $\mathbb{F}_4$ are examples of such sets. For this chapter, we will be interested in distinguishing between positive and negative interactions, and so we will define these concepts now. Though this was defined in the previous chapter, we will repeat it here for self-containment, as it is one of the central concepts in this chapter.

Definition 6.2. Let $f = (f_1, \ldots, f_n)$ be a local model over $\mathbb{F} = \mathbb{F}_p$. If for some $a < b$ (respectively, $a > b$),

$$f_j(x_1, \ldots, x_{i-1}, a, x_{i+1}, \ldots, x_n) \leq f_j(x_1, \ldots, x_{i-1}, b, x_{i+1}, \ldots, x_n) \tag{6.2}$$

with equality not always holding, then we say that x_i *positively* (respectively, *negatively*) affects x_j. Note that the effect of x_i on x_j could be positive, negative, both, or none.

The wiring diagram is a directed graph that encodes this information. There is an edge from i to j if x_i affects x_j. If desired, we can use "signed edges" to distinguish between positive and negative interactions.

Definition 6.3. Let $f = (f_1, \ldots, f_n)$ be a local model over $\mathbb{F} = \mathbb{F}_p$. The (signed) *wiring diagram* of f is the signed directed graph with vertex set $x_1, \ldots, x_n$ (or just $1, \ldots, n$) and an

- edge $x_i \longrightarrow x_j$ if x_i positively affects x_j,
- edge $x_i \dashv x_j$ if x_i negatively affects x_j.

When we do not want or need to distinguish between positive or negative interactions, we either use ordered pairs (x_i, x_j) or just regular arrows, $x_i \to x_j$ and make it clear that they are "unsigned."

Wiring diagram for local models over nonprime fields can be defined similarly, but they have to be unsigned because $\mathbb{F}_{p^k}$ does not have a natural order if $k > 1$. An example of the different types of interactions are following. For ease of notation, we are using both Boolean logical symbols (first two examples) and standard arithmetic (third example).

$f_j = x_i \wedge x_k$	$f_j = \overline{x_i} \wedge x_k$	$f_j = x_i + x_k$
$x_i \longrightarrow x_j$	$x_i \dashv x_j$	x_i (→ and ⊣ arcs) x_j
"x_i activates x_j"	"x_i inhibits x_j"	"x_i affects x_j positively and negatively"

Note that the wiring diagram does not determine the Boolean functions. For example, if there are positive edges at node 1 from nodes 2 and 3, then there is no way to tell just from the wiring diagram whether $f_1 = x_2 \wedge x_3$ or $f_1 = x_2 \vee x_3$.

Since this work arises from biology, we are primarily interested in functions that are "biologically meaningful." Most interactions in a molecular network are simple activations or inhibitions [5]. This just means that every edge in the wiring diagram is either positive or negative. Such a function is said to be *unate*. In the Boolean setting, a function $\mathbb{F}_2^n \to \mathbb{F}_2$ is unate if it can be written in logical form so that no variable x_i and its negation $\overline{x_i}$ both appear. A special case of these are *monomials*, which (over $\mathbb{F}_2$) are just products of a subset of variables. In Boolean logical form, this is just $f = x_{i_1} \wedge \cdots \wedge x_{i_k}$ and called a *conjunction*. Similarly, a logical *disjunction* has the form $f = x_{i_1} \vee \cdots \vee x_{i_k}$.

6.1.2 A Motivating Example of Reverse Engineering

The mathematical problem central to this chapter is to infer the wiring diagram given partial data. This is best motivated by an example. Suppose we have an unknown Boolean function $f_i \colon \mathbb{F}_2^3 \to \mathbb{F}_2$ that satisfies the following:

$$f_i(1,1,1) = 0, \quad f_i(0,0,0) = 0, \quad f_i(1,1,0) = 1.$$

Here, we are intentionally not providing many details about the context. For example, i could be 1 (i.e., $x_i = x_1$), in which case this could be a function in a three-node BN. Or perhaps, $3 < i \leq n$, and the function $f_i \colon \mathbb{F}_2^n \to \mathbb{F}_2$ only depends on the variables x_1, x_2, x_3. Regardless of the setting, we are looking for all Boolean functions on three variables whose truth tables look like the following, where we are writing $x = x_1x_2x_3$ for short instead of $x = (x_1, x_2, x_3)$.

$x_1x_2x_3$	111	110	101	100	011	010	001	000
$f_i(x)$	0	1	?	?	?	?	?	0

(6.3)

Each entry in the truth table can be 0 or 1, and so there are $2^8 = 256$ possible truth tables. Specifying three of the eight output values leaves exactly $2^{8-3} = 32$ possibilities for f_i. We say that these 32 functions *fit the data*, and the set of all of them is called the *model space*.

Exercise 6.1. Find as many of the 32 functions that fit the data from Eq. (6.3) by hand. This is a useful "warm-up" exercise to become more familiar with the type of mathematical objects and problems considered in this chapter.

Remark 6.1. Given a set of partial data, the model space has a nice algebraic structure, much like a coset or affine space. Specifically, it has the form $h + I$, where I is the ideal of functions that vanish on the data, and $h(x)$ is any one particular function that satisfies it. There are algorithms to compute the model space using computational algebra; see Chapter 3 of [6] for a gentle survey.

Now, suppose that our unknown function f_i arose from a model of a molecular network. In other words, the function describes how the nodes v_1, v_2, and v_3, which represent gene products, affect the state of node v_i. In this case, only 1 of these 32 functions best describes the biological interactions. Finding "the correct" function can be a daunting task. Instead, we will focus on simply asking the natural question:

On which node(s) can the function f_i depend?

Of course, one way to answer this question is to compute the entire model space, $h + I$, and record the wiring diagram of each function. However, this can be extremely cumbersome, if not computationally prohibitive. The algorithms described in this chapter provide efficient ways to answer this question in a nonbrute-force method. One natural first step is to eliminate functions that are likely not biologically meaningful, and to just consider unate functions. Of the 32 possible functions for f_i, only 4 are unate, and they are

$$x_1 \wedge \overline{x_3}, \qquad x_2 \wedge \overline{x_3}, \qquad x_1 \wedge x_2 \wedge \overline{x_3}, \qquad (x_1 \vee x_2) \wedge \overline{x_3}.$$

The wiring diagrams of these four functions are following.

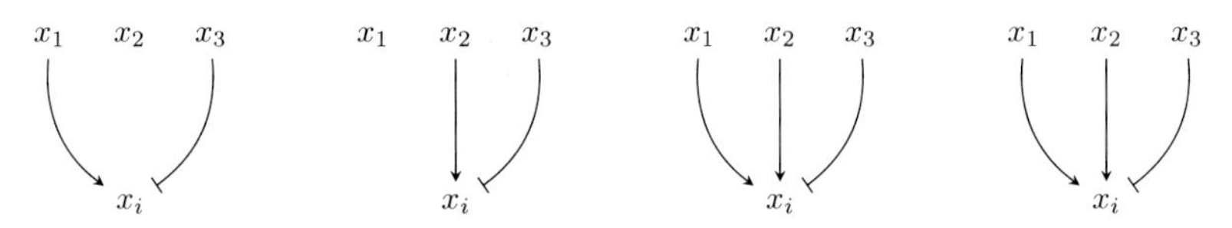

If we just focus on the incoming nodes to vertex x_i, then there are several simple ways to encode this information, whose relative utility depends on the context. For example, we can describe these edges as a triple $(r_1, r_2, r_3) \in \mathbb{F}_3^3$, where $\mathbb{F}_3 = \{0, 1, -1\}$. Specifically, the four wiring diagrams above can be described by the following vectors:

$$(1, 0, -1), \qquad (0, 1, -1), \qquad (1, 1, -1), \qquad (1, 1, -1).$$

We say that the first two sets of signed edges (or vectors) are *minimal sets*, or "*min-sets*" because the signed edges in the last two wiring diagrams are both proper supersets of one of these. Another convenient way to represent this information is by writing the a set of variables that x_i depends on, with a negation for every negative interaction. For example, using this notation, the four wiring diagrams above are

$$\{x_1, \overline{x_3}\}, \qquad \{x_2, \overline{x_3}\}, \qquad \{x_1, x_2, \overline{x_3}\}, \qquad \{x_1, x_2, \overline{x_3}\}.$$

To give a name to the problem that we just solve in our example, we say that we have *reverse engineered the wiring diagram* at node x_i of a BN from partial data. This particular example was small enough that all 32 of the functions that fit the data can be found by hand. Obviously, this is not always the case. In fact, it is quite easy to find relatively small examples that are too large for even a supercomputer to handle. For example, if we had specified three values in the truth table of an unknown *ternary* function $f_i\colon \mathbb{F}_3^3 \to \mathbb{F}_3$, then there would be $3^{3^3-3} \approx 2.82 \times 10^{11}$ functions that fit the data.

In this chapter, we will learn how to reverse engineer wiring diagrams using computational algebra. This will allow us to quickly and efficiently solve such problems for much large networks than we could ever do by brute force. The mathematics behind the scenes involves an area known as *combinatorial commutative algebra* [7], which blends computational algebra [8], commutative algebra [9], and combinatorics. Specifically, we will use ideals of polynomial rings generated by square-free monomials to encode the partial data. There is a beautiful relationship between these ideals and a combinatorial object called a *simplicial complex*. Additionally, the min-sets can be found by taking the *primary decomposition* of the ideal.

The remainder of this chapter is organized as follows. The mathematical background is presented in Section 6.2. Sections 6.3 and 6.4 can be thought of as a survey of the mathematics developed in [10, 11], respectively. Specifically, in Section 6.3, we use these tools to develop the theory needed to reverse engineer the wiring diagram from partial data, but without distinguishing between positive and negative interactions. In other words, we determine how to deduce which variables the local functions can depend on, that is, how to find all *unsigned min-sets*. This involves working with monomial ideals over $\mathbb{F}_2$, and we do not need to impose the restriction that the functions are unate. In Section 6.4, we take this further and learn how to find all of the *signed min-sets*. For this, we will only consider unate functions. This involves working with monomial ideals over $\mathbb{F}_3 = \{0, 1, -1\}$, because we need both $+1$ and -1 $(= 2)$ to distinguish between positive and negative interactions. In Section 6.5, we will apply these ideas to a biological network from the model organism *Caenorhabditis elegans*. Finally, we will conclude in Section 6.6 with a discussion of how to actually find all possible functions that fit the data. Though this sounds like a more challenging problem, solving it actually involves less advanced computational algebra. That said, the solution to this problem is arguably less useful than for just reverse engineering the wiring diagram, which we will discuss in Section 6.6 as well.

6.2 STANLEY-REISNER THEORY

For the mathematics in this section, $\mathbb{F}$ can be any arbitrary field. It is only when we apply this theory to local models in subsequent sections that we will specialize to finite fields. Let $R = \mathbb{F}[x_1, \ldots, x_n]$ denote the polynomial ring in n variables. Though the variables come with a natural order, $x_1, \ldots, x_n$, if $n = 2$ or $n = 3$, we will often use x, y or x, y, z for convenience. A quick review of polynomial rings and ideals, especially the definitions and results relevant to this material, was provided in Section 4.3.1.

6.2.1 Monomial Ideals

A *monomial* is just a product of variables, possibly with a constant in front. Elements in R are just sums of monomials. Every monomial can be written as cx^α, where $x^\alpha := x_1^{\alpha_1} \cdots x_n^{\alpha_n}$ and $\alpha = (\alpha_1, \ldots, \alpha_n) \in \mathbb{Z}_{\geq 0}^n$ is the *exponent*. This is best seen with an example.

Example 6.1. Consider the polynomial ring $R = \mathbb{F}_3[x_1, x_2, x_3, x_4]$, and the polynomial $f \in R$ below written in several different forms:

$$f = x_1^3x_2x_4^2 + 2x_1x_4^5 = x_1^3x_2^1x_3^0x_4^2 + 2x_1^1x_2^0x_3^0x_4^5 = x^{(3,1,0,2)} + 2x^{(1,0,0,5)}.$$

If I is an ideal of R, then we will write $I \leq R$, and use angle brackets $\langle , \rangle$ rather than parentheses $(,)$ when speaking of a generating set. The majority of ideals that we will encounter here are appropriately called "monomial ideals." There is an extensive theory of these ideals; see the recent book [9] for more information.

Definition 6.4. A *monomial ideal* $I \leq \mathbb{F}[x_1, \dots, x_n]$ is an ideal generated by monomials. We write this as $I = \langle x^\alpha, x^\beta, \dots \rangle$.

Let $\mathcal{M}(I)$ be the set of monomials in I. The following result is surprisingly useful given how straightforward it is.

Proposition 6.1. *If I is a monomial ideal, then $I = \langle \mathcal{M}(I) \rangle$.*

Exercise 6.2. Prove Proposition 6.1.

By Proposition 6.1, every monomial ideal is completely determined by the set of monomials it contains. This can be visualized by a *staircase diagram*, which is also best seen by an example.

Example 6.2. Consider the monomial ideal $I = \langle y^3, xy^2, x^3y^2, x^4 \rangle$ in $\mathbb{F}[x, y]$. To construct the staircase diagram, we draw the nonnegative integer grid in the xy-plane, as shown in Fig. 6.1. The point with coordinates (i, j) represents the monomial x^iy^j. Every monomial that lies above and to the right of x^iy^j is in I, and so we shade this region for each generator of I. Note that the monic (i.e.,

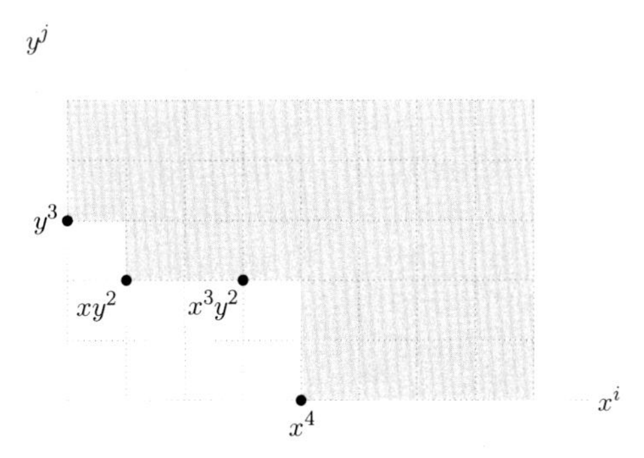

FIG. 6.1 The staircase diagram of the monomial ideal $I = \langle y^3, xy^2, x^3y^2, x^4 \rangle$ in $\mathbb{F}[x, y]$. The monomials x^iy^j that are *not* in I are those corresponding to the integer lattice points (i, j) that are not in the *shaded region*.

having coefficient 1) monomials that are *not* in I are those that are not in the shaded region. In this example, those are

$$1, x, x^2, x^3, y, xy, x^2y, x^3y, y^2.$$

It should be clear from Fig. 6.1 that the monomial x^3y^2 is not needed to generate I, and thus

$$I = \langle y^3, xy^2, x^3y^2, x^4 \rangle = \langle y^3, xy^2, x^4 \rangle.$$

Staircase diagrams are only easy to draw when the polynomial ring has two variables, or in some simple cases, three. However, they are useful because they provide a "visual proof" of Proposition 6.1, how every monomial ideal is uniquely determined by the set of monomials in I. Said differently, every monomial ideal is uniquely determined by its staircase. It is also clear that there is a unique minimal monomial generating set for I; just take the monomials corresponding to the "corners" of the staircase.

This example illustrates the connection between monomial ideals and combinatorics. This interplay will only get stronger, especially as we focus on a special type of monomial ideal in the next section that arises in our biological application.

Exercise 6.3. Draw the staircase diagram of the monomial ideal $I = \langle y^2, x^2y^3, x^3y^2, x^3y, x^5 \rangle$ of $\mathbb{F}[x, y]$ and find a minimal set of generators. Which (monic) monomials are *not* in I?

Exercise 6.4. Suppose I and J are two monomial ideals in $\mathbb{F}[x, y]$. Describe how to construct the staircase diagrams of $I \cap J$ and $\langle I \cup J \rangle$ from the staircase diagrams of I and J.

6.2.2 Square-Free Monomial Ideals

A special type of monomial ideal that arises in a number of applications are those that are *square-free*. These are defined exactly as their name suggests: the monomial x^α is square-free if each $\alpha_i \in \{0, 1\}$, and a monomial ideal I is square-free if can be generated by square-free monomials. These are also called *Stanley-Reisner ideals*. Clearly, the exponent vector $\alpha = (\alpha_1, \ldots, \alpha_n)$ canonically determines a subset of $[n] = \{1, \ldots, n\}$; namely, the coordinates i whose entry α_i is nonzero. Though it is a slight abuse of notation, we will often speak of α as a *subset* of $[n]$ rather than a vector for convenience.

Remark 6.2. For many of our examples, we will work in the ring $\mathbb{F}[x, y, z]$ instead of $\mathbb{F}[x_1, x_2, x_3]$; the variables x, y, z are canonically identified with x_1, x_2, x_3. In this setting, it is more convenient to write subsets as strings instead of vectors. For example, xz is short for $\{x, z\}$—we can express both the subset and the monomial with $\alpha = xz$. This has the added advantage that we no longer need to distinguish between subsets, exponent vectors, and monomials.

For example, in $\mathbb{F}[x_1, x_2, x_3]$, we would write a monomial as $x^\alpha = x^{(1,0,1)} = x_1x_3$, where $\alpha = \{1, 3\} \subseteq [3]$ is a subset, or $\alpha = (1, 0, 1)$ is an exponent vector. In $\mathbb{F}[x, y, z]$, we just write $\alpha = xz$ for both the subset and the monomial, which leaves the exponent vector unnecessary. Though this is a slight abuse of notation, it works well and greatly simplifies things. It should always be clear from the context what α represents.

To summarize, every square-free monomial ideal is minimally generated by a unique collection of monomials, each of which is described by a subset $\alpha \subseteq [n]$. In other words, every square-free monomial ideal can be uniquely described by a canonical collection of subsets of $[n]$. Moreover, this collection of subsets is *not* arbitrary; it has the following key properties, which we will return to soon.

Proposition 6.2. *Let I be a square-free monomial of $\mathbb{F}[x_1 \dots, x_n]$, and $\alpha, \beta \subseteq [n]$. Then*

$$x^\alpha \in I \quad \text{and} \quad x^\beta \in I \quad \Longrightarrow \quad x^{\alpha\cup\beta} \in I,$$
$$x^\alpha \notin I \quad \text{and} \quad x^\beta \notin I \quad \Longrightarrow \quad x^{\alpha\cap\beta} \notin I.$$

Exercise 6.5. Prove Proposition 6.2.

As a direct consequence of Proposition 6.2, every square-free monomial ideal can be expressed uniquely using a very simple combinatorial diagram called a *simplicial complex*. One can think of a simplicial complex as a structure built with triangular "faces" in various dimensions, which we call its *geometric realization*. A k-dimensional face is called a *k-face*, but for small k, these have more familiar names:

- A 0-dimensional face is a vertex, or node.
- A 1-dimensional face is an edge.
- A 2-dimensional face is a triangle.
- A 3-dimensional face is a (solid) triangular pyramid.

If a simplicial complex has a two-face (a triangle), then it must have the three edges and three vertices contained in that triangle. Formally, a k-face is just any size-$(k+1)$ subset, and simplicies must be closed under taking subsets. This is the only restriction; the formal definition is following.

Definition 6.5. An abstract *simplicial complex* over a finite set X is a collection Δ of subsets of X that are closed under the operation of taking subsets. That is, if $\beta \in \Delta$ and $\alpha \subset \beta$, then $\alpha \in \Delta$. The elements in Δ are called *simplices* or *faces*.

Most simplicial complexes in this chapter will be over $[n]$ or over a set of letters (e.g., $\{a, b, c, d, e\}$). Recall that we usually prefer to write subsets as strings, for example,

$$\{\emptyset, a, b, c, d, ab, ac, bc, abc\} \quad \text{instead of} \quad \{\emptyset, \{a\}, \{b\}, \{c\}, \{d\}, \{a, b\}, \{a, c\}, \{b, c\}, \{a, b, c\}\}.$$

Clearly, in this context, the order of the letters in a string does not matter, so $abc = acb = \cdots = cba$.

We will now go over two examples: one on six variables that is large enough to see some intricacies, and a three-variable example that is small enough that we can visualize key features on the three-variable Boolean lattice.

Running Example 6.1. Consider the simplicial complex Δ on $X = \{a, b, c, d, e, f\}$ shown below in two ways: as a collection of strings (subsets) at left, and its geometric realization at right.

$$\Delta = \{\emptyset, a, b, c, d, e, f, bc, cd, ce, de, cde, df, ef\}$$

The triangle cde is shaded to indicate that $cde \in \Delta$, whereas the unshaded triangle def means that $def \notin \Delta$.

For reasons that will become clear soon, we will often be interested in the *nonfaces* of a simplicial complex. These are the subsets that are *not* in Δ. Clearly, if Δ is a simplicial complex over X, then the collection 2^X of all subsets of X is the disjoint union $2^X = \Delta \sqcup \Delta^c$, where $\Delta^c := 2^X \setminus \Delta$ is the *set complement* of Δ. Thus, every subset of X is either a face or a nonface. The following easy observation is crucial.

Remark 6.3. Let Δ be a simplicial complex.

(i) Faces of Δ are closed under intersection: $\alpha, \beta \in \Delta \Rightarrow \alpha \cap \beta \in \Delta$.
(ii) Nonfaces of Δ are closed under unions: $\alpha, \beta \in \Delta^c \Rightarrow \alpha \cup \beta \in \Delta^c$.

Due to Remark 6.3, we do not need to specify all of the faces (or nonfaces) of Δ to completely determine it. It should be clear that every simplicial complex is uniquely defined by its *maximal faces*, also called *facets*. Formally, we say that a face $\alpha \subseteq [n]$ of Δ is *maximal* if there is no $\beta \supsetneq \alpha$ in Δ. In terms of the geometric realization, the facets of a simplicial complex are what one would use if building them from physical materials.

Just as Δ is determined by its maximal faces, Δ^c is determined by its *minimal nonfaces*. Formally, a nonface $\alpha \subseteq [n]$ of Δ^c is *minimal* if there is no $\beta \subsetneq \alpha$ in Δ^c. The following three-node example should help illustrate these concepts.

Running Example 6.2. Consider the following simplicial complex Δ over $X = \{x, y, z\}$, where the geometric realization is shown on the right.

$$\text{Faces: } \Delta = \{\emptyset, x, y, z, xz\} \quad (\text{maximal: } y, xz)$$

$$\text{Nonfaces: } \Delta^c = \{xy, yz, xyz\} \quad (\text{minimal: } xy, yz)$$

Since this example is small enough, it can be nicely visualized on the 3D Boolean lattice, as shown in Fig. 6.2. On the left is the full lattice with the faces in Δ circled, and the maximal faces shaded. In the middle diagram, the nonfaces are boxed with the minimal nonfaces shaded. The edges connecting the faces with nonfaces are dotted to emphasize the partition of the vertices into $\Delta \sqcup \Delta^c$.

We added arrows in Fig. 6.2 to emphasize how the faces are precisely those vertices that lie on a downward path from the maximal faces, y and xz. Similarly, the nonfaces are precisely those that lie on an upward path from the minimal nonfaces, xy and yz, and this is shown in the middle diagram of Fig. 6.2. Formally, we say that the faces form a *down-set* of the Boolean lattice 2^X generated by $\{xz, y\}$, and the nonfaces form an *up-set* generated by $\{xy, yz\}$.

The diagram on the right of Fig. 6.2 shows the complements $\overline{\alpha} := [n] \setminus \alpha$ of the faces $\alpha \in \Delta$. Soon, we will return to this diagram when we learn how to encode this information into an ideal of $\mathbb{F}[x, y, z]$. Note that we are using different notations to help distinguish between the two different

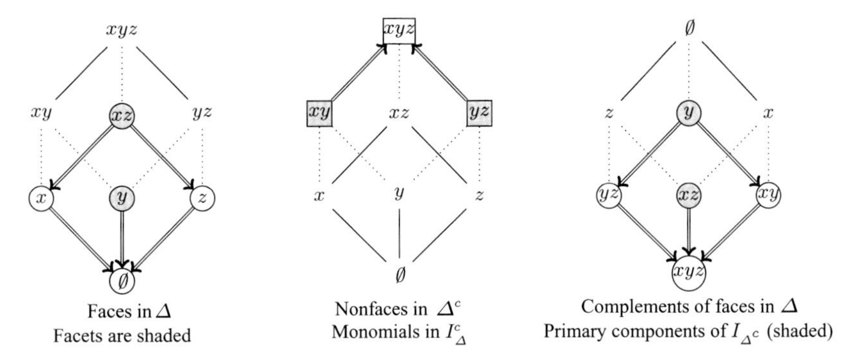

FIG. 6.2 The partition of the Boolean lattice on $X = \{x, y, z\}$ into the simplicial complex $\Delta = \{\emptyset, x, y, z, xz\}$ (a down-set) and its complement $\Delta^C = \{xy, yz, xyz\}$ (an up-set). *At right* are the complements of the faces of Δ.

set-complements that arise in this chapter: we write the complement of a face $\alpha \in \Delta$ as $\overline{\alpha} := [n] \setminus \alpha$, and the complement of a simplicial complex $\Delta \subseteq 2^X$ as $\Delta^c := 2^X \setminus \Delta$.

We will now return to our first running example and find the maximal faces and minimal nonfaces. The same structure of the Boolean lattice 2^X being partitioned into an up-set and a down-set holds, but it is too large to easily draw, as it would contain 64 vertices.

Running Example 6.1 (continued). Let $\Delta = \{\emptyset, a, b, c, d, e, f, bc, cd, ce, de, cde, df, ef\}$. There are $2^6 = 64$ subsets of a six-element set, and Δ has 14 faces. Hence, there are 50 nonfaces in Δ^c. The maximal faces (facets) and minimal nonfaces are listed below.

- *Maximal faces*: a, bc, cde, df, ef.
- *Minimal nonfaces*: $ab, ac, ad, ae, af, bd, be, bf, cf, def$.

Note that by construction, every proper subset of a minimal nonface is in Δ, and every proper superset of a maximal face is in Δ^c.

Now, let us turn back to square-free monomial ideals. Recall that every monomial ideal in $R = \mathbb{F}[x_1, \ldots, x_n]$ is uniquely determined by the (monic) monomials x^α it contains, and so every square-free monomial ideal I is uniquely determined by a collection of subsets of $[n]$.

Remark 6.4. Let I be a square-free monomial ideal, which is completely determined by the subsets α for which $x^\alpha \in I$.

- If $\alpha \subseteq \beta$ and $x^\alpha \in I$, then $x^\beta \in I$.
- If $\alpha \subseteq \beta$ and $x^\beta \notin I$, then $x^\alpha \notin I$.

In other words,

(i) As subsets, exponents of square-free monomials in I are closed under unions.
(ii) As subsets, exponents of square-free monomials *not* in I are closed under intersections.

By Remarks 6.3 and 6.4, we can describe a square-free monomial ideal I combinatorially as a collection of subsets that is closed under intersections. These subsets have two interpretations, one algebraic and one combinatorial.

- *algebraically*: the monomials x^α not in I; and
- *combinatorially*: the faces α of a simplicial complex, that we will denote by Δ_{I^c}.

To summarize, every square-free monomial ideal I canonically defines a simplicial complex Δ_{I^c}, where the (monic) monomials in I correspond to the nonfaces of Δ_{I^c}, that is, the elements of $\Delta_{I^c}^c$. Or equivalently, the faces of Δ_{I^c} correspond to the (monic) monomials *not* in I. This correspondence is bijective, as every simplicial complex Δ canonically defines a square-free monomial ideal. The formal definition is below.

Definition 6.6. Given an ideal I, define the simplicial complex Δ_{I^c} as

$$\Delta_{I^c} = \left\{\alpha \mid x^\alpha \notin I\right\}.$$

Given a simplicial complex Δ on $[n]$, define a square-free monomial ideal I_{Δ^c} as

$$I_{\Delta^c} = \left\langle x^\alpha \mid \alpha \notin \Delta\right\rangle.$$

This is called the *Stanley-Reisner ideal* of Δ. The following theorem highlights the interplay between the combinatorial and algebraic structure of these objects. See [7, Section 1.1] for more details, including proofs.

Theorem 6.1. *The correspondence $I \mapsto \Delta_{I^c}$ and $\Delta \mapsto I_{\Delta^c}$ is a bijection between:*

(i) *simplicial complexes on $[n] = \{1, \ldots, n\}$, and*
(ii) *square-free monomial ideals in $\mathbb{F}[x_1, \ldots, x_n]$.*

Remark 6.5. Our notation in Definition 6.6 is nonstandard. Usually, the correspondence is denoted $I \mapsto \Delta_I$ and $\Delta \mapsto I_\Delta$. We choose to retain the small c to remind the reader that one must take the complement; this is sometimes called *Alexander duality*. The authors remember first-hand how confusing this can be initially! However, one can read right off of our notation that:

- I_{Δ^c} is the ideal generated by the monomials not in Δ.
- Δ_{I^c} is the simplicial complex whose faces are the monomials not in I.

At this point, we will return to our first running example.

Running Example 6.1 (continued). Consider the simplicial complex Δ on $X = \{a, b, c, d, e, f\}$ shown below in two ways: as a collection of strings (subsets) at left, and its geometric realization at right.

$\Delta = \{\emptyset, a, b, c, d, e, f, bc, cd, ce, de, cde, df, ef\}$

Maximal faces: a, bc, cde, df, ef

Minimal nonfaces: ab, ac, ad, ae, af, bd, be, bf, cf, def

$I_{\Delta^c} = \langle ab, ac, ad, ae, af, bd, be, bf, cf, def\rangle$

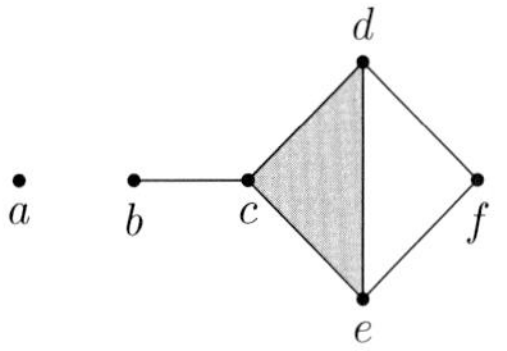

6.2.3 Primary Decompositions

A common theme in mathematics is to decompose, or break up, complicated objects into smaller manageable pieces. This often leads to insight about the structure of the original object. For example, every positive integer can be factored into a product of prime numbers. Given a simplicial complex such as the ones in our Running Examples, it is visually obvious how to "decompose" it into smaller pieces. Your eyes do this for you when you look at the geometric realization and see the maximal faces. Our next goal is to decompose the algebraic object I_{Δ^c} into simpler pieces, called *primary components*. We will begin with the punchline: these pieces correspond to the maximal faces α in Δ, but we have to take the complement $\overline{\alpha}$ of each one first. The following example should motivate this.

Running Example 6.2 (continued)**.** Recall the simplicial complex $\Delta = \{\emptyset,\ x, y, z, xz\}$. The Stanley-Reisner ideal I_{Δ^c} of $R = \mathbb{F}[x, y, z]$ is generated by the minimal nonfaces:

$$I_{\Delta^c} = \langle xy, yz\rangle = \{xy \cdot h_1(x, y, z) + yz \cdot h_2(x, y, z) \mid h_1, h_2 \in R\}.$$

Note that both of the terms in the expression above contain y. Thus, we can factor y out of it and write

$$xy \cdot h_1(x, y, z) + yz \cdot h_2(y, z) = y\,(x \cdot h_1(x, y, z) + z \cdot h_2(x, y, z)) \in \langle y\rangle \cap \langle x, z\rangle.$$

This shows that $\langle xy, yz\rangle \subseteq \langle y\rangle \cap \langle x, z\rangle$, and it is not hard to show that these two sets are in fact equal. Notice how these arose from the maximal faces of Δ, which are y and xz.

The expression of the ideal $\langle xy, yz\rangle$ as an intersection of the two ideals $\langle y\rangle$ and $\langle x, z\rangle$ is called a *primary decomposition*. An ideal $I \subset \mathbb{F}[x_1, \ldots, x_n]$ is *prime* if $fg \in I$ implies either $f \in I$ or $g \in I$. Unlike how integers can be factored into primes, it is not the case that any ideal can be decomposed into (an intersection of) prime ideals. However, something similar is true, though it requires a generalization of the notion of an ideal being prime. An ideal I is *primary* if $fg \in I$ implies either $f \in I$ or $g^m \in I$ for some power $m > 0$. In the ring $\mathbb{Z}$ of integers, prime ideals are of the form $\langle p\rangle = p\mathbb{Z}$, and primary ideals are of the form $\langle p^k\rangle = p^k\mathbb{Z}$, where p is a prime number. Just like how an integer can be decomposed into a product of prime powers, an ideal I can be decomposed into an intersection of primary ideals, called a *primary decomposition* of I.

Theorem 6.2 (Lasker-Noether Theorem)**.** *Every ideal I of $\mathbb{F}[x_1, \ldots, x_n]$ can be written as $I = \bigcap_{i=1}^{r} \mathfrak{p}_i$, where $\mathfrak{p}_i$ is a primary ideal.*

More information about the primary decomposition, including a proof of the Lasker-Noether theorem can be found in [8, Section 4.7]. It actually holds not just for polynomial rings, but a larger class called *Noetherian rings*, which we will not define here. The question of how to compute a primary decomposition is a difficult problem from algebraic geometry, and is also outside of the scope of this chapter. Also, it is generally *not* unique. The important point here is that for square-free monomial ideals, the answer has a simple combinatorial description. To establish the result, we must first provide the following well-known concepts for varieties of an ideal, that is, the set of common zeros to of all of the polynomials.

Definition 6.7. Given an ideal $I \leq \mathbb{F}[x_1, \ldots, x_n]$, the *variety* of I is its set of common zeros:

$$V(I) = \{x \in \mathbb{F}^n \mid f(x) = 0 \text{ for all } f \in I\}.$$

Even though computing the variety of a general ideal can be difficult, it is quite straightforward for square-free monomial ideals. Our focus here is a little different; however, we want to decompose ideals into intersections of simpler ideals. For example, the following result is straightforward.

Lemma 6.1. *For any two varieties V_1 and V_2 in $\mathbb{F}^n$,*

$$I(V_1 \cup V_2) = I(V_1) \cap I(V_2).$$

Our problem will basically be a special case of this for square-free monomial ideals. For any $\alpha \subseteq [n]$, define $\mathfrak{p}^{\alpha} = \langle x_i \mid i \in \alpha \rangle$ and $\mathfrak{p}^{\overline{\alpha}} = \mathfrak{p}^{[n]-\alpha} = \langle x_i \mid i \notin \alpha \rangle$. It is easily checked that the ideal $\mathfrak{p}^{\alpha}$ (and thus $\mathfrak{p}^{\overline{\alpha}}$ as well) is prime, and therefore primary.

Proposition 6.3. *Let Δ be a simplicial complex over $[n]$. The Stanley-Reisner ideal of Δ in $R = \mathbb{F}[x_1, \ldots, x_n]$ is*

$$I_{\Delta^c} = \bigcap_{\alpha \in \Delta} \mathfrak{p}^{\overline{\alpha}} = \bigcap_{\substack{\alpha \in \Delta \\ \text{maximal}}} \mathfrak{p}^{\overline{\alpha}}. \tag{6.4}$$

Proof. See [8]. □

The intersection on the right of Eq. (6.4) is called the *primary decomposition*[1] of I_{Δ^c}, and the ideals $\mathfrak{p}^{\overline{\alpha}}$ are its *primary components*, or *minimal primes*.

1. Though primary decompositions are not unique in general, for square-free monomial ideals they can be considered to be unique up to scalar multiplication.

We will conclude this section by computing the primary decomposition of our two running familiar examples.

Running Example 6.2 (continued). Recall the simplicial complex $\Delta = \{\emptyset, x, z, y, xz\}$, whose maximal faces are y and xz. The Stanley-Reisner ideal in $R = \mathbb{F}[x, y, z]$ is generated by the nonfaces:

$$I_{\Delta^c} = \langle xy, yz, xyz\rangle = \langle xy, yz\rangle.$$

The primary decomposition of I_{Δ^c} is generated by the individual *complements* of the five faces in Δ. Note the subtle difference between this and the *set complement* of Δ, which are the three nonfaces $\Delta^c = \{xy, yz, xyz\}$. The complements of the faces are

$$\{\overline{\emptyset}, \overline{x}, \overline{z}, \overline{y}, \overline{xz}\} = \{xyz, yz, xy, xz, y\}.$$

By Proposition 6.3, the primary decomposition of I_{Δ^c} is

$$\begin{aligned} I_{\Delta^c} = \langle xy, yz\rangle = \bigcap_{\alpha\in\Delta} \mathfrak{p}^{\overline{\alpha}} &= \mathfrak{p}^{\overline{\emptyset}} \cap \mathfrak{p}^{\overline{x}} \cap \mathfrak{p}^{\overline{z}} \cap \mathfrak{p}^{\overline{y}} \cap \mathfrak{p}^{\overline{xz}} \\ &= \mathfrak{p}^{xyz} \cap \mathfrak{p}^{yz} \cap \mathfrak{p}^{xy} \cap \mathfrak{p}^{xz} \cap \mathfrak{p}^{y} \\ &= \underbrace{\langle x, y, z\rangle \cap \langle y, z\rangle \cap \langle x, y\rangle}_{\text{unnecessary}} \cap \langle x, z\rangle \cap \langle y\rangle \\ &= \langle x, z\rangle \cap \langle y\rangle = \bigcap_{\substack{\alpha\in\Delta \\ \text{maximal}}} \mathfrak{p}^{\overline{\alpha}}. \end{aligned}$$

Note that the three ideals that arose from the complements of nonmaximal faces are unnecessary. In this example, the set of complements of the maximal faces happens to be the same as the maximal faces themselves, though this does not typically happen.

Running Example 6.1 (continued). Recall the simplicial complex $\Delta = \{\emptyset, a, b, c, d, e, f, bc, cd, ce, de, cde, df, ef\}$ over $X = \{a, b, c, d, e, f\}$. The maximal nonfaces and their complements are the following.

- *Maximal faces*: a, bc, cde, df, ef.
- *Complements of maximal faces*: $bcdef$, $adef$, abf, $abce$, $abcd$.
- *Minimal nonfaces*: ab, ac, ad, ae, af, bd, be, bf, cf, def.

The Stanley-Reisner ideal I_{Δ^c} is generated by the (minimal) nonfaces. The primary decomposition is the intersection of ideals, one for the complement of each maximal face:

$$\begin{aligned} I_{\Delta^c} &= \langle ab, ac, ad, ae, af, bd, be, bf, cf, def\rangle = \bigcap_{\substack{\alpha\in\Delta \\ \text{maximal}}} \mathfrak{p}^{\overline{\alpha}} \\ &= \langle b, c, d, e, f\rangle \cap \langle a, d, e, f\rangle \cap \langle a, b, f\rangle \cap \langle a, b, c, e\rangle \cap \langle a, b, c, d\rangle. \end{aligned}$$

The following exercise involves a simplicial complex that will arise later in this chapter when we revisit our original BN example from Section 6.1.2.

Exercise 6.6. In this exercise, you will repeat what we did for our two running examples but with the simplicial complex $\Delta = \{\emptyset, x, y\}$ over the three-element set $X = \{x, y, z\}$.

(i) Sketch the simplicial complex and find the maximal faces.
(ii) Circle each node in the Boolean lattice 2^X corresponding to a face of Δ, and additionally shade in those faces that are maximal.
(iii) Box each node that corresponds to a nonface, and shade in those that are minimal.
(iv) Find the Stanley-Reisner ideal I_{Δ^c} and compute its primary decomposition.

Exercise 6.7. Repeat Exercise 6.6 for the simplicial complex $\Delta = \{\emptyset, a, b, c, d, e, ab, ac, bc, bd, cd, abc\}$ over the set $X = \{a, b, c, d, e\}$. Fig. 6.3 shows the five-dimensional Boolean lattice 2^X.

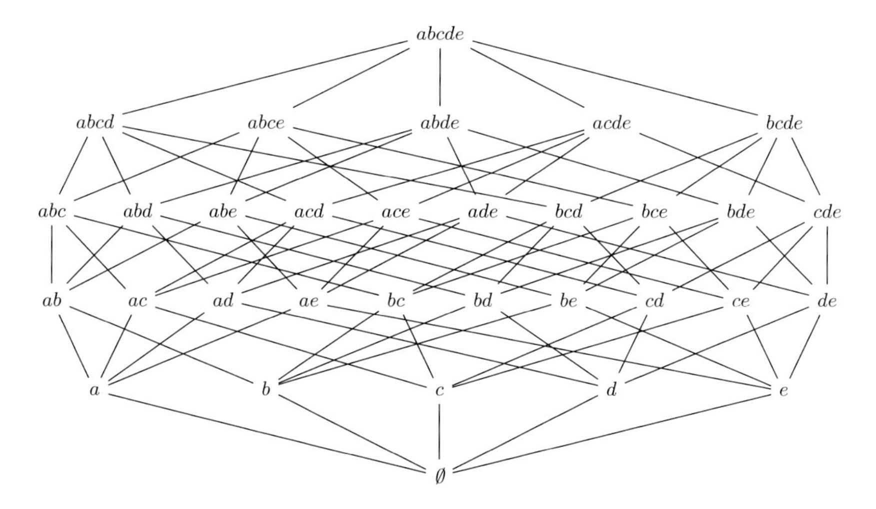

FIG. 6.3 The five-dimensional Boolean lattice for Exercise 6.7.

6.3 FINDING MIN-SETS OF LOCAL MODELS

6.3.1 Wiring Diagrams

In this section, we will apply the algebraic tools introduced in the previous section to the problem of reverse-engineering a wiring diagram of a local model. To summarize the process, we will consider every pair of input vectors that give a different output, and for each one, take the square-free monomial x^{α}, where $\alpha \subseteq [n]$ is the set of variables in which the entries differ. Note that the variable(s) that caused the change in the output are contained in α. The primary decomposition of the ideal generated by these monomials will encode all possible minimal wiring diagrams.

The wiring diagram of a local model $f = (f_1, \ldots, f_n)$ is a collection of signed edges, and it is the disjoint union of the incoming edges at each node. The following example should make this clear.

Example 6.3. Consider the following Boolean local model $f = (f_1, f_2, f_3)$.

$$\begin{cases} f_1 = x_2 \\ f_2 = x_2 \wedge x_3 \\ f_3 = x_1 \vee \overline{x_2} \end{cases}$$

x_2 x_1 x_3 = x_2 x_1 x_3 ∪ x_2 x_1 x_3 ∪ x_2 x_1 x_3

Its wiring diagram can be decomposed into three directed graphs, where each component corresponds to one of the coordinate functions. The first component represents f_1, the second f_2, and the third f_3. For example, since f_1 (the function describing the behavior of x_1) is written in terms of x_2, then its component contains the single edge $x_2 \to x_1$.

It suffices to solve the problem for each component separately. For example, if we wish to reverse engineer the wiring diagram of a local model $f: \mathbb{F}^n \to \mathbb{F}^n$, where $f = (f_1, \ldots, f_n)$, then all we need to do is solve the problem for each $f_1, \ldots, f_n$ separately. Thus, we will consider a function $f: \mathbb{F}^n \to \mathbb{F}$ with partial data, and attempt to reverse-engineer its wiring diagram.

Exercise 6.8. For the BN in Example 6.3, express the variables that each f_i depends on in two different ways: (i) as a vector in $\mathbb{F}_3^3 = \{0, 1, -1\}^3$, and (ii) as a signed min-set, that is, a subset of $\{x_1, x_2, x_3, \overline{x_1}, \overline{x_2}, \overline{x_3}\}$. See the example in Section 6.1.2 for a reminder of how to do this.

6.3.2 Feasible and Disposable Sets of Variables

In the remainder of this section, we will see how to find the unsigned min-sets (i.e., reverse-engineering the unsigned wiring diagrams) of a function. In Section 6.4, we will extend this to signed min-sets. Let $f: \mathbb{F}^n \to \mathbb{F}$ be a function, where $\mathbb{F} = \mathbb{F}_p$. Consider a set

$$\mathcal{D} = \{(\mathbf{s}_1, t_1), \ldots, (\mathbf{s}_m, t_m)\}$$

of state-transition pairs, where all input vectors $\mathbf{s}_i \in \mathbb{F}^n$ are distinct, though the output values $t_i \in \mathbb{F}$ need not be. We call such a set *data*, and we say that f *fits the data* $\mathcal{D}$ if

$$f(\mathbf{s}_i) = f(s_{i1}, \ldots, s_{in}) = t_i, \qquad \text{for all} \quad i = 1, \ldots, m.$$

The *model space* of $\mathcal{D}$ is the set $\mathsf{Mod}(\mathcal{D})$ of all functions that fit the data, that is,

$$\mathsf{Mod}(\mathcal{D}) = \left\{ f : \mathbb{F}^n \to \mathbb{F} \mid f(\mathbf{s}_i) = t_i \qquad \text{for all} \quad i = 1, \ldots, m \right\}.$$

For any f in $\mathsf{Mod}(\mathcal{D})$, the *support* of f, denoted supp(f), is the set of variables on which f depends. Under a slight abuse of notation, we can think of the support as a subset of $\{x_1, \ldots, x_n\}$ or as a subset $\alpha \subseteq [n] = \{1, \ldots, n\}$, which as before, we usually write as a string.

Definition 6.8. With respect to a set $\mathcal{D}$ of data, a set $\alpha \subseteq [n]$ is:

- *feasible* if there is some $f \in \mathsf{Mod}(\mathcal{D})$ for which supp(f) $\subseteq \alpha$.
- *disposable* if there is some $f \in \mathsf{Mod}(\mathcal{D})$ for which supp(f) $\cap\, \alpha = \emptyset$.

Note that a set α is feasible if and only if its complement $\overline{\alpha} := [n] - \alpha$ is disposable.

We chose the names "feasible" and "disposable" to best represent what they mean in plain English. For example, the set α is feasible with respect to $\mathcal{D}$ if there is some f that fits the data and depends only on the variables from α. Similarly, α is disposable if there is some function f that fits the data, none of whose variables are from α. Note that these are *not* opposite concepts, that is, a set can be both feasible and disposable, or neither. We will see an example of this soon. We will use the terms *infeasible* and *nondisposable* in the obvious way. The following result is straight-forward.

Proposition 6.4. *Let $\mathcal{D}$ be a set of data, and $\alpha, \beta \subseteq [n]$.*

(i) *If α and β are feasible with respect to $\mathcal{D}$, then so is $\alpha \cup \beta$.*
(ii) *If α and β are disposable with respect to $\mathcal{D}$, then so is $\alpha \cap \beta$.*

Since the disposable sets of $\mathcal{D}$ are closed under intersection, they form a simplicial complex $\Delta_{\mathcal{D}}$.

Recall that our goal is to find all *min-sets*; minimal sets of variables on which the function f can depend. This is easiest to define in terms of complements of disposable sets.

Definition 6.9. A subset $\alpha \subseteq [n]$ is a *min-set* of $\mathcal{D}$ if its complement $\overline{\alpha} := [n] - \alpha$ is a maximal disposable set of $\mathcal{D}$.

We will find all min-sets by employing Stanley-Reisner theory, which in this setting says the following.

Proposition 6.5. *There is a bijective correspondence between:*

- *the simplicial complex $\Delta_{\mathcal{D}}$ of disposable sets, and*
- *the square-free monomial ideal $I_{\Delta^c_{\mathcal{D}}}$ in $\mathbb{F}[x_1, \ldots, x_n]$ of nondisposable sets.*

In other words, α is a min-set of $\mathcal{D}$ if and only if $\overline{\alpha}$ is a maximal disposable set, and

$$x^\alpha \in I_{\Delta^c_{\mathcal{D}}} \quad \text{if and only if } \alpha \text{ is nondisposable.}$$

The next task is determining which monomials lie in $I_{\Delta^c_{\mathcal{D}}}$. Clearly, this is generated by x^α such that α is a *minimal* nondisposable set. As such, it suffices to check each pair of input vectors $\mathbf{s}$ and $\mathbf{s}'$ in $\mathcal{D}$ that have different outputs, $t \neq t'$ (i.e., $f(\mathbf{s}) \neq f(\mathbf{s}')$). For each coordinate in which $\mathbf{s}$ and $\mathbf{s}'$ differ, we cannot eliminate the possibility that f depends on that variable. We encode this by defining the monomial

$$m(\mathbf{s}, \mathbf{s}') := \prod_{s_i \neq s'_i} x_i.$$

Note that by construction, $\text{supp}(m(\mathbf{s}, \mathbf{s}'))$ is nondisposable. Moreover, all nondisposable sets arise in this manner.

Proposition 6.6. *The ideal of nondisposable sets is the ideal in $\mathbb{F}_2[x_1, \ldots, x_n]$ defined by*

$$I_{\Delta^c_{\mathcal{D}}} = \langle m(\mathbf{s}, \mathbf{s}') \mid t \neq t' \rangle.$$

Remark 6.6. Since we are only interested in which monomials x^α lie in $I_{\Delta^c_{\mathcal{D}}}$, it does not matter what field we work over. Thus, we can safely work over $\mathbb{F}_2$, even if the local model $f: \mathbb{F}^n \to \mathbb{F}$ we are trying to reverse-engineer is over a different field. The Boolean fields particularly convenient because a monomial is monic if and only if it is nonzero.

The ideal of nondisposable sets $I_{\Delta^c_{\mathcal{D}}}$ is the Stanley-Reisner ideal of the simplicial complex $\Delta_{\mathcal{D}}$ of disposable sets. As such, the primary decomposition gives the complements of the maximal faces in $\Delta_{\mathcal{D}}$, which by definition are the min-sets.

Proposition 6.7. *The generators of the primary components of the ideal of nondisposable sets $I_{\Delta^c_{\mathcal{D}}}$ are the min-sets of $\mathcal{D}$ (see Proposition 6.3).*

At this point, we will pause to do several examples. Both of these examples have been carefully chosen to match examples we have previously seen in this chapter. For each of these, we are working in the ring $\mathbb{F}_2[x, y, z]$. Also, recall that we frequently write subsets as strings, for example, xy means $\{x, y\}$. We will begin with an example that we first saw in Section 6.1.2, and also arose in Exercise 6.6. As such, we will call it Running Example 6.3.

Running Example 6.3. Consider a three-variable Boolean function $f\colon \mathbb{F}_2^3 \to \mathbb{F}_2$ with the following partial data:

xyz	111	000	110
f(x,y,z)	0	0	1

Using our notation, the *data* $\mathcal{D}$, grouped by output value, is

$$\mathcal{D} = \{(\mathbf{s}_1, t_1), (\mathbf{s}_2, t_2), (\mathbf{s}_3, t_3)\} = \{(111, 0), (000, 0), (110, 1)\}.$$

There are $2^5 = 32$ functions that fit this data. Note that $t_1 = t_2 \neq t_3$, so we need to compute $m(\mathbf{s}_1, \mathbf{s}_3)$ and $m(\mathbf{s}_2, \mathbf{s}_3)$. Since $\mathbf{s}_1$ and $\mathbf{s}_3$ differ in only the z-coordinate, $m(\mathbf{s}_1, \mathbf{s}_3) = z$. Since $\mathbf{s}_2$ and $\mathbf{s}_3$ differ in the first two coordinates, $m(\mathbf{s}_2, \mathbf{s}_3) = xy$. The ideal of nondisposable sets for this data is thus

$$I_{\Delta_{\mathcal{D}}^c} = \langle m(\mathbf{s}_1, \mathbf{s}_3),\ m(\mathbf{s}_2, \mathbf{s}_3)\rangle = \langle z,\ xy\rangle\,. \tag{6.5}$$

To find the min-sets, we need to compute a primary decomposition of this ideal. The easiest way to do this is using a computer algebra package. There are several that are freely available, such as Macaulay2 [12], Sage [13], and Singular [14]. We will use Macaulay2, which can be downloaded or run online at http://web.macaulay2.com/. The following Macaulay2 commands define the ideal $I_{\Delta_{\mathcal{D}}^c}$ of nondisposable sets from Eq. (6.5) and compute its primary decomposition.

```
R = ZZ/2[x,y,z];
I_nonDisp = ideal(z, x*y);
primaryDecomposition I_nonDisp
```

These commands can be entered one at a time, or all at once. Adding a semicolon at the end of a line suppresses the output of that command. We encourage the reader to try the above commands individually, without semicolons. However, the output of these three commands as they are above will be

```
{ideal (x, z), ideal (y, z)}
```

This means that $I_{\Delta_{\mathcal{D}}^c} = \langle x, z\rangle \cap \langle y, z\rangle$, that is, the primary components of $I_{\Delta_{\mathcal{D}}^c}$, or equivalently the min-sets of $\mathcal{D}$, are $\{x, z\}$ and $\{y, z\}$ (which we write as xz and yz). Take a moment to compare this to the example in the text back in Section 6.1.2.

Next, let us compare this example to the Stanley-Reisner theory presented in the previous section. The simplicial complex of disposable sets is $\Delta_{\mathcal{D}} = \{\emptyset, x, y\}$, shown at left in Fig. 6.4. The ideal of nondisposable sets $I_{\Delta_{\mathcal{D}}^c}$ is generated by the five nondisposable sets,

$$I_{\Delta_{\mathcal{D}}^c} = \langle z, xy, xz, yz, xyz\rangle = \langle z, xy\rangle.$$

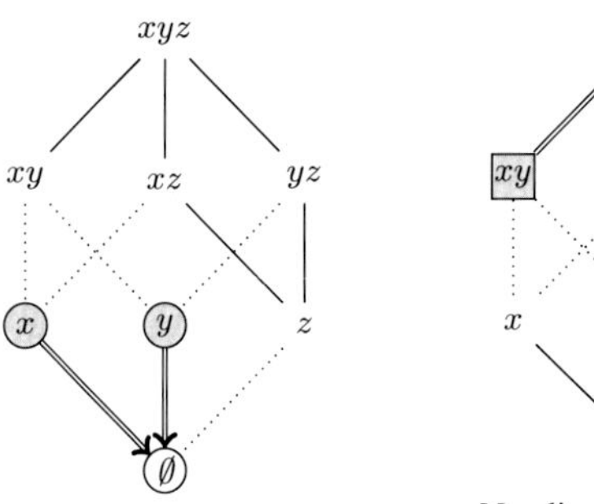

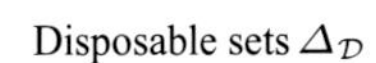
Disposable sets $\Delta_{\mathcal{D}}$

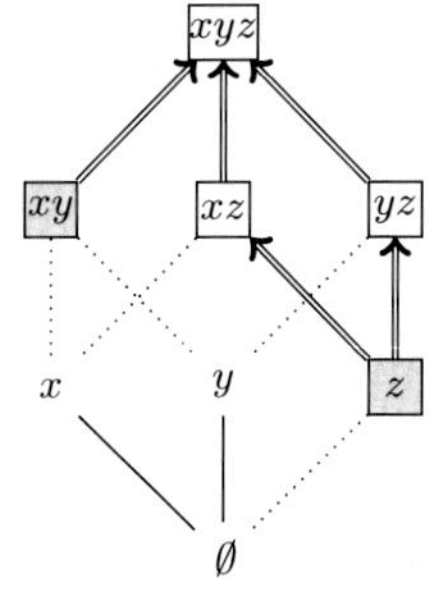

Nondisposable sets $\Delta^c_{\mathcal{D}}$;
Monomials in $I_{\Delta^c_{\mathcal{D}}}$

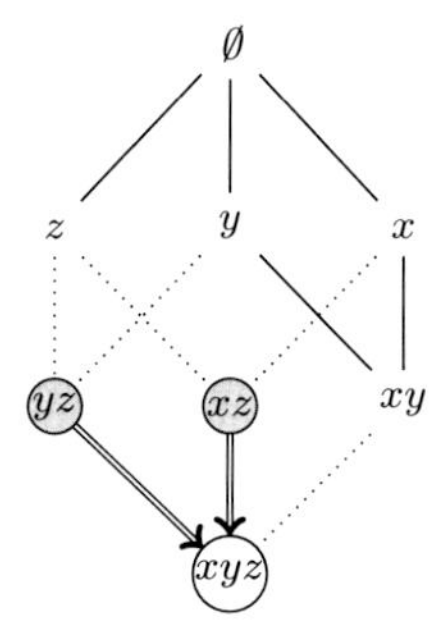

Feasible sets of $\mathcal{D}$
The min-sets are shaded

FIG. 6.4 The disposable sets from Running Example 6.3 form a simplicial complex $\Delta_{\mathcal{D}} = \{\emptyset, x, y\}$. The ideal of nondisposable sets $I_{\Delta^c_{\mathcal{D}}} = \langle z, xy \rangle$ has primary decomposition $\langle y, z \rangle \cap \langle x, z \rangle$, which describes the min-sets of $\mathcal{D}$.

These five nondisposable sets are boxed in the middle diagram of Fig. 6.4. The two shaded ones are the *minimal* nondisposable sets, which are the only ones needed to generate this ideal. These are precisely the monomials $m(\mathbf{s}_1, \mathbf{s}_2) = z$ and $m(\mathbf{s}_2, \mathbf{s}_3) = xy$. On the right of Fig. 6.4 are the individual complements of the faces in the simplicial complex $\Delta_{\mathcal{D}}$ of disposable sets. The minimal complements (i.e., complements of the maximal faces) are the generators of the primary components of $I_{\Delta^c_{\mathcal{D}}}$, which are the min-sets of $\mathcal{D}$. Finally, note how the set z $(= \{z\})$ is neither disposable nor feasible in this example.

Our next example is constructed so that the min-sets are $\{x, z\}$, $\{y\}$, just like one of the running examples from the previous section, and so we will name it accordingly.

Running Example 6.2 (continued)**.** Consider a three-variable Boolean functions with the following partial data:

xyz	101	000	110
f(x,y,z)	0	0	1

Using our notation, the data $\mathcal{D}$ is the set

$$\mathcal{D} = \{(\mathbf{s}_1, t_1), (\mathbf{s}_2, t_2), (\mathbf{s}_3, t_3)\} = \{(101, 0), (000, 0), (110, 1)\}.$$

Like the previous example, there are $2^5 = 32$ functions that fit this data. The ideal of nondisposable sets is generated by the $m(\mathbf{s}_i, \mathbf{s}_j)$ monomials for each $t_i \neq t_j$, and thus

$$I_{\Delta^c_{\mathcal{D}}} = \langle m(\mathbf{s}_1, \mathbf{s}_3), m(\mathbf{s}_2, \mathbf{s}_3) \rangle = \langle yz,\ xy \rangle.$$

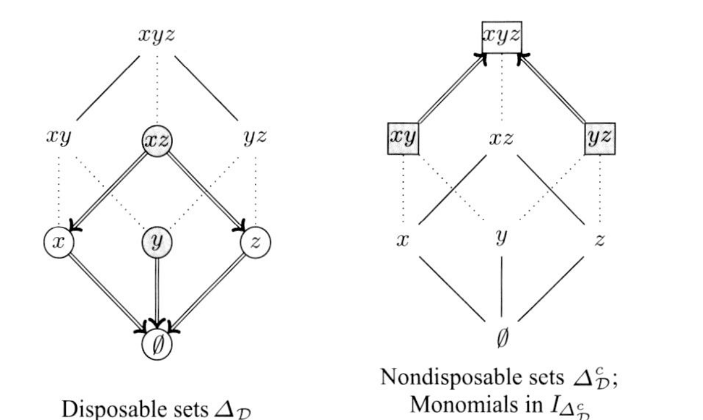

FIG. 6.5 The disposable sets $\Delta_{\mathcal{D}}$, the nondisposable sets $\Delta_{\mathcal{D}}^c$, and the feasible sets from Running Example 6.2.

Once again, we can compute the primary decomposition of this ideal in Macaulay2 with the following commands.

```
R = ZZ/2[x,y,z];
I_nonDisp = ideal(y*z, x*y);
primaryDecomposition I_nonDisp
```

The output is

```
{ideal y, ideal (x, z)}
```

This means that $I_{\Delta_{\mathcal{D}}^c} = \langle y \rangle \cap \langle x, z \rangle$. In other words, there are two min-sets: $\{y\}$ and $\{x, z\}$. Several of the 32 functions in $\mathsf{Mod}(\mathcal{D})$ are listed below:

$$\mathsf{Mod}(\mathcal{D}) = \{y, \quad x \wedge \bar{z}, \quad x \wedge y, \quad y \wedge \bar{z}, \quad x \wedge y \wedge \bar{z}, \quad \ldots\}.$$

It is easy to see that the supports of the first two functions earlier are these (unsigned) min-sets. See Fig. 6.5 for a visual of this, and compare this to Fig. 6.2.

Exercise 6.9. Construct an example of a set $\mathcal{D} = \{(\mathbf{s}_1, t_1), \ldots, (\mathbf{s}_m, t_m)\}$ of data from a Boolean function $f : \mathbb{F}_2^3 \to \mathbb{F}_2$ whose min-sets are $\{x, y\}$, $\{x, z\}$, and $\{y, z\}$. Carry out all of the steps from Running Examples 6.2 and 6.3. Additionally, for each feasible set, find a function whose support is that set.

The following table summarizes the correspondence between the combinatorial structures in the BN problem to classic Stanley-Reisner theory and Alexander duality.

Reverse Engineering of Local Models	Stanley-Reisner Theory
Disposable sets of $\mathcal{D}$	Faces of the simplicial complex $\Delta_{\mathcal{D}}$
Nondisposable sets of $\mathcal{D}$	The nonfaces, $\Delta^c_{\mathcal{D}}$
The ideal $\langle m(\mathbf{s},\mathbf{s}') \mid t \neq t' \rangle$ of nondisposable sets	The Stanley-Reisner ideal $I_{\Delta^c_{\mathcal{D}}}$
Feasible sets of $\mathcal{D}$	Complements of faces of $\Delta_{\mathcal{D}}$
Min-sets of $\mathcal{D}$	Complements of maximal faces of $\Delta_{\mathcal{D}}$ $\leftrightarrow$ primary components of $I_{\Delta^c_{\mathcal{D}}}$

6.3.3 Min-Sets Over Non-Boolean Fields

We will conclude this section with an example of computing the min-sets from a set of partial data $\mathcal{D}$ for a local model over a nonbinary field. The process is basically the same, and we even compute the ideal in a polynomial ring over $\mathbb{F}_2$, for reasons mentioned in Remark 6.6.

Consider the input data $\mathcal{D} = \{(\mathbf{s}_1, t_1), \ldots, (\mathbf{s}_5, t_5)\}$ shown below over $\mathbb{F}_5 = \{0, 1, 2, 3, 4\}$. We align the data vertically so it is easier to see the coordinates in which they differ. Once again, we write vectors $(x_1, x_2, x_3, x_4, x_5)$ as strings $x_1x_2x_3x_4x_5$.

$$\begin{aligned}
(\mathbf{s}_1, t_1) &= (01210, 0)\,, \\
(\mathbf{s}_2, t_2) &= (01211, 0)\,, \\
(\mathbf{s}_3, t_3) &= (01214, 1)\,, \\
(\mathbf{s}_4, t_4) &= (30000, 3)\,, \\
(\mathbf{s}_5, t_5) &= (11113, 4)\,.
\end{aligned}$$

Since $t_i \neq t_j$ except for $t_1 = t_2$, we need to compute $m(\mathbf{s}_i, \mathbf{s}_j)$ for the nine pairs $i < j$ not equal to 1 and 2. Since $\mathbf{s}_3$ and $\mathbf{s}_4$ differ in all five coordinates, as do $\mathbf{s}_4$ and $\mathbf{s}_5$,

$$m(\mathbf{s}_3, \mathbf{s}_4) = m(\mathbf{s}_4, \mathbf{s}_5) = x_1x_2x_3x_4x_5.$$

Next, $\mathbf{s}_5$ differs with $\mathbf{s}_1$, $\mathbf{s}_2$, and $\mathbf{s}_3$ in each of the first, third, and fifth coordinates, and so

$$m(\mathbf{s}_1, \mathbf{s}_5) = m(\mathbf{s}_2, \mathbf{s}_5) = m(\mathbf{s}_3, \mathbf{s}_5) = x_1x_3x_5.$$

Finally, it is easy to check that

$$m(\mathbf{s}_1, \mathbf{s}_4) = x_1x_2x_3x_4, \qquad m(\mathbf{s}_1, \mathbf{s}_3) = m(\mathbf{s}_2, \mathbf{s}_3) = x_5.$$

Even though the data $\mathcal{D}$ is over $\mathbb{F}_5$, the monomial ideal of nondisposable sets is the following ideal in $\mathbb{F}_2[x_1, x_2, x_3, x_4, x_5]$:

$$I_{\Delta^c_{\mathcal{D}}} = \langle m(\mathbf{s}_i, \mathbf{s}_j) \mid t_i \neq t_j \rangle = \langle x_1x_2x_3x_4x_5,\ x_1x_3x_5,\ x_1x_2x_3x_4,\ x_5 \rangle.$$

Recall that if $\alpha \subseteq \beta$, and $x^\alpha \in I$, then $x^\beta \in I$. Since $x_1x_2x_3x_4x_5$ and $x_1x_3x_5$ both contain x_5, we can eliminate them from the generating set above, and just write

$$I_{\Delta^c_{\mathcal{D}}} = \langle x_1x_2x_3x_4,\ x_5 \rangle.$$

The min-sets can be read from the primary decomposition of $I_{\Delta^c_{\mathcal{D}}}$, and we can compute this using the following commands in Macaulay2:

```
R = ZZ/2[x1,x2,x3,x4,x5];
I_nonDisp = ideal(x5, x1*x2*x3*x4);
primaryDecomposition I_nonDisp
```

The output is

```
{ideal (x1, x5), ideal(x2, x5), ideal(x3, x5),
ideal(x4, x5)}
```

This means that the primary decomposition is

$$I_{\Delta^c_{\mathcal{D}}} = \langle x_1, x_5 \rangle \cap \langle x_2, x_5 \rangle \cap \langle x_3, x_5 \rangle \cap \langle x_4, x_5 \rangle.$$

Hence, there are four (unsigned) min-sets,

$$\{x_1, x_5\}, \qquad \{x_2, x_5\}, \qquad \{x_3, x_5\}, \qquad \{x_4, x_5\}.$$

Exercise 6.10. Repeat Exercise 6.7 for the five-variable example over $\mathbb{F}_5$ from this section. Use the five-dimensional Boolean lattice from Fig. 6.3.

6.4 FINDING SIGNED MIN-SETS OF LOCAL MODELS

6.4.1 The Pseudo-Monomial Ideal of Signed Nondisposable Sets

The methods from the previous chapter allowed us to find all possible min-sets given a set of data $\mathcal{D}$. However, what was missing the information of whether the interactions were positive, negative, or both. In this section, we will extend those techniques to partially answer that question. To do this, we need to use $+1$ and -1 to encode positive and negative interactions, respectively. Therefore, the monomial ideal that we construct will be over $\mathbb{F}_3$ instead of $\mathbb{F}_2$. However, we will only consider unate functions. In other words, we will only consider min-sets of functions whose interactions are either positive or negative; not both. As mentioned in the beginning of this section, this is not a big deal, because most functions that arise in molecular networks are unate. Finally, we should mention that it is possible to drop this requirement, but we will not do that here. The interested reader is encouraged to read the final section of [11] for a discussion on how to do this.

As in the previous section, we need to look at every pair in input vectors $\mathbf{s}$ and $\mathbf{s}'$ in $\mathcal{D}$ that have different outputs, $t \neq t'$. This time, we will record this with a so-called *pseudo-monomial* $p(\mathbf{s}, \mathbf{s}')$. To begin, re-order the data so the output values are nondecreasing (i.e., $t_1 \leq \cdots \leq t_m$).

Next, for each pair $\mathbf{s}$ and $\mathbf{s}'$ such that $t < t'$, we look at every coordinate in which $\mathbf{s}$ and $\mathbf{s}'$ differ. If $s_i < s'_i$, then we encode this with the term $(x_i - 1)$, denoting a potential positive interaction. In other words, we are saying that an increase in the ith input coordinate from s_i to s'_i could have resulted in a increase of the output from t_i to t_i'. Similarly, if $s_i > s_i'$, then we encode this with the term $(x_i + 1)$, denoting a potential negative interaction. Finally, we define $p(\mathbf{s}, \mathbf{s}')$ to be the product of each of these terms, over all coordinates in which $s_i \neq s_i'$.

Notice that in the process described earlier, we are simply multiplying $(x_i - \text{sign}(s_i' - s_i))$ for each coordinate i in which $\mathbf{s}$ and $\mathbf{s}'$ differ. In other words, we define

$$p(\mathbf{s}, \mathbf{s}') := \prod_{s_i \neq s_i'} \left(x_i - \text{sign}(s_i' - s_i)\right).$$

Though $p(\mathbf{s}, \mathbf{s}')$ is not a monomial in $\mathbb{F}_3[x_1, \ldots, x_n]$, we call it a *pseudo-monomial*. Throughout this chapter we will continue to use m for a monomial and p for a pseudo-monomial. For monomials, Proposition 6.6 says that the monomials $m(\mathbf{s}_i, \mathbf{s}_i')$ for which $t_i \neq t_i'$ generate the Stanley-Reisner ideal of the simplicial complex $\Delta_{\mathcal{D}}$ of disposable sets. The analog of this concept for pseudo-monomials has to be a definition, rather than a proposition.

Definition 6.10. The ideal of signed nondisposable sets is the ideal in $\mathbb{F}_3[x_1, \ldots, x_n]$ defined by

$$J_{\Delta_{\mathcal{D}}^c} = \left\langle p(\mathbf{s}_i, \mathbf{s}_j) \mid i < j,\ t_i \neq t_j \right\rangle.$$

Note that we use $J_{\Delta_{\mathcal{D}}^c}$ to denote that the nondisposable sets are signed, whereas the unsigned version is denoted $I_{\Delta_{\mathcal{D}}^c}$. Due to the lack of a developed theory of pseudo-monomials,[2] it does not immediately follow from Stanley-Reisner theory that the primary decomposition of this ideal necessarily gives anything useful, let along the (signed) min-sets, the way it does for the unsigned version in Proposition 6.6. However, such a correspondence fortunately does hold, though it had to be proven directly in [11].

Theorem 6.3. *Let $J_{\Delta_{\mathcal{D}}^c}$ be the ideal of signed nondisposable sets in $\mathbb{F}_3[x_1, \ldots, x_n]$, where $\mathbb{F}_3 = \{0, 1, -1\}$. Each minimal prime $\mathfrak{p}$ in the primary decomposition of $J_{\Delta_{\mathcal{D}}^c}$ has the form*

$$\mathfrak{p} = \langle x_i - \alpha_i \mid \alpha_i \neq 0 \rangle$$

for some vector $\alpha = (\alpha_1, \ldots, \alpha_n) \in \{0, 1, -1\}^n$. Moreover, these α's are precisely the signed min-sets of $\mathcal{D}$.

We will revisit our examples from the previous section right away.

2. Not long before the publication of this book, some work along these lines has appeared on the arXiv preprint server [15], involving pseudo-monomials in neural ideals, which are very similar to the pseudo-monomials ideals in this chapter. See Chapter 7 of this book for more on neural ideals.

Running Example 6.3 (continued)**.** Let us reverse-engineer the minimal signed wiring diagrams of a Boolean function $f\colon \mathbb{F}_2^3 \to \mathbb{F}_2$ with the following partial data:

xyz	111	000	110
f(x,y,z)	0	0	1

Recall that we order the input data so the output values are nondecreasing (i.e., $t_1 \leq t_2 \leq t_3$). Using our notation, the *data* $\mathcal{D}$ are

$$\mathcal{D} = \{(\mathbf{s}_1, t_1), (\mathbf{s}_2, t_2), (\mathbf{s}_3, t_3)\} = \{(111, 0), (000, 0), (110, 1)\}.$$

Since $t_1 \neq t_3$ and $t_2 \neq t_3$, we need to compute $p(\mathbf{s}_1, \mathbf{s}_3)$ and $p(\mathbf{s}_2, \mathbf{s}_3)$. Comparing $\mathbf{s}_1$ and $\mathbf{s}_3$, we see a decrease in the z-coordinate, $1 = s_{13} > s_{33} = 0$, but an increase in the output, $0 = t_1 < t_3 = 1$. Therefore,

$$p(\mathbf{s}_1, \mathbf{s}_3) = z - (\text{sign}(s_{33} - s_{13})) = z + 1.$$

Next, $\mathbf{s}_2$ and $\mathbf{s}_3$ differ in the first two coordinates: $s_{21} < s_{31}$ and $s_{22} < s_{32}$, which means that

$$p(\mathbf{s}_2, \mathbf{s}_3) = (x - 1)(y - 1).$$

The ideal of signed nondisposable sets for $\mathcal{D}$ is thus

$$J_{\Delta^c_{\mathcal{D}}} = \langle p(\mathbf{s}_1, \mathbf{s}_2),\ p(\mathbf{s}_2, \mathbf{s}_3)\rangle = \langle z + 1,\ (x - 1)(y - 1)\rangle.$$

The following Macaulay2 commands compute the primary decomposition of $J_{\Delta^c_{\mathcal{D}}}$:

```
R = ZZ/3[x,y,z];
J_nonDisp = ideal(z+1, (x-1)*(y-1));
primaryDecomposition J_nonDisp
```

The output is

```
{ideal (z + 1, y - 1), ideal (z + 1, x - 1)}
```

This means that the primary decomposition of the ideal $J_{\Delta^c_{\mathcal{D}}}$ of signed nondisposable sets is

$$J_{\Delta^c_{\mathcal{D}}} = \langle x - 1, z + 1\rangle \cap \langle y - 1, z + 1\rangle,$$

and hence the signed min-sets are $\{x, \overline{z}\}$ and $\{y, \overline{z}\}$. In other words, every unate function $f\colon \mathbb{F}_2^3 \to \mathbb{F}_2$ in the model space $\text{Mod}(\mathcal{D})$ must either involve x and $\overline{z}$, or y and $\overline{z}$. Recall from Section 6.1.2 that of the 32 functions in $\text{Mod}(\mathcal{D})$, only the following four were unate:

$$x_1 \wedge \overline{x_3}, \qquad x_2 \wedge \overline{x_3}, \qquad x_1 \wedge x_2 \wedge \overline{x_3}, \qquad (x_1 \vee x_2) \wedge \overline{x_3},$$

and these first two correspond to the two signed min-sets.

Exercise 6.11. Compute the signed min-sets from Running Example 6.2 by repeating the steps in the example done in this section.

6.4.2 A Non-Boolean Example

We will now return to the non-Boolean example from Section 6.3.3, but compute the signed min-sets. Recall the following input data over $\mathbb{F}_5 = \{0, 1, 2, 3, 4\}$:

$$(\mathbf{s}_1, t_1) = (01210, 0),$$
$$(\mathbf{s}_2, t_2) = (01211, 0),$$
$$(\mathbf{s}_3, t_3) = (01214, 1),$$
$$(\mathbf{s}_4, t_4) = (30000, 3),$$
$$(\mathbf{s}_5, t_5) = (11113, 4).$$

As before, since only $t_1 = t_2$, we need to compute the pseudo-monomial $p(\mathbf{s}_i, \mathbf{s}_j)$ for the other nine pairs $i < j$ not equal to 1 and 2. However, this time we need to keep track of whether the difference is positive or negative. We will begin with $\mathbf{s}_2$ and $\mathbf{s}_4$, because they differ in every coordinate, and

$$s_{21} < s_{41}, \quad s_{22} > s_{42}, \quad s_{23} > s_{43}, \quad s_{24} > s_{44}, \quad s_{25} > s_{45}.$$

We encode each $s_{2i} < s_{4i}$ with $x_i - 1$ and $s_{2i} > s_{4i}$ with $x_i + 1$. Therefore, the corresponding pseudo-monomial is

$$p(\mathbf{s}_2, \mathbf{s}_4) = (x_1 - 1)(x_2 + 1)(x_3 + 1)(x_4 + 1)(x_5 + 1).$$

It is easy to check that the same inequalities hold when comparing $\mathbf{s}_3$ and $\mathbf{s}_4$, and so $p(\mathbf{s}_3, \mathbf{s}_4) = p(\mathbf{s}_4, \mathbf{s}_5)$. However, all five inequalities are reversed when comparing $\mathbf{s}_4$ with $\mathbf{s}_5$, and so

$$p(\mathbf{s}_4, \mathbf{s}_5) = (x_1 + 1)(x_2 - 1)(x_3 - 1)(x_4 - 1)(x_5 - 1).$$

Next, recall that $\mathbf{s}_5$ differs with $\mathbf{s}_1$, $\mathbf{s}_2$, and $\mathbf{s}_3$ in each of the first, third, and fifth coordinates. However, since we are keeping track of signs, we need to compute these pseudo-monomials separately. We start with $\mathbf{s}_1$ versus $\mathbf{s}_5$:

$$s_{11} < s_{51}, \quad s_{13} > s_{53}, \quad s_{15} < s_{55}, \quad p(\mathbf{s}_1, \mathbf{s}_5) = (x_1 - 1)(x_3 + 1)(x_5 - 1).$$

It is easy to check that these same relations hold for $\mathbf{s}_2$ versus $\mathbf{s}_5$, and so $p(\mathbf{s}_1, \mathbf{s}_5) = p(\mathbf{s}_2, \mathbf{s}_5)$. However, there is a difference in the fifth coordinate when comparing $\mathbf{s}_3$ with $\mathbf{s}_5$:

$$s_{31} < s_{51}, \quad s_{33} > s_{53}, \quad s_{35} > s_{55}, \quad p(\mathbf{s}_3, \mathbf{s}_5) = (x_1 - 1)(x_3 + 1)(x_5 + 1).$$

Next, in comparing $\mathbf{s}_1$ with $\mathbf{s}_4$, we get that

$$s_{11} < s_{41}, \quad s_{12} > s_{42}, \quad s_{13} > s_{43}, \quad s_{14} > s_{44},$$
$$p(\mathbf{s}_1, \mathbf{s}_4) = (x_1 - 1)(x_2 + 1)(x_3 + 1)(x_4 + 1).$$

Finally, $\mathbf{s}_3$ differs with $\mathbf{s}_1$ and $\mathbf{s}_2$ in only the last coordinate, so

$$s_{15} < s_{35}, \quad s_{25} < s_{35}, \quad p(\mathbf{s}_1, \mathbf{s}_3) = p(\mathbf{s}_2, \mathbf{s}_3) = x_5 - 1.$$

The ideal of signed nondisposable sets is generated by all of these pseudo-monomials $p(\mathbf{s}_i, \mathbf{s}_j)$:

$$\begin{aligned} J_{\Delta_{\mathcal{D}}^c} = \langle &(x_1 - 1)(x_2 + 1)(x_3 + 1)(x_4 + 1)(x_5 + 1), \\ &(x_1 + 1)(x_2 - 1)(x_3 - 1)(x_4 - 1)(x_5 - 1), \\ &(x_1 - 1)(x_3 + 1)(x_5 - 1), (x_1 - 1)(x_3 + 1)(x_5 + 1), \\ &(x_1 - 1)(x_2 + 1)(x_3 + 1)(x_4 + 1), x_5 - 1\rangle . \end{aligned}$$

This ideal can be simplified. We can eliminate the second and third pseudo-monomials because they are multiples of the last pseudo-monomial, $x_5 - 1$. Similarly, the first pseudo-monomials are unnecessary because it is a multiple of the fourth. Therefore, our ideal is just

$$J_{\Delta_{\mathcal{D}}^c} = \langle (x_1 - 1)(x_3 + 1)(x_5 + 1),\ (x_1 - 1)(x_2 + 1)(x_3 + 1)(x_4 + 1),\ x_5 - 1\rangle .$$

Once again, we can use Macaulay2 to find the primary decomposition, but we have to work over $\mathbb{F}_3$.

```
R = ZZ/3[x1,x2,x3,x4,x5];
J = ideal((x1-1)*(x3+1)*(x5+1),
(x1-1)*(x2+1)*(x3+1)*(x4+1), x5-1);
primaryDecomposition J
```

The output is

```
{ideal (x5-1, x3+1), ideal(x5-1, x1-1)}
```

This means that there are two signed min-sets,

$$\{x_1, x_5\}, \quad \{\overline{x_3}, x_5\}.$$

Let us take a moment to compare this result with that of the previous section, where we computed unsigned min-sets for this same set $\mathcal{D}$ of data. Then, we got two additional (unsigned) min-sets, $\{x_2, x_5\}$ and $\{x_4, x_5\}$. Together, this means that every function $f : \mathbb{F}_5^5 \to \mathbb{F}_5$ in the model space $\mathrm{Mod}(\mathcal{D})$ depends on at least one of the pairs of variables x_1, x_5, or x_2, x_5, or x_3, x_5, or x_4, x_5. However, upon computing the signed min-sets, we conclude that any unate function in $\mathrm{Mod}(\mathcal{D})$ must depend on either the first or third pair. Moreover, if it involves the third pair, then x_3 negatively affects f but x_5 positively affects f. If it involves the first pair, then both x_1 and x_5 positively affect f.

6.5 APPLICATIONS TO A REAL GENE NETWORK

In this section, we will apply the theory developed in this chapter to the reconstruction of a developmental network in the microscopic roundworm *C. elegans*, a common model organism in biology. It was the first multicellular organism to have its full genome sequenced, as well as its nervous system (called its *connectome*) completely mapped. The latter consists of just 302 neurons and approximately 7000 synapses.

	x_1	x_2	x_3	x_4	x_5	x_6	x_7	x_8	x_9	x_{10}	x_{11}	x_{12}	x_{13}	x_{14}	x_{15}	x_{16}	x_{17}	x_{18}	x_{19}	x_{20}
$\mathbf{s}_1$	4	6	5	0	3	0	0	0	0	1	0	0	0	1	0	1	0	0	0	0
$\mathbf{s}_2 = \mathbf{t}_1$	3	6	5	0	2	1	1	1	0	0	0	0	1	1	0	0	0	1	0	0
$\mathbf{s}_3 = \mathbf{t}_2$	1	3	1	0	2	1	1	1	0	1	0	0	1	1	0	1	0	1	0	1
$\mathbf{s}_4 = \mathbf{t}_3$	1	3	1	2	2	1	1	1	1	1	0	0	0	1	0	0	0	1	2	1
$\mathbf{s}_5 = \mathbf{t}_4$	0	1	1	2	2	1	1	1	1	1	0	0	1	1	0	1	0	1	2	1
$\mathbf{s}_6 = \mathbf{t}_5$	0	2	1	4	6	4	1	3	1	1	0	0	1	2	0	1	0	1	1	1
$\mathbf{s}_7 = \mathbf{t}_6$	0	3	1	6	5	5	1	4	2	1	0	0	1	1	1	2	1	1	1	0
$\mathbf{s}_8 = \mathbf{t}_7$	1	3	1	4	2	6	1	4	2	3	1	1	3	2	4	4	0	3	3	0
$\mathbf{s}_9 = \mathbf{t}_8$	1	3	1	6	2	5	1	5	1	5	2	5	6	2	5	5	0	4	4	0
$\mathbf{t}_9$	0	2	1	4	2	3	1	3	1	4	1	3	4	2	5	3	1	5	5	2

	x_1	x_2	x_3	x_4	x_5	x_6	x_7	x_8	x_9	x_{10}	x_{11}	x_{12}	x_{13}	x_{14}	x_{15}	x_{16}	x_{17}	x_{18}	x_{19}	x_{20}
$\mathbf{u}_1$	4	3	3	0	1	0	0	0	1	1	1	0	1	0	0	1	0	0	0	0
$\mathbf{u}_2 = \mathbf{v}_1$	4	0	0	0	0	0	1	0	0	1	1	0	1	2	1	1	0	0	0	0
$\mathbf{u}_3 = \mathbf{v}_2$	5	3	2	0	1	0	0	0	0	1	0	1	0	2	0	1	0	0	0	0
$\mathbf{u}_4 = \mathbf{v}_3$	4	4	3	0	2	0	1	1	1	1	0	1	1	1	0	1	0	0	1	1
$\mathbf{u}_5 = \mathbf{v}_4$	1	2	1	1	2	0	1	1	1	1	0	1	2	0	0	1	0	1	0	1
$\mathbf{u}_6 = \mathbf{v}_5$	2	3	1	2	4	2	2	2	3	1	0	0	2	1	0	1	0	1	1	1
$\mathbf{u}_7 = \mathbf{v}_6$	5	3	1	3	2	2	3	3	5	2	0	1	2	3	1	1	1	1	0	1
$\mathbf{u}_8 = \mathbf{v}_7$	6	5	6	5	4	5	6	4	6	1	0	4	2	2	3	2	1	2	2	0
$\mathbf{u}_9 = \mathbf{v}_8$	3	3	1	4	2	2	4	2	4	3	0	4	5	0	3	2	2	2	4	0
$\mathbf{v}_9$	4	5	4	6	2	3	5	6	2	6	2	6	5	2	6	6	1	6	6	3

FIG. 6.6 The time series data from the *C. elegans* developmental network in [16]. These data come from two experiments intended to produce "wildtype" expression.

The network that we will consider has 20 genes involved in the embryonal development of *C. elegans*. These data are described in [16] and consist of 2 time series of 10 data points, $\mathbf{s}_1, \ldots, \mathbf{s}_{10}$ discretized to 7 states (i.e., $\mathbf{s}_i \in \mathbb{F}_7^{20}$). The first input state is $\mathbf{s}_1$ and the first output state is $\mathbf{t}_1 = f(\mathbf{s}_1) = \mathbf{s}_2$, where $f\colon \mathbb{F}_7^{20} \to \mathbb{F}_7^{20}$ is the FDS map arising from the unknown local model over $\mathbb{F}_7$. More generally, the ith input state is $\mathbf{s}_i$ and the ith output state is $\mathbf{t}_i = f(\mathbf{s}_i) = \mathbf{s}_{i+1}$. Note that this means that the 20 points in $\mathbb{F}_7^{20}$ describe 18 input-output pairs in 2 time series, which are shown in Fig. 6.6.

We aim to reconstruct a wiring diagram for the subnetwork of three genes responsible for body wall (mesodermal) tissue development. These genes are known to be regulated by the maternally controlled *pal-1* gene. While all three regulate the development of a single tissue type in *C. elegans*, some vertebrates have homologous transcription factors related to these genes that function in one of three different muscle types (see Fig. 6.5 in [17]). Understanding their regulatory interactions has implications in human muscle development and disease. The gene names, the variable labels from [16], and the associated muscle types in vertebrates are listed below.

Gene	Variable	Muscle Type
hlh-1	x_8	Skeletal
hnd-1	x_{18}	Cardiac
unc-120	x_{19}	Cardiac, smooth, skeletal

For each gene j of interest ($j = 8, 18, 19$), we extract a set $\mathcal{D}_j$ of data consisting of 18 input-output pairs $(\mathbf{s}_i, t_{ij})$ and $(\mathbf{u}_i, v_{ij})$ where t_{ij} and v_{ij} are the corresponding coordinates of the output vectors. For example, the data for the *hlh-1* gene are the set

$$\mathcal{D}_8 = \{(\mathbf{s}_1, t_{18}), (\mathbf{s}_2, t_{28}), \ldots, (\mathbf{s}_9, t_{98}), (\mathbf{u}_1, v_{18}), (\mathbf{u}_2, v_{28}), \ldots, (\mathbf{u}_9, v_{98})\}$$

taking corresponding pairs from both time series. The primary decomposition of the ideal $I_{\mathcal{D}_8^c}$ of nondisposable sets for the *hlh-1* gene can be computed using Macaulay2. To get a sense for what this entails, we encourage the interested reader to explore this on their own; see Exercise 6.12 and the data from Fig. 6.6. Though this is called an exercise, it is more of a mini-project. The size-18 data set $\mathcal{D}_8$ has $\binom{18}{2} = 153$ pairs of data. Though 28 of these pairs have the same output value t_{i8} (or v_{i8}), that still leaves 125 monomials to generate the ideal $I_{\mathcal{D}_8^c}$. Even though some of these monomials are repeats, computing the primary decomposition of $I_{\mathcal{D}_8^c}$ with Macaulay2 spits out a daunting list of 483 primary components.

Exercise 6.12. Give an explicit generating set for the ideal

$$I_{\mathcal{D}_8^c} = \left\langle \{m(\mathbf{s}_i, \mathbf{s}_j) \mid t_{i8} \neq t_{j8}\} \cup \{m(\mathbf{u}_i, \mathbf{u}_j) \mid v_{i8} \neq v_{j8}\} \cup \{m(\mathbf{s}_i, \mathbf{u}_j) \mid t_{i8} \neq v_{j8}\} \right\rangle.$$

of nondisposable sets for the *hlh-1* gene.

Since $I_{\mathcal{D}_8^c}$ has 483 primary components, we have 483 possible min-sets. Choosing "the correct" one initially appears to be an impossible task. However, it is known experimentally that *hlh-1* is controlled by the *pal-1* genes, which are represented by the variables x_1, x_2, x_3. Therefore, we can safely disregard all of the min-sets that involve none of these variables, and this happens to be 481 of them. This leaves just two candidates for the min-sets of *hlh-1*, which are

$$\{x_2, x_3, x_8, x_{18}\} \quad \text{and} \quad \{x_2, x_3, x_8, x_{19}\}.$$

To summarize, the interpretation of this is that there are just two ways to construct the *hlh-1* component in a minimal wiring diagram of the three-gene subnetwork using only the variables of interest. Furthermore, we use the scientific knowledge that there is no way to construct a minimal wiring diagram without the controller *pal-1*; that is, the subnetwork of the three mesodermal genes does not function without direct input/regulation from *pal-1*. If we merge the three variables that represent the *pal-1* genes into a single node P, then there are two possibilities for the wiring diagrams at the *hlh-1* node, which are shown below.

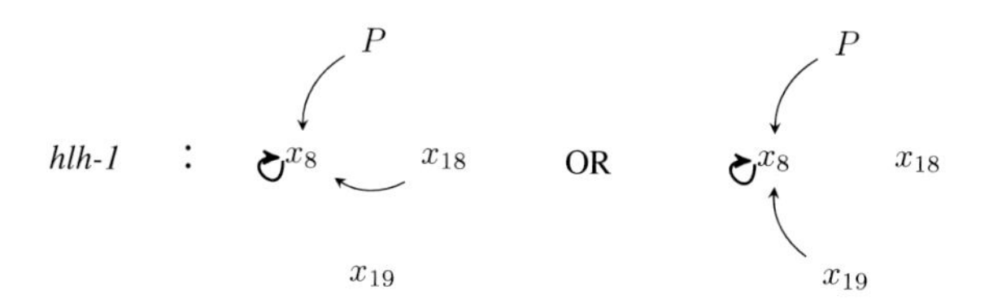

TABLE 6.1 Biologically Feasible Min-Sets for *hlh-1*, *hnd-1*, and *unc-120*

hlh-1 (x_8)	*hnd-1* (x_{18})	*unc-120* (x_{19})
$\{x_2, x_3, x_8, x_{18}\}$	$\{x_2, x_8, x_{18}\}$	$\{x_2, x_3, x_8, x_{18}\}$
$\{x_2, x_3, x_8, x_{19}\}$	$\{x_2, x_8, x_{19}\}$	$\{x_2, x_3, x_8, x_{19}\}$
	$\{x_3, x_8, x_{19}\}$	$\{x_2, x_8, x_9, x_{19}\}$
	$\{x_3, x_8, x_9, x_{18}\}$	

Notes: These were selected from the complete sets of 483, 580, and 498, respectively, so that each min-set contains (i) at least one of the pal-1 *variables, (ii) as many of the variables representing* hlh-1, hnd-1, *and* unc-120 *as possible, and (iii) no other variables.*

Applying a similar process for the other two genes gives a total of 580 min-sets for *hnd-1* and 498 for *unc-120*. Like with the *hlh-1* genes, these lists can be drastically reduced by discarding the min-sets that do not contain any of the *pal-1* genes. The resulting min-sets are listed in Table 6.1.

The wiring diagrams of the subnetwork of $\{x_8, x_{18}, x_{19}\}$ are the unions of the min-sets of x_8, x_{18}, and x_{19}. There are $2 \cdot 4 \cdot 3 = 24$ ways to do this using the min-sets from Table 6.1. However, if we merge the *pal-1* genes x_1, x_2, x_3 into a single node P as before, then we get only two possible wiring diagrams at each of the three genes *hlh-1*, *hnd-1*, and *unc-120*, as shown in Table 6.2. The two min-sets for *hlh-1* were already shown, and the two for the genes *hnd-1* and *unc-120* are shown below. Together, this gives $2^3 = 8$ possible minimal wiring diagrams of the subnetwork.

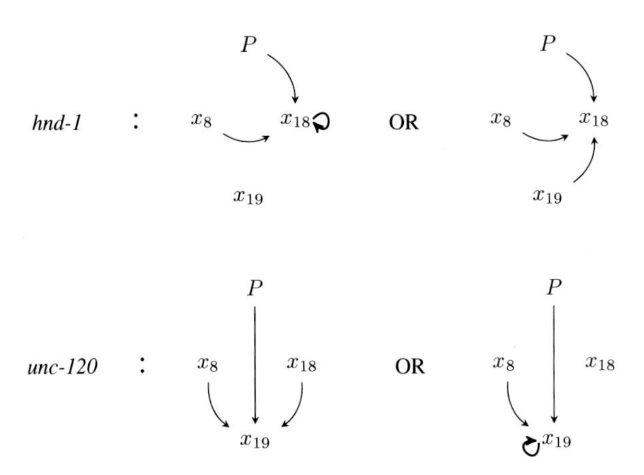

TABLE 6.2 Simplified Min-Sets for *hlh-1*, *hnd-1*, and *unc-120*

hlh-1 (x_8)	*hnd-1* (x_{18})	*unc-120* (x_{19})
$\{P, x_8, x_{18}\}$	$\{P, x_8, x_{18}\}$	$\{P, x_8, x_{18}\}$
$\{P, x_8, x_{19}\}$	$\{P, x_8, x_{19}\}$	$\{P, x_8, x_{19}\}$

Notes: Here P represents the pal-1 *variables,* x_1, x_2, x_3. *We removed the simplified min-sets* $\{P, x_8, x_9, x_{18}\}$ *from the* hnd-1 *column and* $\{P, x_8, x_9, x_{19}\}$ *from the* unc-120 *column as they are no longer minimal: removal of a variable from the set is another (simplified) min-set.*

6.6 CONCLUDING REMARKS

In this chapter, we considered the problem of having experimental discrete data from a biological network, and wanting to know which nodes or sets of nodes influenced others. To tackle this problem, we assumed that the actual biological system could be modeled by a *local model* $f = (f_1, \ldots, f_n)$, and as such, the data are a set of input-output pairs

$$\left\{(\mathbf{s}_1, \mathbf{t}_1), \ldots, (\mathbf{s}_m, \mathbf{t}_m) \mid \mathbf{s}_i, \mathbf{t}_i \in \mathbb{F}^n,\ f(\mathbf{s}_i) = \mathbf{t}_i\right\} \tag{6.6}$$

that represent edges in the phase space. There are many possible local models that fit this data, and the set of all of them is called the *model space*. Through this lens, our original problem is simply to find all possible wiring diagrams of elements in the model space, and it suffices to find the minimal ones. We will conclude this chapter here with a summary of what we learned, as well as some details on the algebraic structure of the model space that were not covered in this chapter. This should both solidify the main concepts and shine a light on the relevance and bigger context.

A key simplification of this problem arose from the simple observation that finding all functions $(f_1, \ldots, f_n)$ that fit the data can be done coordinate-by-coordinate. In other words, if F_j is the set of all possible f_j's, then the full model space is simply the direct product $F_1 \times \cdots \times F_n$. Because of this, we focused on the problem of finding each F_j separately. The restriction to this coordinate means that the *data* is the following set of input-output pairs

$$\mathcal{D}_j = \left\{(\mathbf{s}_1, t_1), \ldots, (\mathbf{s}_m, t_m) \mid \mathbf{s}_i \in \mathbb{F}^n,\ t_i \in \mathbb{F},\ f_j(\mathbf{s}_i) = t_i\right\},$$

where t_i is the jth coordinate of the vector $\mathbf{t}_i$ from Eq. (6.6). Recall that in Sections 6.3 and 6.4, we dropped these subscripts and wrote $\mathcal{D} = \mathcal{D}_j$ and $f = f_j$. Now that we are looking at the bigger picture and multiple coordinates, we will need to retain the subscripts. To unify this with the notation we used earlier, $F_j = \mathsf{Mod}(\mathcal{D}_j)$. Though the elements of this set are functions rather than local models, we will continue to refer this object as the model space of $\mathcal{D}_j$.

It is well known that F_j has an affine-like structure, much like the solutions to inhomogeneous linear differential equations, or to systems of linear equations, for example, $\mathbf{Ax} = \mathbf{b}$ [18]. Specifically, for any particular model f_j that fits the data, that is, any $f_j \in F_j$,

$$F_j = \mathrm{Mod}(\mathcal{D}_j) = f_j + I := \{f_j + h \mid h \in I\},$$

where I is the set (in fact, ideal) of all functions $h\colon \mathbb{F}^n \to \mathbb{F}$ that vanish on the data, that is, $h(\mathbf{s}_i) = 0$ for all $(\mathbf{s}_i, t_i) \in \mathcal{D}_j$. Even better news is that it is not difficult to compute F_j. We will summarize this now; see [6, Chapter 3] for more details, which includes helpful examples and exercises. There are many algorithms that can construct a function f_j that fits the data. Lagrange interpolation is one, and another more algebraic method that uses the Chinese remainder theorem was developed in [10]. Finding the vanishing ideal I is also straightforward. If I_k is the set of functions that vanish on $\mathbf{s}_k$, then clearly, the set of functions that vanish on *all* input vectors is simply $I = I_1 \cap \cdots \cap I_m$. Moreover, each I_k is easy to compute. The linear function $x_i - s_{ki}$ clearly vanishes on $\mathbf{s}_k$, and thus any multiple of it does as well. Therefore, I_k is the ideal generated by these n functions, that is,

$$I_k = \langle x_1 - s_{k1}, \ldots, x_n - s_{kn} \rangle = \{(x_1 - s_{k1})h_1(x) + \cdots + (s_n - s_{kn})h_n(x) \mid h_i \in \mathbb{F}[x_1, \ldots, x_n]\},$$

where $x := (x_1, \ldots, x_n)$. Computing these ideals and their intersection is an easy task for a computational algebra software package. For example, in Macaulay2, the following commands compute the vanishing ideal of the data $\mathcal{D}_j = \{(101, 0), (000, 0), (110, 1)\}$ from Running Example 6.2.

```
R = ZZ/2[x1,x2,x3];
I1 = ideal(x1-1, x2, x3-1);
I2 = ideal(x1, x2, x3);
I3 = ideal(x1-1, x2-1, x3);
I = intersect{I1,I2,I3};
gens gb I
```

Only the last line should be potentially unfamiliar—it computes a *Gröbner basis* $\mathcal{G}$ for the ideal I, which is a particularly "nice" choice of generating set, for reasons that do not need to be discussed here; see [8]. The output of the above code is

```
| x1+x2+x3 x3^2+x3 x2x3 x2^2+x2 |
```

which means that $\mathcal{G} = \{x_1 + x_2 + x_3, x_3^2 + x_3, x_2x_3, x_2^2 + x_2\}$. However, note that $x^2 = x$ and $x + x = 0$ over $\mathbb{F}_2$, and so the second and fourth elements in $\mathcal{G}$, though they are nonzero *polynomials* (elements of the ring $\mathbb{F}_2[x_1, x_2, x_3]$), are actually identically zero as Boolean *functions*, and thus can be removed from our generating set. Simplifying a Gröbner basis by hand in this manner can be avoided altogether by entering the original code into Macaulay2 but with the first line changed to

```
Q = ZZ/2[x1,x2,x3] / ideal(x1^2-x1,x2^2-x2,x3^2-x3);
```

By taking the *quotient* of R by the ideal $\langle x_1^2-x_1, x_2^2-x_2, x_3^2-x_3\rangle$, we are basically declaring that $x_i^2 - x_i = 0$, and so the subsequent computations are done in the quotient ring Q of $2^{2^3} = 256$ *Boolean functions*, rather than the infinite ring $R = \mathbb{F}_2[x_1, x_2, x_3]$ of *Boolean polynomials*. This important technicality was unnecessary in the prior parts of this chapter because we were working with square-free monomial ideals. Indeed, if the first line of the original Macaulay2 code is tweaked in the manner described above, then the output becomes

```
| x1+x2+x3 x2x3 |
```

This means that our Gröbner basis is $\mathcal{G} = \{x_1 + x_2 + x_3, x_2x_3\}$, and so the vanishing ideal I consists of the sums of all multiples of these two polynomials, that is,

$$I = \langle x_1 + x_2 + x_3, x_2x_3\rangle = \{(x_1 + x_2 + x_3)h_1(x) + x_2x_3h_2(x) \mid h_1, h_2 \in Q\}.$$

To summarize, the set of all local models $f = (f_1, \ldots, f_n)$ that fit the data from Eq. (6.6) has the form

$$F_1 \times \cdots \times F_n = (f_1 + I, \ldots, f_n + I),$$

where f_j is any particular model in $\mathsf{Mod}(\mathcal{D}_j)$ and I is the vanishing ideal. This is seemingly a much more powerful result than what was done in this chapter—simply finding the minimal sets that these functions can depend on. However, in the next paragraph, we will argue why the min-set approach is often more useful, and it mostly comes down to the enormous size of I. Counting the number of local models and the size of the model space is a straightforward exercise in enumerative combinatorics, and is discussed in this chapter on local models. Here, we will assume this knowledge, as the details of the derivation are not needed to see the punchline.

Turning our attention back to the 20-node *C. elegans* developmental network in Section 6.5, let us suppose that we had wanted to find the entire model space from the two given time-series. From a naïve point of view, finding the coordinate functions for each of the mesodermal genes requires computations in 20 variables, and there are $7^{7^{20}}$ possible functions : $\mathbb{F}_7^{20} \to \mathbb{F}_7$. Moreover, there are $(7^{7^{20}})^{20} = 7^{20\cdot 7^{20}}$ local models $(f_1, \ldots, f_{20})$ over $\mathbb{F}_7$, and since the data consist of 18 input-output pairs, $(7^{20})^{7^{20}-18}$ of these fit the data. Since the model space has the form $F_1 \times \cdots \times F_{20}$, then each $F_j = f_j + I$ has cardinality $7^{7^{20}-18}$. This number has approximately 6.74×10^{16} *digits*.

Now, recall that the largest min-set in Table 6.1 has size four, so this reduces the search space for each of the three genes from $7^{7^{20}}$ functions $f_j \colon \mathbb{F}_7^{20} \to \mathbb{F}_7$ to a "meager" 7^{7^4} functions $f_j \colon \mathbb{F}_7^4 \to \mathbb{F}_7$, a number that at least has only 2029 digits. Despite such a massive reduction, it is still not feasible to select "the correct" model from such a large space, something which is called the *model selection problem*. Moreover, in the spirit the popular quote of George Box, "all models are wrong, but some models are useful" [19], one can make the argument that even "the best" local model that fits the data, if one could actually come up with such a thing, is still very artificial. In some cases, such as a model over $\mathbb{F}_2$ whose functions use fewer variables (most biological networks are quite sparse), one can still massively reduce the model space by requiring that the functions chosen be of a particular type that naturally arise in biology, for example, threshold, canalizing, monotone, or unate. In some cases, this is enough to make an accurate prediction of the local model. In other cases, people have proposed using *Gröbner normal forms*, which is in some sense, a way to "maximally reduce" an element of F_j by "modding out" by the vanishing ideal, much like how one can reduce 149 down to 4, modulo the ideal $5\mathbb{Z}$. However, this depends on the Gröbner basis, which depends on a choice of a *monomial ordering*, which is needed to carry out computational algorithms such as multivariate polynomial long division. Resolutions of this technical and completely nonbiological relic have included using a *universal Gröbner basis*, or using a *Gröbner fan* and choosing the Gröbner basis corresponding to the largest polytopal region [20].

One take-home message of this chapter should be that even though on the surface, the method to find all min-sets seems much weaker than finding all local models, for the *C. elegans* network, the former has lead to new biological insight that was not predicted experimentally [16], whereas the latter does not provide much useful information. More broadly, the ideas in this chapter demonstrate how nontraditional topics such as computational commutative algebra can lead to new methods and knowledge in scientific fields like molecular biology and genetics. This is a wonderful example of the emerging field of *algebraic biology*. Moreover, not only is mathematics helping biology in a new way, but the biological applications we saw have lead to new mathematical questions, theorems, and even areas. For example, nobody had any reason to study pseudo-monomial ideals, their primary decompositions, and their Gröbner bases, until these topics arose in mathematical biology. Remarkably, these objects have also arisen recently in mathematical neuroscience, where they (and related objects) are called *neural ideals*. This is the topic of Chapter 9, as well as being an active area of pure mathematical research [21, 22], all which would not have existed without the biological questions which these tools were developed to tackle. It answers a classic question of Bernd Sturmfels from 2005: *Can biology lead to new theorems* [23], and it provides explicit evidence supporting Joel Cohen's bold claim a year earlier that *Mathematics is biology's next microscope, only better; biology is mathematics' next physics, only better.*

REFERENCES

[1] E. Goles, M. Noual, Block-sequential update schedules and Boolean automata circuits, in: Automata 2010—16th Intl. Workshop on CA and DCS, Discrete Math. Theor. Comput. Sci., 2010, pp. 41–50.

[2] J. Liang, J. Han, Stochastic Boolean networks: an efficient approach to modeling gene regulatory networks, BMC Syst. Biol. 6 (2012) 113.
[3] D. Murrugarra, A. Veliz-Cuba, B. Aguilar, S. Arat, R. Laubenbacher, Modeling stochasticity and variability in gene regulatory networks, EURASIP J. Bioinform. Sys. Biol. 2012 (1) (2012) 1–11.
[4] H.S. Mortveit, C.M. Reidys, An Introduction to Sequential Dynamical Systems, Universitext, Springer Verlag, New York, NY, 2007.
[5] L. Raeymaekers, Dynamics of Boolean networks controlled by biologically meaningful functions, J. Theor. Biol. 218 (3) (2002) 331–341.
[6] R. Robeva, T. Hodge, Mathematical Concepts and Methods in Modern Biology: Using Modern Discrete Models, Academic Press, London, 2013.
[7] E. Miller, B. Sturmfels, Combinatorial Commutative Algebra, vol. 227, Springer Science & Business Media, New York, NY, 2004.
[8] D. Cox, J. Little, D. O'Shea, Ideals, Varieties, and Algorithms: An Introduction to Computational Algebraic Geometry and Commutative Algebra, Springer International Publishing, Fourth Edition, 2015.
[9] W.F. Moore, M. Rogers, S. Sather-Wagstaff, Monomial Ideals and Their Decompositions, Springer, New York, NY, 2017.
[10] A.S. Jarrah, R. Laubenbacher, B. Stigler, M. Stillman, Reverse-engineering of polynomial dynamical systems, Adv. Appl. Math. 39 (4) (2007) 477–489.
[11] A. Veliz-Cuba, An algebraic approach to reverse engineering finite dynamical systems arising from biology, SIAM J. Appl. Dyn. Syst. 11 (1) (2012) 31–48.
[12] D. Grayson, M. Stillman, Macaulay2, a software system for research in algebraic geometry. Available from: https://faculty.math.illinois.edu/Macaulay2/. (Accessed November 1, 2017).
[13] I. SageMath, CoCalc Collaborative Computation Online. Available from: https://cocalc.com/. (Accessed November 1, 2017).
[14] W. Decker, G.-M. Greuel, G. Pfister, H. Schönemann, SINGULAR 4-1-0: a computer algebra system for polynomial computations. Available from: http://www.singular.uni-kl.de. (Accessed November 1, 2017).
[15] S. Gunturkun, J. Jeffries, J. Sun, Polarization of Neural Rings, ArXiv preprint arXiv:1706.08559, 2017.
[16] B. Stigler, H.M. Chamberlin, A regulatory network modeled from wild-type gene expression data guides functional predictions in *Caenorhabditis elegans* development, BMC Syst. Biol. 6 (77) (2012) 1–15.
[17] T. Fukushige, T.M. Brodigan, L.A. Schriefer, R.H. Waterston, M. Krause, Defining the transcriptional redundancy of early bodywall muscle development in *C. elegans*: evidence for a unified theory of animal muscle development, Genes Dev. 20 (24) (2006) 3395–3406.
[18] G. Strang, Differential Equations and Linear Algebra, Wellesley-Cambridge Press, Wellesley, MA, USA, 2014.
[19] G.E.P. Box, N.R. Draper, Empirical Model-Building and Response Surfaces, John Wiley & Sons, London, 1987.
[20] E.S. Dimitrova, A.S. Jarrah, R. Laubenbacher, B. Stigler, A Gröbner fan method for biochemical network modeling, in: Proc. Int. Symposium Symb. Algebraic Comput., ACM, 2007, pp. 122–126.
[21] C. Curto, E. Gross, J. Jeffries, K. Morrison, M. Omar, Z. Rosen, A. Shiu, N. Youngs, What makes a neural code convex?, SIAM J. Appl. Algebra Geom. 1 (1) (2017) 222–238.
[22] R. Garcia, L.D. García-Puente, R. Kruse, J. Liu, D. Miyata, E. Petersen, K. Phillipson, A. Shiu, Gröbner Bases of Neural Ideals, Int. J. Algebra Comput. 28 (4) (2017) 1–19.
[23] B. Sturmfels, Can biology lead to new theorems, Annu. Rep. Clay Math. Inst. (2005) 13–26.

Chapter 7

Analysis of Combinatorial Neural Codes: An Algebraic Approach

Carina Curto*, Alan Veliz-Cuba† and Nora Youngs‡
*Department of Mathematics, The Pennsylvania State University, University Park, PA, United States, †Department of Mathematics, University of Dayton, Dayton, OH, United States, ‡Department of Mathematics and Statistics, Colby College, Waterville, ME, United States

7.1 INTRODUCTION

7.1.1 Biological Motivation: Neurons With Receptive Fields

A major challenge in neuroscience is to understand how the brain encodes such an enormous amount of information about the outside world. By observing the behavior of neurons in response to various stimuli, we can hypothesize about the function of groups of neurons, and try to create a dictionary which can predict how the brain will respond to different stimuli. However, to understand how the brain uses the neural response to stimuli to create its own picture of the outside world, we must consider how the firing of a population of neurons intrinsically encodes information about the space of possible stimuli.

One specific example of a population of neurons which provides an internal representation of an external structure is that of *place cells*, discovered by O'Keefe and Dostrovsky in the early 1970s [1]. Place cells are neurons which code for a particular region of an animal's (2D) spatial environment. In this example, the stimulus space is the set of possible locations in the environment. Each place cell is associated with a particular region of the environment (a *place field*); when the animal is within that region, the associated cell fires. Fig. 7.1 shows the activity of three different place cells as an animal passes through their respective place fields.

From an external perspective, determining the place field associated with a particular place cell is simply a matter of recording the activity of that cell while simultaneously observing the animal as it moves about its environment. However, for the animal itself to obtain useful information from the firing of

Algebraic and Combinatorial Computational Biology. https://doi.org/10.1016/B978-0-12-814066-6.00007-6

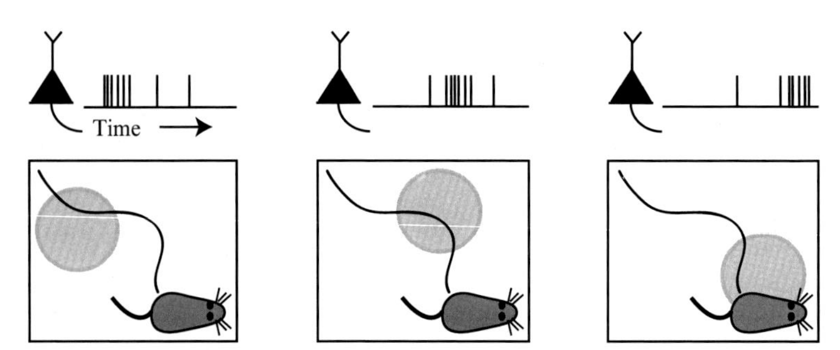

FIG. 7.1 Three place cells (represented by the *triangular* "neurons"), and the firing times of each one as an animal passes through their respective place fields (represented by *shaded gray regions*). *(Figure first published in C. Curto, What can topology tells us about the neural code?, Bull. AMS 54 (1) (2017) 63–78, © 2017 American Mathematical Society.)*

a place cell without the bird's eye view of the researcher, there would have to be some spatial information intrinsic to the firing of these cells. Considered individually, these cells give very little information about the shape of the animal's environment. However, by considering the collective behavior of the neurons, considerably more information can be determined [2].

Place cells are not the only neurons whose firing behavior can be associated with regions of stimulus. Other cells with such *receptive fields*, as they are called, are not uncommon; some examples include cells which code for angle of a visual stimulus, or 3D place fields in bats [3, 4]. With each of these examples, there is a space of stimulus in some dimension—one dimension for angles, and three dimensions for bat place fields. Given a set of any of these cells, we could record their firing behavior for various stimuli, obtaining what is called a (combinatorial) neural code.

A *neural code* $\mathcal{C}$ on n neurons is simply a subset of $\{0,1\}^n$ assumed to represent the behavior of the neurons in the set $\{1,\ldots,n\} = [n]$. We associate each codeword $\mathbf{c} \in \mathcal{C}$ to its support, representing the set of neurons which the codeword indicates were firing.

Definition 7.1. Let $\mathbf{c} \in \{0,1\}^n$. The *support* of $\mathbf{c}$ is the set $\text{supp}(\mathbf{c}) = \{i \in [n] \mid \mathbf{c}_i = 1\}$. Likewise, we can associate the entire code $\mathcal{C}$ to the collection of subsets of neurons observed in the code: $\text{supp}(\mathcal{C}) = \{\sigma \subset [n] \mid \sigma = \text{supp}(\mathbf{c}) \text{ for some } \mathbf{c} \in \mathcal{C}\}$.

The presence of the codeword $\mathbf{c}$ in $\mathcal{C}$ indicates that, at some point, the neurons in $\text{supp}(\mathbf{c})$ were firing, while the other neurons were silent. Thus, each individual codeword indicates a set of neurons which were observed to respond a common stimulus; the entire code can be seen as providing a relatively complete picture of which combinations of neurons (and thus receptive fields) are possible within this particular set of neurons and receptive fields.

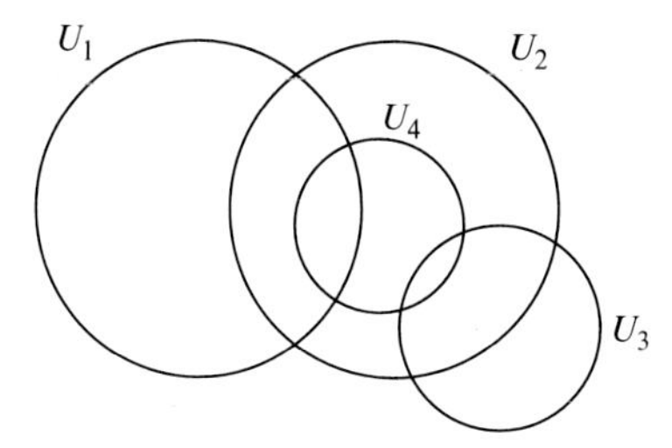

FIG. 7.2 A collection $\mathcal{U} = \{U_1, U_2, U_3, U_4\}$ of four receptive fields in $\mathbb{R}^2$.

Given knowledge of the receptive fields (and assuming that a broad range of stimuli in the space is experienced), we can predict the neural code associated with these receptive fields by considering the possible combinations of receptive fields a stimulus could come from.

Definition 7.2. Let X be a stimulus space. Given a collection of receptive fields $\mathcal{U} = \{U_1, \ldots, U_n\}$ with $U_i \subset X$ the receptive field for neuron i, we define the associated code $\mathcal{C}(\mathcal{U})$ as follows:

$$\mathcal{C}(\mathcal{U}) \stackrel{\text{def}}{=} \left\{ \mathbf{c} \in \{0,1\}^n \mid \left(\bigcap_{i \in \text{supp}(\mathbf{c})} U_i \right) \Big\backslash \left(\bigcup_{i \notin \text{supp}(\mathbf{c})} U_i \right) \neq \emptyset \right\}.$$

That is, $\mathcal{C}(\mathcal{U})$ encodes the set of nonempty regions formed by the sets in $\mathcal{U}$. If $\mathcal{C} = \mathcal{C}(\mathcal{U})$ for some collection of sets $\mathcal{U}$, we say $\mathcal{U}$ is a *realization* of $\mathcal{C}$ in the space X.

Example 7.1. Consider the collection of receptive fields $\mathcal{U} = \{U_1, U_2, U_3, U_4\}$ in $\mathbb{R}^2$ as represented in Fig. 7.2.

The code associated with this collection of sets is

$$\mathcal{C}(\mathcal{U}) = \{0000, 1000, 0100, 0010, 1100, 0110, 0101, 1101, 0111\};$$

and we can say that $\mathcal{U}$ is a realization of this code. Note that other realizations of the same code may exist.

To obtain the code associated with a known arrangement $\mathcal{U}$ of n receptive fields, we must simply determine which of the possible 2^n regions actually occur, and thus which combinations of neurons are possible. However, in a neural data context where the receptive fields are usually not known to us, we are more interested in the reverse question: given a set of neural data recorded in response to a certain type of stimulus, what can we say about the receptive fields, the relationships between them, and the structure of the space of stimuli they represent? We frame this question as follows:

Question 1. Given a neural code $\mathcal{C} \subset \{0, 1\}^n$, can we find a realization for it in a particular stimulus space X? If so, how can we find such a realization, and what can it tell us about the stimulus space X?

The next few sections will describe one approach to answering this question, using algebraic structures to encapsulate the information given by the code and extract relevant combinatorial features. These methods can be applied to any combinatorial code, not only to neural codes. However, in order to interpret this information to obtain meaningful results in the neural context, it will be necessary to make additional assumptions about the receptive fields U_i. We now briefly describe why this is so.

If no restrictions (e.g., convex, connected, etc.) of any kind are placed on the receptive fields U_i, then we can find a realization of any code, in nearly any space. In particular, there is no particular dimensional or structural information about the space associated with the code.

Exercise 7.1. Show that any arbitrary code $\mathcal{C}$ has a realization $\mathcal{U}$ in $\mathbb{R}$.

Such generic realizations are not satisfying for a neural code, as the receptive fields involved in the realization bear no resemblance to those seen in the neuroscience context and thus give no insight about the structure of the stimulus space. In Exercise 7.1, we see that without any assumptions on the receptive fields we could conclude every stimulus space is structurally equivalent to $\mathbb{R}$. However, by making a few reasonable assumptions about the nature of the receptive fields, we can drastically change the amount we can learn about the stimulus space from a neural code. Our main assumption will be to assume that the receptive fields U_i are *convex*; Section 7.4 gives more information about the results which can be obtained when we make this assumption. These results will rely heavily upon the information the code implies about the relationships between the receptive fields themselves.

7.1.2 Receptive Field Relationships

The neural code captures relationships between the firing behavior of different neurons, and thus captures ideas about the interactions between their receptive fields. For example, if there is a codeword $\mathbf{c} \in \mathcal{C}$ where $i, j \in \text{supp}(\mathbf{c})$ (so $\mathbf{c}_i = \mathbf{c}_j = 1$), we would note that at some point, neurons i and j are both firing. Thus their receptive fields overlap, or equivalently, $U_i \cap U_j$ is nonempty. However, if no such codeword appears, then we must assume that $U_i \cap U_j = \emptyset$. Similarly, we might assume that $U_i \subset U_j$ if we observe no codewords where neuron i is firing and neuron j is not. In general, a relationship of the form

$$\bigcap_{i \in \sigma} U_i \subset \bigcup_{j \in \tau} U_j$$

is referred to as a *receptive field (RF) relationship*, where we use the conventions that $\bigcap_{i\in\emptyset} U_i = X$, the entire stimulus space, and $\bigcup_{j\in\emptyset} U_j = \emptyset$.

Example 7.2. Consider the code $\mathcal{C} = \{000, 100, 010, 011\}$. Since there is no codeword where all three neurons fire together, we note the RF relationship $U_1 \cap U_2 \cap U_3 = \emptyset$ must hold in any realization of $\mathcal{C}$. We also note that there is no codeword where both neuron 1 and neuron 2 fire, so the RF relationship $U_1 \cap U_2 = \emptyset$ holds. Moreover, the former RF relationship we found is a consequence of the latter. Similarly, we note that neuron 3 fires only when neuron 2 is also firing, so $U_3 \subset U_2$ in any realization of $\mathcal{C}$.

Exercise 7.2. Extract as many receptive field relationships as you can from the following code:

$$\mathcal{C} = \{0000, 0001, 0011, 0111, 1001\}.$$

Are any of these RF relationships consequences of others on the list?

Importantly, the RF relationships extracted from a code show relationships which must hold in *any* realization of the code $\mathcal{C}$; they are features of the code, rather than of any one particular realization $\mathcal{U}$.

Exercise 7.3. Let $\mathcal{C}$ be the code from Exercise 7.2. Draw two different realizations of $\mathcal{C}$ in $\mathbb{R}^2$. Verify that the set of RF relationships you found in that exercise hold for both realizations.

While RF relationships are associated directly with a code $\mathcal{C}$, there are some RF relationships which will hold for *any* code $\mathcal{C}$, as the following exercise shows.

Exercise 7.4. Let $\mathcal{C}$ be an arbitrary code. Show that if σ and τ are subsets of $[n]$ such that $\sigma \cap \tau \neq \emptyset$, then any receptive field relationship of the form

$$\bigcap_{i\in\sigma} U_i \subset \bigcup_{j\in\tau} U_j$$

must be true in any realization $\mathcal{U}$ of $\mathcal{C}$.

We now give a preview of how RF relationships can be used to infer structure.

Example 7.3. Consider the code $\mathcal{C} = \{000, 010, 001, 110, 101\}$. We observe that in this code, neuron 1 never fires without one of neuron 2 or neuron 3 firing, so in any realization $\mathcal{U}$ of $\mathcal{C}$, we must have $U_1 \subset (U_2 \cup U_3)$. However, we also see that neurons 2 and 3 never fire at the same time, so in any realization, $U_2 \cap U_3 = \emptyset$. Thus, U_1 is contained within two completely disjoint sets.

If we make no assumptions about the properties of the receptive fields U_i, then we have merely made observations about the code itself which must hold in any realization. If, however, we make some assumptions about properties of the receptive fields, then we can determine much more interesting properties of possible realizations. For example, we may find that a realization exhibiting RF relationships is impossible.

Exercise 7.5. Show that if that our sets U_i must be convex and open, then there is no realization of the code $\mathcal{C} = \{000, 010, 001, 110, 101\}$ in $\mathbb{R}^2$ (or, indeed, in $\mathbb{R}^d$ for any d).

This exercise exhibits one example of a *topological obstruction*, a feature which indicates that this code cannot have a realization consisting of convex open sets. Section 7.4 goes into more detail about such obstructions.

For small examples, as in Exercise 7.2, we can list all the receptive field relationships by hand. However, for larger codes, such a list will not be feasible. Nor is it clear that a full list is necessary, since some of these relationships are redundant, or are true for every code, and thus convey no important information about the particular code in question. Thus, we need a method to efficiently extract the important RF relationships from a code.

For this, we will turn to algebraic geometry. We motivate its use by considering the example of the simplicial complex, a related combinatorial/topological object which has a well-known algebraic encoding.

7.1.3 The Simplicial Complex of a Code

As we have noted, a code $\mathcal{C}$ represents a collection supp($\mathcal{C}$) of subsets of $[n]$. A simplicial complex is also a collection of subsets of $[n]$, but with an additional property.

Definition 7.3. An abstract *simplicial complex* Δ on $[n]$ is a collection Δ of subsets of $[n]$, such that whenever $\sigma \in \Delta$ and $\tau \subset \sigma$, we have $\tau \in \Delta$ also.

For a particular code $\mathcal{C}$, supp($\mathcal{C}$) may not be a simplicial complex. However, we can associate a simplicial complex to any code as follows.

Definition 7.4. Define the *simplicial complex of the code* $\mathcal{C} \subset \{0, 1\}^n$ as

$$\Delta(\mathcal{C}) \stackrel{\text{def}}{=} \{\sigma \subset [n] \mid \sigma \subset \text{supp}(\mathbf{c}) \text{ for some } \mathbf{c} \in \mathcal{C}\}.$$

Exercise 7.6. Confirm that for any code $\mathcal{C}$, the set $\Delta(\mathcal{C})$ as defined earlier is actually a simplicial complex.

Exercise 7.7. Let $\mathcal{C}$ be the code from Example 7.1. Compute the simplicial complex $\Delta(\mathcal{C})$.

Exercise 7.8. Show that the following three codes have the same simplicial complex; that is, that $\Delta(\mathcal{C}_1) = \Delta(\mathcal{C}_2) = \Delta(\mathcal{C}_3)$.

- $\mathcal{C}_1 = \{0000, 1000, 1100, 1101, 1110\}$
- $\mathcal{C}_2 = \{0000, 1000, 0100, 1100, 1010, 0110, 1110, 1101\}$
- $\mathcal{C}_3 = \{0000, 1000, 0100, 1010, 0110, 1110, 1101\}$

As we see in Exercise 7.8, distinct codes may have the same simplicial complex, as long as any set of neurons which fires together in one code also fires together in the other. This co-firing information would be captured in a realization by noting that the respective receptive fields overlap. We now introduce the *nerve*, a set which encodes the set of nonempty intersections of any potential realization.

Definition 7.5. Let $\mathcal{U} = \{U_1, \ldots, U_n\}$ be a collection of sets in $\mathbb{R}^d$. The *nerve* of $\mathcal{U}$ is

$$\mathcal{N}(\mathcal{U}) = \left\{ \sigma \subset [n] \mid \bigcap_{i \in \sigma} U_i \neq \emptyset \right\}.$$

The following two exercises illustrate the relationship between the simplicial complex of a code $\mathcal{C}$ and the nerve of a realization $\mathcal{U}$ of $\mathcal{C}$.

Exercise 7.9. Let $\mathcal{U} = \{U_1, U_2, U_3, U_4\}$ be the collection of sets from Example 7.1. Compute the nerve $\mathcal{N}(\mathcal{U})$, and show that it is the same as the simplicial complex $\Delta(\mathcal{C})$ computed in Exercise 7.7.

Exercise 7.10. Show that for any code $\mathcal{C} \subset \{0, 1\}^n$ and any realization $\mathcal{U}$ of $\mathcal{C}$, we have

$$\Delta(\mathcal{C}) = \mathcal{N}(\mathcal{U}).$$

We will now discuss an algebraic method for encoding a simplicial complex, and as such, a brief note about algebra is in order. In the rest of this chapter, $k[x_1, \ldots, x_n]$ will denote the ring of polynomials in the variables $x_1, \ldots, x_n$ with coefficients in the field k. An ideal is a nonempty subset I of $k[x_1, \ldots, x_n]$ that is closed under addition (if $a, b \in I$ then $a + b \in I$) and under multiplication by elements in $k[x_1, \ldots, x_n]$ (if $a \in I$ and $f \in k[x_1, \ldots, x_n]$, then $fa \in I$). We will also often refer to an ideal *generated by* a set A, which we will denote $I = \langle A \rangle$; this is the set of elements which can be written as finite combinations of elements in I. For more on these definitions, see for example [5, 6].

Simplicial complexes can be encoded algebraically through the Stanley-Reisner ideal. For a set $\sigma \subset [n]$, denote $x_\sigma = \prod_{i\in\sigma} x_i$. The *Stanley-Reisner ideal* over the field k for the simplicial complex Δ on $[n]$ is defined as

$$I_{\Delta^c} \stackrel{\text{def}}{=} \langle x_\sigma \mid \sigma \notin \Delta \rangle \subset k[x_1, \ldots, x_n].$$

For a more thorough treatment of the Stanley-Reisner ideal, see [7] or [8, Chapter 6]. As the Stanley-Reisner ideal encodes simplicial complexes, it can be used to capture the simplicial complex of a code and as Exercise 7.10 shows, it therefore implies certain intersection information for receptive fields.

Example 7.4. Let $\mathcal{C}$ be the code from Example 7.1. The Stanley-Reisner ideal of $\Delta(\mathcal{C})$ is $I_{\Delta(\mathcal{C})} = \langle x_1x_3 \rangle$. This captures the fact that $\{1,3\} \notin \Delta(\mathcal{C})$, as $\mathcal{C}$ does not exhibit any codewords where neurons 1 and 3 both fire. In any realization of $\mathcal{C}$, we would see the RF relationship $U_1 \cap U_3 = \emptyset$.

The simplicial complex and the nerve capture information about RF relationships of the form $\bigcap_{i\in\sigma} U_i = \emptyset$. However, neither structure captures the information about relationships which show containment; that is, RF relationships of the form $\bigcap_{i\in\sigma} U_i \subset \bigcup_{j\in\tau} U_j$.

Since the simplicial complex information is insufficient to understand a code, we will now introduce an algebraic structure similar to the Stanley-Reisner ideal which encodes *all* RF relationships of a code. This will be the *neural ideal.*

7.2 THE NEURAL IDEAL

Except where otherwise noted, the material in Sections 2 and 3 comes from [9], where the idea of the neural ideal was introduced. As we consider only binary codes, we will work over the field of two elements, $\mathbb{F}_2 = \{0, 1\}$ and in particular, over the polynomial ring $\mathbb{F}_2[x_1, \ldots, x_n]$.

Definition 7.6. Suppose I is an ideal of a polynomial ring $k[x_1, \ldots, x_n]$. The *variety* of I, denoted $V(I)$, is defined to be the set of common zeros of the elements of I; that is,

$$V(I) = \{\mathbf{v} \in k^n \mid f(\mathbf{v}) = 0 \text{ for all } f \in I\}.$$

For an ideal $I \subset \mathbb{F}_2[x_1, \ldots, x_n]$, the variety $V(I)$ is a subset of $\mathbb{F}_2^n$. The neural ideal for a code $\mathcal{C}$ is designed to be an ideal whose variety is precisely $\mathcal{C}$. Before introducing the neural ideal, we consider an important related ideal called the Boolean ideal.

Exercise 7.11. Show that the set of polynomials

$$\mathcal{B} = \{f \in \mathbb{F}_2[x_1, \ldots, x_n] \mid f(\mathbf{v}) = 0 \text{ for all } \mathbf{v} \in \mathbb{F}_2^n\}$$

forms an ideal, and furthermore that $V(\mathcal{B}) = \mathbb{F}_2^n$. $\mathcal{B}$ is called the *Boolean ideal.*

Remark 7.1. Since we will be working over $\mathbb{F}_2$ for the remainder of the chapter, we note that $1 = -1$ in $\mathbb{F}_2$, and thus that $1 + x_i = 1 - x_i$. We usually choose to use the latter notation.

Exercise 7.12. Show that $\mathcal{B} = \langle x_i(1-x_i) \mid i \in [n] \rangle$; that is, that $\mathcal{B}$ is generated by the set of Boolean polynomials $x_i(1 - x_i)$.

The next few examples and exercises provide additional practice with ideals and varieties in $\mathbb{F}_2[x_1, \ldots, x_n]$.

Example 7.5. Consider the ideal $I = \langle f_1, f_2, f_3 \rangle \subset \mathbb{F}_2[x_1, x_2, x_3, x_4]$ where $f_1 = (1-x_1)x_2, f_2 = x_1(1-x_3)$, and $f_3 = x_3x_4$. This ideal contains polynomials such as $h = x_1x_3x_4$ (since $h = x_1f_3$), $g = x_1x_2x_3 - x_2$ (since $g = 1f_1 + x_2f_2$), and $w = x_3^3x_4$ (since $w = x_3^2f_3$).

The variety of this ideal is the set of binary strings that satisfy the system of equations

$$(x_1 - 1)x_2 = 0, \quad x_1(x_3 - 1) = 0, \quad x_3x_4 = 0.$$

An example of such a binary string is 0001. To find all the elements of the variety we can proceed as follows: Since $x_i \in \mathbb{F}_2$, the first equation implies that $x_1 = 1$ or $x_2 = 0$. In the case $x_1 = 1$, the other equations become $x_3 = 1$ and $x_4 = 0$. This gives the elements 1010 and 1110. In the case $x_1 = 0$, then $x_2 = 0$ and $x_3x_4 = 0$, so this gives the elements 0000, 0001, and 0010. This shows that the variety of the ideal is $V = \{1010, 1110, 0000, 0001, 0010\}$. This variety can also be said to have elements of the form 1*10, 000*, 00*0, where the asterisk means that any value is allowed.

Exercise 7.13. Consider the ideal $I = \langle x_1(1-x_3), x_3x_4 \rangle \subset \mathbb{F}_2[x_1, x_2, x_3, x_4]$. Compute all of the elements in the variety of I and compare with the variety found in Example 7.5.

Exercise 7.14. Consider the ideal $I = \langle x_1(1 - x_2), x_2(1 - x_3), x_1(1 - x_3) \rangle$. Show that $x_1(1 - x_3) \in \langle x_1(1 - x_2), x_2(1 - x_3) \rangle$ and use this to show that $I = \langle x_1(1 - x_2), x_2(1 - x_3) \rangle$. Compute all of the elements in the variety of I.

7.2.1 Definition of the Neural Ideal

We now define our main object of study, the neural ideal of a code, which we will associate to the code a collection of polynomials. As we will see, these polynomials can be used to extract information about the RF relationships associated with $\mathcal{C}$. Recall that since $\mathbb{F}_2 = \{0, 1\}$, a code $\mathcal{C}$ can be thought of as a subset of $\mathbb{F}_2^n$.

Let $\mathbf{v} \in \mathbb{F}_2^n$. The *characteristic polynomial* of $\mathbf{v}$ is the polynomial $\rho_{\mathbf{v}} \in \mathbb{F}_2[x_1, \ldots, x_n]$ defined by

$$\rho_{\mathbf{v}} = \prod_{i \in \text{supp}(\mathbf{v})} x_i \prod_{j \notin \text{supp}(\mathbf{v})} (1 - x_j).$$

Exercise 7.15. Show that $\rho_{\mathbf{v}}(\mathbf{c}) = \begin{cases} 1 & \mathbf{v} = \mathbf{c} \\ 0 & \mathbf{v} \neq \mathbf{c}. \end{cases}$

We define two ideals associated with each neural code.

Definition 7.7. Let $\mathcal{C} \subset \mathbb{F}_2^n$. The *vanishing ideal* of $\mathcal{C}$ is defined as

$$I_{\mathcal{C}} = \{f \in \mathbb{F}_2[x_1, \ldots, x_n] \,|\, f(\mathbf{c}) = 0 \text{ for all } \mathbf{c} \in \mathcal{C}\}.$$

Informally, $I_{\mathcal{C}}$ is the set of polynomials in $\mathbb{F}_2[x_1, \ldots, x_n]$ which vanish on the entire code.

The *neural ideal* of the code is the ideal

$$J_{\mathcal{C}} = \langle \rho_{\mathbf{v}} \,|\, \mathbf{v} \in \mathbb{F}_2^n \backslash \mathcal{C} \rangle.$$

That is, $J_{\mathcal{C}}$ is the ideal generated by the characteristic polynomials of noncodewords.

While $J_{\mathcal{C}}$ and $I_{\mathcal{C}}$ are defined quite differently, they capture similar information. The following exercises illustrate the relationship between the neural ideal $J_{\mathcal{C}}$ and the vanishing ideal $I_{\mathcal{C}}$.

Exercise 7.16. Show that every element of $J_{\mathcal{C}}$ will vanish on the entire code, that is, show that $J_{\mathcal{C}} \subset I_{\mathcal{C}}$.

Exercise 7.17. Show that

$$I_{\mathcal{C}} = J_{\mathcal{C}} + \mathcal{B}.$$

Exercise 7.17 shows that the polynomials in $I_{\mathcal{C}}$ and $J_{\mathcal{C}}$ differ only by a polynomial in $\mathcal{B}$ that vanishes on every vector in $\mathbb{F}_2^n$.

Exercise 7.18. Prove that both $I_{\mathcal{C}}$ and $J_{\mathcal{C}}$ have the same variety, $\mathcal{C}$, that is,

$$V(I_{\mathcal{C}}) = V(J_{\mathcal{C}}) = \mathcal{C}.$$

We now give a specific example of the ideal $J_{\mathcal{C}}$ for a code $\mathcal{C}$, and note how the relationship between $J_{\mathcal{C}}$ and $I_{\mathcal{C}}$ can be used to learn about the RF structure of the code.

Example 7.6. Consider the code $\mathcal{C} = \{000, 010, 110, 011\}$. Then,

$$\begin{aligned} J_{\mathcal{C}} &= \langle \rho_v \,|\, v \in \{001, 100, 101, 111\} \rangle \\ &= \langle (1 - x_1)(1 - x_2)x_3, x_1(1 - x_2)(1 - x_3), x_1(1 - x_2)x_3, x_1x_2x_3 \rangle. \end{aligned}$$

This ideal has four generators, but one can reduce the number of generators needed as follows: Since the sum of the last two generators is x_1x_3, it is in $J_{\mathcal{C}}$. Similarly, by adding the first and third generators, we can see that $(1-x_2)x_3 \in J_{\mathcal{C}}$, and by adding the second and third we obtain that $x_1(1-x_2) \in J_{\mathcal{C}}$. It follows that $J_{\mathcal{C}} = \langle x_1x_3, (1-x_2)x_3, x_1(1-x_2)\rangle$ (see Exercise 7.19).

Beyond describing which codewords are not in $\mathcal{C}$, the polynomials in $J_{\mathcal{C}}$ tell us further information about the code. For example, we know the polynomial $x_1x_3 \in J_{\mathcal{C}}$. Since $J_{\mathcal{C}} \subset I_{\mathcal{C}}$, this means that for any $\mathbf{c} \in \mathcal{C}$, we have $c_1c_3 = 0$, and thus the first and third neurons never fire in the same codeword. This leads us to conclude that the RF relationship $U_1 \cap U_3 = \emptyset$ must hold for $\mathcal{C}$.

This example illustrates how we might find RF relationships from the polynomials in $J_{\mathcal{C}}$, and the following section will elaborate on this idea. This example also shows that while our definition for the neural ideal provides a specific set of generators, it is often possible to use a different, shorter list of polynomials as generators for $J_{\mathcal{C}}$. The following exercises illustrate this idea.

Exercise 7.19. Let $\mathcal{C}$ be as in Example 7.6, and let $K = \langle x_1x_3, (1-x_2)x_3, x_1(1-x_2)\rangle$. Show that $J_{\mathcal{C}} = K$ and that $V(K) = V(J_{\mathcal{C}}) = \mathcal{C}$.

Exercise 7.20. Show that the ideals $J = \langle x_1(1-x_2)x_3, x_1x_2x_3\rangle$ and $K = \langle x_1x_3\rangle$ are equal. Show that there is a unique code $\mathcal{C} \subset \mathbb{F}_2^3$ such that $J = J_{\mathcal{C}}$.

Exercise 7.21. Suppose $J_{\mathcal{C}} = \langle x_1(1-x_2), x_1x_2, (1-x_1)(1-x_2), (1-x_1)x_2\rangle$. Show that $J_{\mathcal{C}} = \langle 1\rangle$. What would the associated code $\mathcal{C} \subset \mathbb{F}_2^2$ be in this case?

Exercise 7.22. Consider the code $\mathcal{C} = \{000, 111, 011, 001\}$. Compute $J_{\mathcal{C}}$ and show that $x_1(1-x_3)$ is in $J_{\mathcal{C}}$.

Exercise 7.23. Consider the code $\mathcal{C} = \{000, 100, 110, 010, 011, 001, 101\}$. Compute $J_{\mathcal{C}}$.

Exercise 7.24. Consider the code $\mathcal{C} = \{000, 100, 110, 010, 011, 001\}$. Compute $J_{\mathcal{C}}$, and compare this result with the previous exercise.

7.2.2 The Neural Ideal and Receptive Field Relationships

Since the variety of both ideals $I_{\mathcal{C}}$ and $J_{\mathcal{C}}$ is precisely the code $\mathcal{C}$, information about the elements of the ideals $J_{\mathcal{C}}$ and $I_{\mathcal{C}}$ can be translated into information about the code itself. In particular, we can often conclude RF relationships hold by showing that certain polynomials exist in $I_{\mathcal{C}}$. Example 7.6 gave one example; the following exercise considers another.

Exercise 7.25. Suppose $\mathcal{C} \subset \mathbb{F}_2^n$ is a code and $x_1(1 - x_2) \in J_{\mathcal{C}}$. Show that the RF relationship $U_1 \subset U_2$ must hold for $\mathcal{C}$.

We now formalize this connection between the polynomials in the ideal $I_{\mathcal{C}}$ and the RF relationships associated with $\mathcal{C}$. Throughout this and the following sections, we will use the convention that $U_\sigma = \bigcap_{i \in \sigma} U_i$ and $x_\sigma = \prod_{i \in \sigma} x_i$.

Theorem 7.1 (Lemma 4.2 from [9]). *Let $\mathcal{C} \subset \mathbb{F}_2^n$ be a neural code, and let $\mathcal{U} = \{U_1, \dots, U_n\}$ be any collection of sets in a stimulus space X such that $\mathcal{C} = \mathcal{C}(\mathcal{U})$. Then, for any sets $\sigma, \tau \subset [n]$,*

$$x_\sigma \prod_{i \in \tau} (1 - x_i) \in I_{\mathcal{C}} \Leftrightarrow U_\sigma \subset \bigcup_{i \in \tau} U_i.$$

Exercise 7.26. Use Exercises 7.12 and 7.17 to prove that if $\sigma \cap \tau \neq \emptyset$, then $x_\sigma \prod_{i \in \tau} (1 - x_i) \in I_{\mathcal{C}}$. Apply Theorem 7.1 to give a new proof of Exercise 7.4.

The previous exercise shows that when $\sigma \cap \tau \neq \emptyset$, Theorem 7.1 is a consequence of the Boolean relationships implied by $\mathcal{B}$ and provides trivial information about receptive fields, and in particular, information which is not dependent on the actual code $\mathcal{C}$. Thus, we look to $J_{\mathcal{C}}$ to obtain the RF relationships specific to the code.

Proposition 7.1. *If $\sigma \cap \tau = \emptyset$, we have*

$$x_\sigma \prod_{i \in \tau} (1 - x_i) \in J_{\mathcal{C}} \Leftrightarrow U_\sigma \subset \bigcup_{i \in \tau} U_i.$$

Assuming that $\sigma \cap \tau = \emptyset$, there are three major types[1] of relationships we observe, depending on whether σ or τ or neither is empty.

Monomial: $(\sigma \neq \emptyset, \tau = \emptyset)$: $\quad x_\sigma \in J_{\mathcal{C}} \Leftrightarrow \bigcap_{i \in \sigma} U_i = \emptyset$

Mixed monomial: $(\sigma \neq \emptyset, \tau \neq \emptyset)$: $\quad x_\sigma \prod_{j \in \tau} (1 - x_j) \in J_{\mathcal{C}} \Leftrightarrow U_\sigma \subset \bigcup_{j \in \tau} U_j$

Negative monomial: $(\sigma = \emptyset, \tau \neq \emptyset)$: $\quad \prod_{j \in \tau} (1 - x_j) \in J_{\mathcal{C}} \Leftrightarrow X \subset \bigcup_{j \in \tau} U_j$

We use the convention that $\bigcap_{i \in \emptyset} U_i = X$, the entire stimulus space, and that $x_\emptyset = 1$.

1. In previous work, such as [9], these have been referred to as Type 1, 2, and 3 relationships.

We usually assume that our codes contain the all-zero codeword **0**, representing an instance when none of the neurons were firing. In this case, we find that negative monomial relationships are impossible.

Exercise 7.27. Show that if $\mathbf{0} \in \mathcal{C}$ then it is impossible to have any negative monomials in $J_{\mathcal{C}}$.

The monomial relationships give the same information as the Stanley-Reisner ideal; namely, they describe the simplicial complex of the code.

Exercise 7.28. Show that if $\mathcal{C}_1, \mathcal{C}_2 \subset \mathbb{F}_2^n$ are two codes such that $J_{\mathcal{C}_1}$ and $J_{\mathcal{C}_2}$ contain the same set of monomials, then $\Delta(\mathcal{C}_1) = \Delta(\mathcal{C}_2)$.

We now give some examples and exercises using Theorem 7.1 and its consequence Proposition 7.1 to find the nontrivial RF relationships associated with a code by using its neural ideal.

Example 7.7. Consider the neural ideal $J = \langle x_1x_2x_3, x_1(1-x_2)x_3\rangle$; Exercise 7.20 shows that $J = J_{\mathcal{C}}$ for $\mathcal{C} = \{000, 100, 010, 001, 110, 011\}$. The generator $x_1x_3(1-x_2)$ encodes the information that the intersection of the receptive fields of neuron 1 and neuron 3 is contained in the receptive field of neuron 2 (i.e., $U_1 \cap U_3 \subseteq U_2$). Similarly, the generator $x_1x_2x_3$ encodes that $U_1 \cap U_2 \cap U_3 = \emptyset$. These two pieces of information imply that $U_1 \cap U_3$ is contained in both U_2 and its complement; thus, $U_1 \cap U_3 = \emptyset$. Since this information is encoded by the polynomial x_1x_3, it follows that $x_1x_3 \in J$. Indeed, x_1x_3 is the sum of the two generators of J, and Exercise 7.20 shows that $J = \langle x_1x_3\rangle$.

Example 7.8. Consider the neural ideal $J = \langle x_1(1-x_2), x_2(1-x_3)\rangle \subset \mathbb{F}_2[x_1, x_2, x_3])$. The generator $x_1(1-x_2)$ encodes the information that $U_1 \subseteq U_2$ and $x_2(1-x_3)$ encodes $U_2 \subseteq U_3$. This implies that $U_1 \subseteq U_3$, so the polynomial $x_1(1-x_3)$ is in J (see Exercise 7.14).

Exercise 7.29. Identify the RF relationships of the code with ideal $J_{\mathcal{C}} = \langle x_1x_2x_3, (1-x_1)x_2x_3, (1-x_1)x_2(1-x_3), (1-x_1)(1-x_2)x_3\rangle$. Then, use basic set theory to show that the same RF relationships can be determined from the generators of the ideal $K = \langle (1-x_1)x_3, (1-x_1)x_2, x_2x_3\rangle$. Does $J_{\mathcal{C}} = K$?

Exercise 7.30. Identify the RF relationships of the code with ideal $J_{\mathcal{C}} = \langle x_1x_2x_3, (1-x_1)(1-x_2)x_3, x_1(1-x_2)x_3\rangle$. Then, use basic set theory to show that the same information can be determined from the generators of the ideal $K = \langle (1-x_2)x_3, x_1x_3\rangle$. Does $J_{\mathcal{C}} = K$?

7.3 THE CANONICAL FORM

As the preceding examples and exercises emphasize, it is often the case that the receptive field relationships we discover from polynomials in $I_{\mathcal{C}}$ or $J_{\mathcal{C}}$ give redundant information. In Example 7.7 and Exercise 7.20, once we know that x_1x_3 is in $J_{\mathcal{C}}$ and thus have the RF relationship $U_1 \cap U_3 = \emptyset$, we can infer through basic set theory that $U_1 \cap U_2 \cap U_3 = \emptyset$. Thus, the polynomial $x_1x_2x_3 \in J_{\mathcal{C}}$ provides redundant information about the receptive fields. This polynomial is also in a sense algebraically redundant, since once we know $x_1x_3 \in J_{\mathcal{C}}$, then we know that $x_1x_2x_3 \in J_{\mathcal{C}}$, as it is a multiple of x_1x_3.

We now introduce a set of polynomials which captures the important RF relationship information, while excluding as much as possible those polynomials which give redundant information.

Definition 7.8. A *pseudo-monomial* is a polynomial of the form $x_\sigma \prod_{j\in\tau}(1-x_j)$, where $\sigma \cap \tau = \emptyset$.

Observe that for any $\mathbf{v} \in \mathbb{F}_2^n$, the characteristic polynomial $\rho_{\mathbf{v}}$ is a pseudo-monomial, but that the Boolean polynomial $x_i(1-x_i)$ is not a pseudo-monomial. Monomials, mixed monomials, and negative monomials are all examples of pseudo-monomials.

Definition 7.9. Let $J \subset \mathbb{F}_2[x_1, \ldots, x_n]$ be an ideal. A pseudo-monomial $f \in J$ is *minimal* in J if there is no pseudo-monomial $g \in J$ with $\deg(g) < \deg(f)$ such that $f = gh$ for some $h \in \mathbb{F}_2[x_1, \ldots, x_n]$.

Definition 7.10. Let $\mathcal{C} \subset \mathbb{F}_2^n$ be a code, and let $J_{\mathcal{C}}$ be the associated neural ideal. The *canonical form* of $J_{\mathcal{C}}$, denoted $CF(J_{\mathcal{C}})$, is the set of all minimal pseudo-monomials in $J_{\mathcal{C}}$.

Example 7.9. Let $\mathcal{C}$ be as in Example 7.7, so $J_{\mathcal{C}} = \langle x_1(1-x_2)x_3, x_1x_2x_3\rangle$. The canonical form is $CF(J_{\mathcal{C}}) = \{x_1x_3\}$, and as we have seen, $J_{\mathcal{C}} = \langle x_1x_3\rangle$.

Exercise 7.31. Show that the canonical form for the neural ideal generates the neural ideal, that is, that

$$J_{\mathcal{C}} = \langle CF(J_{\mathcal{C}})\rangle.$$

The canonical form collects *all* of the minimal pseudo-monomials in $J_{\mathcal{C}}$; as such, the term "minimal" comes with a warning. There are many cases in which the set of all minimal pseudo-monomials is not the smallest possible set of pseudo-monomials which generate the ideal, as in the following example.

Example 7.10. Consider the code $\mathcal{C} = \{000, 111, 011, 001\}$. Then $J_{\mathcal{C}} = \langle (1-x_1)x_2(1-x_3), x_1(1-x_2)(1-x_3), x_1(1-x_2)x_3, x_1x_2(1-x_3)\rangle$. The canonical form is $CF(J_{\mathcal{C}}) = \{x_1(1-x_2), x_2(1-x_3), x_1(1-x_3)\}$, but note that $J_{\mathcal{C}} = \langle x_1(1-x_2), x_2(1-x_3)\rangle$ (see Exercise 7.14). Thus, the canonical form does not provide the smallest possible set of pseudo-monomial generators.

Exercise 7.32. Draw a realization of the code with ideal $J_{\mathcal{C}} = \langle x_1x_2x_3, (1-x_1)x_2x_3, (1-x_1)x_2(1-x_3), (1-x_1)(1-x_2)x_3\rangle$. Extract the RF relationships from the canonical form $CF(J_{\mathcal{C}}) = \{x_3(1-x_1), x_2(1-x_1), x_2x_3\}$ and note that these are sufficient to draw a realization (see Exercise 7.29).

Exercise 7.33. Draw a realization of the code with ideal $J_{\mathcal{C}} = \langle x_1x_2x_3, x_1(1-x_2)x_3, (1-x_1)(1-x_2)x_3\rangle$. Extract the RF relationships from the canonical form $CF(J_{\mathcal{C}}) = \langle (1-x_2)x_3, x_1x_3\rangle$, and note that these are sufficient to draw the realization (see Exercise 7.30).

As these exercises illustrate, the canonical form gives a simpler set of information than the original presentation of $J_{\mathcal{C}}$, but still implies all of the same RF relationships. Thus, rather than using $J_{\mathcal{C}}$ as originally presented, we usually compute only the canonical form for $J_{\mathcal{C}}$. The next sections will show two methods for finding the canonical form of a code.

7.3.1 Computing the Canonical Form

Computing the canonical form for the neural ideal code $\mathcal{C}$ can be done iteratively by codeword. Writing down the canonical form for the ideal consisting of no codewords is simple; thereafter, we add the codewords one at a time, updating the canonical form at each step.

Here, we describe an algorithm which takes the canonical form for a given code $\mathcal{C} \subset \{0,1\}^n$, and a codeword $\mathbf{c} \in \{0,1\}^n$, and outputs the canonical form for $\mathcal{C} \cup \{\mathbf{c}\}$. It is generally assumed that $\mathbf{c} \notin \mathcal{C}$ since otherwise the canonical form will not change; however, the success of the algorithm does not depend on this assumption. This algorithm can then be iterated to build the canonical form for a code.

Algorithm 7.1

Input: The canonical form $CF(J_{\mathcal{C}})$ for a code $\mathbf{C} \subset \{0,1\}^n$, and a word $\mathbf{c} \in \{0,1\}^n$.
Output: The canonical form $CF(J_{\mathcal{C}\cup\{\mathbf{c}\}})$.
Step 0: Initialize empty lists L, M, and N.
Step 1: For $f \in CF(J_{\mathcal{C}})$, if $f(\mathbf{c}) = 0$, add f to L; otherwise, add f to M.
Repeat for all polynomials $f \in CF(J_{\mathcal{C}})$.
Step 2: For each polynomial $g \in M$ and each $1 \leq i \leq n$, if
(i) neither x_i nor $(1-x_i)$ divide g, and

(ii) $(x_i - \mathbf{c}_i)g$ is not a multiple of some polynomial in L,
then add $(x_i - \mathbf{c}_i)g$ to N. Repeat for all polynomials $g \in M$ and $1 \leq i \leq n$.
Step 3: Output $L \cup N$.

A brief explanation of the algorithm: the polynomials in the original canonical form $CF(J_{\mathcal{C}})$ will either vanish on the new codeword $\mathbf{c}$, or they will not. In Step 1, we sort the polynomials by this feature. If the polynomial f does vanish on $\mathbf{c}$, then f will remain in the new canonical form (add f to L); if not, we add f to the list M to potentially adjust f so that it will vanish on $\mathbf{c}$. In Step 2, we perform this adjustment to each f, multiplying by a linear term which will certainly vanish on $\mathbf{c}$; namely, $(x_i - \mathbf{c}_i)$. If the result of this multiplication is not a pseudo-monomial, or if it is a multiple of some pseudo-monomial in L and therefore redundant, then we will not retain it; otherwise, we add it to the list of new pseudo-monomials to add to $CF(J_{\mathcal{C}})$ (list N). To ensure that we obtain the entire new canonical form and not only a subset, it is necessary to consider all such linear terms $(x_i - \mathbf{c}_i)$ and add all the valid possible adjustments to the new canonical form. Finally, we output the collection of the old polynomials which are still valid (list L) and the newly-created polynomials (list N) as the new canonical form.

A proof that this process will result in the correct canonical form for the new code can be found in [10].

Example 7.11. Let $\mathcal{C} = \{000, 100, 010\}$. Observe that in this code we have the RF relationships $U_3 = \emptyset$ since neuron 3 never fires, and $U_1 \cap U_2 = \emptyset$ since neurons 1 and 2 never fire together. The canonical form of $J_{\mathcal{C}}$ is $CF(J_{\mathcal{C}}) = \{x_1x_2, x_3\}$.

We will compute $CF(\mathcal{C}')$, where $\mathcal{C}' = \mathcal{C} \cup \{110\}$, inputting $CF(\mathcal{C}) = \{x_1x_2, x_3\}$ and $\mathbf{c} = 110$ into Algorithm 7.1:

Step 0: Set $L = M = N = \emptyset$.
Step 1 for x_3, x_1x_2: Since $f = x_3$ satisfies $f(110) = 0$, add x_3 to L. Since $f = x_1x_2$ does not satisfy $f(110) = 0$, add x_1x_2 to M. Then, $M = \{x_1x_2\}$ and $L = \{x_3\}$.
Step 2 for x_1x_2 and $i = 1, 2, 3$: It is not true that neither x_1 nor $1 - x_1$ divide x_1x_2. So N does not change.

It is not true that neither x_2 nor $1 - x_2$ divide x_1x_2. So N does not change.
It is true that neither x_3 nor $1 - x_3$ divide x_1x_2. However, $(x_3 - 0)x_1x_2$ is a multiple of a polynomial in L. So N does not change.

Step 3: The output is $L \cup N = \{x_3\}$. This ends Algorithm 7.1.

Thus, we find that $CF(\mathcal{C}') = \{x_3\}$. This result makes sense, since our new code $\mathcal{C}' = \{000, 100, 010, 110\}$ is characterized by the RF relationship $U_3 = \emptyset$.

Exercise 7.34. Let $\mathcal{C}' = \{000, 100, 010, 110\}$ as in Example 7.11. Define $\mathcal{C}'' = \mathcal{C}' \cup \{011\}$. Use Algorithm 7.1 to show that $CF(\mathcal{C}'') = \{x_1x_3, x_3(1 - x_2)\}$.

Algorithm 7.1 shows how to adjust a given canonical form to include one new codeword; computing the canonical form for a given code $\mathcal{C}$ is simply an iteration of this process. We note that the canonical form for the empty code $\mathcal{C}_0 = \emptyset$ is $CF(J_{\mathcal{C}_0}) = \{1\}$. Then, we order our code $\mathcal{C}$, and add in the codewords one at a time, using Algorithm 7.1 each time to compute the new canonical form. Once all codewords have been added, the result is the canonical form for the complete code $\mathcal{C}$. Algorithm 7.2 describes this process.

Algorithm 7.2

Input: A code $\mathcal{C} \subset \{0,1\}^n$.
Output: The canonical form $CF(J_{\mathcal{C}})$.
Step 0: Arbitrarily order the codewords of $\mathcal{C}$ (so $\mathcal{C} = \{\mathbf{c}^1, \mathbf{c}^2, \ldots, \mathbf{c}^d\}$).
Step 1: Define the code $\mathcal{C}_0 = \emptyset$. The canonical form of $\mathcal{C}_0$ is $CF(J_{\mathcal{C}_0}) = \{1\}$. Set $j = 1$.
Step 2: Define $\mathcal{C}_j = \mathcal{C}_{j-1} \cup \{\mathbf{c}^j\}$. Input $CF(J_{\mathcal{C}_{j-1}})$ and $\mathbf{c}^j$ into Algorithm 7.1; the output will be $CF(J_{\mathcal{C}_j})$. If $j = d$, output $CF(J_{\mathcal{C}_d})) = CF(J_{\mathcal{C}})$. Otherwise, set $j = j + 1$ and repeat Step 2.

Exercise 7.35. Using the iterative process described in Algorithm 7.2 and $\mathcal{C} = \{000, 100, 010\}$ as in Example 7.11, show that $CF(J_{\mathcal{C}}) = \{x_3, x_1x_2\}$. Then, reorder the codewords in a new way, recompute $CF(J_{\mathcal{C}})$, and show that the result is the same.

7.3.2 Alternative Computation Method: The Primary Decomposition

One well-known way to break down an ideal into manageable computational pieces is by finding its *primary decomposition*. This strategy has been leveraged with considerable success for the Stanley-Reisner ideal; see Ref. [8, Chapter 6] for a description. A *primary decomposition* of the neural ideal is

$$J_{\mathcal{C}} = \mathfrak{p}_1 \cap \cdots \cap \mathfrak{p}_r,$$

where each ideal $\mathfrak{p}_i$ is a primary ideal. We will not go into much detail about primary decompositions here, but we note that the primary decomposition of the neural ideal provides an alternative method to compute the canonical form, as well as providing combinatorial information about the code in its own right.

The primary decomposition gives information about the code which is complementary to that provided by the canonical form. Each ideal in the primary decomposition gives an interval in the Boolean lattice $\mathbb{F}_2^n$ covered by the code; on the other hand, the polynomials in the canonical form can be interpreted to describe intervals of *non*-codewords in the Boolean lattice. The primary ideals used in the decomposition described earlier can be written as ideals with linear generators; recombining these linear generators in their various possible combinations gives us the canonical form. This algorithm and the Boolean lattice interpretation are described in detail in [9].

Example 7.12. Consider the code $\mathcal{C} = \{000, 100, 010, 001, 101\}$. The neural ideal for this code is

$$J_{\mathcal{C}} = \langle (1 - x_1)x_2x_3, x_1x_2(1 - x_3), x_1x_2x_3 \rangle.$$

The canonical form of $J_{\mathcal{C}}$ is $CF(J_{\mathcal{C}}) = \{x_1x_2, x_2x_3\}$, where the pseudo-monomials correspond to the intervals $\{110, 111\}$ and $\{011, 111\}$ among the noncodewords. A primary decomposition of $J_{\mathcal{C}}$ is $J_{\mathcal{C}} = \langle x_1, x_3 \rangle \cap \langle x_2 \rangle$, where $\langle x_1, x_3 \rangle$ has as its variety the Boolean lattice interval $\{000, 010\}$ of codewords, and $\langle x_2 \rangle$ has the interval $\{000, 100, 001, 101\}$ as its variety. Note that polynomials obtained by taking one generator from each ideal in the primary decomposition result in precisely the generators of the canonical form.

7.3.3 Sage Code for Computations

As we have shown, it is possible to compute the canonical form by hand, but for large examples it quickly becomes tedious. Fortunately, the iterative algorithm we have described has been coded for computation in SageMath [11]. To use this code, first download the packages for neural computations from https://github.com/e6-1/NeuralIdeals. If necessary, rename the folder `NeuralIdeals-master` as `NeuralIdeals`. The following example will describe how to use the code to compute the canonical form and extract the RF relationships. For a more detailed guide, see [10].

Once the packages have downloaded, run the following to compile and load them.

```
load("NeuralIdeals/iterative_canonical.spyx")
load("NeuralIdeals/neuralcode.py")
load("NeuralIdeals/examples.py")
```

The code above should give the message

```
Compiling ./NeuralIdeals/iterative_canonical.spyx...
```

Example 7.13. We will use the Sage code to compute the canonical form for the neural code $\mathcal{C} = \{100, 010, 001, 101, 011, 111, 000\}$.

To define the neural code, run the following:

```
neuralCode = NeuralCode(['100','010','001','101','
             011','111','000'])
```

To compute the canonical form, run `CF=neuralCode.canonical()` and `CF` to see the result. This will give the following output.

```
Ideal (x0*x1*x2+x0*x1) of Multivariate Polynomial Ring in
x0, x1, x2 over Finite Field of size 2
```

To have an output with factored generators, we run `CF=neuralCode.factored_canonical()` and `CF` to see the result. The output is `[(x2 + 1) * x1 * x0]`.

Furthermore, we can find the receptive field structure by running the following.

```
RF=neuralCode.canonical_RF_structure()

RF
```

This prints the output, which in this case will be:

```
Intersection of U_['1', '0'] is a subset of Union of
  U_['2']
```

Thus, we obtain the following receptive field structure: $(U_0 \cap U_1) \subseteq U_2$. A possible receptive field in $\mathbb{R}^2$ is shown (on the left) in Fig. 7.3.

Note that we can also obtain a one-dimensional realization using the intervals $U_0 = (0, 3)$, $U_1 = (2, 5)$, and $U_2 = (1, 4)$.

Example 7.14. We will compute the canonical form of $\mathcal{C} = \{1010, 1110, 0000, 0001, 0010\}$ and attempt to draw a convex realization. We proceed as in the previous example.

```
neuralCode = NeuralCode(['1010','1110','0000','0001','
0010'])
CF=neuralCode.factored_canonical()
CF
```

This will give the output

```
[x1*(x0+1),(x2+1)*x0,(x2+1)*x1,x3*x0,x3*x1,x3*x2]
```

To obtain the receptive field structure we use `neuralCode.canonical_RF_structure()` and obtain

```
Intersection of U_['1'] is a subset of Union of U_['0']
Intersection of U_['0'] is a subset of Union of U_['2']
Intersection of U_['1'] is a subset of Union of U_['2']
Intersection of U_['3', '0'] is empty
Intersection of U_['3', '1'] is empty
Intersection of U_['3', '2'] is empty
```

Thus, we obtain the following neural field structure: $U_1 \subseteq U_0 \subseteq U_2$ and $U_3 \cap U_0 = U_3 \cap U_1 = U_3 \cap U_2 = \emptyset$. A possible receptive field in $\mathbb{R}^2$ is shown (on the right) in Fig. 7.3.

Note that we can also obtain a one-dimensional realization using the intervals $U_0 = (1, 4)$, $U_1 = (2, 3)$, $U_2 = (0, 5)$, and $U_3 = (6, 7)$.

Exercise 7.36. For the following codes, use Sage to compute the canonical form of $\mathcal{C}$. Use the RF relationships you obtain to draw a realization of $\mathcal{C}$, using convex sets if possible.

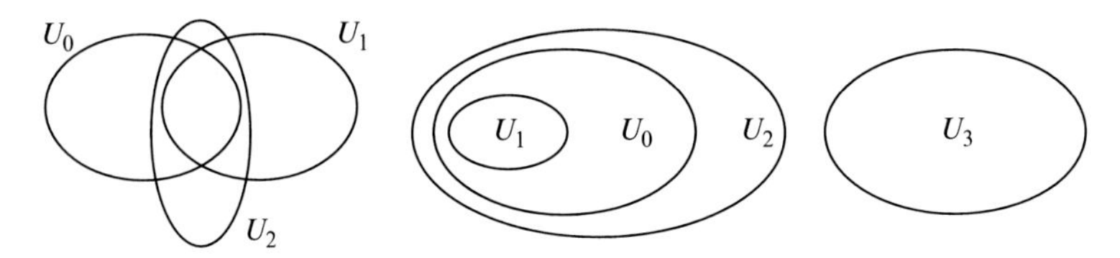

FIG. 7.3 A realization of the code in Example 7.13 (*left*), and one of the code in Example 7.14 (*right*).

1. $\mathcal{C} = \{000, 010, 110, 011\}$
2. $\mathcal{C} = \{000, 100, 110, 010, 011, 001, 101\}$
3. $\mathcal{C} = \{000, 100, 110, 010, 011, 001\}$
4. $\mathcal{C} = \{000, 111, 011, 001\}$
5. $\mathcal{C} = \{000, 100, 010, 101, 011\}$
6. $\mathcal{C} = \{0000, 1000, 0100, 1100, 0010, 1010, 0110, 1110, 0011, 0001\}$
7. $\mathcal{C} = \{0000, 1000, 1100, 0100, 0110, 0010, 0011, 0001\}$
8. $\mathcal{C} = \{0000, 1000, 0100, 1100, 0010, 1010, 0110, 1110, 0001, 1001, 0101, 1101, 0011, 1011, 0111\}$
9. $\mathcal{C} = \{0000, 1000, 0100, 1010, 0110, 0010, 0011\}$

Note: Some of the computations for this book chapter were done using the Ohio Supercomputer Center [12].

7.4 APPLICATIONS: USING THE NEURAL IDEAL

7.4.1 Convex Realizability

One of the main questions motivating the construction of the neural ideal is to determine which codes have realizations using *convex* sets, and which do not. Many types of neurons with receptive fields, including both 2D and 3D place cells, have natural convex receptive fields. Structural features of the code are known which either guarantee or prohibit the existence of a realization with convex sets, and the neural ideal and canonical form can often be used to detect these features. We will refer to a code which has a realization using open convex receptive fields in $\mathbb{R}^d$ for some d as a *convex* code.

As Exercise 7.8 shows, the simplicial complex alone is insufficient to characterize a code. In addition to the fact that different codes may have the same simplicial complex, codes with the same simplicial complex may have very different properties with respect to convex realizability.

Exercise 7.37. Let $\mathcal{C}_1, \mathcal{C}_2$, and $\mathcal{C}_3$ be the three codes from Exercise 7.8. Show that $\mathcal{C}_1$ has a realization with convex open sets in $\mathbb{R}$, that $\mathcal{C}_2$ does not have such a realization, but can be realized with convex open sets in $\mathbb{R}^2$, and that $\mathcal{C}_3$ cannot be realized with convex open sets in $\mathbb{R}^d$ for any dimension d.

As seen in Exercise 7.37 and previously in Exercise 7.5, there exist codes which are not convex. The obstructions to convexity in these exercises are topological in nature, and are specific examples of a broad class of obstructions known as *local obstructions*, defined in [13, 14]. These obstructions follow from an application of the Nerve lemma.

Lemma 7.1 (Nerve Lemma). *If the sets $\mathcal{U} = \{U_1 \ldots, U_n\}$ are convex, then the homotopy type of $\bigcup_{i=1}^n U_i$ is equal to the homotopy type of the nerve $\mathcal{N}(\mathcal{U})$. In particular, $\bigcup_{i=1}^n U_i$ and $N(\mathcal{U})$ have exactly the same homology groups.*

The Nerve lemma is a consequence of [15, Corollary 4G.3]. Local obstructions arise in instances where features of the code dictate that any realization of the code using convex sets would violate the Nerve lemma; thus, a code which can be realized with convex sets must have no local obstructions.

Determining if a code $\mathcal{C}$ has local obstructions to convexity can be reduced to the question of determining whether $\mathcal{C}$ contains a particular minimal code, as Theorem 7.2 will show. However, we first require a definition.

Definition 7.11. Let Δ be a simplicial complex. For any $\sigma \in \Delta$, the *link* of σ in Δ is

$$\mathrm{Lk}_\sigma(\Delta) \stackrel{\text{def}}{=} \{\omega \in \Delta \mid \sigma \cap \omega = \emptyset \text{ and } \sigma \cup \omega \in \Delta\}.$$

Example 7.15. Consider the simplicial complex

$$\Delta = \{\emptyset, \{1\}, \{2\}, \{3\}, \{4\}, \{1,2\}, \{1,3\}, \{1,4\}, \{2,3\}, \{3,4\}, \{1,2,3\}\}.$$

Then, $\mathrm{Lk}_{\{1\}}(\Delta) = \{\emptyset, \{2\}, \{3\}, \{4\}, \{2,3\}\}$, whereas $\mathrm{Lk}_{\{2\}}(\Delta) = \{\emptyset, \{1\}, \{3\}, \{1,3\}\}$ and $\mathrm{Lk}_{\{1,3\}}(\Delta) = \{\emptyset, \{2\}\}$ (See Fig. 7.4).

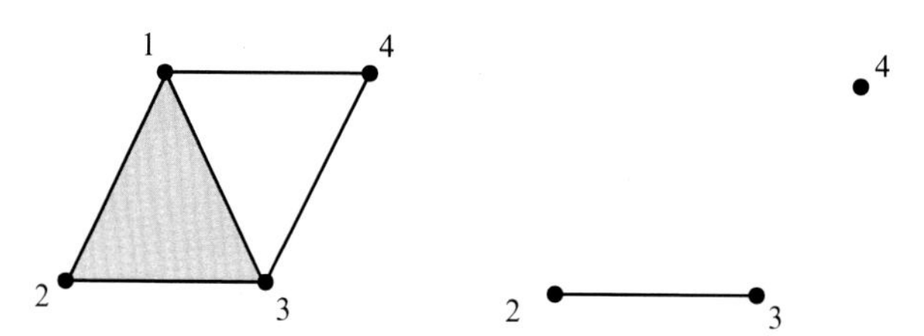

FIG. 7.4 The simplicial complex Δ from Example 7.15 (at *left*) and the link $\mathrm{Lk}_{\{1\}}(\Delta)$ (at *right*). Here, $\mathrm{Lk}_{\{1\}}(\Delta)$ is disconnected, and thus not contractible.

The necessary condition for a code to have no local obstructions will involve determining whether links are contractible. Informally, a topological space (such as a simplicial complex) is *contractible* if it is connected and has no "holes"; more formally, a space is contractible if it is homotopy-equivalent to a single point. In the previous example, $\text{Lk}_{\{1\}}(\Delta)$ is disconnected, and thus is not contractible, but the other two links are both contractible.

Theorem 7.2 (Theorem 1.3 from [13]). *For each simplicial complex Δ, there is a unique minimal code $\mathcal{C}_{\min}(\Delta)$ with the following properties:*

1. *the simplicial complex of $\mathcal{C}_{\min}(\Delta)$ is Δ, and*
2. *any code $\mathcal{C}$ satisfying $\Delta(\mathcal{C}) = \Delta$ has no local obstructions if and only if $\mathcal{C} \supseteq \mathcal{C}_{\min}(\Delta)$.*

Determining this minimal code depends only on the simplicial complex $\Delta(\mathcal{C})$:

$$\mathcal{C}_{\min}(\Delta) = \{\sigma \in \Delta \mid \text{Lk}_{\sigma}(\Delta) \text{ is noncontractible}\} \cup \{\emptyset\}.$$

Exercise 7.38. Recall code $\mathcal{C} = \mathcal{C}_3$ from Exercises 7.8 and 7.37. With $\Delta = \Delta(\mathcal{C})$ as computed in that exercise, show that $\text{Lk}_{\{1,2\}}(\Delta)$ is not contractible. Conclude that $1100 \in \mathcal{C}_{\min}$, and that since $\mathcal{C}$ does not contain 1100, it cannot be convex.

Exercise 7.39. Recall $\mathcal{C} = \{000, 010, 001, 110, 101\}$, the code from Exercise 7.5. Find $\Delta(\mathcal{C})$. Then, identify a set σ such that $\text{Lk}_{\sigma}(\Delta)$ is not contractible, and the associated codeword $\mathbf{c}$ with $\text{supp}(\mathbf{c}) = \sigma$ is not in $\mathcal{C}$. Conclude that $\mathcal{C}$ is not convex.

While it can be difficult in general to detect these local obstructions from the canonical form, it is sometimes possible. Exercise 7.38 exhibits a simple example of such a situation.

Example 7.16. The code $\mathcal{C}_3$ from Exercises 7.8 and 7.38 has canonical form

$$CF(J_{\mathcal{C}_3}) = \{x_3x_4, x_1x_2(1-x_3)(1-x_4), x_3(1-x_1)(1-x_2), x_4(1-x_2), x_4(1-x_1)\}.$$

The first two polynomials together tell us that $(U_1 \cap U_2) \subset (U_3 \cup U_4)$ and $U_3 \cap U_4 = \emptyset$. This gives us a clue that we may have an obstruction using $\sigma = \{1, 2\}$, since in any convex realization, $U_1 \cap U_2$ should also be convex and thus connected. However, we just found that it is contained in the union of two disjoint sets, so it cannot be connected.

We can show that examples such as this one provide a pattern for similar local obstructions.

Exercise 7.40. Suppose $\mathcal{C} \subset \mathbb{F}_2^n$. Show that if there is some set $\sigma \subset [n]$ such that both of the following conditions hold:

1. $x_\sigma(1 - x_i)(1 - x_j) \in CF(J_\mathcal{C})$
2. $x_\sigma x_i x_j \in J_\mathcal{C}$

then $\mathcal{C}$ is not convex, as we can use σ to find a local obstruction.

Demonstrating that a code has a local obstruction proves immediately that the code cannot be convex. However, the converse does not hold; there are codes which have no local obstructions but still have no open convex realization, as in the following example.

Example 7.17 (Lienkaemper et al. [16]). The following code on five neurons has no local obstructions, but has no open convex realization:

$$\mathcal{C} = \{00000, 00100, 00010, 10100, 10010, 01100, 00110, 00011, 11100, 10110, 10011, 01111\}.$$

The obstruction to convexity in this case is more geometric in its flavor—one can use convexity show that the straight line which connects points in two different regions has only a limited set of possible regions it can pass through, and only in certain orders, and this can be shown to be impossible in all cases.

Thus, local obstructions can only prove nonconvexity; we cannot assume that a code without local obstructions is necessarily convex. That said, several large classes of codes are known to have convex realizations. We will now show some examples of these classes, and give algebraic signatures from the canonical form which can tell us immediately if a code fits into one of these classes. The first such class is simplicial complexes.

Theorem 7.3 (Curto et al. [13]). *If $\mathcal{C}$ is a simplicial complex, then $\mathcal{C}$ has a convex realization.*

Proposition 7.2. *Let $\mathcal{C}$ be a neural code. Then $\mathcal{C}$ is a simplicial complex if and only if $CF(J_\mathcal{C})$ consists only of monomials.*

Exercise 7.41. Prove Proposition 7.2.

Theorem 7.3 and Proposition 7.2 combined show that by computing the canonical form $CF(J_\mathcal{C})$, we can immediately detect if $\mathcal{C}$ is a simplicial complex, and if so, conclude that $\mathcal{C}$ is convex. However, it is quite specific to require that a code be a simplicial complex, and many codes which are quite clearly realizable (including any code where one receptive field is covered by others, as in Example 7.1) are not simplicial complexes. Fortunately, there is a relaxation of the simplicial complex condition which is also known to be convex.

Definition 7.12. A code $\mathcal{C}$ is *intersection-complete* if for any $\mathbf{c}, \mathbf{d} \in \mathcal{C}$, there is a codeword $\mathbf{v} \in \mathcal{C}$ such that $\text{supp}(\mathbf{v}) = \text{supp}(\mathbf{c}) \cap \text{supp}(\mathbf{d})$.

Exercise 7.42. Show that codes which are simplicial complexes are intersection-complete. Give an example to show that the converse does not hold.

The larger class of intersection-complete codes is also known to be convex realizable.

Theorem 7.4 (Cruz et al. [17]). *If $\mathcal{C}$ is intersection-complete, then $\mathcal{C}$ has a convex realization.*

Intersection-complete codes are more general than simplicial complex codes, but we can use the canonical form to detect them as well, as the following result indicates.

Theorem 7.5 (Curto et al. [18]). *A code $\mathcal{C}$ containing the all-zero word $\mathbf{0}$ is intersection-complete if and only if $CF(J_{\mathcal{C}})$ consists only of monomials, and mixed monomials of the form*

$$\prod_{i \in \sigma} x_i(1 - x_j).$$

From the previous two results, we see that codes whose canonical forms are entirely monomial (as with simplicial complexes) or allow only the most basic mixed monomials (intersection-complete codes) are convex. This might lead us to suspect that having mostly monomial-like relationships is somehow necessary for convexity. However, this is not the case, as codes with no monomials at all in their canonical form are also convex.

Theorem 7.6 (Curto et al. [13]). *If $\mathcal{C}$ is a neural code and $CF(J_{\mathcal{C}})$ contains no monomials, then $\mathcal{C}$ has a convex realization.*

Furthermore, codes with no monomial relationships are also known to have convex realizations in very low dimensions (1 or 2); see Fig. 7.5 and related discussion. This property also has a simple characterization in terms of the canonical form, as seen in the following exercise.

Exercise 7.43. Show that $CF(J_{\mathcal{C}})$ contains no monomials if and only if $\mathcal{C}$ contains the codeword $\mathbf{1} = 111\ldots1$.

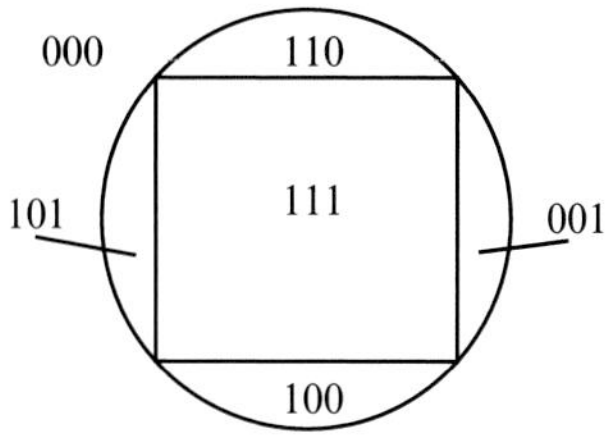

FIG. 7.5 Constructing a 2D realization of $\mathcal{C} = \{000, 100, 001, 101, 110, 111\}$. The codeword 111 is placed at the center of a polygon inscribed within a *circle*; the remaining codewords are assigned to the spaces created around the edges. We then take U_i to be the union of the regions labeled by codewords with neuron i firing.

7.4.2 Dimension

The question of realizability in general can tell us which codes can be represented with convex receptive fields in some dimension. However, the question of which dimension is appropriate is equally difficult, but still quite important from a biological perspective, as the set of neurons may be associated to a space of stimulus with a particular dimension. In our motivating example of place fields, the stimulus space is 2D. However, other famous examples of neurons with receptive fields include neurons in visual cortex which code for the 1D angle of a stimulus [3], and the more recently discovered example of 3D place fields in bats [4]. The dimension is one of the most basic features of the stimulus space which a code should capture.

We will provide some partial answers to the question of in which dimension a code can be realized with convex receptive fields; recall from Exercise 7.1 that without any assumptions on the receptive fields, all codes can be realized in $\mathbb{R}$. Mathematically, the question of the lowest dimension where a code is convex-realizable requires a different approach than the question of realizability. Many of the constructions which provide realizations for the classes of realizable codes seen in the previous section are in very high dimension, and in particular, may not be minimal [13].

Definition 7.13. The *dimension* of a code $\mathcal{C}$ is the minimum $d \in \mathbb{N}$ such that $\mathcal{C}$ has a convex open realization in $\mathbb{R}^d$. If $\mathcal{C}$ is not realizable in any dimension, we say $d = \infty$.

As a basic example, consider the codes containing the codeword $\mathbf{1}$, as described in Theorem 7.6. The proof that all such codes are realizable is constructive, and shows how to realize such codes in two dimensions. Hence, any code containing the codeword $\mathbf{1}$ has dimension $d = 1$ or 2. Fig. 7.5 shows an example of the construction for a simple code.

Obtaining some basic lower bounds on the dimension of a convex realization can be done using the simplicial complex information in the code via the Nerve lemma. A simplicial complex Δ is said to be *d-representable* if there exists a collection of convex sets $\mathcal{U} = \{U_1, \ldots, U_n\}$ in $\mathbb{R}^d$ such that $\Delta = \mathcal{N}(\mathcal{U})$. Thus, for any code $\mathcal{C}$, the dimension of $\mathcal{C}$ is at least as high as the minimal d such that $\Delta(\mathcal{C})$ is d-representable [18].

Some of this basic dimensional information can be extracted very quickly using well-known theorems from convex geometry, such as Helly's theorem.

Theorem 7.7 (Helly's Theorem). *Suppose $U_1, \ldots, U_k \subset \mathbb{R}^d$ are convex sets with $d < k$. If the intersection of every $d + 1$ of these sets is nonempty, then the full intersection $\bigcap_{i=1}^{k} U_i$ is nonempty.*

Since monomial relationships capture intersection information, we can obtain dimension bounds from the canonical form via Helly's theorem.

Exercise 7.44. Apply Helly's theorem to prove that if $\mathcal{C}$ is a code and $CF(J_{\mathcal{C}})$ contains a monomial of degree $\geq d$, then $\mathcal{C}$ cannot be realized in any dimension less than d.

All of the results outlined here rely on information about $\Delta(\mathcal{C})$ to obtain dimension bounds. However, when considering dimension, the simplicial complex does not tell the whole story. In Exercises 7.8 and 7.37, we saw three codes $\mathcal{C}_1, \mathcal{C}_2$, and $\mathcal{C}_3$, all with the same simplicial complex, where $d(\mathcal{C}_1) = 1$, $d(\mathcal{C}_2) = 2$, and $d(\mathcal{C}_3) = \infty$. Using specifics of the canonical form beyond the monomial/simplicial complex information to compute dimension is still a very open problem.

7.5 CONCLUDING REMARKS

We conclude by giving a description of the ongoing work in this area, highlighting those areas where little is known, in the hopes that the reader will consider taking on these challenges.

In the previous section, we saw partial answers to questions about when a code has a convex realization, and in what dimension such a realization might be possible. In each of these situations, we used the algebraic structure of the neural ideal to recognize the RF relationships which informed us how to apply the result. Further work extracting algebraic signatures of convexity continues [18]. However, many related questions about convexity and dimension remain unanswered. Even for known results, it is not immediately clear how to apply the algebraic characterization, as in the following example.

A code $\mathcal{C}$ is *max intersection-complete* if for any collection of facets (maximal sets) $\rho_i \in \Delta(\mathcal{C})$, we have $\bigcap_{i=1}^{k} \rho_i \in \mathcal{C}$.

Theorem 7.8 (Cruz et al. [17]). *If $\mathcal{C}$ is max intersection-complete, then $\mathcal{C}$ is convex.*

Max intersection-complete is a weakening of the condition that $\mathcal{C}$ be intersection-complete. However, unlike intersection-complete codes, an algebraic signature of max intersection-complete codes in $J_{\mathcal{C}}$ is not known.

Similarly, although nonlocal obstructions to convexity have been found (such as those in [16]), it is not known how to detect those obstructions algebraically. The search for other types of obstructions to (or guarantees of) convexity, whether detectable through the neural ideal or not, is still ongoing.

The question of the dimension of a code is likewise still very much open. We have shown that the combinatorial information encoded in the simplicial complex can be used to find a weak upper bound on the dimension, and that Helly's theorem can give us a weak lower bound. Results related to Helly's theorem can be used to find other lower bounds, and in the very specific case of a code containing the all-1s word, we can bound the dimension by 2 [13]. All of this information is based on the information about $\Delta(\mathcal{C})$, however. While this means it can be detected algebraically from the monomial relations, we have also seen that this information by no means sufficient to characterize dimension. In recent work, Zhang and Rosen have characterized the codes of dimension 1 [19], but no such characterization is known for higher dimensions, and no algebraic signature of this characterization is known.

Even in those cases where the code is known to have a convex realization in a low dimension, it is not always clear how to actually construct a realization. In specific cases, a realization can be constructed using the theory of Euler diagrams [20], but the more general question of how to algorithmically construct a convex realization is still open.

Finally, while the results highlighted in this chapter have focused on using the neural ideal under the assumption of convexity, we could also consider loosening this assumption. For example, some recent work has considered the question of dimension of codes under the assumption of connectedness [21]. The machinery of the neural ideal and canonical form is not specific to convexity, and can be used in any context where RF relationships provide useful information.

ACKNOWLEDGMENTS

The authors thank Matthew Macaulay for many helpful suggestions. Carina Curto was supported by NSF DMS-1225666/1537228, NSF DMS-1516881, and an Alfred P. Sloan Research Fellowship. Alan Veliz-Cuba was supported by the Simons Foundation (award ID 516088) and the Mathematical Biosciences Institute (funding through NSF DMS-1440386). Nora Youngs was supported by the Luce Foundation through the Clare Boothe Luce Program.

REFERENCES

[1] J. O'Keefe, J. Dostrovsky, The hippocampus as a spatial map. Preliminary evidence from unit activity in the freely-moving rat, Brain Res. 34 (1) (1971) 171–175.

[2] C. Curto, V. Itskov, Cell groups reveal structure of stimulus space, PLoS Comput. Biol. 4 (10) (2008) e1000205.

[3] D.H. Hubel, T.N. Wiesel, Place fields of single neurons in the cat's striate cortex, J. Physiol. 148 (3) (1959) 574–591.
[4] M. Yartsev, N. Ulanovsky, Representation of three-dimensional space in the hippocampus of flying bats, Science 340 (6130) (2013) 367–372.
[5] D.A. Cox, J. Little, D. O'Shea, Ideals, varieties, and algorithms: an introduction to computational algebraic geometry and commutative algebra, in: Undergraduate Texts in Mathematics, fourth ed., Springer-Verlag, Berlin, 2015.
[6] D.S. Dummit, R.M. Foote, Abstract Algebra, third ed., John Wiley & Sons, New York, NY, 2004.
[7] E. Miller, B. Sturmfels, Combinatorial commutative algebra, in: Graduate Texts in Mathematics, vol. 227, Springer-Verlag, New York, NY, 2005.
[8] R. Robeva, M.E. Macauley, Algebraic and Combinatorial Computational Biology, Elsevier, Amsterdam, 2018.
[9] C. Curto, V. Itskov, A. Veliz-Cuba, N. Youngs, The neural ring: an algebraic tool for analyzing the intrinsic structure of neural codes, Bull. Math. Biol. 75 (9) (2013) 1571–1611.
[10] L.D. Garica Puente, R. Garcia, R. Kruse, D. Miyata, E. Petersen, N. Youngs, Neural ideals in SageMath, In: Davenport, J.H., Kauers, M., Labahn, G., Urban, J. (eds) Mathematical Software – ICMS 2018. ICMS 2018. Lecture Notes in Computer Science, vol 10931, Springer, 2018.
[11] SageMath, The Sage Mathematics Software System (Version 8.2), Available from: http://www.sagemath.org. (Accessed July 12, 2018).
[12] Ohio Supercomputer Center, Columbus, OH, 1987, Available from: http://osc.edu/ark:/19495/f5s1ph73. (Accessed July 10, 2018).
[13] C. Curto, E. Gross, J. Jeffries, K. Morrison, M. Omar, Z. Rosen, A. Shiu, N. Youngs, What makes a neural code convex?, SIAM J. Appl. Algebr. Geom. 1 (1) (2017) 222–238.
[14] C. Giusti, V. Itskov, A NO-GO theorem for one-layer feedforward networks, Neural Comput. 26 (11) (2014) 2527–2540.
[15] A. Hatcher, Algebraic Topology, Cambridge University Press, Cambridge, UK, 2002.
[16] C. Lienkaemper, A. Shiu, Z. Woodstock, Obstructions to convexity in neural codes, Adv. Appl. Math. 85 (2017) 31–59.
[17] J. Cruz, C. Giusti, V. Itskov, W. Kronholm, On open and closed convex codes, arxiv.org/abs/1609.03502.
[18] C. Curto, E. Gross, J. Jeffries, K. Morrison, Z. Rosen, A. Shiu, N. Youngs, Algebraic signatures of convex and non-convex codes, arxiv.org/abs/1807.02741.
[19] Z. Rosen, Y.X. Zhang, Convex neural codes in dimension 1, arxiv.org/abs/1702.06907.
[20] E. Gross, N. Kazi Obatake, N. Youngs, Neural ideals and stimulus space visualization, Adv. Appl. Math. 95 (2018) 65–95.
[21] R. Mulas, N.M. Tran, Minimal embedding dimensions of connected neural codes, arxiv.org/abs/1706.03999.

FURTHER READING

[22] C. Curto, What can topology tells us about the neural code?, Bull. AMS 54 (1) (2017) 63–78.
[23] W.A. Stein, et al., Sage Mathematics Software (Version 8.2), 2016, Available from: http://www.sagemath.org. (Accessed July 12, 2018).

Chapter 8

Predicting Neural Network Dynamics via Graphical Analysis

Katherine Morrison*[*] and Carina Curto[†]
[*]School of Mathematical Sciences, University of Northern Colorado, Greeley, CO, United States, [†]Department of Mathematics, The Pennsylvania State University, University Park, PA, United States

8.1 INTRODUCTION

8.1.1 Neuroscience Background and Motivation

Neurons in the brain have intricate patterns of nonrandom connections between them. Indeed, the complexity of connectivity is one of the most important features setting neurons apart from other types of cells in the body. How does this connectivity shape dynamics? This question is of particular interest in the study of local recurrent networks, which contain collections of neurons with similar functional properties. Such networks are found in cortical areas like the mammalian hippocampus and visual cortex, and the role of recurrent—as opposed to feedforward [1]—connectivity serves to shape neural responses into meaningful patterns of activity. Even in simple models, however, the effects of connectivity on neural activity are poorly understood.

In this chapter, we focus on the Combinatorial Threshold-Linear Network (CTLN) model, first introduced in 2016 [2]. This is a simplified mathematical model of neural networks that allows us to focus specifically on connectivity as the key ingredient controlling the dynamics. The emergent dynamics, however, are nonlinear and complex, exhibiting many of the features believed to underlie information processing in the brain. For example, CTLNs can be multistable, meaning that the network possesses multiple steady states (a.k.a. stable fixed points). Depending on the initial condition, the activity will evolve to one steady state or another, mimicking decision making and memory retrieval in the brain. In this manner, CTLNs are similar to Hopfield networks and other classical attractor neural networks that are popular models for associative memory [3, 4]. Because of their mathematical tractability, however, CTLNs provide a new window into understanding how detailed connectivity influences these processes in the brain.

Algebraic and Combinatorial Computational Biology. https://doi.org/10.1016/B978-0-12-814066-6.00008-8

CTLNs also exhibit other aspects of nonlinear dynamics that play a functional role in the nervous system. For example, a network can possess multiple limit cycles or even multiple chaotic attractors. Limit cycles, in particular, have long been used to model central pattern generators (CPGs) controlling animal locomotion, breathing, or other periodic behaviors [5–7]. The activity of neurons in a limit cycle is often sequential, with neurons taking turns firing in an orderly sequence of activation. Such sequences have also been observed in higher-level areas, such as the mammalian cortex and hippocampus [8–11]. As an example, consider the problem of remembering a seven-digit phone number, such as 867-5309. Many people will repeat the number over and over again in their working memory, a process that can be modeled as selecting a limit cycle in a network where neurons representing the various digits fire in a repeating sequence. How does the connectivity of a network support these kinds of neural functions? Can one predict the emergent sequences from the structure of the underlying graph?

In this chapter, we will introduce CTLNs and make some of our motivating neuroscience questions more precise. Next, we will explore how CTLNs can be analyzed as a patchwork of linear systems of ordinary differential equations (ODEs), with the nonlinear behavior emerging from the transitions between adjacent linear regimes. After that we will develop a graph-theoretic analysis that enables us to predict various features of the dynamics directly from the underlying connectivity graph. These results greatly simplify the fixed point analysis from the previous section, and also reveal the remarkable degree to which the combinatorial structure of the graph controls dynamics, irrespective of the model's other parameters. Finally, we will use these findings to predict sequences from the graph, and study the effect of symmetry on a network's attractors. The mathematical topics we will visit along the way include concepts from linear algebra, differential equations, dynamical systems, graph theory, and a bit of group theory (in disguise).

8.1.2 The CTLN Model

The dynamics of threshold-linear networks (TLNs) are governed by the following system of ODEs,

$$\frac{dx_i}{dt} = -x_i + \left[\sum_{j=1}^{n} W_{ij}x_j + \theta\right]_+, \quad i = 1, \ldots, n, \tag{8.1}$$

where n is the number of neurons. The dynamic variable $x_i(t) \in \mathbb{R}_{\geq 0}$ is the activity level (or "firing rate") of the ith neuron, and $\theta > 0$ is a constant external input. The values W_{ij} are entries of an $n \times n$ matrix of real-valued connection strengths. The threshold nonlinearity $[\cdot]_+ \stackrel{\text{def}}{=} \max\{0, \cdot\}$ is critical for the model to produce nonlinear dynamics; without it, the system would be linear (see Appendix for brief review of linear systems of ODEs).

CTLNs are a special case of TLNs, where we restrict to connection strengths W_{ij} that are obtained from a simple[1] directed graph G in the following way:

$$W_{ij} = \begin{cases} 0 & \text{if } i = j, \\ -1 + \varepsilon & \text{if } i \leftarrow j \text{ in } G, \\ -1 - \delta & \text{if } i \not\leftarrow j \text{ in } G, \end{cases} \tag{8.2}$$

where $i \leftarrow j$ indicates that there is an edge from j to i in the graph G, and $i \not\leftarrow j$ indicates that there is no such edge. A CTLN is thus completely specified by the choice of a directed graph G, along with three positive real parameters: ε, δ, and θ. We additionally require that $\delta > 0$, and $0 < \varepsilon < \frac{\delta}{\delta+1}$; when these conditions are met, we say the parameters are within the *legal range*.

The rate of change dx_i/dt consists of two parts: a leak term, $-x_i$, and a thresholded term. In the thresholded term, $\sum_{j=1}^{n} W_{ij}x_j$ is the sum of the weighted synaptic inputs to neuron i. Note that since $x_j \geq 0$, this sum is negative; however, the external drive $\theta > 0$ allows the net input to be positive when the inhibitory inputs to the neuron are not too strong. The leak term ensures that in the absence of a net positive input, i.e., when the thresholded term is zero, a neuron's firing rate will decay exponentially to zero.

Notice that $W_{ij} < 0$ whenever $i \neq j$. We interpret the CTLN as modeling a network of n excitatory neurons, whose net interactions are inhibitory due to a background inhibition that does not enter explicitly into the model (see Fig. 8.1A). When $j \not\to i$, we say j *strongly inhibits* i; when $j \to i$, we say j *weakly inhibits* i, and we interpret the weak inhibition as the sum of an excitatory connection with the background inhibition. Note that because $-1 - \delta < -1 < -1 + \varepsilon$, when $j \not\to i$, neuron j inhibits i *more* than it "inhibits itself" via its leak term; when $j \to i$, neuron j inhibits i *less* than it inhibits itself. These differences in inhibition strength cause the activity to follow the arrows of the graph (see Fig. 8.1C).

For fixed parameters, only the graph G varies between networks. Thus, we can attribute all differences in dynamics to differences in connectivity, providing insight into how neural connectivity shapes emergent dynamics. For all simulations in this chapter, we fix the parameters at $\theta = 1$, $\varepsilon = 0.25$, and $\delta = 0.5$, unless otherwise noted. We refer to these values as the *standard parameters*.

Variety of Dynamics of CTLNs

Despite the simplicity of the nonlinearity, CTLNs exhibit the full range of nonlinear dynamic phenomena: multistability, limit cycles, quasiperiodic attractors, and chaos. Multistability (i.e., the coexistence of multiple stable fixed points)

1. A graph is *simple* if it does not have loops or multiple edges in the same direction between a pair of nodes.

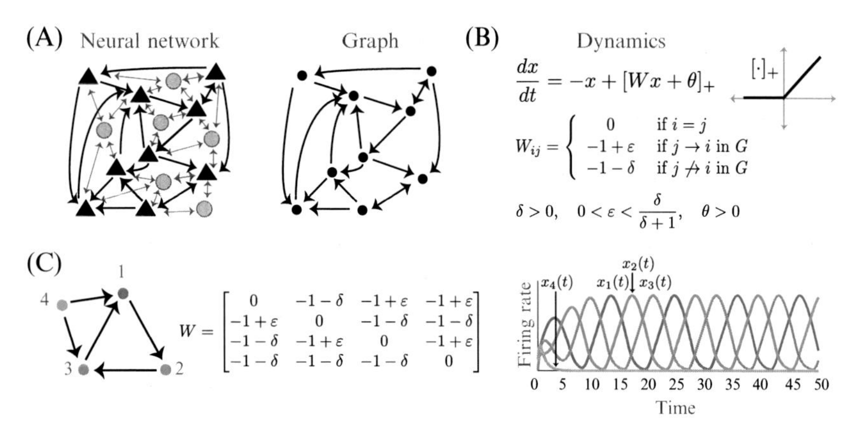

FIG. 8.1 (A) A neural network with excitatory pyramidal neurons (*triangles*) and a background network of inhibitory interneurons (*gray circles*) that produce global inhibition. The corresponding graph (*right*) retains only the excitatory neurons and their connections. (B) Equations for the CTLN model. (C) A directed graph (*left*), and its corresponding connection strength matrix W (*middle*). (*Right*) The periodic firing pattern produced by the CTLN with standard parameters. Firing rate curves are *color-coded* to match the corresponding neuron in the graph.

is the only nonlinear behavior that occurs in the case where W is symmetric [12, 13]. For nonsymmetric W, limit cycles, chaotic, and quasiperiodic attractors can also occur. As an example of limit cycle behavior, consider the CTLN in Fig. 8.1C. Notice that the firing rate curve for the *source* neuron 4 (yellow) quickly decays to 0, and the network settles into a sequential firing pattern following the 3-cycle (123) in the graph. This limit cycle emerges for every initial condition, and is thus a *global attractor*.

Fig. 8.2A and B show two graphs with matching *in-degree* and *out-degree* across the nodes (see box of graph theory terminology in Section 8.3.1), and yet they exhibit significantly different dynamics, with A producing a single limit cycle that is a global attractor, while B has two limit cycles. This shows that the CTLN model exhibits dynamics that are truly *emergent*, as the difference cannot be explained by local properties of the nodes. Notice that the high-firing neurons in these limit cycles correspond to three cycles in the graph. In graph A, the 3-cycle is (235); in graph B, they are (125) and (253). Interestingly, both graphs A and B contain an additional 3-cycle (145) that does not have a corresponding limit cycle.

To any limit cycle, we can associate a sequence of neural firing as shown in Fig. 8.2D. These sequences are shaped by cycles in the graph, but neurons not involved in a cycle still participate in the sequence, and a given CTLN can produce multiple sequences of different lengths. Finally, Fig. 8.2C shows that multiple stable fixed points can arise in the same network. Moreover, the sets of active neurons (i.e., the *supports* of the fixed points) can have different sizes.

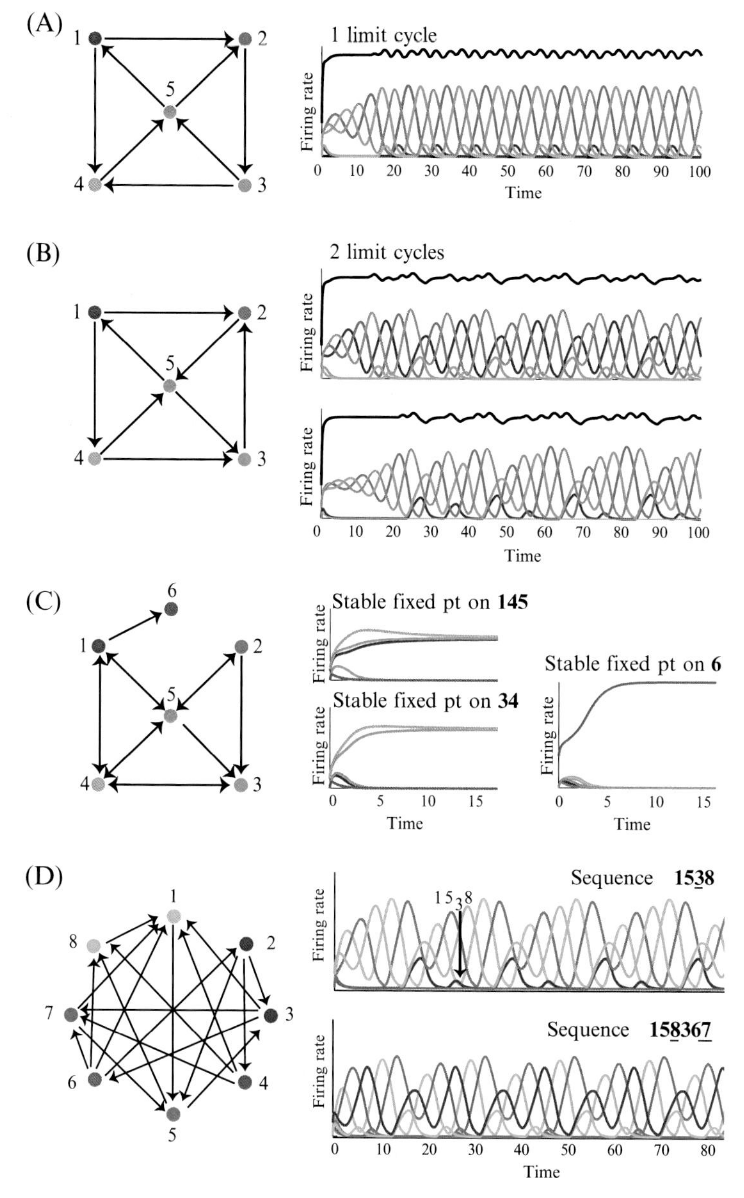

FIG. 8.2 Four graphs and all attractors of the corresponding CTLNs. Firing rate curves are color-coded to match the neuron in each graph. The black curves in A and B show total population activity, obtained by summing all the individual firing rates. (A) A graph whose CTLN has a single global attractor, with high-firing neurons matching the 3-cycle (235). (B) A graph with same set of in-degrees and out-degrees as the graph in A, but with two limit cycle attractors corresponding to the 3-cycles (125) and (253). (C) A graph whose CTLN has three stable fixed points, corresponding to some – but not all – of the cliques in the graph. (D) A graph whose CTLN produces two limit cycles, labeled with the sequence of peaks of neural firing.

Note that each fixed point support corresponds to a *clique* in the graph, but not every clique has a corresponding fixed point.

This variety of dynamic behaviors motivates a number of questions. Which graph structures correspond to stable fixed points of the network? When will limit cycles, chaotic, or quasiperiodic attractors emerge? Why do some three cycles in a graph have corresponding limit cycles, but not others? What determines the sequence of neural firing in a dynamic attractor? The primary goal of this chapter is to introduce methods for analyzing the underlying graph in order to predict features of the dynamics. This will allow us to directly relate a network's connectivity to its dynamics.

8.2 A CTLN AS A PATCHWORK OF LINEAR SYSTEMS

The dynamics in Eq. (8.1) can be written more compactly as $\frac{d\mathbf{x}}{dt} = -\mathbf{x} + [W\mathbf{x} + \boldsymbol{\theta}]_+$. If the threshold nonlinearity were dropped, this would yield the linear system $\frac{d\mathbf{x}}{dt} = (-I + W)\mathbf{x} + \boldsymbol{\theta}$. Assuming $-I + W$ is invertible, this system has a unique fixed point that is stable if all eigenvalues of $-I + W$ have negative real part, and is unstable otherwise (see Appendix for a brief review of fixed points of linear systems). While the threshold is crucial for producing the nonlinear dynamics observed in CTLNs, the fact that the nonlinearity is piecewise linear allows us to analyze these networks as a patchwork of linear systems that partition the positive orthant. In particular, we can identify and classify the fixed points of a CTLN by analyzing the fixed points of each linear system in the patchwork.

Let

$$y_i \stackrel{\text{def}}{=} \sum_{j=1}^{n} W_{ij}x_j + \theta,$$

and rewrite $\frac{dx_i}{dt} = -x_i + [y_i]_+$. When $y_i \leq 0$, we obtain $\frac{dx_i}{dt} = -x_i$ for neuron i. When $y_i > 0$, we have $\frac{dx_i}{dt} = -x_i + y_i$. Thus the set of hyperplanes $\{y_i = 0\}$ partitions the positive orthant into chambers where purely linear systems of ODEs apply.[2] We identify each chamber by a corresponding subset $\sigma \stackrel{\text{def}}{=} \{i \mid y_i > 0\}$; there are up to 2^n possible chambers in the positive orthant. The linear system for each chamber has a fixed point $\mathbf{x}^*$; this fixed point has the form $x_k^* = 0$ for all $k \notin \sigma$ and $\mathbf{x}_\sigma^* = (I - W_\sigma)^{-1}\boldsymbol{\theta}_\sigma$, where the subscript σ indicates restricting the vector/matrix to only the entries indexed by σ. Note that $\mathbf{x}^*$ may or may not be located inside the chamber in which the linear system applies. Thus, for each σ, the corresponding fixed point $\mathbf{x}^*$ is only a true fixed point of the CTLN if it resides in the appropriate chamber. For example, the fixed point $\mathbf{x}^* = \mathbf{0}$, corresponding to $\sigma = \emptyset$, will never lie in its corresponding chamber because $\theta > 0$, and thus is never a fixed point of a CTLN. Note a

2. The hyperplanes $y_i = 0$ should not be confused with the *nullclines* $\frac{dx_i}{dt} = 0$.

fixed point of a CTLN is stable precisely when all eigenvalues of $-I + W_\sigma$ have negative real part.

At a fixed point of the CTLN we must have $x_i = [y_i]_+$ for each $i \in [n]$, so for a fixed point $\mathbf{x}^*$ of the linear system associated with σ to be in its correct chamber, the system must satisfy (i) $x_i^* > 0$ for all $i \in \sigma$, and (ii) $y_k^* \leq 0$ for all $k \notin \sigma$, where y_k^* is obtained by evaluating y_k at $\mathbf{x}^*$. We refer to (i) and (ii) as the "ON"- and "OFF"-neuron conditions, respectively. When these conditions are all satisfied, we say the CTLN has a fixed point with *support* σ.

Example 8.1. Consider the graph G on two neurons with a single directed edge $1 \to 2$, and the corresponding CTLN (see Fig. 8.3). For this network, $y_1 = (-1-\delta)x_2 + \theta$ and $y_2 = (-1+\varepsilon)x_1 + \theta$. Thus the hyperplane $y_1 = 0$ is the horizontal line $x_2 = \frac{\theta}{1+\delta}$, and the hyperplane $y_2 = 0$ is the vertical line $x_1 = \frac{\theta}{1-\varepsilon}$. This cuts the first quadrant into four chambers with a different linear system of ODEs holding in each one. Each chamber has an associated $\sigma = \{i \mid y_i > 0\}$.

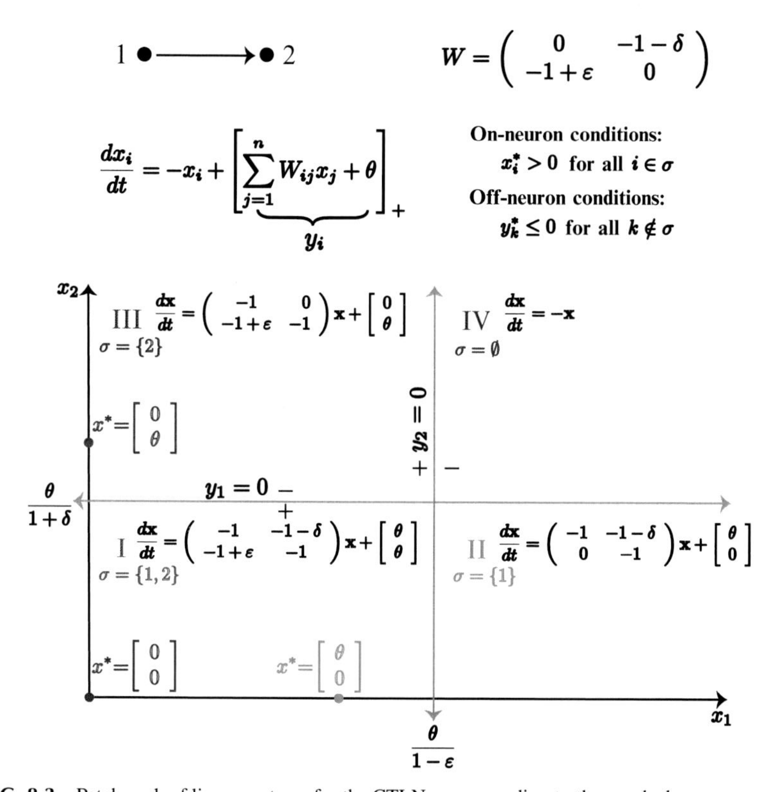

FIG. 8.3 Patchwork of linear systems for the CTLN corresponding to the graph shown.

To find the fixed points of the CTLN, we solve for the fixed point $\mathbf{x}^*$ of each linear system and determine whether it lives in its corresponding chamber. For chamber I, since $y_1, y_2 > 0$, we obtain $\frac{dx_i}{dt} = -x_i + W_{ij}x_j + \theta$ for $i = 1, 2$. Solving $\frac{d\mathbf{x}}{dt} = 0$ yields the fixed point

$$\mathbf{x}^* = (I - W)^{-1}\begin{bmatrix} \theta \\ \theta \end{bmatrix} = \frac{1}{\delta - \varepsilon(\delta + 1)}\begin{bmatrix} \delta\theta \\ -\varepsilon\theta \end{bmatrix}.$$

Within the legal parameter regime, $x_1^* > 0$ and $x_2^* < 0$, and so this fixed point violates the on-neuron conditions and lies outside of chamber I. We conclude that the fixed point for the linear system in chamber I is *not* a fixed point of the CTLN.

By contrast, for chamber III, the fixed point $\mathbf{x}^* = [0,\ \theta]^\top$ of the linear system *does* lie in its chamber (see Fig. 8.3), and is thus a fixed point of the CTLN. By analyzing the remaining two linear systems for chambers II and IV, as in the exercise below, we see that $[0,\ \theta]^\top$ is in fact the unique fixed point for the CTLN. Furthermore, the eigenvalues of the associated matrix for chamber III are both -1, and so this fixed point is stable.

Exercise 8.1. Verify that the linear systems given for chambers II and IV are those shown in Fig. 8.3, and then find the fixed points of those systems. Show that these fixed points do not lie in their corresponding chambers. Conclude that these are not fixed points of the CTLN.

Exercise 8.2. Let G be the graph on two neurons with no edges between them. Analyze the corresponding CTLN to verify that the network has exactly two stable fixed points, $[\theta,\ 0]^\top$ and $[0,\ \theta]^\top$, and one unstable fixed point $\frac{1}{\delta+2}[\theta, \theta]^\top$.

Exercise 8.3. Let G be the graph on two neurons with a bidirectional edge between them. Analyze the corresponding CTLN to verify that the network has exactly one fixed point, $\frac{1}{2-\varepsilon}[\theta,\ \theta]^\top$, which is stable.

8.2.1 How Graph Structure Affects Fixed Points

To build intuition for how the graph structure affects fixed points, we will compute the fixed points for each of the CTLNs defined by the five graphs in Fig. 8.4, using the same strategy that we used in Example 8.1. Each graph in this sequence is obtained from the previous one by adding a single edge. Thus, this analysis will illustrate the impact of individual edges on the collection of CTLN fixed points.

For the CTLN corresponding to graph A, the relevant chambers are defined by the hyperplanes $x_j + x_k = \frac{\theta}{\delta+1}$ for distinct $j, k \in \{1, 2, 3\}$. In Fig. 8.5 we

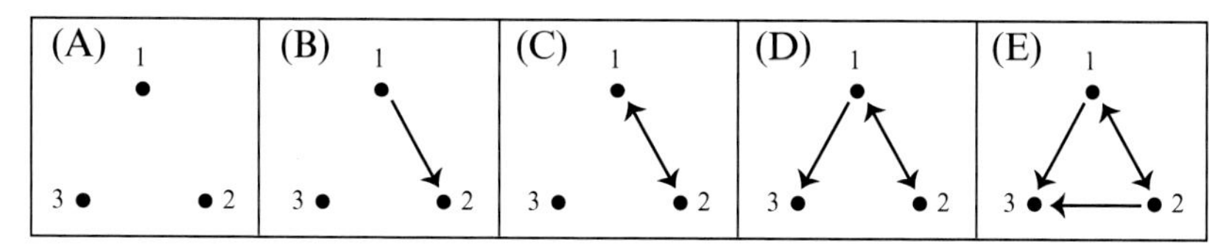

FIG. 8.4 A sequence of graphs where consecutive graphs differ only by a single edge.

depict slices of these chambers corresponding to the x_1x_2-plane and the x_2x_3-plane. Since $x_3 = 0$ in the x_1x_2-plane, the hyperplane $y_1 = 0$ projects to the horizontal line $x_2 = \frac{\theta}{\delta+1}$, with $y_1 > 0$ below the line and $y_1 < 0$ above it. The hyperplane $y_2 = 0$ projects to the vertical line $x_1 = \frac{\theta}{\delta+1}$, and $y_3 = 0$ projects to the line $x_2 = -x_1 + \frac{\theta}{\delta+1}$. Each chamber has a corresponding $\sigma = \{i \mid y_i > 0\}$ that prescribes for which i we can set $[y_i]_+ = y_i$ in the corresponding linear system, while $[y_k]_+ = 0$ for all $k \notin \sigma$. The analogous picture for the x_2x_3-plane is also shown in Fig. 8.5 (top right).

Table 8.1 contains the value of the fixed point for each linear system in the CTLN for graph A, indexed by σ. Note that each fixed point can be obtained by solving its corresponding linear system, as in Example 8.1. Alternatively, the fixed point can be found as

$$x_k^* = 0 \text{ for all } k \notin \sigma, \text{ and } \mathbf{x}_\sigma^* = (I - W_\sigma)^{-1}\boldsymbol{\theta}_\sigma,$$

where the subscript σ indicates restricting the vector/matrix to only the entries indexed by σ. From the top two panels in Fig. 8.5, we see each fixed point, other than $[0, 0, 0]^\top$, lies in its defining chamber for the chambers shown. Similarly, the fixed points for $\{1, 2, 3\}$ and $\{1, 3\}$ also lie in their respective chambers (not shown). Thus, each of these is a fixed point of the CTLN.

To check stability of these fixed points, it is sufficient to check the eigenvalues of $-I + W_\sigma$, since the remaining eigenvalues of the full matrix for the σ system are all -1. Computing these eigenvalues, we find that the fixed points corresponding to the singletons $\{1\}, \{2\}$, and $\{3\}$ are stable, while the matrices $-I + W_\sigma$ for all other systems have at least one positive eigenvalue, and thus their fixed points are unstable.

Exercise 8.4. Analyze the CTLN for graph B in Fig. 8.4, and verify that the network has exactly two stable fixed points, corresponding to $\{2\}, \{3\}$, and one unstable fixed point, corresponding to $\{2, 3\}$.

Exercise 8.5. Analyze the CTLN for graph C in Fig. 8.4 and verify that the network has exactly two stable fixed points, corresponding to $\{1, 2\}, \{3\}$, and one unstable fixed point, corresponding to $\{1, 2, 3\}$.

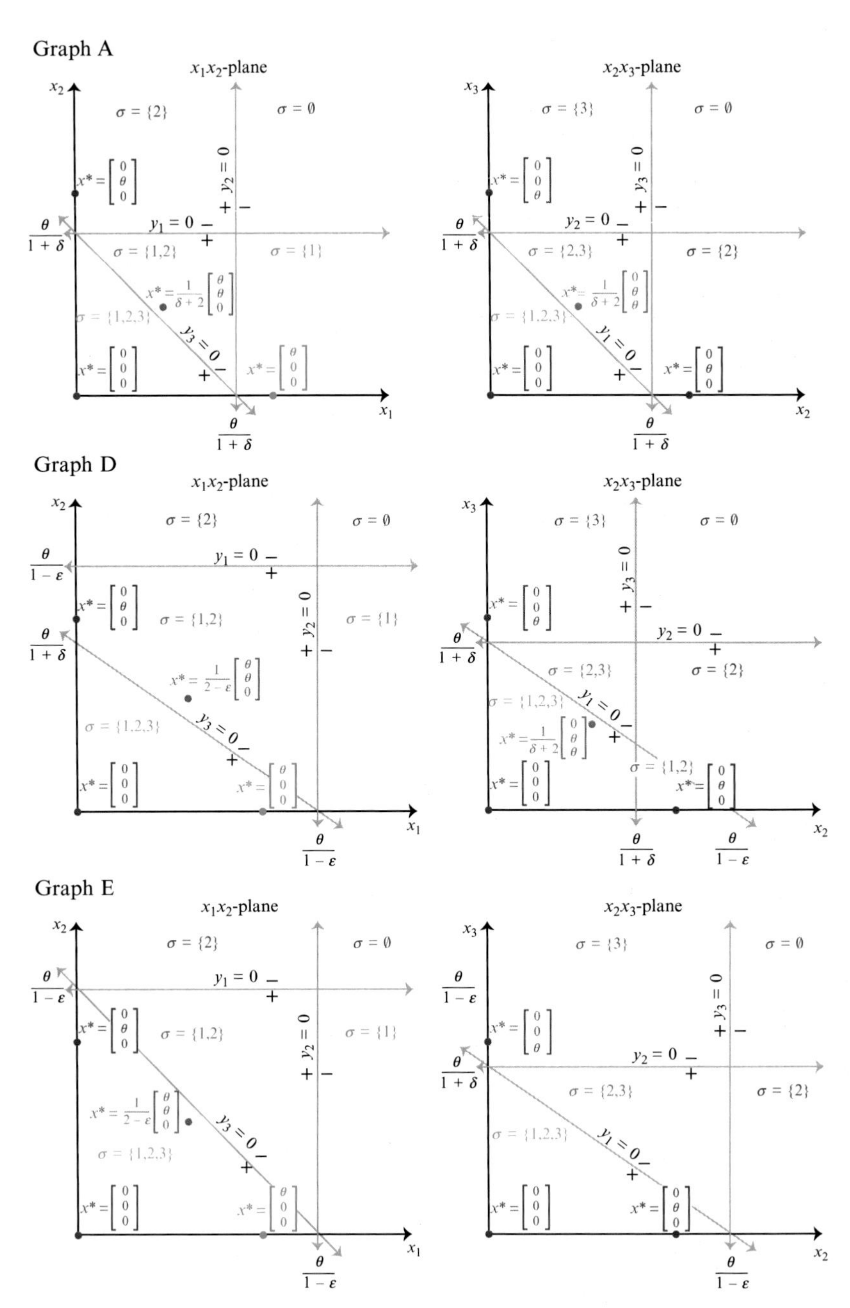

FIG. 8.5 Projections of chambers of linear systems with their corresponding fixed points for the CTLNs defined by graphs A, D, and E from Fig. 8.4.

TABLE 8.1 Fixed Points Corresponding to Each Chamber in the Patchwork of Linear Systems for the CTLN Defined by Graph A in Fig. 8.4

Graph A								
σ	$\{1,2,3\}^*$	$\{1,2\}^*$	$\{1,3\}^*$	$\{2,3\}^*$	$\{1\}^*$	$\{2\}^*$	$\{3\}^*$	$\emptyset$
x^*	$\frac{1}{2\delta+3}\begin{bmatrix}\theta\\\theta\\\theta\end{bmatrix}$	$\frac{1}{\delta+2}\begin{bmatrix}\theta\\\theta\\0\end{bmatrix}$	$\frac{1}{\delta+2}\begin{bmatrix}\theta\\0\\\theta\end{bmatrix}$	$\frac{1}{\delta+2}\begin{bmatrix}0\\\theta\\\theta\end{bmatrix}$	$\begin{bmatrix}\theta\\0\\0\end{bmatrix}$	$\begin{bmatrix}0\\\theta\\0\end{bmatrix}$	$\begin{bmatrix}0\\0\\\theta\end{bmatrix}$	$\begin{bmatrix}0\\0\\0\end{bmatrix}$

This is also a fixed point of the CTLN.

TABLE 8.2 Fixed Points of the Linear Systems of the CTLN for Graph D in Fig. 8.4

Graph D					
σ	$\{1,2,3\}^*$	$\{1,2\}^*$	$\{1,3\}$	$\{2,3\}$	$\{3\}^*$
x^*	$\frac{1}{3+\delta-\varepsilon}\begin{bmatrix}\theta\\\theta\\\theta\end{bmatrix}$	$\frac{1}{2-\varepsilon}\begin{bmatrix}\theta\\\theta\\0\end{bmatrix}$	$\frac{1}{\delta-\varepsilon(\delta+1)}\begin{bmatrix}\delta\theta\\0\\-\varepsilon\theta\end{bmatrix}$	$\frac{1}{\delta+2}\begin{bmatrix}0\\\theta\\\theta\end{bmatrix}$	$\begin{bmatrix}0\\0\\\theta\end{bmatrix}$

**Fixed points of the CTLN. Fixed points for $\sigma = \{1\}, \{2\}, \emptyset$ (not shown) are not fixed points of the CTLN.*

To further investigate the effect of adding edges on the set of CTLN fixed points, we next consider graph D from Fig. 8.4. This graph differs from graph C only by the addition of the $1 \to 3$ edge. The CTLN for graph C has three fixed points corresponding to the subsets $\{1, 2\}$, $\{3\}$, and $\{1, 2, 3\}$ (see Exercise 8.5). Similarly, as shown in Table 8.2 and the plots in Fig. 8.5, the CTLN for graph D has fixed points corresponding to the same supports. In fact, the values of the fixed points for $\{1, 2\}$ and $\{3\}$ are identical to those from graph C, since the $I - W_\sigma$ matrices are identical; while the value of the fixed point for $\{1, 2, 3\}$ differs. Thus, the addition of the $1 \to 3$ edge did not change the fixed point supports. One can also check that the stability of each fixed point remains the same.

Finally, consider graph E, which differs from graph D only by the addition of the $2 \to 3$ edge. Since the submatrix $I - W_\sigma$ for $\sigma = \{1, 2\}$ does not change, the value of the fixed point for this linear system is the same as for graph D (observe that this fixed point is in the same location on the phase plane plots in Fig. 8.5). However, for graph E this is not a fixed point of the CTLN, because the $y_3 = 0$ hyperplane is shifted as a result of the added $2 \to 3$ edge. Additionally, the added edge changes the value of the fixed point for $\{1, 2, 3\}$ such that it is no

TABLE 8.3 Fixed Points of the Linear Systems of the CTLN for Graph E in Fig. 8.4

	Graph E				
σ	$\{1,2,3\}$	$\{1,2\}$	$\{1,3\}$	$\{2,3\}$	$\{3\}^*$
x^*	$\frac{1}{2\delta-2\delta\varepsilon-\varepsilon}\begin{bmatrix}\delta\theta\\ \delta\theta\\ -\varepsilon\theta\end{bmatrix}$	$\frac{1}{2-\varepsilon}\begin{bmatrix}\theta\\ \theta\\ 0\end{bmatrix}$	$\frac{1}{\delta-\varepsilon(\delta+1)}\begin{bmatrix}\delta\theta\\ 0\\ -\varepsilon\theta\end{bmatrix}$	$\frac{1}{\delta-\varepsilon(\delta+1)}\begin{bmatrix}0\\ \delta\theta\\ -\varepsilon\theta\end{bmatrix}$	$\begin{bmatrix}0\\ 0\\ \theta\end{bmatrix}$

**Fixed points of the CTLN. Fixed points for $\sigma = \{1\}, \{2\}, \emptyset$ (not shown) are not fixed points of the CTLN.*

longer a fixed point support for the new graph E network. In fact, $\{3\}$ is the only fixed point support of the CTLN for graph E (Table 8.3).

From the above analyses, we see that the addition of the $1 \to 3$ edge from graph C to graph D does not affect the fixed point supports. By contrast, the addition of the $2 \to 3$ edge from graph D to graph E dramatically alters the set of CTLN fixed point supports. Another important point is that the fixed point supports in each case are independent of the values of ε and δ, provided these fall within the legal range.

Which subgraphs correspond to fixed point supports? How does the way a subgraph is embedded in the larger graph determine whether or not a fixed point for a linear system survives to be a fixed point of the CTLN? In the next section, we will focus on the development of graph rules for predicting fixed point supports, thus eliminating the need to do an (often tedious) analysis of the patchwork of linear systems, as we have done here.

8.3 GRAPHICAL ANALYSIS OF STABLE AND UNSTABLE FIXED POINTS

The previous section illustrated how the graph structure controls the collection of CTLN fixed point supports. Furthermore, the values of ε and δ did not affect the set of supports, it only affects the values of the fixed points themselves. But these values can be immediately computed once the collection of fixed point supports is known, since for each support σ the corresponding fixed point is given by $\mathbf{x}_\sigma^* = (I - W_\sigma)^{-1}\boldsymbol{\theta}_\sigma$ and $x_k^* = 0$ for all $k \notin \sigma$. We can thus restrict our attention to finding the collection of (stable and unstable) fixed point supports given a graph G, which we denote:

$$\text{FP}(G) = \text{FP}(G, \varepsilon, \delta) \stackrel{\text{def}}{=} \{\sigma \subseteq [n] \mid \sigma \text{ is the support of a fixed point of the CTLN } W(G, \varepsilon, \delta)\}.$$

We will use the notation FP(G) because all of the following results on fixed point supports are independent of the actual values of ε and δ, provided these parameters lie within the legal range.

The rest of this section is dedicated to developing tools for finding FP(G) through graphical analysis alone, without appealing to computations such as the ones performed in Section 8.2. All of the mathematical results presented here are contained in [14] and [2].

8.3.1 Graph Theory Concepts

To aid in the graphical analysis of fixed point supports, the following table reviews background graph theory terminology and some useful new graph concepts.

Graph Theory Terminology

Let G be a simple directed graph on n nodes.

- The *in-degree* of a node is the number of incoming edges it receives.
- The *out-degree* of a node is the number of outgoing edges it projects.
- A node is a *sink* if it has out-degree 0.
- A node is a *source* if it has in-degree 0. A source is called *proper* if it has at least one outgoing edge.
- A node is *isolated* if it has in-degree 0 and out-degree 0.
- G is *oriented* if it contains no bidirectional edges.
- For a subset σ, the *induced subgraph* $G|_\sigma$ (read as "G restricted to σ") is the subgraph consisting solely of nodes in σ and the edges between those nodes.
- A subset σ is an *independent set* if there are no edges between any pair of nodes in σ, that is, if every node is isolated in $G|_\sigma$.
- A subset σ is a *cycle* if there is an ordering of the nodes $1, \ldots, |\sigma|$ such that $G|_\sigma$ has $|\sigma|$ edges, all of the form $i \to i+1 \pmod{|\sigma|}$.
- A subset σ is a *clique* if every pair of nodes in σ has a bidirectional edge between them in $G|_\sigma$. A clique is called *maximal* if it is not contained in any larger clique.
- A node $k \notin \sigma$ is called a *target of a clique* σ if it receives an edge from every node in σ. A clique with no targets is called *target-free*.
- A graph G has *uniform in-degree d* if every vertex has in-degree d. We say that a subset σ has *uniform in-degree d* if $G|_\sigma$ has uniform in-degree d.

Example 8.2. Consider the graphs in Fig. 8.4. In graph B, node 1 is a proper source, node 2 is a sink, and node 3 is isolated (it is also a source and a sink). Graphs A and B are oriented. In graph A, $\{1, 2, 3\}$ and all of its nonempty subsets are independent sets, while in graph B, $\{1, 3\}$ and $\{2, 3\}$ (and trivially the singletons $\{i\}$) are independent sets. Furthermore, node 3 of graph D is not a target of the maximal clique $\{1, 2\}$, whereas in graph E, node 3 is a target of $\{1, 2\}$.

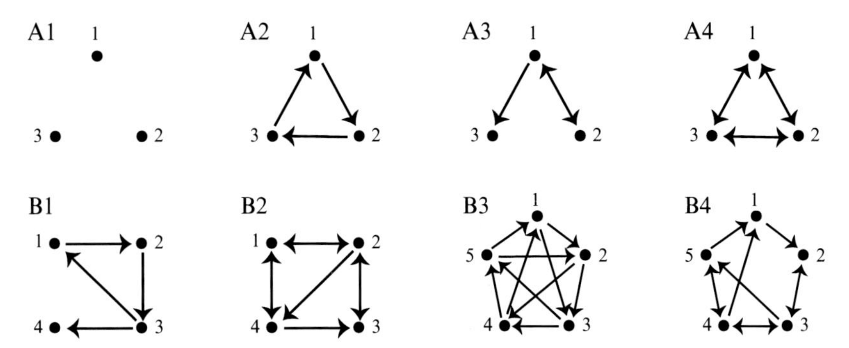

FIG. 8.6 (A1–A4) All uniform in-degree graphs on $n = 3$ nodes. (B1–B4) Some other uniform in-degree graphs on $n = 4, 5$ nodes.

Example 8.3. Consider the graphs in Fig. 8.6, graph A1 is an independent set and graph A2 is a cycle. In graph A4, for $\sigma = \{1, 2\}$ the induced subgraph $G|_\sigma$ is the pair of nodes 1 and 2 with the double edge between them. In A4, all subsets of the vertex set $\{1, 2, 3\}$ are cliques; but only $\{1, 2, 3\}$ is a maximal clique.

Exercise 8.6 (Maximal vs. target-free cliques)**.**

(a) Prove that all target-free cliques are maximal.
(b) Draw an example graph with a maximal clique that is not target-free.

Conclude that the number of target-free cliques in a directed graph G is less than or equal to the number of maximal cliques in G.

Moon and Moser [15] proved that in an undirected graph on n vertices, an upper bound on the number of maximal cliques is given by

$$\text{max \# of maximal cliques} = \begin{cases} 3^{n/3} & \text{if } n \equiv 0 \pmod 3 \\ 4 \cdot 3^{\lfloor n/3 \rfloor - 1} & \text{if } n \equiv 1 \pmod 3 \\ 2 \cdot 3^{\lfloor n/3 \rfloor} & \text{if } n \equiv 2 \pmod 3. \end{cases}$$

Given an undirected graph, one can create a corresponding directed graph by replacing each edge with a bidirectional edge. In this case, every maximal clique of the undirected graph becomes a target-free clique of the directed graph. Thus, the upper bound on the number of target-free cliques in a directed graph is at least as large as the upper bound on the number of maximal cliques in an undirected graph.

Exercise 8.7. For any directed graph, show that there is a corresponding undirected graph such that each target-free clique of the directed graph becomes a maximal clique of the undirected graph. Conclude that the upper bound on target-free cliques in a directed graph equals the above upper bound on the number of maximal cliques in an undirected graph.

Example 8.4. Uniform in-degree subgraphs will prove particularly important in the analysis of fixed point supports. As examples, if σ is a clique, it has uniform in-degree $d = |\sigma| - 1$; at the other extreme, if σ is an independent set, then it is uniform in-degree with $d = 0$. A cycle has uniform in-degree 1. Fig. 8.6 shows the examples of other types of graphs with varying uniform in-degree, including all four uniform in-degree graphs on three nodes. Notice that in a uniform in-degree graph, it is not necessary that the graph be symmetric or even that every node has the same out-degree. Additionally, a subgraph $G|_\sigma$ can have uniform in-degree without every node in σ having the same in-degree in G; it is only necessary that the nodes have identical in-degrees in the induced subgraph.

Exercise 8.8. Draw all graphs on four nodes that have uniform in-degree. *Hint*: There are 14 graphs.

Exercise 8.9. Identify all the uniform in-degree induced subgraphs of graph B4 in Fig. 8.6. Which of the cliques have targets? Which are target-free?

8.3.2 Stable Fixed Points

The stable fixed points of a network correspond to steady states, or stable equilibria, of the system. The population activity $\mathbf{x}(t)$ can converge to any of the stable fixed points, but which (if any) is selected depends on the initial conditions. As is typical in attractor neural networks, these fixed points represent stored memory patterns; the process of evolving from an initial condition into one of the stable fixed points is a standard model for *pattern completion* [13]. The set of all stable fixed points is the collection of all static memories stored in the network.

In this section, we will learn how to infer stable fixed points directly from the graph of a CTLN. As noted before, we restrict our attention to fixed point supports; recovering the actual values of $\mathbf{x}^*(t)$ is straightforward once the supports are known.

Recall that Fig. 8.2C showed how a variety of subgraphs of different sizes could support stable fixed points. Furthermore, analyzing the networks in Fig. 8.4 showed that the interaction of a given subgraph with other nodes in the network affects whether that subgraph corresponds to a fixed point support; for example, the $\{1, 2\}$ clique in Fig. 8.4D supports a stable fixed point, while that same clique in Fig. 8.4E does not.

Opening Exploration: Stable Fixed Point Supports

Fig. 8.7 shows 15 graphs together with the supports of their stable fixed points. Carefully analyze these graphs to conjecture which graph structures give rise to stable fixed points. Be sure to check your conjecture against the full collection of graphs provided.

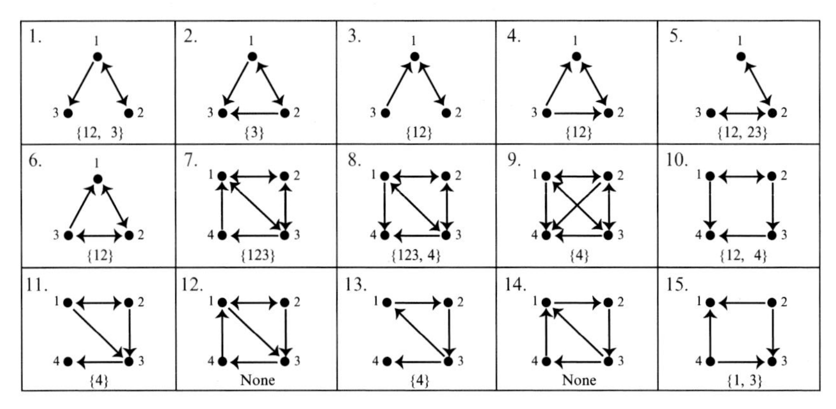

FIG. 8.7 Graphs for opening exploration. Below each graph is the set of stable fixed point supports for the corresponding CTLN. For example, the stable supports for graph 1 are $\{1,2\}$ and $\{3\}$; we denote the set of these supports as $\{12,3\}$ for brevity.

From the graphs in Fig. 8.7, we see that the only time a single neuron supports a stable fixed point is when it has no outgoing edges (i.e., it is a sink). This fact is actually true more broadly: a single node i is the support of a fixed point for a graph G if and only if i is a sink in G; in this case the fixed point is stable.

When do larger subsets support stable fixed points? For σ to be the support of a stable fixed point, it must be a maximal clique. However, not every maximal clique is the support of a stable fixed point, as seen in graphs 2, 6, 9, 11, and 12 of Fig. 8.7. In each of these cases, the cliques that did not support fixed points have a target.

Fact. (Morrison et al. [2]). A clique σ is the support of a fixed point if and only if it is target-free. In this case, the fixed point is stable.

Note that a singleton is trivially a clique and is target-free precisely when it is a sink. Thus the earlier result on sinks is actually a special case of the target-free cliques result. These are the only subgraphs that have been observed to correspond to stable fixed points, and so we have the following conjecture.

Conjecture 8.1 (Morrison et al. [2]). *A subset $\sigma \subseteq [n]$ is the support of a stable fixed point if and only if σ is a target-free clique.*

Exercise 8.10. For each of the following sets of possible stable fixed point supports, create a graph whose CTLN has these supports, or explain why no such graph can exist.
Stable fixed point supports:

(a) $\{1,2,3\}$, $\{3,4\}$ (b) $\{1,2,3\}$, $\{1,2,4\}$ (c) $\{1,2,3\}$, $\{1,2,4\}$, $\{3,4\}$
(d) $\{1,2,3,4\}$ $\{1,5\}$ (e) $\{1,2,3,4\}$ $\{1,5\}$, $\{4,5\}$ (f) $\{1,2\}$, $\{1,3\}$, $\{1,4\}$, $\{2,3\}$

As a special case of the conjecture, we can prove that no stable fixed points emerge for certain classes of graphs guaranteed to have no target-free cliques.

Theorem 8.1 (Morrison et al. [2]). *If G is an oriented graph with no sinks, then the corresponding CTLN has no stable fixed points. Furthermore, the network dynamics are bounded, and thus are guaranteed to be oscillatory or chaotic.*

8.3.3 Unstable Fixed Points

Thus far we have focused on stable fixed points, as these produce attractors that may have computational functions such as associative memory storage and retrieval. But Theorem 8.1 ensures that if G is an oriented graph with no sinks, then the CTLN has no stable fixed point attractors and can thus exhibit only dynamic attractors: limit cycles, quasiperiodic, or chaotic attractors. What shapes these attractors?

In Fig. 8.2, the graphs in A and B are oriented with no sinks, and the attractors displayed by these networks are limit cycles. For the network in Fig. 8.2A, there is a single attractor whose high-firing neurons correspond to the 3-cycle (235). The graph contains a second 3-cycle (145), yet there is no corresponding attractor. Meanwhile the network in B, which differs from that in A only by the orientation of the (235) cycle, has attractors corresponding to the 3-cycles (125) and (253), but does not have an attractor corresponding to the 3-cycle (145). What distinguishes the (145) cycle from the other 3-cycles? It turns out this can be explained by which 3-cycles support *unstable fixed points*. Thus, to predict the presence of dynamic attractors, it is essential that we understand what graph structures produce unstable fixed points. The rest of this section is focused on developing rules for analyzing graphs to find the full collection of fixed point supports $\text{FP}(G)$.

Recall that for σ to support a fixed point, on-neuron conditions must be satisfied for all $i \in \sigma$ and off-neuron conditions must hold for all $k \in [n] \setminus \sigma$. The on-neuron conditions are independent of any nodes outside of σ, and thus σ satisfies the on-neuron conditions if and only if $\sigma \in \text{FP}(G|_\sigma)$. Since the off-neuron condition for a given $k \notin \sigma$ can be checked independently of any other nodes outside σ, this condition is equivalent to checking $\sigma \in \text{FP}(G|_{\sigma \cup \{k\}})$. This recasting of the fixed point conditions gives Rule 0a in the summary table of rules given below.

8.3.3.1 Parity

Recall that for each support σ, the fixed point can be computed as $\mathbf{x}^*_\sigma = (I - W_\sigma)^{-1}\boldsymbol{\theta}_\sigma$. Since each subset has a unique associated fixed point, we have the immediate upper bound $|\text{FP}(G)| \leq 2^n - 1$ simply because a network on n nodes has $2^n - 1$ nonempty subsets. This upper bound is attained when G is an independent set (see Exercise 8.12).

Furthermore, as a straightforward consequence of the Poincaré-Hopf theorem from differential topology [16], every CTLN must satisfy the following parity condition:

$$\sum_{\sigma \in \mathrm{FP}(G)} \operatorname{sgn} \det(I - W_\sigma) = 1. \tag{8.3}$$

In particular, since each term in the sum is either $+1$ or -1, there must be an odd number of terms and thus $|\mathrm{FP}(G)|$ is odd (Rule 0b: parity). This fact is particularly useful for determining if there is a fixed point of full support $[n]$ once all proper subgraphs have been analyzed; specifically, $[n] \in \mathrm{FP}(G)$ precisely when there are an even number of smaller subsets that support fixed points. Eq. (8.3) also yields an upper bound on the number of stable fixed points of a CTLN.

Exercise 8.11. Use Eq. (8.3) to prove that a CTLN on n neurons has at most 2^{n-1} stable fixed points. *Hint*: Recall that a fixed point σ is stable if all the eigenvalues of $-I + W_\sigma$ have negative real part. How are these related to eigenvalues of $I - W_\sigma$ and to its determinant?

8.3.3.2 Sinks and Sources

To graphically characterize $\mathrm{FP}(G)$, we begin with some of the simplest graph structures to identify, namely *sinks* and *sources*. Recall that a singleton $\{i\}$ supports a fixed point if and only if it is a sink; in this case the fixed point is stable. Additionally, an independent set is the support of a fixed point precisely when it is a union of sinks (see Exercise 8.15). Finally, a sink k can be added to an existing fixed point support σ to create a larger fixed point, as long as k does not "kill" σ. In other words, if $\sigma \in \mathrm{FP}(G)$, so that the sink k satisfied the off-neuron conditions and did not kill σ, then $\sigma \cup \{k\} \in \mathrm{FP}(G)$ as well. The converse of this also holds, yielding Rule 1.

Exercise 8.12. Prove that if G is an independent set on n nodes, then it has $2^n - 1$ fixed points.

While sinks are involved in many types of fixed point supports, it turns out that a proper source is never involved in a fixed point support. This holds even when a node is not a source in the full graph, but acts as a proper source in a restricted subgraph. Specifically, if there exists an $i \in \sigma$ such that i is a proper source in $G|_\sigma$, then $\sigma \notin \mathrm{FP}(G)$. In fact, if i is a proper source in G, then $\mathrm{FP}(G) = \mathrm{FP}(G \setminus \{i\})$ (see Exercise 8.18).

Exercise 8.13. Prove that if $\sigma = \{i,j\}$ and $i \to j$ but $j \not\to i$ (i.e., there is just a unidirectional edge from i to j), then $\sigma \notin \mathrm{FP}(G)$.

Rules for identifying fixed points from graphs (adapted from [14])

0. *Fixed point conditions and parity.*
   a. A subset $\sigma \in \text{FP}(G) \Leftrightarrow \sigma \in \text{FP}(G|_\sigma)$ and $\sigma \in \text{FP}(G|_{\sigma \cup \{k\}})$ for every $k \notin \sigma$.
   b. The total number of (stable and unstable) fixed points $|\text{FP}(G)|$ is odd.

1. *Sinks.*
   a. A singleton $\{i\} \in \text{FP}(G) \Leftrightarrow i$ is a sink in G.
   b. An independent set $\sigma \in \text{FP}(G) \Leftrightarrow \sigma$ is a union of sinks.
   c. If k is a sink in G, then $\sigma \cup \{k\} \in \text{FP}(G) \Leftrightarrow \sigma \in \text{FP}(G)$.

2. *Sources.*
   a. If $i \in \sigma$ is a proper source in $G|_\sigma$, then $\sigma \notin \text{FP}(G)$.
   b. If i is a proper source in G, then $\text{FP}(G) = \text{FP}(G \setminus \{i\})$.

3. *Uniform in-degree.* Suppose σ has uniform in-degree. Then $\sigma \in \text{FP}(G) \Leftrightarrow \sigma$ is target-free.
   If σ is a target-free clique, then σ supports a stable fixed point.

4. *Domination.*
   a. If there exists $j, k \in \sigma$ such that k dominates j w.r.t σ, then $\sigma \notin \text{FP}(G|_\sigma)$, and so $\sigma \notin \text{FP}(G)$.
   b. If there exists $j \in \sigma$ and $k \notin \sigma$ such that k dominates j w.r.t σ, then $\sigma \notin \text{FP}(G|_{\sigma \cup \{k\}})$, and so $\sigma \notin \text{FP}(G)$.
   c. For $j \notin \sigma$, if there exists $k \in \sigma$ such that k dominates j w.r.t σ, then $\sigma \in \text{FP}(G|_{\sigma \cup \{j\}}) \Leftrightarrow \sigma \in \text{FP}(G|_\sigma)$.

8.3.3.3 Uniform In-Degree Subgraphs

We now turn to analyzing when uniform in-degree subgraphs support fixed points. Recall that a cycle has uniform in-degree, and Fig. 8.2A and B showed that some 3-cycles have a corresponding limit cycle, but the 3-cycle (145) never has a corresponding attractor. This can be explained by the presence/absence of a corresponding unstable fixed point. To understand when a uniform in-degree subgraph is a fixed point support, we must generalize the notion of target, first introduced for the special case of cliques in Section 8.3.1. Suppose σ has uniform in-degree d and $k \notin \sigma$; we say that k is a *target* of σ if k receives at least $d + 1$ edges from nodes in σ, and σ is said to be *target-free* if it has no targets in G.

Theorem 8.2 (Curto et al. [14]). *Suppose $\sigma \subseteq [n]$ has uniform in-degree d in G. Then $\sigma \in \text{FP}(G)$ if and only if there is no $k \notin \sigma$ receiving at least $d + 1$ incoming edges from σ; in other words,*

$$\sigma \in \text{FP}(G) \Leftrightarrow \sigma \textit{ is target-free in } G.$$

If $d < \frac{|\sigma|}{2}$, then the fixed point is unstable. If $d = |\sigma| - 1$, that is, if σ is a clique, then it is stable.

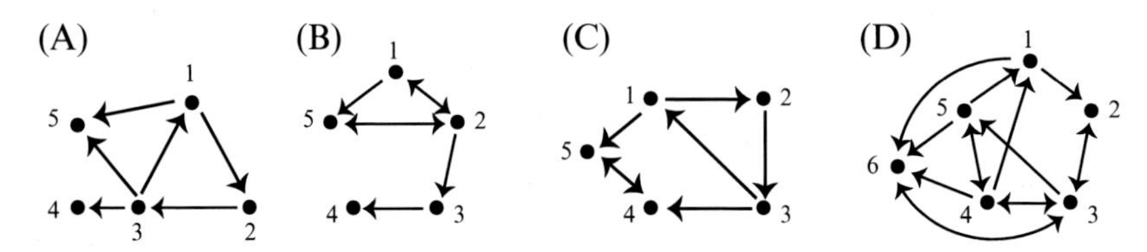

FIG. 8.8 (A–D) Example graphs containing uniform in-degree subgraphs with targets; see Example 8.5 and Exercise 8.14 for details.

The intuition behind a target killing a uniform in-degree fixed point with support σ is that if all the neurons in σ are "on" at the fixed point, this will force the target node to also turn on since it receives a higher input than any of the nodes in σ.[3] Thus there cannot be a fixed point with only the nodes in σ firing.

Example 8.5 (Targets of Uniform In-Degree Subsets)**.** Fig. 8.8A contains a 3-cycle (123), which has uniform in-degree $d = 1$. Node 5 is a target of $\{1, 2, 3\}$ since it receives at least $d + 1 = 2$ edges; in contrast, node 4 is not a target since it only receives d edges. Thus, $\{1, 2, 3\}$ is not a fixed point support of G, although it is a fixed point of $G|_{\{1,2,3,4\}}$ (note also that $\{1, 2, 3, 4\}$ actually has uniform in-degree 1, so it is a fixed point of that restricted subgraph as well). In Fig. 8.8B, both $\{1, 2, 3\}$ and $\{1, 2, 3, 4\}$ have uniform in-degree 1, and node 5 is a target of both of these sets, guaranteeing they do not support fixed points. Note that the edge from 5 back to 2 is irrelevant to the uniform in-degree of the induced subgraphs $G|_{\{1,2,3\}}$ and $G|_{\{1,2,3,4\}}$ and is irrelevant to node 5 being a target. The independent set $\{2, 4\}$ has uniform in-degree 0, but since node 2 is not a sink, $\{2, 4\}$ does not support a fixed point (see Exercise 8.15).

Example 8.6 (FP(G) for the butterfly graph)**.** Consider the *butterfly graph* shown in Fig. 8.9A. We will identify all the uniform in-degree subgraphs and

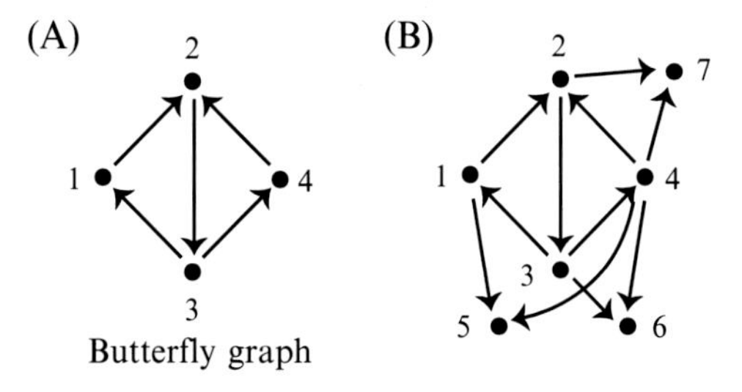

FIG. 8.9 (A) Butterfly graph. (B) Example nodes added to the butterfly graph.

3. Note that the firing rates of all "on" neurons in σ are equal at the fixed point if σ has uniform in-degree.

use this to determine the full set of fixed point supports. There are two 3-cycles, (123) and (234), which are both uniform in-degree 1. Neither has an external node receiving two or more edges, so they are both target-free and thus support fixed points. There is one other uniform in-degree subgraph, the independent set $\{1,4\}$ with $d = 0$. Node 2 is a target of this set, since it receives at least 1 edge from $\{1,4\}$; thus $\{1,4\}$ is not a fixed point support.

Note that no singletons can support a fixed point since there are no sinks (Rule 1a). Additionally, no pair of nodes can support a fixed point because every pair other than $\{1,4\}$ has just a unidirectional edge (see Exercise 8.13). Every set of three nodes, other than the 3-cycles, contains a node that is a proper source in the subgraph, and so cannot be a fixed point support (Rule 2a). Thus the only proper subsets that support fixed points are $\{1,2,3\}$ and $\{2,3,4\}$. But by parity (Rule 0b), $|\,\text{FP}(G)|$ must be odd, so we conclude that the full support $\{1,2,3,4\}$ must also yield a fixed point. Thus for the butterfly graph, $\text{FP}(G) = \{123, 234, 1234\}$; note that for simplicity we drop the set notation for each support contained in $\text{FP}(G)$.

Exercise 8.14 (Uniform In-Degree and Targets).

(a) Find all uniform in-degree subsets of the graph in Fig. 8.8C. Determine which of these are fixed point supports.

(b) Find all uniform in-degree subsets of the graph in Fig. 8.8D. Determine which of these are fixed point supports.

Exercise 8.15 (Unions of Sinks).

(a) Using Theorem 8.2, prove Rule 1b showing that if $\sigma \subseteq [n]$ is an independent set, then $\sigma \in \text{FP}(G)$ if and only if every node $i \in \sigma$ is a sink in G.

(b) Prove that if a graph G has s sinks, then $|\,\text{FP}(G)| \geq 2^s - 1$.

8.3.3.4 Domination

The intuition behind Theorem 8.2, of why a target node would be turned on in the presence of a fixed point of uniform in-degree, can be extended to other scenarios. This leads us to the concept of *domination*, where a node receives the same inputs as another node, and possibly more.

Definition 8.1.[4] We saythat k *dominates* j *with respect to* σ, and write $k >_\sigma j$, if $\sigma \cap \{j,k\} \neq \emptyset$ and the following three conditions hold:

4. In [14], this concept is referred to as *graphical domination*.

(1) for each $i \in \sigma \setminus \{j, k\}$, if $i \to j$ then $i \to k$,
(2) if $j \in \sigma$, then $j \to k$, and
(3) if $k \in \sigma$, then $k \not\to j$.

Note that if $k >_\sigma j$, then $j \not>_\sigma k$, and thus $>_\sigma$ is an *antisymmetric* relation.

Exercise 8.16. Prove that domination $>_\sigma$ is *transitive* for $j, k, \ell \in \sigma$; in other words, if $\ell >_\sigma k$ and $k >_\sigma j$, then $\ell >_\sigma j$.

The following theorem shows how domination can be used to rule in or rule out certain fixed point supports. This gives us Rule 4 in the table of graph rules.

Theorem 8.3 (Curto et al. [14]). *Suppose k dominates j with respect to σ. The following statements all hold:*

(a) *[inside-in] If $j, k \in \sigma$, then $\sigma \notin \mathrm{FP}(G|_\sigma)$, and so $\sigma \notin \mathrm{FP}(G)$.*
(b) *[outside-in] If $j \in \sigma$ and $k \notin \sigma$, then $\sigma \notin \mathrm{FP}(G|_{\sigma \cup \{k\}})$, and so $\sigma \notin \mathrm{FP}(G)$.*
(c) *[inside-out] If $j \notin \sigma$ and $k \in \sigma$, then $\sigma \in \mathrm{FP}(G|_{\sigma \cup \{j\}})$ if and only if $\sigma \in \mathrm{FP}(G|_\sigma)$.*

Fig. 8.10 illustrates the three cases of domination. In the first panel, both k and ℓ receive all inputs that node j receives (as well as possibly other inputs), and so condition (1) of the definition of domination is satisfied. Since $j, k, \ell \in \sigma$ we also need $j \to k$ and $k \not\to j$ for domination to hold. This is true for k, and so k dominates j; however, this does not hold for ℓ since $\ell \to j$, and so ℓ does not dominate j. A single inside-in domination relationship is sufficient to rule out a fixed point, though, and thus by Theorem 8.3(a), σ is not a fixed point support. The second panel in Fig. 8.10 illustrates outside-in domination. In this case, both k_1 and k_2 dominate j since both receive all the inputs that j receives from σ and both receive an edge from j; since $k_1, k_2 \notin \sigma$ it is not necessary that $k \not\to j$. By Theorem 8.3(b) we conclude that σ cannot support a fixed point. Finally, the third panel shows inside-out domination: $k_1, k_2 \in \sigma$ both receive all inputs that j receives, and there is no edge $k \to j$; thus both k_1 and k_2 dominate j.

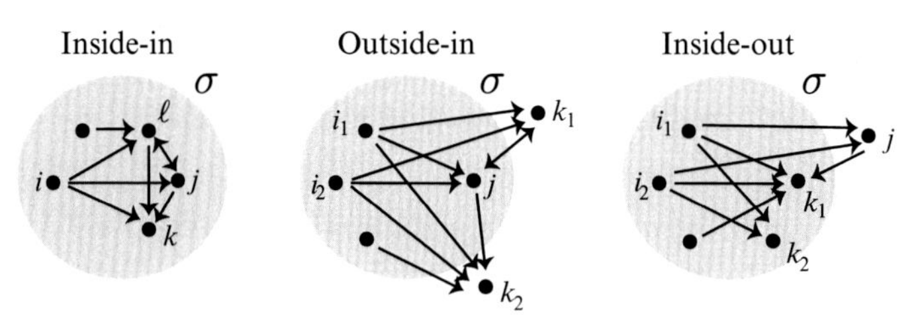

FIG. 8.10 Three cases of domination.

Note that it is not necessary that $j \to k$ since $j \notin \sigma$. Theorem 8.3(c) guarantees that σ survives the addition of node j, that is, $\sigma \in \text{FP}(G|_{\sigma \cup \{j\}})$, precisely when $\sigma \in \text{FP}(G|_\sigma)$.

Example 8.7 (Butterfly graph survival rules)**.** Returning to the butterfly graph, we examine when the fixed point $\sigma = \{1, 2, 3, 4\}$ survives the addition of a single node. Note that since σ is a fixed point of the butterfly graph, it cannot contain any inside-in domination relationships by Theorem 8.3(a); thus we need only determine whether inside-out or outside-in domination arise from the addition of the single node.

In Fig. 8.9B, consider $G|_{\{1,2,3,4,5\}}$, where node 5 receives from 1 and 4. Node 5 does not dominate any nodes in σ since 1 and 4 both receive from node 3 while 5 does not. In fact, node 5 is dominated by node 2, since 2 receives from 1 and 4 and $2 \not\to 5$. Thus, by Theorem 8.3(c), inside-out domination guarantees that $\sigma \in \text{FP}(G|_{\{1,2,3,4,5\}})$ since σ was a fixed point of the restricted butterfly graph.

Next, consider $G|_{\{1,2,3,4,6\}}$ where node 6 receives from nodes 3 and 4. In this case, 6 dominates 4 since 6 receives from 3 and $4 \to 6$. Theorem 8.3(b) guarantees that $\sigma \notin \text{FP}(G|_{\{1,2,3,4,6\}})$ by outside-in domination.

The survival of σ when a single node is added can actually be determined via domination in every case (see Exercise 8.17), except when the added node receives from 1 and 2, or equivalently (by symmetry) from 2 and 4, as in the case of node 7 in Fig. 8.9B. In this case, there are no domination relationships of any type. Thus an explicit computation is necessary to check whether the added node satisfies the off-neuron condition for σ. It turns out this condition is in fact satisfied and $\sigma \in \text{FP}(G|_{\{1,2,3,4,7\}})$. Furthermore, the off-neuron condition holds for every value of ε and δ in the legal range, and thus the survival rules for the butterfly graph are parameter-independent.

Exercise 8.17. Let G be the butterfly graph union a single node k, and let $\sigma = \{1, 2, 3, 4\}$ so that $G|_\sigma$ is the butterfly graph. Prove that $\sigma \in \text{FP}(G)$ if and only if k receives at most one edge from σ or k receives two edges from σ from among the nodes 1, 2, and 4.

Hint. Use domination for each of the remaining cases not covered in Example 8.7.

Exercise 8.18. Use domination to prove Rule 2:

(a) If $i \in \sigma$ is a proper source in $G|_\sigma$, then $\sigma \notin \text{FP}(G)$.
(b) If i is a proper source in G, then $\text{FP}(G) = \text{FP}(G \setminus \{i\})$.

In addition to not participating in fixed point supports, proper sources appear to "die" in dynamic attractors as well. For example, in Fig. 8.1C node 4 is a proper source, and this node is not active once the network falls into its global attractor corresponding to the 3-cycle (123).

Conjecture 8.2. *Proper sources always die and are never active in an attractor of a network.*

Exercise 8.19 (Mini project)**.** Consider graph B from Fig. 8.4. Node 1 is a proper source and thus is not involved in any attractors of the CTLN, so the attractors of graph B are identical to those of the independent set of neurons 2 and 3. Interestingly, though, the presence of the source 1 affects the basins of attraction of those attractors.

Using the Matlab code provided with this chapter (see Section 8.4), study the CTLN for the independent set on neurons 2 and 3 to confirm that half of the space of initial conditions evolves to the stable fixed point supported on $\{2\}$ while the other half evolves to $\{3\}$, and so the two basins of attraction have the same size. Next, explore the CTLN for graph B to see how the addition of 1 as a source affects the size of the basins of attraction of $\{2\}$ and $\{3\}$.

Exercise 8.20. Suppose $i \in \sigma$ is an isolated node in $G|_\sigma$. Using domination, prove that if i is not a sink in G, then $\sigma \notin \mathrm{FP}(G)$.

Exercise 8.21. Prove that an n-cycle has a unique fixed point.

Exercise 8.22. Explain why if σ has uniform in-degree, then there can be no $j, k \in \sigma$ such that k dominates j.

8.3.3.5 *Using the Rules to Compute* FP(G)

As a culmination of the rules in this section, we will demonstrate how to find FP(G) for the two graphs in Fig. 8.11.

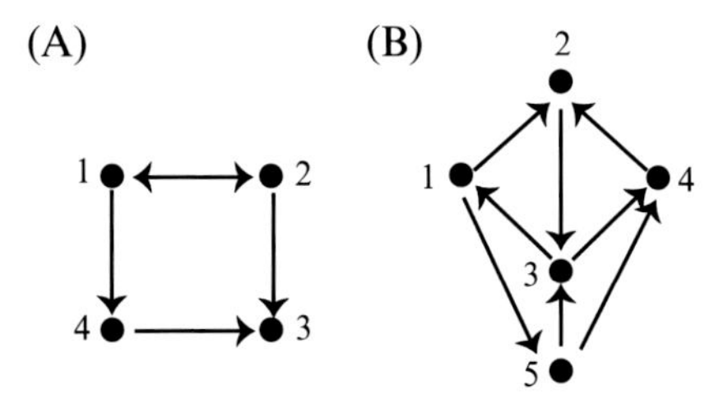

FIG. 8.11 (A–B) Graphs for Example 8.8.

Example 8.8 (Finding FP(G)).

(A) Let G be the graph in Fig. 8.11A. Node 3 is the only sink, and so by Rule 1a, $\{3\}$ is the only singleton fixed point support. The clique $\{1,2\}$ is target-free, and thus supports a fixed point by Rule 3. The independent sets $\{1,3\}$ and $\{2,4\}$ are not unions of sinks, and thus do not support fixed points by Rule 1b. Every other pair of nodes has a unidirectional edge, yielding a proper source, and so cannot be fixed point supports (Rule 2a). The subset $\{1,2,3\}$ has uniform in-degree 1, and 4 is not a target, hence $\{1,2,3\} \in \text{FP}(G)$ by Rule 3. Note that $\{1,2,3\}$ is also the union of a fixed point $\{1,2\}$ with a sink, and so it is a fixed point support by Rule 1c. The subset $\{1,2,4\}$ also has uniform in-degree 1, but 3 is a target, and so $\{1,2,4\} \notin \text{FP}(G)$. Additionally, $\{2,3,4\} \notin \text{FP}(G)$ by Rule 2a since both 2 and 4 are proper sources in $G|_{\{2,3,4\}}$. Thus, FP(G) contains three proper subsets; by Rule 8.3.3.2b, $|\,\text{FP}(G)|$ is odd, and so the full set $\{1,2,3,4\} \notin \text{FP}(G)$. Thus, $\text{FP}(G) = \{3, 12, 123\}$.

(B) Let G be the graph in Fig. 8.11B. G has no sinks, so no singletons and no independent sets can be fixed point supports (Rule 1). Since G is an oriented graph, every pair of nodes that is not an independent set must have a unidirectional edge, and so contains a proper source and cannot be a fixed point support. Among the triples, $\{1,2,3\}$ and $\{2,3,4\}$ both have uniform in-degree 1 and are target-free, so they are fixed point supports. By contrast, $\{1,3,5\}$ has node 4 as a target, and so $\{1,3,5\} \notin \text{FP}(G)$. All other triples contain a proper source in the restricted subgraph, and thus cannot support fixed points. Both $\{1,2,3,4\}$ and $\{1,2,3,5\}$ are butterfly graphs, and by the survival rules in Exercise 8.17, $\{1,2,3,4\} \in \text{FP}(G)$ while $\{1,2,3,5\} \notin \text{FP}(G)$. Both $\{1,2,4,5\}$ and $\{2,3,4,5\}$ contain proper sources in the restricted subgraphs and so do not support fixed points. The subset $\{1,3,4,5\} \notin \text{FP}(G)$ since 4 dominates 3; additionally, $\{1,3,4,5\}$ is the union of a nonfixed point and a sink, and so cannot support a fixed point by Rule 1c. Thus, FP(G) contains four proper subsets, and so by parity (Rule 8.3.3.2b) the full set $\{1,2,3,4,5\} \in \text{FP}(G)$. Thus, $\text{FP}(G) = \{123, 234, 1234, 1235, 12345\}$.

Exercise 8.23. Return to the graphs in the Opening Exploration of Section 8.3.2 (see Fig. 8.7). Use the graph rules summarized in the table to find FP(G) for each graph.

8.4 PREDICTING DYNAMIC ATTRACTORS VIA GRAPH STRUCTURE

In addition to static memory patterns, which are given by stable fixed points, neural networks also encode dynamic patterns of neural activity. These are

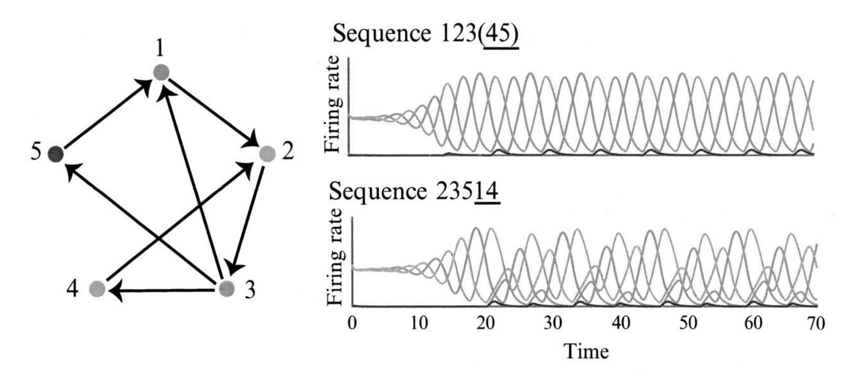

FIG. 8.12 Graph for a CTLN with two limit cycles, and the corresponding sequences. In the top limit cycle, neurons 4 and 5 fire synchronously, and so only the firing rate curve for neuron 5 (*red*) is visible. These two limit cycles are the attractors for parameters $\varepsilon = 0.35$ and $\delta = 0.9$. The top limit cycle has a much larger basin of attraction, while the bottom limit cycle occurs for a smaller set of initial conditions. It can be obtained by initializing the CTLN at or near the unstable fixed point corresponding to $\sigma = \{2, 3, 4\}$.

typically modeled by periodic attractors, such as limit cycles, that represent repeating patterns of neural activation. Such patterns often take the form of sequences, in which neurons fire in a repeatable order that is functionally meaningful. Such sequences have been observed in the mammalian cortex, hippocampus, and CPG circuits [5, 6, 8–10]. They model everything from episodic memories (i.e., sequences of places or events) to rhythmic locomotion [7, 11]. The problem of remembering a phone number, first described in the introduction, is an example of a sequence being maintained in working memory.

Recall from Theorem 8.1 that if G is an oriented graph with no sinks, then the attractors of the corresponding CTLN will be oscillatory or chaotic. Typically, given such a dynamic attractor, it is possible to associate a sequence of neural firing based on the order in which neurons achieve their peak firing rate. Consider the graph in Fig. 8.12 with its two limit cycle attractors, shown on the right. To associate a sequence to the first attractor, notice that starting at a blue peak (neuron 1), next is always a green peak (neuron 2), then a gray peak (neuron 3), followed by a low red peak (neuron 5) that is simultaneous with a yellow peak (neuron 4, not visible). The corresponding sequence is thus $123(\underline{45})$, where the (45) indicates that neurons 4 and 5 fire synchronously, and the underlining denotes low firing. The second limit cycle in Fig. 8.12 has sequence $23\underline{5}1 4$, with neurons 5 and 1 being low firing.

Matlab Exploration: Sequences of Attractors

The Matlab package CTLN Basic, available at: https://github.com/nebneuron/CTLN-bookchapter contains Matlab code to run simulations for CTLNs obtained from any directed graph. Graphs can be coded by the user into the executable file, run_CTLN_model_script.m, in the form of a binary adjacency

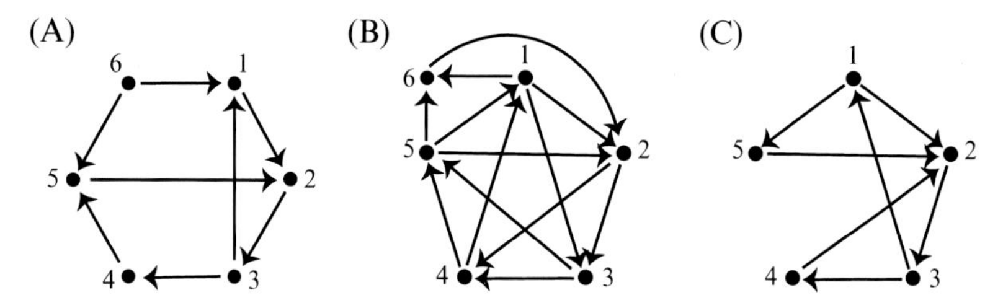

FIG. 8.13 (A–C) Graphs for exploratory Matlab activity.

matrix called sA (see README file for instructions). The parameters ε, δ, and θ can also be adjusted in this file, with defaults matching those in this chapter. Initial conditions may be chosen at random, or specified by the user. This package was used to produce the firing rate plots seen in Fig. 8.12, as well as those in earlier figures.

Using CTLN Basic, code the graph in Fig. 8.13A. Try a variety of initial conditions to find the two limit cycle attractors of the corresponding CTLN. Note that it may be helpful to try initial conditions where the nonzero entries correspond to a cycle in the graph. Record the sequence of neural firing in each of the limit cycles. Repeat this process for the two remaining graphs in Fig. 8.13. *Hint*: Graph B has one attractor while graph C has two attractors.

8.4.1 Sequence Prediction Algorithm

Using the sequence prediction algorithm summarized in the following table, it is often possible to predict the sequences of attractors directly from the graph. This algorithm was first introduced in [17], but has been updated here. To understand the algorithm, we first need some terminology.

Definition 8.2. A node k in G is said to be *freely removable* if removing k from G

(a) does not create a sink in $G \setminus \{k\}$ and

(b) does not create a target-free 3-cycle in $G \setminus \{k\}$ if k was otherwise a target of that 3-cycle.

We say a graph G is *irreducible* if it is cyclically symmetric or has no freely removable nodes. If G is cyclically symmetric, it is also called a *core cycle*.

The sequence prediction algorithm consists of two phases; in the first phase, the graph G is deconstructed by removing nodes according to the rules in the table until an irreducible subgraph is reached. There may be choices as to which node to remove at a given step; each of these choices must be separately pursued to determine if different irreducible subgraphs result. At the end of phase I, after pursuing all choices of nodes, a list of irreducible subgraphs is produced. For each of these subgraphs that is a core cycle, the algorithm proceeds to phase II. A sequence is then constructed for each core cycle as described in the previous table. For each resulting sequence, we expect to see a corresponding attractor of the CTLN defined by G.

Sequence Prediction Algorithm (for an Oriented Graph G With No Sinks)

Phase I: Deconstruct graph

If a node is removed at any step, return to Step 0 with the subgraph remaining after the node's removal. If no node can be removed, then proceed to the next step. Terminate these steps when the remaining subgraph is irreducible. If choices in node removal were made at any step, repeat Phase I with alternate choices. When all irreducible subgraphs are obtained, proceed to Output step, and then to Phase II.

0. Remove all sources (in-degree 0) from the graph.

1. Remove a node of in-degree 1 whose removal does not create a sink in the resulting subgraph.

2. Remove a node of lowest in-degree that is *freely removable*.

Output step: For each irreducible subgraph that is a core cycle, output the corresponding cycle. If a final subgraph is irreducible, but not a core cycle, then declare algorithm failure for that subgraph.

Phase II: Reconstruct sequence

Each core cycle yields a sequence, which is obtained as follows.

0. List the core-cycle nodes in the order they cyclically appear. These are the high-firing nodes of the sequence. The inserted nodes (next step) will be low firing and thus underlined in the sequence.

1. For each node i not in the core cycle, insert it into the sequence only if it receives at least one edge from the core cycle, as follows:

- **a.** Consider the induced subgraph $G|_\omega$ of core-cycle nodes ω that are inputs to node i. Insert i in the sequence after each core-cycle node that is a sink in $G|_\omega$.
- **b.** If two nodes i and j are to be inserted in the same place, check how they interact. If $i \to j$ in G, then insert i before j, and vice versa if $j \to i$. If there is no edge between i and j, then the nodes will fire synchronously, denoted (ij).

Example 8.9. As an illustration of the sequence prediction algorithm, first consider the graph in Fig. 8.13A. The algorithm begins by removing the source node 6 (see Fig. 8.14). Then nodes 1, 3, 4, and 5 have in-degree 1; but neither 3 nor 5 can be removed because they would cause nodes 2 and 4, respectively, to become sinks. When node 1 is removed, the remaining subgraph is a 4-cycle, which is cyclically symmetric, so (2345) is irreducible and is a core cycle. Alternatively, when node 4 is removed, node 5 becomes a source, and must be removed next. Then the cycle (123) remains, and this is a core cycle. Thus, phase I outputs two core cycles (2345) and (123).

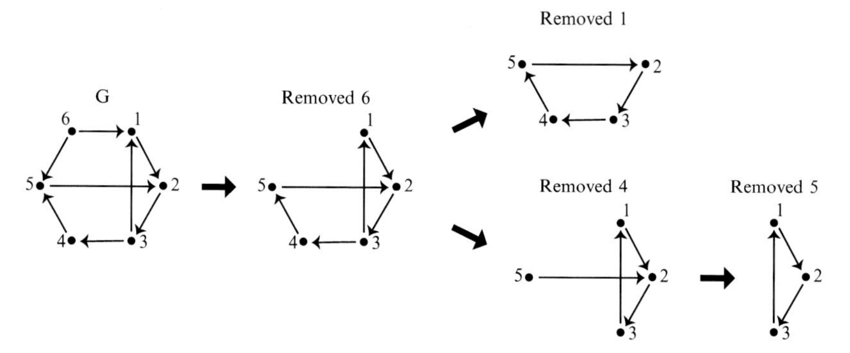

FIG. 8.14 The graph in Fig. 8.13A, deconstructed via phase I of the sequence prediction algorithm.

We now proceed to phase II. Since node 6 was a source, it does not receive from any core cycle and is thus not inserted into any sequence. For the core cycle (2345), the high-firing nodes are, in order, 2345 (or a cyclically equivalent ordering). Node 1 receives from the core-cycle node 3 only, and so it is inserted after 3 and underlined to indicate low firing. The predicted sequence is thus 23$\underline{1}$45. For the core cycle (123), the high-firing part of the sequence is 123. Node 4 receives from 3 only, and so is inserted after 3 in the sequence. Node 5 does not receive from this core cycle, so it is <u>not</u> inserted in this sequence. The second predicted sequence is thus 123$\underline{4}$.

Next, consider the graph in Fig. 8.13B. There are no sources and no nodes of in-degree 1, so we proceed to Step 2. Every node other than 2 has in-degree 2, and every node is a target of a 3-cycle (e.g., node 1 is a target of (245)). Recall that a target of a 3-cycle is only freely removable if the 3-cycle has at least one other target. The only 3-cycle with two or more targets is (135), whose targets are nodes 2 and 6. Since these nodes are not targets for any other 3-cycle, they are both freely removable. We remove node 6, however, because it has lower in-degree. The remaining subgraph is cyclically symmetric, and thus is irreducible and a core cycle. In phase II, we obtain the high-firing sequence 12345 from the core cycle. Node 6 receives from the core-cycle nodes $\omega = \{1, 5\}$, but only 1 is a sink of $G|_{\omega}$, so node 6 is inserted after 1 only. This produces the sequence 1$\underline{6}$2345.

Note that the sequences obtained above match those observed in the Matlab exploration exercise.

Exercise 8.24. Perform the sequence prediction algorithm on graph C of Fig. 8.13 to obtain the two sequences found in the opening Matlab exploration.

Finally, consider the graph in Fig. 8.12. There are no sources to remove, but nodes 3, 4, and 5 all have in-degree 1. Removing node 3 would cause node 2 to become a sink, but both nodes 4 and 5 are valid candidates for removal. If node

4 is removed first, then node 5 must be removed next, yielding the core cycle (123). If instead, node 5 is removed first, then either 1 or 4 can be removed next. Removing node 4 produces the same (123) core cycle already observed. By contrast, removing node 1 produces a second core cycle (234).

In phase II, for the core cycle (123), 123 is the sequence of high-firing neurons. Both nodes 4 and 5 must be inserted into the sequence following node 3. Since there is no edge between 4 and 5, these nodes will fire synchronously, producing the sequence $123(\underline{45})$. For the core cycle (234), 234 is the sequence of high-firing neurons. Nodes 1 and 5 must be inserted into the sequence after node 3. Since $5 \to 1$ in G, node 5 is inserted first, yielding $23\underline{51}4$. Note that both these sequences have corresponding limit cycle attractors when $\varepsilon = 0.35$ and $\delta = 0.9$, as observed in Fig. 8.12. Interestingly, only the first attractor with sequence $123(\underline{45})$ is observed for the standard parameters ($\varepsilon = 0.25$ and $\delta = 0.5$).

Exercise 8.25. Perform the sequence prediction algorithm on the graph in Fig. 8.2D to obtain the two sequences shown there.

The sequence prediction algorithm has been tested on all 160 permutation-inequivalent oriented graphs with no sinks on $n \leq 5$ neurons [17]. With the standard parameters, the algorithm correctly predicted the sequences corresponding to all the attractors of the CTLNs for 152 of the 160 graphs. There were four graphs for which multiple sequences were predicted, but each had one sequence that was not observed with the standard parameters; for example, the second sequence in Fig. 8.12. Nevertheless, for each of these graphs there was an alternative set of legal parameters where all predicted sequences were observed. Thus, taken across the full legal parameter range, the algorithm was successful for 156 of the 160 graphs [17].

There were four graphs, however, for which the algorithm consistently failed to predict the sequences; these are shown in Fig. 8.15 along with the actual observed sequences. The attractors for these graphs all have synchrony that

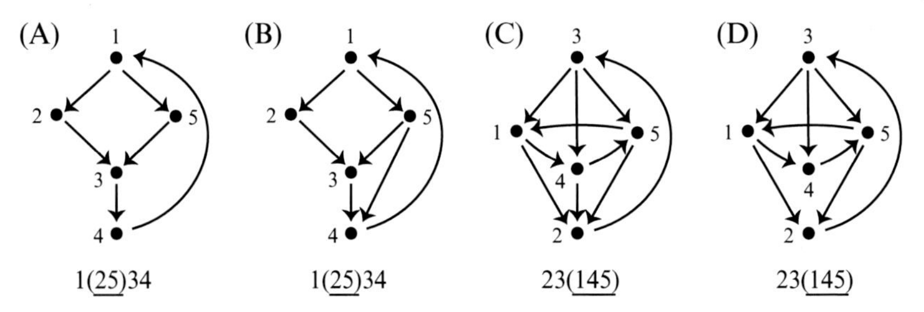

FIG. 8.15 (A–D) Graphs for which the sequence prediction algorithm fails across all legal parameter values. Below each graph is the sequence of the single attractor that is observable across all legal parameters.

is not predicted by the algorithm, and which would require merging multiple core cycles. This unexpected synchrony is the result of *graph automorphisms* in graphs A and C, while graphs B and D are one edge away from having a graph automorphism. The problem with the sequence prediction, therefore, appears to arise from symmetry that is not taken into account by the algorithm. In the next section, we explore graph automorphisms in more detail and examine their impact on the set of attractors of a network.

8.4.2 Symmetry of Graphs Acting on the Space of Attractors

Informally, a graph automorphism is a bijective map from a graph to itself that reflects its symmetry. More precisely, for a graph G with vertex set $V(G)$, a bijection $\alpha\colon V(G) \to V(G)$ is a *graph automorphism* if it preserves the edges of G, that is, $i \to j$ in G if and only if $\alpha(i) \to \alpha(j)$.

Example 8.10. The butterfly graph from Fig. 8.9A has a graph automorphism α that interchanges nodes 1 and 4 and fixes nodes 2 and 3. This is an automorphism since the map sends the edges $1 \to 2$ and $4 \to 2$ to each other and similarly sends the edges $3 \to 1$ and $3 \to 4$ to each other, while fixing the edge $2 \to 3$.

Example 8.11. For graph A in Fig. 8.15, the map α that sends node 2 to 5 and vice versa, while fixing all other nodes, is a graph automorphism. This same map is not an automorphism of graph B since it sends the edge $5 \to 4$ to the edge $2 \to 4$, which is not present in the original graph.

For graph C in Fig. 8.15, the map α that sends node 1 to 4, 4 to 5, and 5 to 1, while fixing nodes 2 and 3, is a graph automorphism. (Verify this!) Note that α is not an automorphism of graph D because the edge $1 \to 2$ is mapped to $4 \to 2$, which is not an edge of the original graph.

Exercise 8.26. Return to the graphs in Fig. 8.7. Identify which graphs have nontrivial graph automorphisms and find the automorphisms of these graphs.

How does a graph automorphism affect the corresponding CTLN? Since the map permutes the neurons of a network in a way that preserves the graph, the resulting CTLN must have the same set of attractors. In other words, the automorphism induces a bijection on the set of attractors. A single attractor may be fixed, or sent to another attractor that differs from it only by permuting neuron labels. For example, the attractor of a 3-cycle (123) has neurons 1, 2, and 3 periodically firing in sequence, and the graph automorphism that sends node 1 to 2, 2 to 3, and 3 to 1 fixes the attractor. Another way an attractor can be fixed is if the exchanged nodes in the automorphism fire synchronously. For example, the attractor produced by the graph in Fig. 8.15A has neurons 2 and 5 firing synchronously, and thus the automorphism that exchanges nodes 2 and 5 maps this attractor to itself. Similarly, the automorphism of the graph in Fig. 8.15C fixes the attractor, since the exchanged nodes 1, 4, and 5 fire synchronously.

Exercise 8.27 (Extended Research Project)**.** Investigate graphs with automorphisms and find the set of attractors of their CTLNs using the Matlab package provided. It is easiest to find all attractors by choosing initial conditions that are small perturbations around the unstable fixed points of the network. Compare the sequences of attractors observed to the outputs of the sequence prediction algorithm. Investigate possible modifications to the sequence prediction algorithm to improve its predictive power for graphs with various types of automorphisms. Test your improved algorithm on the graphs in Fig. 8.15.

When an automorphism does not fix an attractor, it must send it to another one of the same type. For example, the butterfly graph in Fig. 8.9A has an automorphism that exchanges nodes 1 and 4. The two attractors of this network are identical up to permutation, and have sequences $123\underline{4}$ and $423\underline{1}$. It is easy to see that the automorphism sends these attractors to each other. Note that if we had only discovered one of them, the automorphism would tell us that the other must exist. In this way, the presence of a graph automorphism can aid in predicting new attractors in a network once a (nonfixed) one has been observed.

Example 8.12. Consider the graph in Fig. 8.16, with m neurons (nodes) in each layer, where the layers wrap around in a cylindrical fashion. This graph has two types of automorphisms. The first type consists of all permutations of the nodes that keep each node inside its original layer. If we consider the limit cycle displayed on the right, we see that each such automorphism produces another limit cycle with a different set of five neurons firing in sequence. The existence of one such limit cycle, together with the m^5 automorphisms, thus predicts that this network has at least m^5 sequential attractors. Note that this network

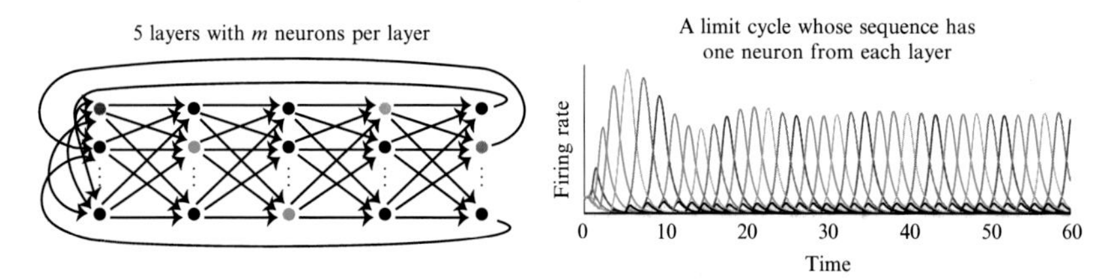

FIG. 8.16 (*Left*) A cyclically structured graph with m neurons per layer, and all m^2 feedforward connections from one layer to the next. (*Right*) A limit cycle for the corresponding CTLN (with parameters $\varepsilon = 0.75$, $\delta = 4$) in which five neurons fire in a repeating sequence, with one neuron from each layer.

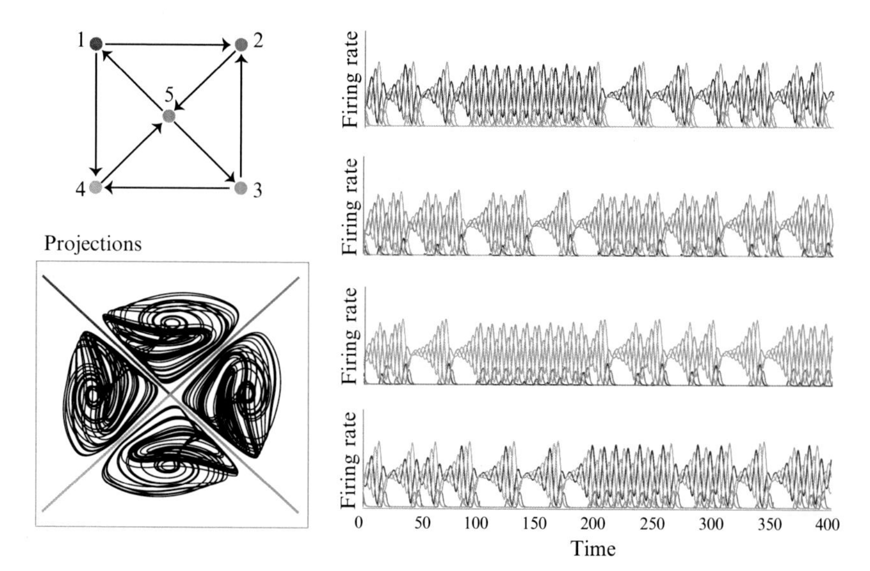

FIG. 8.17 Graph of CTLN with four chaotic attractors—one attractor for each 3-cycle. Note that this graph differs from the graph in Fig. 8.2B only by flipping the $3 \rightarrow 4$ edge.

architecture, with seven layers, could serve as a mechanism for storing phone numbers in working memory ($m = 10$ for digits 0–9). The phone number is stored as a sequence that is repeated indefinitely, with different initial conditions producing different phone number sequences. (Can you find another type of automorphism for this network? What does it do to the attractors?)

Graph automorphisms may also play a role in producing more exotic attractors. For example, in Fig. 8.17 we see chaotic attractors in a network of only five nodes, with four perfectly symmetric overlapping 3-cycles. There is an attractor for each 3-cycle, and these attractors are chaotic for the standard parameters. Interestingly, for different (legal) parameter values the chaotic attractors may become limit cycles. The graph automorphisms ($1 \leftrightarrow 3$ and $2 \leftrightarrow 4$) permute these attractors without fixing any.

We also see quasiperiodic attractors in small networks with symmetry. A *quasiperiodic* attractor is one that is nearly periodic, but whose trajectory does not perfectly repeat. The attractor thus has the shape of a torus, rather than a circle as in the case of a limit cycle. Fig. 8.18 displays a CTLN from a cyclically symmetric graph on $n = 7$ nodes. For standard parameters, this network has two attractors: one limit cycle and one quasiperiodic. Interestingly, the quasiperiodic attractor only emerges for a portion of the legal parameter range, while the limit cycle is present for the full range. The emergence of the quasiperiodic attractor is particularly surprising because it does not have a corresponding unstable fixed point. In fact, this network has exactly one unstable fixed point, and initial

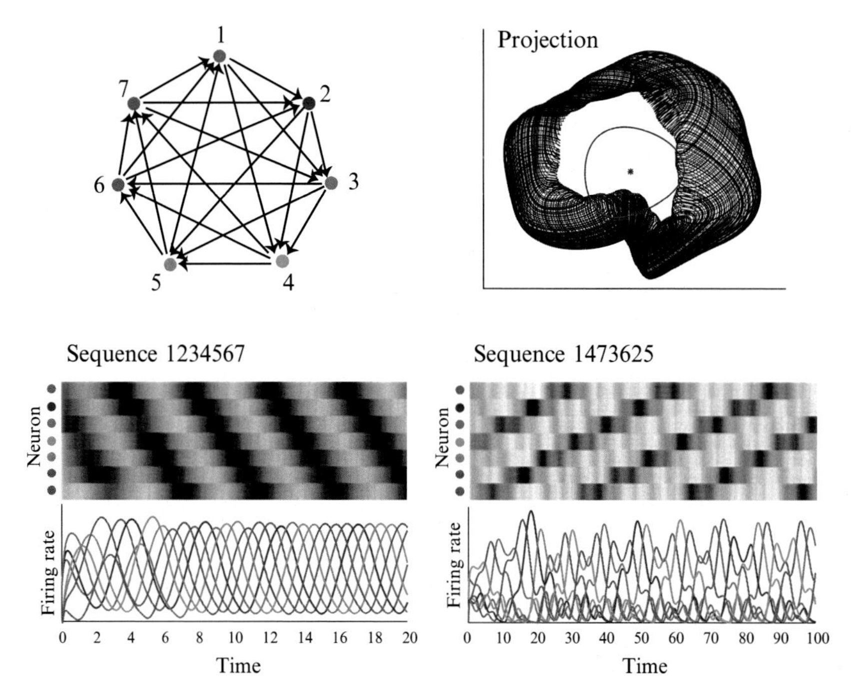

FIG. 8.18 (Top left) Graph of a cyclically symmetric CTLN with two attractors: one limit cycle and one quasiperiodic. (Bottom row) The limit cycle has the expected sequence 1234567, while the quasiperiodic attractor has a sequence corresponding to another cycle in the graph. (Top right) A 2-dimensional projection of the two attractors, and the unique (unstable) fixed point. The limit cycle and fixed point are shown in red, while the quasiperiodic attractor is the torus-like trajectory shown in black.

conditions near this fixed point always yield the limit cycle. The quasiperiodic attractor can be obtained from a much smaller set of initial conditions, including $[0.1, 0, 0, 0.1, 0, 0, 0]$.

Exercise 8.28 (Mini project, for those familiar with bifurcation theory)**.** Conduct a bifurcation analysis for the graph in Fig. 8.12. Fix $\varepsilon > 0$ and choose a set of δs ranging from $-\varepsilon$ to 2. For each choice of parameters, find the collection of attractors of the CTLN by sampling a variety of initial conditions, including perturbations of all unstable fixed points. Identify the parameter values where bifurcations occur by finding when the set of dynamic attractors qualitatively changes.

ACKNOWLEDGMENTS

Katherine Morrison and Carina Curto were supported by NIH R01 EB022862. Carina Curto was also supported by NSF DMS-1516881. We thank David Falk, an undergraduate student, for his help in developing various exercises and examples.

APPENDIX

A.1 REVIEW OF LINEAR SYSTEMS OF ODES

In this appendix, we briefly review linear systems of differential equations with constant coefficients; for further details, see any textbook on ODEs.

Consider a linear system with constant real-valued coefficients of the form

$$\frac{dx_1}{dt} = a_{11}x_1 + a_{12}x_2 + \cdots + a_{1n}x_n + b_1$$
$$\frac{dx_2}{dt} = a_{21}x_1 + a_{22}x_2 + \cdots + a_{2n}x_n + b_2$$
$$\vdots$$
$$\frac{dx_n}{dt} = a_{n1}x_1 + a_{n2}x_2 + \cdots + a_{nn}x_n + b_n$$

which can be written compactly as

$$\frac{d\mathbf{x}}{dt} = A\mathbf{x} + \mathbf{b} \tag{A.1}$$

where $\mathbf{x}$ and $\mathbf{b}$ are column vectors and A is the matrix of coefficients. A *fixed point* of such a system is a point $\mathbf{x}^*$ where each derivative dx_i/dt is zero, that is, $\frac{d\mathbf{x}}{dt}|_{\mathbf{x}=\mathbf{x}^*} = 0$. Assuming the matrix A is invertible, the system (A.1) has a unique fixed point $\mathbf{x}^* = -A^{-1}\mathbf{b}$. The behavior of the system around this fixed point is dictated by the eigenvalues $\{\lambda_i\}$ and corresponding eigenvectors $\{\mathbf{v}_i\}$ of the matrix A.

For simplicity, consider the system (A.1) with just two variables $\mathbf{x} = \begin{bmatrix} x_1 \\ x_2 \end{bmatrix}$ and 2×2 matrix A, with eigenvalues λ_1 and λ_2 corresponding to eigenvectors $\mathbf{v}_1$ and $\mathbf{v}_2$, respectively. When the eigenvalues are distinct, the solutions to the system have the form

$$\mathbf{x}(t) = C_1\mathbf{v}_1e^{\lambda_1 t} + C_2\mathbf{v}_2e^{\lambda_2 t} - A^{-1}\mathbf{b},$$

where C_1, C_2 are constants determined by the initial conditions. Observe that if λ_1 and λ_2 are both real and negative, then the first two terms decay toward 0 as $t \to \infty$ and all trajectories converge to the fixed point $\mathbf{x}^* = -A^{-1}\mathbf{b}$. In this case, the fixed point is *stable* and is an *attractor* of the network. If the eigenvalues are both real and positive, then solutions will tend toward infinity as $t \to \infty$, and the fixed point $\mathbf{x}^*$ is called *unstable*. In contrast, if one eigenvalue is positive while the other is negative, then the fixed point is a *saddle*. The top row of Fig. A.1 shows vector fields centered at the fixed point with axes corresponding to the eigenvectors for each of these three cases. For ease of drawing, we have assumed that the eigenvalues are distinct and the eigenvectors are orthogonal.

When the eigenvalues are complex, $e^{\lambda_i t}$ can be rewritten as $e^{\mathrm{Re}(\lambda_i)t}[\cos(\mathrm{Im}(\lambda_i)t) + i\sin(\mathrm{Im}(\lambda_i)t)]$ using Euler's formula. This produces spiral-like behavior that converges toward the fixed point when $\mathrm{Re}(\lambda_i) < 0$ and diverges when

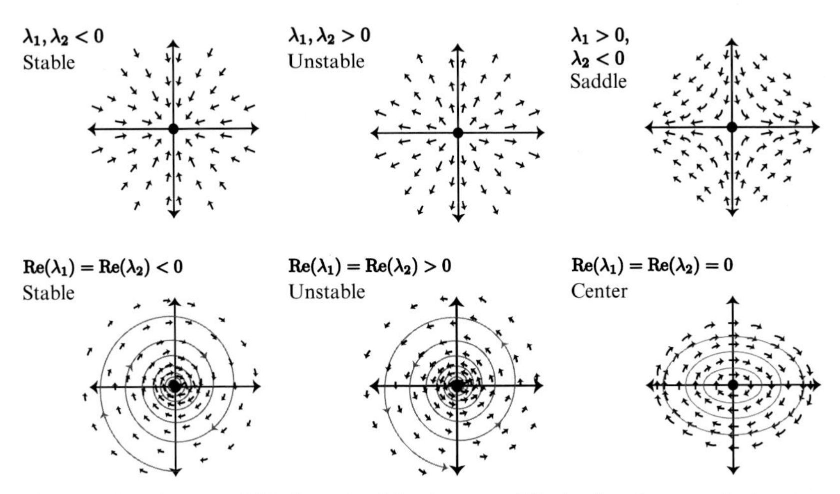

FIG. A.1 Sample vector fields for each of the six types of fixed points that can arise in a two-dimensional linear system.

$\mathrm{Re}(\lambda_i) > 0$. (Note that $\mathrm{Re}(\lambda_1) = \mathrm{Re}(\lambda_2)$, as the eigenvalues must be complex conjugate pairs.) In the former case, the fixed point is again called *stable*, while in the latter case it is *unstable*. Finally, when $\mathrm{Re}(\lambda_i) = 0$, the system produces perfectly periodic orbits around the fixed point, but these are unstable, as small changes in initial conditions result in different periodic trajectories. Vector fields corresponding to these three cases are shown in the bottom panel of Fig. A.1.

In higher-dimensional linear systems, these same behaviors occur around fixed points, and the fixed point's type is again dictated by the eigenvalues of the matrix A. A fixed point of the system (A.1) is *stable* when all eigenvalues have negative real part, and is *unstable* if at least one eigenvalue has positive real part.

REFERENCES

[1] H.S. Seung, R. Yuste, Appendix E: neural networks, in: Principles of Neural Science fifth ed., McGraw-Hill Education/Medical, New York, 2012, pp. 1581–1600.

[2] K. Morrison, J. Geneson, C. Parmelee, C. Langdon, A. Degeratu, V. Itskov, and C. Curto, Emergent dynamics from network connectivity: a minimal model, *In preparation.* Earlier version available online at https://arxiv.org/abs/1605.04463.

[3] J.J. Hopfield, Neural networks and physical systems with emergent collective computational abilities, Proc. Natl Acad. Sci. 79 (8) (1982) 2554–2558.

[4] D.J. Amit, Modeling Brain Function: The World of Attractor Neural Networks, Cambridge University Press, Cambridge, MA, 1989.

[5] E. Marder, D. Bucher, Central pattern generators and the control of rhythmic movements, Curr. Biol. 11 (23) (2001) R986–R996.

[6] J.J. Collins, I.N. Stewart, Coupled nonlinear oscillators and the symmetries of animal gates, J. Nonlinear Sci. 3 (1993) 349–392.

[7] G.B. Ermentrout, D.H. Terman, Mathematical Foundations of Neuroscience, Springer-Verlag, New York, NY, 2010.
[8] E. Stark, L. Roux, R. Eichler, G. Buzsáki, Local generation of multineuronal spike sequences in the hippocampal CA1 region, Proc. Natl Acad. Sci. 112 (33) (2015) 10521–10526.
[9] E. Pastalkova, V. Itskov, A. Amarasingham, G. Buzsáki, Internally generated cell assembly sequences in the rat hippocampus, Science 321 (5894) (2008) 1322–1327.
[10] A. Luczak, P. Barthó, S.L. Marguet, G. Buzsáki, K.D. Harris, Sequential structure of neocortical spontaneous activity *in vivo*, Proc. Natl Acad. Sci. 104 (1) (2007) 347–352.
[11] G. Buzsáki, Rhythms of the Brain, Oxford University Press, New York, NY, 2011.
[12] R.H. Hahnloser, H.S. Seung, J.J. Slotine, Permitted and forbidden sets in symmetric threshold-linear networks, Neural Comput. 15 (3) (2003) 621–638.
[13] C. Curto, K. Morrison, Pattern completion in symmetric threshold-linear networks, Neural Comput. 28 (2016) 2825–2852.
[14] C. Curto, J. Geneson, and K. Morrison, Fixed points of competitive threshold-linear networks, Available online at https://arxiv.org/abs/1804.00794.
[15] J.W. Moon, L. Moser, On cliques in graphs, Israel J. Math. 3 (1965) 23–28.
[16] J.W. Milnor, Topology From the Differentiable Viewpoint, Princeton Landmarks in Mathematics, Princeton University Press, Princeton, NJ, 1997.
[17] C. Parmelee, Applications of Discrete Mathematics for Understanding Dynamics of Synapses and Networks in Neuroscience (Ph.D. thesis), University of Nebraska-Lincoln, 2016.

Chapter 9

Multistationarity in Biochemical Networks: Results, Analysis, and Examples

Carsten Conradi* and Casian Pantea†
**Hochschule für Technik und Wirtschaft Berlin, Berlin, Germany,* †*West Virginia University, Morgantown, WV, United States*

9.1 INTRODUCTION

Biochemical networks are complex objects, almost always containing nonlinear interactions among a usually large number of chemical species. The smallest activation or inhibition interactions involve two components (a gene and its protein, or an inhibitor and an enzyme), are well understood biologically, and have simple stable long-term behavior that can be inferred straight from the interaction diagram [1] (Fig. 9.1A). But if the interaction involves a larger number of components, or if feedback loops are present, then the dynamics may become far more complicated, and understanding it requires strategies more subtle than chasing paths in the interaction diagram. Instabilities of different kinds are possible even for small biological structures and are associated with signaling events [2–5]. One key class of instabilities is those leading to multiple positive steady states (Fig. 9.1B), also known as *multistationarity*, and seen experimentally as irreversible switch-like behavior. There is significant theoretical evidence, backed by experiment, that important pathways may exhibit multistationarity as response to chemical signaling. This phenomenon is particularly relevant in crucial cell behaviors, including generating sustained oscillatory responses, remembering transitory stimuli, differentiation, or apoptosis [2, 4, 6–9]. Multistationarity occurs in chemistry and chemical engineering as well, but is much less common. There is, in fact, a great deal of stable behavior in networks of chemical reactions, and (to a lesser degree) in biological networks. This can be explained in part by the fact that the possibility of exotic behavior places rather delicate constraints on the structure of an interaction network [10–18]. A seminal remark is due to Thomas [19] who conjectured that positive feedbacks in the logical structure of an interaction network are necessary for multistationarity. Much theoretical and simulation

Algebraic and Combinatorial Computational Biology. https://doi.org/10.1016/B978-0-12-814066-6.00009-X

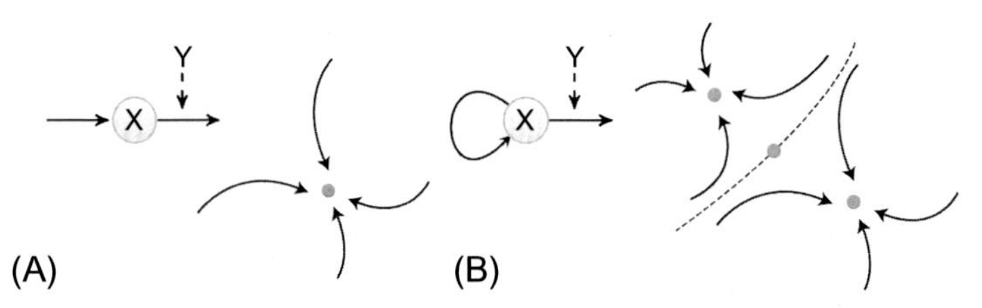

FIG. 9.1 Even simple interactions may authorize complicated behaviors. (A) In reactions of synthesis and degradation the steady-state response X is unique for any value of signal Y [1]. (B) In the presence of positive feedback, as is the case of autocatalysis, certain values of signal Y may generate multiple positive steady states for X [13].

work followed, and proposed a series of increasingly refined design principles for pathways allowing multistationarity [1, 13, 20–23]. Building on this effort, multistationarity has been demonstrated experimentally in bacterial synthetic genetic networks [24].

The structure of reaction networks can be encoded by their reaction graph, Jacobian matrix sign pattern, various stoichiometric matrices, and so on. Recent notable work has established subtle connections between properties of these objects and multistationarity [11, 12, 14–17, 25–29]. In particular, various features of the stoichiometric matrix may allow strong conclusions about the existence of multiple equilibria; this chapter is a review of results along this line. We note that a survey such as this will necessarily be not exhaustive, and we have omitted a number of multistationarity methods, including powerful algebraic tools well-suited for enzymatic networks [30, 31], as well as a series of results based on degree theory [32, 33]. We also refer the reader to the recent review [34].

The chapter is structured as follows. Section 9.2 is dedicated to terminology and notation; we follow a matrix-theoretical framework, much like in [27, 28]. The next two sections focus on necessary conditions for multistationarity of reaction networks. There is a wealth of results in this direction; we present theorems based on injectivity [26, 27, 35–39] (Section 9.3) and the directed species-reaction (DSR) graph [37, 40] (Section 9.4). We then continue with two sections on sufficient conditions for multistationarity: recent work on inheritance of multistationarity from subnetworks [28, 29] is presented in Section 9.5, and the determinant optimization method [26] is reviewed in Section 9.6. Finally, methods for ruling out multistationarity based on deficiency theory [41–43] are collected in Section 9.7. Exercises are proposed at the end of each section. Some of them are important theoretical facts not included in the main text, while others are applications of the results contained in that section.

We restrict the presentation to mass action, where, informally, the rate of a chemical reaction is proportional to the number of molecular collisions, and therefore proportional to the concentration of each of the reactants. We note, however, that many of the results apply to more general kinds of kinetics.

Our focus is on readability and a unifying presentation of the various results from the literature, rather than on giving the strongest statements possible. For these, the reader is encouraged to consult the references accompanying the theorems shown here.

9.2 REACTION NETWORK TERMINOLOGY AND BACKGROUND

We proceed rather informally at first, using examples to introduce some basic terminology, which is later revisited in a general context. For ease of presentation, some of these examples are chosen somewhat artificially. Others are important systems from biochemistry; for example, throughout this chapter we will often consider versions of the *futile cycle*

$$E + S \rightleftharpoons ES \rightarrow E + P, \quad F + P \rightleftharpoons FP \rightarrow F + S, \tag{9.1}$$

a well-studied structure that serves as building blocks in cellular signaling pathways [20, 22, 44, 45]. The name comes from the fact that, in may cases, the two opposite pathways (from S to P and from P to S) run simultaneously and have no overall effect other than to dissipate energy. An important instance of a futile cycle is the *phosphorylation-dephosphorylation* network, where the substrate (often a protein) S is converted into the product P by adding a phosphoryl group. The process is triggered by an enzyme E, which binds to the substrate, forming an intermediate enzyme-substrate complex ES; this, in turn, dissociates to release E and P, the phosphorylated form of S. The reverse process of dephosphorylation proceeds similarly, catalyzed by the enzyme F. Dynamical properties, and in particular multistationarity of systems of coupled futile cycles, have been studied extensively [2, 4, 46–51]. The methods overviewed here can conclude whether or not many such enzymatic systems are multistationary.

9.2.1 Chemical Reaction Networks and Their Dynamics

A *chemical reaction network* (CRN), also called *biochemical reaction network*, or simply *reaction network* throughout the text is a list of *reactions* involving a finite list of *species*. The futile cycle above involves six reactions among six species E, S, ES, P, F, FP. Likewise, the example in Fig. 9.2 involves seven reactions among five species A, B, C, D, E, taking place in a defined reaction environment, such as a cell or a chemostat. Each arrow in the diagram represents a *reaction* and is interpreted in a natural way: for example, in $A + B \rightarrow 2C$, one molecule of species A combines with one molecule of species B to produce two molecules of C.

A reaction arrow connects formal linear combinations of species called *complexes*: it starts at a *source* complex and it ends at a *product* complex. Degradation or discharge of species E from the system is encoded as an *outflow reaction* $E \rightarrow 0$. The exterior of the reaction environment is represented in the

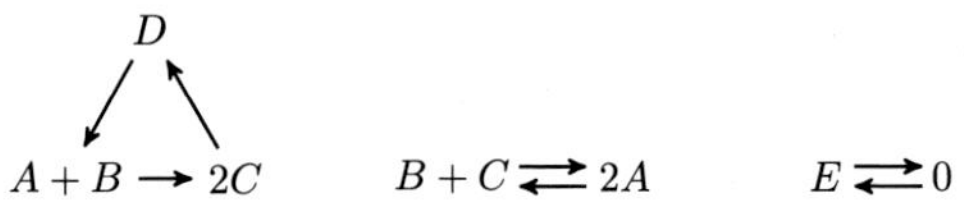

$$A + B \rightarrow 2C \qquad B + C \rightleftarrows 2A \qquad E \rightleftarrows 0$$

(with $2C \rightarrow D$ and $D \rightarrow A + D$ forming a triangle)

FIG. 9.2 A reaction network with seven reactions involving five species A, B, C, D, E and the complexes $A + D, 2C, D, B + C, 2A, E, 0$.

CRN by the *zero complex* 0. Likewise, the *inflow reaction* $0 \rightarrow E$ encodes a constant supply of E into the system. A CRN containing inflow and outflow reactions for all of its species is called *fully open*. Adding to a CRN the missing inflow and outflow reactions ($A \rightleftharpoons 0,\ B \rightleftharpoons 0,\ C \rightleftharpoons 0,\ D \rightleftharpoons 0$ for the example in Fig. 9.2) yields its *fully open extension*.

A pair of reactions that switch their source and product complexes (e.g., $B + C \rightleftharpoons 2A$) is called a *reversible reaction*. One can always view a reversible reaction as two separate, irreversible reactions, and most of our treatment follows this convention. We will clearly state when that is not the case. Species involved in the source complex of a certain reaction are called *reactants*. As Fig. 9.2 illustrates, the CRN can be viewed as a directed graph with complexes as vertices; this is called the *reaction graph* of the CRN.

The molar *concentrations* x_A, x_B, x_C, x_D, x_E of species A, B, C, D, E are nonnegative quantities varying with time according to a system of ordinary differential equations, as follows. The net gain of molecules for each species in a single occurrence of a reaction is encoded in a column vector called its *reaction vector*: for example, setting species A, B, C, D, E in this order, the reaction vector of $A + B \rightarrow 2C$ is $[-1, -1, 2, 0, 0]^t$. Each reaction "pushes" the vector field in the direction of its reaction vector, at a rate proportional to the product of its reactant concentrations (this way of constructing rates is called *mass action*): the *rate* of $A+B \rightarrow 2C$ is equal to $k_1 x_A x_B$, where $k_1 > 0$ is a reaction-dependent quantity called its *rate constant*. The rate constant is usually indicated on top of the reaction arrow, for example $A + B \xrightarrow{k_1} 2C$. Inflow reactions (like $0 \rightarrow E$) have constant rates, equal to their rate constants. Aggregating the contributions of all reactions gives the system of differential equations of the CRN. For the example in Fig. 9.2 this is

$$\begin{bmatrix} \dot{x}_A \\ \dot{x}_B \\ \dot{x}_C \\ \dot{x}_D \\ \dot{x}_E \end{bmatrix} = k_1 x_A x_B \begin{bmatrix} -1 \\ -1 \\ 2 \\ 0 \\ 0 \end{bmatrix} + k_2 x_C^2 \begin{bmatrix} 0 \\ 0 \\ -2 \\ 1 \\ 0 \end{bmatrix} + k_3 x_D \begin{bmatrix} 1 \\ 1 \\ 0 \\ -1 \\ 0 \end{bmatrix} + k_4 x_B x_C \begin{bmatrix} 2 \\ -1 \\ -1 \\ 0 \\ 0 \end{bmatrix}$$
$$+ k_5 x_A^2 \begin{bmatrix} -2 \\ 1 \\ 1 \\ 0 \\ 0 \end{bmatrix} + k_6 x_E \begin{bmatrix} 0 \\ 0 \\ 0 \\ 0 \\ -1 \end{bmatrix} + k_7 \begin{bmatrix} 0 \\ 0 \\ 0 \\ 0 \\ 1 \end{bmatrix},$$

conveniently rewritten as the product of the *stoichiometric matrix* of the network and its *rate vector*.

$$\begin{bmatrix} \dot{x}_A \\ \dot{x}_B \\ \dot{x}_C \\ \dot{x}_D \\ \dot{x}_E \end{bmatrix} = \begin{bmatrix} -1 & 0 & 1 & 2 & -2 & 0 & 0 \\ -1 & 0 & 1 & -1 & 1 & 0 & 0 \\ 2 & -2 & 0 & -1 & 1 & 0 & 0 \\ 0 & 1 & -1 & 0 & 0 & 0 & 0 \\ 0 & 0 & 0 & 0 & 0 & -1 & 1 \end{bmatrix} \begin{bmatrix} k_1 x_A x_B \\ k_2 x_C^2 \\ k_3 x_D \\ k_4 x_B x_C \\ k_5 x_A^2 \\ k_6 x_E \\ k_7 \end{bmatrix}. \quad (9.2)$$

The stoichiometric matrix is made out of reaction vectors arranged as columns, and it is therefore dependent on the ordering of reactions (corresponding to its columns) and the ordering of species (corresponding to its rows). Clearly, this does not change the expression of the vector field (9.2), nor does it have any impact on the results presented here, which are independent of these orderings.

It is easy to see (Exercise 2) that the nonnegative orthant $\mathbb{R}^5_{\geq 0}$ is *forward invariant* with respect to Eq. (9.2), that is, solutions of Eq. (9.2) with nonnegative initial conditions will stay nonnegative. This important remark holds for any reaction network driven by mass action.

9.2.2 Some Useful Notation

Throughout this chapter we denote the stoichiometric matrix by Γ and the rate vector by v. We note that every complex can be viewed as a (column) vector, for example, $A+B$ corresponds to $[1, 1, 0, 0, 0]^t$. Source complexes can be arranged as columns in the *reactant matrix*, denoted by Γ_l. For example, the reactant matrix corresponding to Fig. 9.2 is

$$\Gamma_l = \begin{bmatrix} 1 & 0 & 0 & 0 & 2 & 0 & 0 \\ 1 & 0 & 0 & 1 & 0 & 0 & 0 \\ 0 & 2 & 0 & 1 & 0 & 0 & 0 \\ 0 & 0 & 1 & 0 & 0 & 0 & 0 \\ 0 & 0 & 0 & 0 & 0 & 1 & 0 \end{bmatrix}.$$

A useful notation in the context of mass action is vector exponentiation: for $u, v \in \mathbb{R}^n_{\geq 0}$, u^v is defined as the product of componentwise powers: $u^v = \prod_{i=1}^n u_i^{v_i}$. Mass action rates have convenient expressions using this notation; for example, if $x = [x_A, x_B, x_C, x_D, x_E]^t$ denotes the *concentration vector* for our network in Fig. 9.2, then the reaction rate $k_1 x_A x_B$ of $A + B \to 2C$ is simply $k_1 x^{[1,1,0,0,0]^t}$; note that the exponent is the reactant vector of the reaction, and therefore it figures as a column in Γ_l. We introduce yet another abbreviation, that of exponentiation by a matrix, to write the vector of monomials

$$w(x)^t = [x_Ax_B, x_C^2, x_D, x_Bx_C, x_A^2, x_E, 1] = \begin{bmatrix} x_A \\ x_B \\ x_C \\ x_D \\ x_E \end{bmatrix} \begin{bmatrix} 1 & 0 & 0 & 0 & 2 & 0 & 0 \\ 1 & 0 & 0 & 1 & 0 & 0 & 0 \\ 0 & 2 & 0 & 1 & 0 & 0 & 0 \\ 0 & 0 & 1 & 0 & 0 & 0 & 0 \\ 0 & 0 & 0 & 0 & 0 & 1 & 0 \end{bmatrix} = x^{\Gamma_l}$$

so that the rate vector becomes

$$v(x) = D_k(x^{\Gamma_l})^t,$$

where D_k denotes the diagonal matrix with rate constants k_i on the diagonal.

9.2.3 The Jacobian Matrix

Much of the theory to be presented here relies on Jacobian properties of the CRN vector field. The notation introduced so far allows for a useful expression of the Jacobian matrix, as follows: if $f = \Gamma v$ denotes the mass action vector field, then $Df = \Gamma Dv$, and Dv has a useful factorization, illustrated here for the CRN in Fig. 9.2:

$$(\partial v/\partial x) = \begin{bmatrix} k_1x_B & k_1x_A & 0 & 0 & 0 \\ 0 & 0 & 2k_1x_C & 0 & 0 \\ 0 & 0 & 0 & k_3 & 0 \\ 0 & k_4x_C & k_4x_B & 0 & 0 \\ 2k_5x_A & 0 & 0 & 0 & 0 \\ 0 & 0 & 0 & 0 & k_6 \\ 0 & 0 & 0 & 0 & 0 \end{bmatrix}$$

$$= \begin{bmatrix} k_1x_Ax_B & k_1x_Ax_B & 0 & 0 & 0 \\ 0 & 0 & 2k_1x_C^2 & 0 & 0 \\ 0 & 0 & 0 & k_3x_D & 0 \\ 0 & k_4x_Bx_C & k_4x_Bx_C & 0 & 0 \\ 2k_5x_A^2 & 0 & 0 & 0 & 0 \\ 0 & 0 & 0 & 0 & k_6x_E \\ 0 & 0 & 0 & 0 & 0 \end{bmatrix} \begin{bmatrix} 1/x_A & 0 & 0 & 0 & 0 \\ 0 & 1/x_B & 0 & 0 & 0 \\ 0 & 0 & 1/x_C & 0 & 0 \\ 0 & 0 & 0 & 1/x_D & 0 \\ 0 & 0 & 0 & 0 & 1/x_E \end{bmatrix}$$

$$= \begin{bmatrix} k_1x_Ax_B & 0 & 0 & 0 & 0 & 0 & 0 \\ 0 & k_1x_C^2 & 0 & 0 & 0 & 0 & 0 \\ 0 & 0 & k_3x_D & 0 & 0 & 0 & 0 \\ 0 & 0 & 0 & k_4x_Bx_C & 0 & 0 & 0 \\ 0 & 0 & 0 & 0 & k_5x_A^2 & 0 & 0 \\ 0 & 0 & 0 & 0 & 0 & k_6x_E & 0 \\ 0 & 0 & 0 & 0 & 0 & 0 & k_7 \end{bmatrix} \begin{bmatrix} 1 & 1 & 0 & 0 & 0 \\ 0 & 0 & 2 & 0 & 0 \\ 0 & 0 & 0 & 1 & 0 \\ 0 & 1 & 1 & 0 & 0 \\ 2 & 0 & 0 & 0 & 0 \\ 0 & 0 & 0 & 0 & 1 \\ 0 & 0 & 0 & 0 & 0 \end{bmatrix}$$

$$\begin{bmatrix} 1/x_A & 0 & 0 & 0 & 0 \\ 0 & 1/x_B & 0 & 0 & 0 \\ 0 & 0 & 1/x_C & 0 & 0 \\ 0 & 0 & 0 & 1/x_D & 0 \\ 0 & 0 & 0 & 0 & 1/x_E \end{bmatrix} = D_kD_w\Gamma_l^tD_{1/x}.$$

Here D_k, D_w, and $D_{1/x}$ denote diagonal matrices with entries coming from the vector of rate constants k, monomials $w(x)$, and the vector $[1/x_A, 1/x_B, 1/x_C, 1/x_D]^t$, respectively. This factorization holds for any mass action CRN. Note that although Df is defined everywhere on $\mathbb{R}^5_{\geq 0}$, $D_{1/x}$ is not defined on the boundary of the $\mathbb{R}^5_{\geq 0}$. However, that is not important for the purposes of this chapter, as our Jacobian calculations are restricted to the positive orthant.

9.2.4 Stoichiometry Classes

A simple derivation from Eq. (9.2) shows that $\dot{x}_A + \dot{x}_B + \dot{x}_C + 2\dot{x}_D = 0$, which means that $x_A + x_B + x_C + 2x_D$ stays constant along trajectories. This linear combination of concentrations is called a *conservation law*. A systematic way to find the conservation laws of a CRN is to look for vectors $c \in \ker \Gamma^t$ since then $c \cdot \dot{x} = 0$ (here "$\cdot$" denotes the usual dot product). In our example, $[1, 1, 1, 2, 0]^t$ forms a basis of $\ker \Gamma^t$, and therefore, while the phase space is $\mathbb{R}^5_{\geq 0}$, the trajectories are constrained to four-dimensional affine spaces orthogonal to $[1, 1, 1, 2, 0]^t$, called *stoichiometry classes*. Note that since $(\ker \Gamma^t)^\perp = \operatorname{im} \Gamma$, the stoichiometry classes are affine spaces parallel to the *stoichiometric subspace* $\operatorname{im} \Gamma$. (Throughout the chapter we let $\ker A$ and $\operatorname{im} A$ denote the nullspace and the column space of a matrix A, respectively).

The stoichiometry classes foliate the phase space; the simpler example

$$A \rightleftharpoons 2B \tag{9.3}$$

allows for a helpful picture—see Fig. 9.3. Here the stoichiometric subspace is the span of $[-1, 2]^t$, and translating it by $p \in \mathbb{R}^2_{\geq 0}$ into $\mathbb{R}^2_{\geq 0}$ yields the stoichiometry classes $\left\{p + s \begin{bmatrix} -1 \\ 2 \end{bmatrix} \mid s \in \mathbb{R}\right\} \cap \mathbb{R}^2_{\geq 0}$. Each stoichiometry class corresponds to a choice of "total mass" T and can also be defined as $\left\{[x_A, x_B]^t \in \mathbb{R}^2_{\geq 0} \mid 2x_A + x_B = T\right\}$. For this example, there is only one independent conservation law $[2, 1]^t$, and it spans $\ker \Gamma^t$.

As trajectories are confined to stoichiometry classes, these, rather than the whole phase space, are the relevant spaces where dynamical behaviors of CRNs (like existence or uniqueness of equilibria) are studied. We will come back to this in the next section.

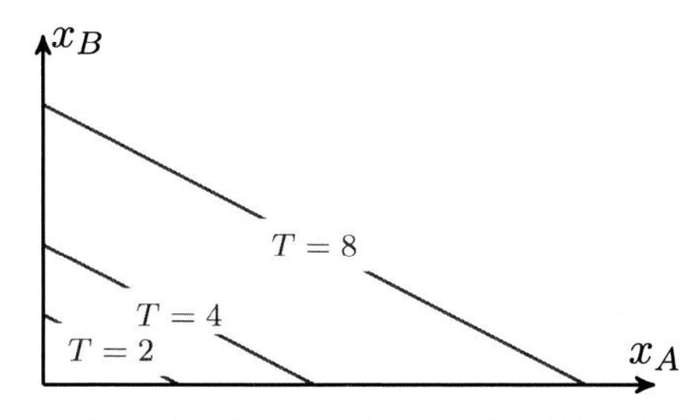

FIG. 9.3 Stoichiometry classes for the network $A \rightleftharpoons 2B$. Although the phase space is two-dimensional, every trajectory is either stuck at the origin, or constrained to one of the one-dimensional stoichiometry classes.

9.2.5 Equilibria

An *equilibrium* (or *steady state*) of a reaction network is a point in the phase space where its vector field vanishes (i.e., a single-point trajectory of the CRN dynamics). In this chapter, we focus on the study of equilibria with positive coordinates (from now on called *positive equilibria*), and in particular, on their uniqueness. As described in the introduction, the existence of two or more positive equilibria is of central importance in the study of many important cellular processes.

Equilibrium points sit at the intersection of *nullclines*, which are manifolds defined by the vanishing of the derivative of one of the variables. In the case of mass action, these are algebraic varieties cut out by a single multivariate polynomial. For example, the network

$$2A \xrightarrow{1} B,\ B \xrightarrow{1} A,\ A \xrightarrow{1} A + B \tag{9.4}$$

with all rate constants chosen to be equal to 1, gives rise to the mass action system

$$\begin{bmatrix} \dot{x}_A \\ \dot{x}_B \end{bmatrix} = \begin{bmatrix} -2 & 1 & 0 \\ 1 & -1 & 1 \end{bmatrix} \begin{bmatrix} x_A^2 \\ x_B \\ x_A \end{bmatrix}$$

and therefore the equilibria sit at the intersection of

$$\dot{x}_A = -2x_A^2 + x_B = 0 \text{ and } \dot{x}_B = x_A^2 - x_B + x_A = 0$$

(see Fig. 9.4A). A quick calculation shows that there is only one positive equilibrium at $(1, 2)$. Note that the stoichiometric matrix here has rank 2, and so there are no conservation laws. When these are present, the discussion on

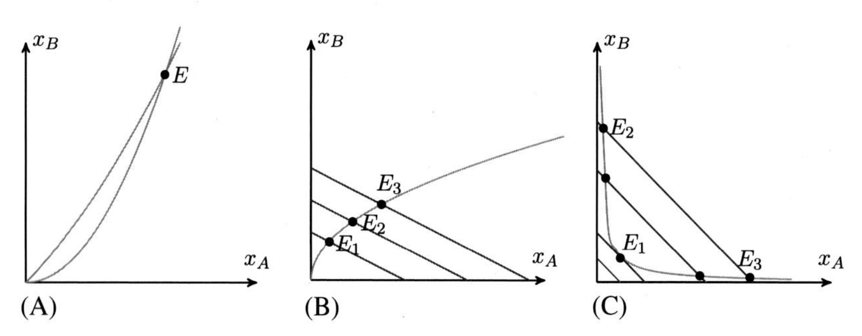

FIG. 9.4 (A) Nullclines for Eq. (9.4). The unique positive equilibrium is $(1, 2)$. (B) The two nullclines for Eq. (9.5) coincide. There is a unique equilibrium in each positive stoichiometry class, and it is nondegenerate. (C) The two nullclines for Eq. (9.6) coincide. Some stoichiometry classes contain two nondegenerate equilibria, some of them contain no equilibria, while the stoichiometry class $x_A + x_B = 2$ has a single degenerate equilibrium.

equilibria becomes a little more subtle; for example, let us revisit network (Eq. 9.3)

$$A \underset{1}{\overset{1}{\rightleftharpoons}} 2B \tag{9.5}$$

with both rate constants set to 1. The steady-state equations read

$$\dot{x}_A = -x_A + x_B^2 = 0, \quad \dot{x}_B = 2x_A - 2x_B^2 = 0,$$

the two polynomials are multiple of each other, and therefore we really only have one equation in two variables. This produces a continuum of equilibria $x_A = x_B^2$ called the *steady-state manifold*—see Fig. 9.4B. However, now we have a conservation law $2x_A + x_B = \text{const}$. When speaking about uniqueness of positive equilibria in networks with conservation laws, the relevant question is whether steady states are unique *within some (same) stoichiometry class*. It is easy to check that for any $T > 0$, the stoichiometry class $2x_A + x_B = T$ contains precisely one equilibrium at $\left(\frac{1+4T-\sqrt{1+8T}}{8}, \frac{-1+\sqrt{1+4T}}{4}\right)$ (see also Fig. 9.4B).

On the other hand, the network

$$2A + B \xrightarrow{1} 3A, \; A \xrightarrow{1} B \tag{9.6}$$

has two positive equilibria within some stoichiometry classes. Indeed, the mass action equations read

$$\begin{bmatrix} \dot{x}_A \\ \dot{x}_B \end{bmatrix} = \begin{bmatrix} 1 & -1 \\ -1 & 1 \end{bmatrix} \begin{bmatrix} x_A^2 x_B \\ x_A \end{bmatrix}$$

and the steady-state equations are

$$\dot{x}_A = x_A^2 x_B - x_A = 0, \quad \dot{x}_B = -x_A^2 x_B + x_A = 0.$$

Note that the two equations are equivalent (linearly dependent), and in the positive quadrant they both reduce to $x_A x_B = 1$. The intersection of this curve with the stoichiometry class $x_A + x_B = T$, $T > 0$ yields two positive equilibria $\left(\frac{T+\sqrt{T^2-4}}{2}, \frac{T-\sqrt{T^2-4}}{2}\right)$ and $\left(\frac{T-\sqrt{T^2-4}}{2}, \frac{T+\sqrt{T^2-4}}{2}\right)$ if $T > 2$, a single equilibrium if $T = 2$, and no equilibria if $T < 2$ (see Fig. 9.4C).

Since there exists at least one stoichiometry class that contains more than one steady state, we say that network (9.6) has multiple positive equilibria (MPE). Here we happened to fix the rate constants, but throughout this chapter that won't be the case. We say that a network has MPE if *for some choice of rate constants* there exists a stoichiometry class with two or more steady states.

Network (9.6) is a favorite example in the multistationarity literature [28, 29, 43, 52]: it is simple enough for calculations to be done by hand, but it is complex enough to have interesting features, like multistationarity and degenerate equilibria. We will often use versions of this network to illustrate various points throughout this chapter.

9.2.6 Nondegenerate Equilibria

An equilibrium is called *nondegenerate* if the various manifolds at whose intersection it lies (linearly independent nullclines and conservation laws, when the latter exist) intersect transversally. This technicality is important in various places during our presentation: for example, we will be interested to see how equilibria survive under small perturbations of the vector field (and therefore of the nullclines), which leads naturally to considering the notion of nondegeneracy. Fig. 9.4 suggests that all equilibria of Eqs. (9.4), (9.5) are nondegenerate, whereas the steady state E_1 of Eq. (9.6) is degenerate. To make this precise, note that nondegeneracy is equivalent to saying that the normal vectors to the (linearly independent) hypersurfaces that cut out an equilibrium point are linearly independent. When there are no conservation laws, these normal vectors are the rows of the Jacobian matrix, and therefore nondegeneracy is equivalent to nonvanishing of the Jacobian at the equilibrium point. The Jacobian matrix of network (9.4) is

$$Df = \begin{bmatrix} -4x_A & 1 \\ 2x_A + 1 & -1 \end{bmatrix}$$

and the Jacobian is $\det(Df) = 2x_A - 1$. At equilibrium $(1, 2)$ this is positive, and nondegeneracy follows.

When conservation laws are present, linearly dependent nullclines are replaced by linearly independent conservation laws, and the Jacobian of the new algebraic system is then computed. For network (9.5), nondegeneracy is checked by computing the Jacobian of the algebraic system

$$-x_A + x_B^2 = 0 \quad 2x_A + x_B - T = 0,$$

that is,

$$\det \begin{bmatrix} -1 & 2x_B \\ 2 & 1 \end{bmatrix} = -1 - 4x_B.$$

This is negative everywhere in the positive orthant, and therefore equilibria are nondegenerate. As for network (9.6), one computes the Jacobian of the system

$$x_A^2 x_B - x_A = 0, \quad x_A + x_B - T = 0$$

to get

$$\det \begin{bmatrix} 2x_A x_B - 1 & x_A^2 \\ 1 & 1 \end{bmatrix} = 2x_A x_B - 1 - x_A^2. \tag{9.7}$$

At a positive equilibrium point (p_A, p_B) we have $p_A p_B = 1$, so the Jacobian is equal to $1 - p_A^2$, which shows that all equilibria are nondegenerate except for when $p_A = 1$, that is, when the steady state is $E_1 = (1, 1)$.

9.2.7 The Reduced Jacobian

Here we illustrate another way to think about nondegeneracy. Let us revisit network (9.6) and the nondegeneracy condition (9.7) for an equilibrium $p \in \mathbb{R}^2$. This condition means that any nonzero vector $v \in \mathbb{R}^2$ satisfying $[2p_Ap_B - 1, p_A^2]^t v = 0$ cannot be orthogonal to $[1,1]^t$, that is, it cannot lie in the stoichiometric subspace span $([-1,1]^t)$. In other words, the Jacobian map of the network at p, $Df(p)$: $\mathbb{R}^2_{\geq 0} \to$ span $([-1,1]^t)$ does not vanish on nonzero vectors of the stoichiometric subspace im Γ, that is, its restriction to im Γ is invertible. To compute this restriction, we use local coordinates on the stoichiometric subspace as follows. For $y \in$ im Γ, let $z \in \mathbb{R}$ denote the coordinate of y in our basis, $y = \begin{bmatrix} 1 \\ -1 \end{bmatrix} z$. Then, recalling that $v(x) = [x_A^2 x_B, x_A]^t$, we have

$$\begin{aligned} Df(p)y = \Gamma Dv(p)\left(\begin{bmatrix} 1 \\ -1 \end{bmatrix} z\right) &= \begin{bmatrix} -1 & 1 \\ 1 & -1 \end{bmatrix} \begin{bmatrix} 2p_Ap_B & p_A^2 \\ 1 & 0 \end{bmatrix} \begin{bmatrix} 1 \\ -1 \end{bmatrix} z \\ &= \begin{bmatrix} 1 \\ -1 \end{bmatrix} [1 \; -1] \begin{bmatrix} 2p_Ap_B & p_A^2 \\ 1 & 0 \end{bmatrix} \begin{bmatrix} 1 \\ -1 \end{bmatrix} z = \begin{bmatrix} 1 \\ -1 \end{bmatrix} (2p_Ap_B - 1 - p_A^2)z \end{aligned} \tag{9.8}$$

and therefore $z \mapsto (2p_Ap_B - 1 - p_A^2)z$ is the action of the Jacobian matrix $Df(p)$ using the basis $\{[1,-1]^t\}$ of the stoichiometric subspace. This is generally a linear map (it happens to be a scalar multiplication in this example since rank $\Gamma = 1$). Its determinant is called *the reduced Jacobian* of the system. Nonvanishing of the reduced Jacobian at p is equivalent to p being nondegenerate. Not coincidently, the reduced Jacobian computed here is the same as the determinant computed in Eq. (9.7). This is revisited in the next section, where the construction of the reduced determinant is presented in full generality.

9.2.8 General Setup and Preliminaries

Here we put the various terminology discussed so far into a general framework. Each notion introduced below is referred to a previous section where examples were discussed informally. We adopt a presentation of matrix-theoretical flavor along the lines of [27]. Throughout this section, we consider a CRN of m reactions on n species, and we fix an order of species and an order of reactions. We can then refer to "species i" or "reaction j." Reactions may be reversible, but we will view these as two separate irreversible reactions. For convenience, a reaction arrow proceeds from left to right: reactants sit on the left, and products on the right.

In what follows, $\mathbb{R}^n_{\geq 0}$ and $\mathbb{R}^n_{>0}$ denote the nonnegative, respectively positive orthants in $\mathbb{R}^n$, i.e. the set of vectors with nonnegative, respectively positive coordinates. The image (column space) of a matrix $A \in \mathbb{R}^{n \times m}$, a linear subspace of $\mathbb{R}^n$, is denoted im A, and the nullspace of A is denoted by ker A. The transpose of A is denoted by A^t. If $x \in \mathbb{R}^n$, we denote by $D_x \in \mathbb{R}^{n\times n}$ the diagonal matrix with elements of x on the diagonal, that is, $(D_x)_{ii} = x_i$. We will use the following notation for submatrices and minors: for a matrix $A \in \mathbb{R}^{n\times m}$ and sets $\alpha, \beta \subseteq \{1, \ldots, n\}$, with $|\alpha| = |\beta|$, $A[\alpha|\beta]$ denotes the minor of A corresponding to rows in α and columns in β. When $\alpha = \beta$, this is simply denoted by $A[\alpha]$.

Definition 9.1 (Reactant Matrix, Product Matrix, Stoichiometric Matrix—Sections 9.2.1 and 9.2.2)**.** The *reactant matrix* $\Gamma_l \in \mathbb{Z}^{n\times m}$ and the *product matrix* $\Gamma_r \in \mathbb{Z}^{n\times m}$ are defined as follows: $(\Gamma_l)_{ij}$ is the number of molecules of species i occurring on the reactant side (left-hand side) of reaction j; $(\Gamma_r)_{ij}$ is the number of molecules of species i occurring on the product side (right-hand side) of reaction j. The *stoichiometric matrix* of the network is defined as $\Gamma = \Gamma_r - \Gamma_l$.

Note that Γ_l and Γ_r have nonnegative entries. The (i,j) entry of the stoichiometric matrix is the net gain of molecules of species i in reaction j. The stoichiometric matrix is not uniquely defined, as it depends on the orderings on the species and reactions. However, these orderings do not impact any of the results we discuss here.

Definition 9.2 (Complexes, Inflow/Outflow Reactions, Reaction Graph—Section 9.2.1)**.** If species are denoted $X_1, \ldots, X_n$, we define formally the vector $X = (X_1, \ldots, X_n)^t$. The formal dot products of columns of Γ_l and Γ_r with X are termed *complexes* of the network [53]. Each reaction converts its *source complex* into its *product complex*. The zero vector (as a column in Γ_l or Γ_r) is called the *zero complex*, and simply denoted by 0. Reactions of the form $0 \to C$ and $C \to 0$ (where C is a nonzero complex) are called *inflow* and *outflow* reactions, respectively. Any reaction can be viewed as an edge in a directed graph whose vertices are the network complexes. This graph fully characterizes the CRN, and is termed the *reaction graph* of the network.

The time evolution of the vector of species concentrations $x = x(t) \in \mathbb{R}^n_{\geq 0}$ is governed by the ODE system

$$\dot{x} = \Gamma v(x) =: f(x), \tag{9.9}$$

where $\dot{x}$ denotes $\frac{dx}{dt}$, $\Gamma \in \mathbb{R}^{n\times m}$ denotes the stoichiometric matrix, and $v \in \mathbb{R}^{m\times 1}$ denotes the vector of reaction rates or *kinetics* of the system. While there are many types of kinetics that arise in practice, we restrict our attention to *mass action*, perhaps the most important one.

It is convenient to introduce the vector power notation: given $x = (x_1, \ldots, x_n)^t, y = (y_1, \ldots, y_n)^t \in \mathbb{R}^n_{\geq 0}$, we let $x^y = \prod_{i=1}^n x_i^{y_i}$. If $M \in \mathbb{R}^{n\times m}$ has columns $M_1, \ldots, M_m$ we use x^M as a convenient abbreviation for the $1 \times m$ vector $(x^{M_1}, \ldots, x^{M_m})$.

Definition 9.3 (Mass Action Kinetics, Rate Constants—Sections 9.2.1 and 9.2.2)**.** Let Γ_l and $\Gamma \in \mathbb{Z}^{n\times m}$ denote the reactant and stoichiometric matrices, respectively. For $k \in \mathbb{R}^n_{>0}$, we say that Eq. (9.9) with $v(x) = D_k(x^{\Gamma_l})^t$ is a *CRN with mass action kinetics* and *rate constants* k. In mass action, the rate v_j of reaction j is proportional to the product of reactant concentrations, and the proportionality factor is k_j. A CRN with mass action is therefore governed by the ODE system

$$\dot{x} = \Gamma D_k(x^{\Gamma_l})^t =: f(x). \tag{9.10}$$

Note that the rate of an inflow reaction $0 \xrightarrow{k} A$ is constant, and equal to k.

It is easy to show that the nonnegative orthant is forward invariant under the dynamics of a CRN with mass action. The proof of this fact is left to the reader (Exercise 2). For $x \in \mathbb{R}^n_{>0}$, we let $1/x = (1/x_1, \ldots, 1/x_n)$. The Jacobian matrix of Eq. (9.9) factors as

$$Df(x) = \Gamma D_{v(x)} \Gamma_l^t D_{1/x}. \tag{9.11}$$

(This is an easy calculation; see Section 9.2.3 for an example.)

Definition 9.4 (Stoichiometric Subspace, Stoichiometry Classes—Section 9.2.4)**.** Given a CRN with stoichiometric matrix $\Gamma \in \mathbb{R}^{n\times m}$, the linear space $\operatorname{im}\Gamma \subseteq \mathbb{R}^n$ is called the *stoichiometric subspace* of the network. For $p \in \mathbb{R}^n_{\geq 0}$, the coset of $\operatorname{im}\Gamma$ containing p and intersected with $\mathbb{R}^n_{\geq 0}$

$$S_p = (p + \operatorname{im}\Gamma) \cap \mathbb{R}^n_{\geq 0} = \left\{ y \in \mathbb{R}^n_{\geq 0} \mid y - p \in \operatorname{im}\Gamma \right\}$$

is called the *stoichiometry class of* p. A stoichiometry class which intersects $\mathbb{R}^n_{>0}$ is called *nontrivial*. The intersection of a stoichiometry class with $\mathbb{R}^n_{>0}$ (if nonempty) is a *positive stoichiometry class*.

Integrating Eq. (9.9) yields

$$x(t) = x(0) + \Gamma \int_0^t v(x(s))ds,$$

which shows that the solution $\{x(t) \mid t \geq 0\}$ of Eq. (9.9) is constrained to the stoichiometry class of $x(0)$. Put another way, trajectories of CRN dynamics satisfy certain linear constraints, called conservation laws.

Definition 9.5 (Conservation Laws—Section 9.2.4)**.** A nonzero element of $\ker \Gamma^t$ is called a *conservation law* of the CRN. A basis of $\ker \Gamma^t$ is called a *complete set of independent conservation laws*.

If v is a conservation law, then Eq. (9.9) yields $v \cdot x(t) = \text{constant}$. Since $\ker \Gamma^t = (\operatorname{im}\Gamma)^\perp$, a set of constants for a complete set of conservation laws uniquely defines a coset of $\operatorname{im}\Gamma$, that is, a stoichiometry class.

Definition 9.6 (Fully Open Systems, Fully Open Extensions—Section 9.2.1)**.** A CRN that contains inflow reactions $A \to 0$ and outflow reactions $0 \to A$ for all species A is called a *fully open network.* A fully open network has stoichiometric subspace equal to $\mathbb{R}^n$, and no conservation laws. Adding all inflow and outflow reactions to a network defines its *fully open extension.*

Definition 9.7 (Equilibria, MPE—Section 9.2.5)**.** A point $p \in \mathbb{R}^n_{\geq 0}$ is called an equilibrium of a CRN if $\Gamma v(p) = 0$. It is called a *positive equilibrium* if $p \in \mathbb{R}^n_{>0}$. A CRN has *the capacity for MPE* [26] if there exist rate constants $k \in \mathbb{R}^n_{>0}$ for which Eq. (9.10) admits two distinct equilibria within the same stoichiometry class, that is, there exist distinct $p_1, p_2 \in \mathbb{R}^n_{>0}$ such that $p_1 - p_2 \in \operatorname{im} \Gamma$ and

$$\Gamma D_k \left(p_1^{\Gamma_l}\right)^t = \Gamma D_k \left(p_2^{\Gamma_l}\right)^t = 0.$$

Let $r > 0$ be the rank of the stoichiometric matrix Γ. Choose any basis for $\operatorname{im} \Gamma$ and arrange its vectors as columns of a matrix $\Gamma_0 \in \mathbb{R}^{n \times r}$. We can write $\Gamma = \Gamma_0 Q$ for some matrix Q. Letting Γ' be a left inverse of Γ_0 we get $\Gamma' \Gamma = Q$, and therefore $\Gamma = \Gamma_0 \Gamma' \Gamma$. We write $x \in \operatorname{im} \Gamma$ in coordinates corresponding to the basis Γ_0 (i.e., $x = \Gamma_0 y$). The action of the Jacobian map $\Gamma D v$ on $x \in \operatorname{im} \Gamma$ is as follows: $\Gamma D v \Gamma_0 y = \Gamma_0 \Gamma' \Gamma D v \Gamma_0 y = \Gamma_0 z$ with $z = \Gamma' \Gamma D v \Gamma_0 y$. Therefore, the Jacobian matrix acts on local coordinates on $\operatorname{im} \Gamma$ according to $y \mapsto \Gamma' \Gamma D v \Gamma_0 y$.

Definition 9.8 (Reduced Jacobian [27]—Section 9.2.7)**.** With the notations introduced above, a *reduced Jacobian matrix* of a CRN is defined as $\Gamma' \Gamma D v \Gamma_0$; its determinant is called the *reduced Jacobian* and denoted by $\det_\Gamma(\Gamma D v)$.

Note that different choices of Γ_0 and Γ' may result in different reduced Jacobian matrices; however, they are all similar matrices (see [27, Appendix A]). Therefore, the reduced Jacobian does not depend on these choices and the notation $\det_\Gamma(\Gamma D v)$ is unambiguous. In fact, one can show that the reduced Jacobian of a CRN is equal to the sum of $r \times r$ principal minors of the Jacobian:

Proposition 9.1 ([27])**.**

$$\det_\Gamma(\Gamma D v) = \sum_{\substack{\alpha \subseteq \{1,\ldots,n\} \\ |\alpha| = r}} (\Gamma D v)[\alpha].$$

In particular, if $\operatorname{rank} \Gamma = n$ (e.g., if the network is fully open), then the reduced Jacobian and the Jacobian are equal.

Definition 9.9 (Nondegenerate Equilibria—Sections 9.2.6 and 9.2.7)**.** An equilibrium $p \in \mathbb{R}^n_{\geq 0}$ of a mass action CRN is called nondegenerate if $\det_\Gamma(\Gamma Dv(p)) \neq 0$. We say that a CRN has *the capacity for multiple positive nondegenerate equilibria (MPNE)* if there exist rate constants $k \in \mathbb{R}^n_{>0}$ for which Eq. (9.10) admits two distinct nondegenerate equilibria within the same positive stoichiometry class, that is, there exist distinct $p_1, p_2 \in \mathbb{R}^n_{>0}$ such that $p_1 - p_2 \in \operatorname{im} \Gamma$,

$$\Gamma D_k \left(p_1^{\Gamma_l}\right)^t = \Gamma D_k \left(p_2^{\Gamma_l}\right)^t = 0,$$

and $\det_\Gamma(\Gamma Dv(p_i)) \neq 0$ for $i = 1, 2$.

Alternatively, an equilibrium is nondegenerate if it sits at a *transversal* intersection of nullclines and hyperplanes defined by conservation laws. To see that this is equivalent to the previous definition, we use the following setup (e.g., [35, 36]). First, we reorder the species (i.e., the rows of Γ) so that a complete set of conservation laws can be arranged as the columns of the block matrix $\begin{bmatrix} W \\ I \end{bmatrix}$ with $I \in \mathbb{R}^{(n-r)\times(n-r)}$ denoting the identity matrix, and $W \in \mathbb{R}^{r\times(n-r)}$. In other words, the last $n-r$ concentrations can be linearly eliminated using conservation laws; their derivatives are linear combinations of the derivatives of the first r concentrations. We form a new algebraic system characterizing the equilibria in some stoichiometric class by removing the equations corresponding to the last $n - r$ concentrations from the steady-state system $\Gamma v(x) = 0$, and replacing them with the $n - r$ conservation laws which define the stoichiometric class $[W^t|I]x = T$ (here $T \in \mathbb{R}^{n-r}$ is the vector of conserved values). Nondegeneracy according to the new geometric interpretation amounts to the nonvanishing of the new algebraic system's Jacobian. It turns out that this is equivalent to the nondegeneracy in Definition 9.9. Indeed, if

$$\Gamma = \begin{bmatrix} \Gamma_1 \\ \Gamma_2 \end{bmatrix}, \tag{9.12}$$

where $\Gamma_1 \in \mathbb{R}^{r\times m}$, then the new algebraic system reads

$$\Gamma_1 v(x) = 0, \quad [W^t|I]x - T = 0 \tag{9.13}$$

and we have the following result generalizing the calculations (9.7), (9.8) in Sections 9.2.6 and 9.2.7.

Proposition 9.2. *The Jacobian of Eq. (9.13) is equal to the reduced Jacobian* $\det_\Gamma(\Gamma Dv)$.

Proof. Note that from Eq. (9.12) and the definition of W we get $\Gamma_2 = -W^t \Gamma_1$. We rearrange reactions (columns of Γ) to get a nonzero principal minor $\Gamma(\{1, \ldots, r\})$, and we write Γ and Dv in corresponding block form

$$\Gamma = \begin{bmatrix} \Gamma_{11} & \Gamma_{12} \\ -W^t\Gamma_{11} & -W^t\Gamma_{12} \end{bmatrix}, \quad Dv = \begin{bmatrix} V_{11} & V_{12} \\ V_{21} & V_{22} \end{bmatrix},$$

where $\Gamma_{11} \in \mathbb{R}^{r\times r}$ is nonsingular and $V_{11} \in \mathbb{R}^{r\times r}$. The Jacobian matrix of Eq. (9.13) is

$$\begin{bmatrix} \Gamma_{11}V_{11} + \Gamma_{12}V_{21} & \Gamma_{11}V_{12} + \Gamma_{12}V_{22} \\ W^t & I \end{bmatrix}. \tag{9.14}$$

To compute the reduced determinant, note that the columns of $\Gamma_0 = \begin{bmatrix} \Gamma_{11} \\ -W^t\Gamma_{11} \end{bmatrix}$ form a basis of $\operatorname{im}\Gamma$ and that $\Gamma' = [\Gamma_{11}^{-1}|0]$ is a left inverse of Γ_0. Then

$$\begin{aligned} \det{}_\Gamma(\Gamma Dv) &= \det(\Gamma'\Gamma Dv\Gamma_0) \\ &= \det\left([\Gamma_{11}^{-1}|0]\begin{bmatrix} \Gamma_{11} & \Gamma_{12} \\ -W^t\Gamma_{11} & -W^t\Gamma_{12} \end{bmatrix}\begin{bmatrix} V_{11} & V_{12} \\ V_{21} & V_{22} \end{bmatrix}\begin{bmatrix} \Gamma_{11} \\ -W^t\Gamma_{11} \end{bmatrix}\right) \\ &= \det\Gamma_{11}^{-1} \det\left([I|0]\begin{bmatrix} \Gamma_{11} & \Gamma_{12} \\ -W^t\Gamma_{11} & -W^t\Gamma_{12} \end{bmatrix}\begin{bmatrix} V_{11} & V_{12} \\ V_{21} & V_{22} \end{bmatrix}\begin{bmatrix} I \\ -W^t \end{bmatrix}\right)\det\Gamma_{11} \\ &= \det(\Gamma_{11}V_{11} - \Gamma_{11}V_{12}W^t + \Gamma_{12}V_{21} - \Gamma_{12}V_{22}W^t), \end{aligned}$$

which equals the determinant of Eq. (9.14) by the Schur determinant formula. □

Some of the results to follow regard not only the existence of MPE of a CRN, but also their linear stability in the following sense.

Definition 9.10 (Linear Stability, Multiple Positive Linearly Stable Equilibria: MPSE)**.** An equilibrium p of Eq. (9.9) will be termed *linearly stable* if it is linearly stable w.r.t. its stoichiometry class, namely all eigenvalues of the reduced Jacobian matrix have negative real parts. A mass action CRN displays *multiple positive linearly stable equilibria* (MPSE) if, for some choice of its rate constants, it has two distinct positive equilibria in some same stoichiometry class, both linearly stable.

Note that if a CRN has MPSE, then it also has MPNE. The network (9.6) shows that the converse is not true in general (Exercise 5).

EXERCISES

1. *(Hungarian Lemma [54])*. Let $f_i(x)$: $\mathbb{R}^n \to \mathbb{R}$, $i \in \{1,\ldots,n\}$ be polynomial functions and $f(x) = (f_1(x),\ldots,f_n(x))$. Show that there exists a CRN whose mass action equations are $\dot{x} = f(x)$ if and only if any monomial with

negative coefficient in f_i contains x_i. (For example, this is not satisfied for $\dot{x}_1 = -2x_1x_2^2 + 1$, $\dot{x}_2 = -x_1^2 + 3x_2$, since the negative monomial $-x_1^2$ in the expression of $\dot{x}_2$ does not contain x_2.)

Remark 9.1. This shows that a large class of polynomial dynamical systems can be realized as mass action, and therefore results for mass action may have implications beyond reaction networks.

2. *(Forward invariance under mass action).* Show that if $x(t)$, $t \geq 0$ is a solution of Eq. (9.10) with $x(0) \in \mathbb{R}^n_{\geq 0}$, then $x(t) \in \mathbb{R}^n_{\geq 0}$ for all $t \geq 0$.

3. *(The futile cycle).* Find a complete set of conservation laws for the following CRN (the *futile cycle (Eq. 9.1)*, sometimes also called the *one-step phosphorylation-dephosphorylation network*), and compute its reduced Jacobian:

$$E + S \rightleftharpoons ES \rightarrow E + P$$
$$F + P \rightleftharpoons FP \rightarrow F + S.$$

4. *(Examples of small networks).* Analyze the capacity of MPE/MPNE/MPSE for the following mass action networks (see [27] for related examples, and [29] for results on two-species networks):

a. $2A + B \rightarrow 3A,\ A \rightarrow B,\ A \rightarrow 0,\ 0 \rightarrow B$
b. $2A + B \rightarrow 3A,\ A \rightarrow B,\ A \rightleftharpoons 0,\ B \rightleftharpoons 0$
c. $2A + B \rightarrow 3A,\ A \rightarrow B,\ A + B \rightleftharpoons 0$
d. $A \rightarrow B,\ 2B \rightarrow 2A,\ 2A + 2B \rightarrow 3A + B$
e. $A + B \rightarrow 2B,\ 2A + B \rightarrow 2B,\ 3B \rightarrow A + 2B$

5. *(MPNE does not imply MPSE).* Show that the CRN (Eq. 9.6) does not have MPSE.

9.3 NECESSARY CONDITIONS FOR MULTISTATIONARITY I: INJECTIVE CRNS

Much of the literature on multistationarity for CRNs have focused on studying necessary conditions for the existence of two or more positive equilibria. A particularly successful approach in this direction has been that of *injective* CRNs, that is, CRNs for which the corresponding vector field $f = \Gamma v$ (Eq. 9.10) is an injective function on positive stoichiometry classes for any choice of rate constants. A very simple, but perhaps illuminating example of a noninjective CRN is $2A \rightleftharpoons A$; for rate constants equal to 1, the mass action vector field is equal to $-x_A^2 + x_A$, which is not injective on the unique stoichiometry class, namely $\mathbb{R}_{>0}$. Clearly, an injective CRN cannot have the capacity for MPE, since the latter requires that two points in the same positive stoichiometry class be both mapped by f to 0. Note, however, that injectivity is *not equivalent* to the lack of capacity for MPE (see Exercise 1).

The study of injective CRNs was started by Craciun and Feinberg for fully open networks [26] and has since been extended by work of various authors [27, 35–39]. These papers have led to complete characterizations of injective CRNs, some of which are gathered in the following theorem. First, we introduce some terminology and notation.

Definition 9.11 (Positive and Negative Matrices)**.** For a real matrix (or vector) A, write $A \geq 0$ to denote the fact that all entries of A are nonnegative, and $A > 0$ to denote the fact that $A \geq 0$ and $A \neq 0$.

Definition 9.12 (Qualitative Class)**.** Given $A \in \mathbb{R}^{n\times m}$, the *qualitative class* $\mathcal{Q}(A) \subseteq \mathbb{R}^{n\times m}$ of A consists of all matrices with the same sign pattern as A, that is, $B \in \mathcal{Q}(A)$ if and only if $(A_{ij} > 0) \Rightarrow (B_{ij} > 0)$; $(A_{ij} < 0) \Rightarrow (B_{ij} < 0)$; and $(A_{ij} = 0) \Rightarrow (B_{ij} = 0)$. If $\mathcal{A}$ is a set of matrices or vectors, we write $\mathcal{Q}(\mathcal{A})$ for $\cup_{A\in\mathcal{A}} \mathcal{Q}(A)$.

Finally, if $A, B \in \mathbb{R}^{n\times m}$ we write $A \circ^r B > 0$ (respectively, $A \circ^r B < 0$) if the product of any pair of corresponding $r \times r$ minors of A and B is nonnegative (nonpositive), and at least one of these products is positive (respectively, negative). To be precise, $A \circ^r B > 0$ if for any $\alpha \subset \{1, \ldots, n\}$, $\beta \subset \{1, \ldots, m\}$ with $|\alpha| = |\beta| = r$ we have

$$A[\alpha|\beta]B[\alpha|\beta] \geq 0$$

and at least one such product is positive.

Consider a CRN with stoichiometric matrix Γ and reactant matrix Γ_l, and let $r = \operatorname{rank}\Gamma$.

Theorem 9.1. *The following are equivalent:*

1. (Injectivity) *The mass action vector field* $\Gamma v(x) := \Gamma D_k(x^{\Gamma_l})^t$ *(as a function of* x*) is injective on any positive stoichiometry class, for any rate constants* $k \in \mathbb{R}^m_{>0}$.
2. (Nonvanishing of reduced Jacobian) *The reduced Jacobian* $\det_\Gamma(\Gamma Dv)$ *(as a polynomial in* x *and* k*) is nonzero and all of its terms have the same sign.*
3. (Concordance) $\Gamma \circ^r \Gamma_l > 0$ *or* $\Gamma \circ^r \Gamma_l < 0$.
4. (Sign condition) $\Gamma_l^t(\mathcal{Q}(im\, \Gamma)\backslash\{0\}) \cap \mathcal{Q}(ker\, \Gamma) = \emptyset$.

For the proofs of these and other related results, the reader is referred to [27, 35]. A network is called *concordant* if Condition 3 in Theorem 9.1 holds. Concordance was first defined in [38] in an equivalent way, and versions of the results above appear in that work. In many cases, the concordance condition may be easiest to check—note that its calculation does not involve the rate constants. On the other hand, the reduced Jacobian is also not hard to compute—recall from Proposition 9.1 that the reduced Jacobian is the sum of $r \times r$ principal minors of the Jacobian matrix ΓDv.

Next, we discuss an example illustrating the applicability of Theorem 9.1. For more examples, the reader is referred to the exercises at the end of the section. Consider the following version of the futile cycle (Eq. 9.1), where metabolite S gets transformed into a product P in a reaction catalyzed by enzyme E, while the reverse process does not require an enzymatic mechanism:

$$E + S \underset{k_2}{\overset{k_1}{\rightleftharpoons}} ES \overset{k_3}{\rightarrow} E + P, \quad P \overset{k_4}{\rightarrow} S. \tag{9.15}$$

Setting species E, S, ES, P in this order, the corresponding stoichiometric matrix Γ, reactant matrix Γ_l, and ODE system are

$$\Gamma = \begin{bmatrix} -1 & 1 & 1 & 0 \\ -1 & 1 & 0 & 1 \\ 1 & -1 & -1 & 0 \\ 0 & 0 & 1 & -1 \end{bmatrix}, \quad \Gamma_l = \begin{bmatrix} 1 & 0 & 0 & 0 \\ 1 & 0 & 0 & 0 \\ 0 & 1 & 1 & 0 \\ 0 & 0 & 0 & 1 \end{bmatrix},$$

$$\dot{x} = \begin{bmatrix} -1 & 1 & 1 & 0 \\ -1 & 1 & 0 & 1 \\ 1 & -1 & -1 & 0 \\ 0 & 0 & 1 & -1 \end{bmatrix} \begin{bmatrix} k_1 x_1 x_2 \\ k_2 x_3 \\ k_3 x_3 \\ k_4 x_4 \end{bmatrix}.$$

The Jacobian matrix can be computed as

$$\Gamma Dv = \begin{bmatrix} -1 & 1 & 1 & 0 \\ -1 & 1 & 0 & 1 \\ 1 & -1 & -1 & 0 \\ 0 & 0 & 1 & -1 \end{bmatrix} \begin{bmatrix} k_1 x_2 & k_1 x_1 & 0 & 0 \\ 0 & 0 & k_2 & 0 \\ 0 & 0 & k_3 & 0 \\ 0 & 0 & 0 & k_4 \end{bmatrix}$$
$$= \begin{bmatrix} -k_1 x_2 & -k_1 x_1 & k_2 + k_3 & 0 \\ -k_1 x_2 & -k_1 x_1 & k_2 & k_4 \\ k_1 x_2 & k_1 x_1 & -k_2 - k_3 & 0 \\ 0 & 0 & k_3 & -k_4 \end{bmatrix},$$

and since rank $\Gamma = 2$, the reduced Jacobian of the system is the sum of all 2×2 principal minors of ΓDv, that is,

$$\det{}_\Gamma(\Gamma Dv) = \begin{vmatrix} -k_1 x_2 & 0 \\ 0 & -k_4 \end{vmatrix} + \begin{vmatrix} -k_1 x_1 & k_2 \\ k_1 x_1 & -k_2 - k_3 \end{vmatrix} + \begin{vmatrix} -k_1 x_1 & k_4 \\ 0 & -k_4 \end{vmatrix}$$
$$+ \begin{vmatrix} -k_2 - k_3 & 0 \\ k_3 & -k_4 \end{vmatrix}$$
$$= k_1 k_3 x_1 + k_1 k_4 x_1 + k_1 k_4 x_2 + k_2 k_4 + k_3 k_4.$$

This polynomial in $x_1, \dots, x_4, k_1, \dots, k_4$ has only positive coefficients, and so Theorem 9.1 implies that the vector field is injective on each positive stoichiometry class. Therefore, the mass action network (9.15) does not have the capacity for MPE.

One may reach the same conclusion by checking concordance of the network, that is, by computing products of corresponding 2×2 minors of Γ and $-\Gamma_l^t$. By inspection, the only nonzero products $\Gamma[\alpha|\beta]\Gamma_l[\alpha|\beta]$ occur for $(\alpha|\beta) \in \{(\{1,4\}|\{1,4\}),(\{2,3\}|\{1,3\}),(\{2,4\}|\{1,4\}),(\{3,4\}|\{2,4\}),(\{3,4\}|\{3,4\})\}$. The corresponding products are

$$\begin{vmatrix} -1 & 0 \\ 0 & -1 \end{vmatrix}\begin{vmatrix} 1 & 0 \\ 0 & 1 \end{vmatrix}, \begin{vmatrix} -1 & 0 \\ 1 & -1 \end{vmatrix}\begin{vmatrix} 1 & 0 \\ 0 & 1 \end{vmatrix}, \begin{vmatrix} -1 & 1 \\ 0 & -1 \end{vmatrix}\begin{vmatrix} 1 & 0 \\ 0 & 1 \end{vmatrix}, \begin{vmatrix} -1 & 0 \\ 0 & -1 \end{vmatrix}\begin{vmatrix} 1 & 0 \\ 0 & 1 \end{vmatrix},$$
$$\begin{vmatrix} -1 & 0 \\ 1 & -1 \end{vmatrix}\begin{vmatrix} 1 & 0 \\ 0 & 1 \end{vmatrix},$$

and are all equal to 1. Therefore, the network is concordant, and injectivity follows. Note that the number of nonzero products of minors (five) equals the number of monomials in the reduced determinant computed above. This is not coincidental; it turns out that in general, the pairwise products of minors from the definition of concordance are precisely the coefficients of the polynomial $\det_\Gamma(\Gamma Dv)$—see, for example, [27] and Exercise 2.

Finally, we illustrate how one checks the sign Condition 4 in the previous theorem. We note that our treatment here is rather rudimentary, and there is a great deal of subtlety that we miss in a short description like this one. For a detailed discussion of the sign condition and for insightful connections with the machinery of oriented matroids, the reader is referred to [55]. Informally, Γ_l^t acts on sign vectors using the commutative addition rules $(+)+(+) = +$, $(-)+(-) = -$, $(+)+(0) = +$, $(-)+(0) = -$, $(+)+(-) \in \{0,+,-\}$.

Choosing the second and fourth column of Γ as basis, we can write $\operatorname{im}\Gamma = \{[x, x+y, -x, -y]^t \mid x,y \in \mathbb{R}\}$. The possible nonzero sign patterns of $\operatorname{im}\Gamma$ are obtained by listing all sign combinations of x and y: $(0,+),(+,0),(+,+),(+,-),(0,-),(-,0),(-,-),(-,+)$. The first three sign combinations of x and y produce precisely one sign pattern and the fourth one, $(+,-)$ produces three more depending on the relative sizes of $|x|$ and $|y|$. The last four sign pairs give rise to the negative of the sign patterns we have so far. Therefore, $\mathcal{Q}(\operatorname{im}\Gamma)\setminus\{0\}$ consists of 12 sign patterns, listed here:

$$\begin{bmatrix} 0 \\ + \\ 0 \\ - \end{bmatrix}, \begin{bmatrix} + \\ + \\ - \\ 0 \end{bmatrix}, \begin{bmatrix} + \\ + \\ - \\ - \end{bmatrix}, \begin{bmatrix} + \\ 0 \\ - \\ + \end{bmatrix}, \begin{bmatrix} + \\ + \\ - \\ + \end{bmatrix}, \begin{bmatrix} + \\ - \\ - \\ + \end{bmatrix}, \begin{bmatrix} 0 \\ - \\ 0 \\ + \end{bmatrix}, \begin{bmatrix} - \\ - \\ + \\ 0 \end{bmatrix}, \begin{bmatrix} - \\ - \\ + \\ + \end{bmatrix}, \begin{bmatrix} - \\ 0 \\ + \\ - \end{bmatrix}, \begin{bmatrix} - \\ - \\ + \\ - \end{bmatrix}, \begin{bmatrix} - \\ + \\ + \\ - \end{bmatrix}.$$

Applying Γ_l^t to all sign vectors earlier, one obtains the sign patterns in $\Gamma_l^t(\mathcal{Q}(\operatorname{im}\Gamma)\setminus\{0\})$ (for simplicity, we only list the combinations with the first nonzero component equal to "+"; the remaining ones are simply the negative of these):

$$\begin{bmatrix} + \\ 0 \\ 0 \\ - \end{bmatrix}, \begin{bmatrix} + \\ - \\ - \\ 0 \end{bmatrix}, \begin{bmatrix} + \\ - \\ - \\ - \end{bmatrix}, \begin{bmatrix} + \\ - \\ - \\ + \end{bmatrix}, \begin{bmatrix} 0 \\ + \\ + \\ - \end{bmatrix}, \begin{bmatrix} + \\ + \\ + \\ - \end{bmatrix}. \tag{9.16}$$

On the other hand, $\ker \Gamma = \{[x+y, x, y, y]^t \mid x, y \in \mathbb{R}\}$ and therefore $\mathcal{Q}(\ker \Gamma)$ has the following nonzero sign patterns (once again, for shortness, we only list those whose first nonzero coordinate is "+"):

$$\begin{bmatrix} + \\ 0 \\ + \\ + \end{bmatrix}, \begin{bmatrix} + \\ + \\ 0 \\ 0 \end{bmatrix}, \begin{bmatrix} + \\ + \\ + \\ + \end{bmatrix}, \begin{bmatrix} 0 \\ + \\ - \\ - \end{bmatrix}, \begin{bmatrix} + \\ + \\ - \\ - \end{bmatrix}, \begin{bmatrix} + \\ - \\ + \\ + \end{bmatrix}. \tag{9.17}$$

Since no sign pattern appears in both Eqs. (9.16), (9.17), the sign condition is satisfied and therefore the network is injective.

Checking the various injectivity conditions can be easily implemented computationally. For example, the software packages CoNtRol [56, 57] and CRNToolbox [58] include this functionality and many other CRN computation tools.

EXERCISES

1. *(1D injective CRNs).* Find all injective one-species mass action CRNs. Give an example of a noninjective CRN without capacity for MPE.
2. *(Coefficients of the reduced determinant).* Show that the coefficients of the reduced Jacobian of a CRN (as a polynomial in x and k) are precisely the products $\Gamma[\alpha|\beta]\Gamma_l[\beta|\alpha]$, with $\alpha \subseteq \{1,\ldots,n\}$, $\beta \subseteq \{1,\ldots,m\}$, $|\alpha| = |\beta| = \operatorname{rank} \Gamma$ (as usual, Γ denotes the stoichiometric matrix of the CRN).

   Hint. Use the Cauchy-Binet formula: Given $A \in \mathbb{R}^{n\times m}$, $B \in \mathbb{R}^{m\times n}$, and any $\alpha, \beta \subseteq \{1,\ldots,n\}$, $|\alpha| = |\beta| = r > 0$, we have

$$(AB)[\alpha|\beta] = \sum_{\substack{\gamma\subseteq\{1,\ldots,m\} \\ |\gamma|=r}} A[\alpha|\gamma]B[\gamma|\beta].$$

3. *Phosphorylation networks.* Are the following networks injective? Use CoNtRol to double check your answer.
   a. *(futile cycle)*

$$E + S \rightleftharpoons ES \rightarrow E + P$$
$$F + P \rightleftharpoons FP \rightarrow F + S.$$

b. *(distributive double phosphorylation)*

$$E + S_1 \rightleftharpoons ES_1 \to E + S_2 \rightleftharpoons ES_2 \to E + S_3$$
$$F + S_3 \rightleftharpoons FS_3 \to F + S_2 \rightleftharpoons FS_2 \to F + S_1.$$

c. *(distributive double phosphorylation without shared enzymes)*

$$E_1 + S_1 \rightleftharpoons E_1S_1 \to E_1 + S_2, \quad E_2 + S_2 \rightleftharpoons E_2S_2 \to E_2 + S_3$$
$$F_1 + S_3 \rightleftharpoons F_1S_3 \to F_1 + S_2, \quad F_2 + S_2 \rightleftharpoons F_2S_2 \to F_2 + S_1.$$

d. *(processive double phosphorylation)*

$$E + S_1 \rightleftharpoons ES_1 \to ES_2 \to E + S_3$$
$$F + S_3 \rightleftharpoons FS_3 \to FS_2 \to F + S_1.$$

Remark 9.2. Injectivity is just one example of how dynamics changes as an effect of enzyme sharing, or by switching from distributive to processive mechanisms. See [46, 59, 60] for further discussion.

9.4 NECESSARY CONDITIONS FOR MULTISTATIONARITY II: THE DSR GRAPH

There is a great deal of stable behavior in networks of chemical reactions and, to a lesser degree, in biological networks. This can be explained in part by the fact that the possibility of exotic behavior (such as multistability) places rather delicate constraints on the structure of an interaction network; a seminal remark is due to Thomas, who noticed that positive feedback in the logical structure of a CRN is necessary for multistationarity [19]. Subsequent theoretical work proved this claim [11]; here we discuss the DSR graph condition, a far-reaching refinement of Thomas' observation. The DSR graph, introduced by Banaji and Craciun [40], is based on earlier work by Craciun and Feinberg [14], and it provides an elegant sufficient condition for injectivity of CRNs. We note that this condition is not also necessary, so that the methods of Section 9.3 are more powerful than the results that follow here. However, the DSR graph is closely related to the typical diagram depicting a biological network, and it offers unique insight into the connection between its structure and its capacity for multiple equilibria.

Throughout this section we consider *nonautocatalytic networks*, that is, networks for which no species occurs on both sides of the same reaction. We note that the DSR theory does not need this restriction. However, the exposition is significantly simpler for nonautocatalytic networks, and moreover, most networks in practice are nonautocatalytic. In what follows, we regard each reversible reactions as one reaction, as opposed to splitting them in two irreversible reactions, and we (arbitrarily) choose a left side and a right side of a reversible reaction. This way, every species that enters a reversible reaction is either a *left reactant* or a *right reactant*. We also recall that species involved in an irreversible reaction are either *reactant species* (inputs) or *product species* (outputs).

Definition 9.13 (DSR Graph, [40]). The DSR graph of a CRN is a labeled bipartite directed multigraph, with nodes corresponding to *species* and *reactions*. The labels are all positive, but the graph will contain *positive* and *negative* edges. Moreover, given a species node S and a reaction node R, two edges $S \to R$ and $R \to S$ of the same sign are by convention merged into one undirected edge $S{-}R$ of the same sign. The DSR is defined in the following way:

1. For every *irreversible reaction* R and every one of its *reactant species* S, we draw an undirected negative edge (depicted as a dashed line) $S{-}R$. The edge is labeled with the *stoichiometric coefficient* of S in R, that is, the number of molecules of S consumed in reaction R.
2. For every *irreversible reaction* R and every one of its *product species* S, we draw a directed positive edge (depicted as a solid arrow) $R \to S$. The edge is labeled with the *stoichiometric coefficient* of S in R, that is, the number of molecules of S produced in reaction R.
3. For every *reversible reaction* R and every one of its *left reactant species* S, we draw an undirected negative edge $S{-}R$. The edge is labeled with the *stoichiometric coefficient* of S in R, that is, the number of molecules of S that enters reaction R.
4. For every *reversible reaction* R and every one of its *right reactant species* S, we draw an undirected positive edge $S{-}R$. The edge is labeled with the *stoichiometric coefficient* of S in R, that is, the number of molecules of S that enters reaction R.

Fig. 9.5 is perhaps illuminating; it illustrates two examples of DSR graphs, one of which corresponds to CRN (Eq. 9.15). By convention, edge labels equal to 1 are omitted from the figure. As we will see following, the way various cycles intersect in the DSR graph may allow conclusions about the lack of multiple equilibria of the CRN's fully open extension. We carry on with a little more terminology.

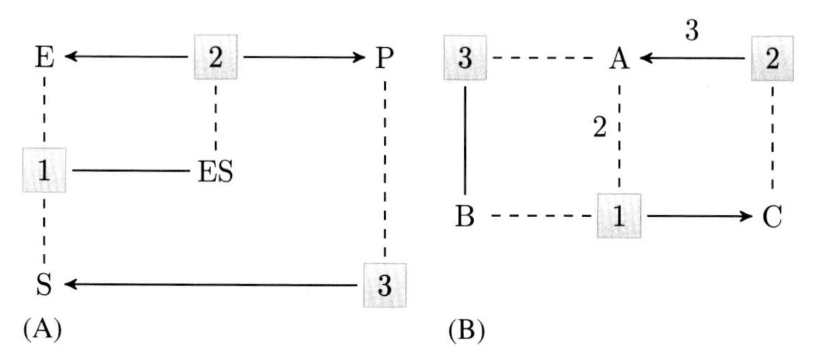

FIG. 9.5 Examples of DSR graphs: (A) $E + S \rightleftharpoons ES \to E + P,\ P \to S$. (B) $2A + B \to C \to 3A,\ A \rightleftharpoons B$.

Recall that a cycle in a directed graph is a path from some vertex to itself which repeats no other vertices, and which respects the orientation of any edges traversed. The unoriented edges in the DSR graph can be viewed as having two orientations, and can be traversed either way. Let $|C|$ denote the length of a cycle in the DSR graph, that is, the number of vertices (or edges) it contains. For an edge e, let $l(e)$ denote its positive label as defined earlier. Recall that e is also assigned a sign, $+1$ (solid) or -1 (dashed). Also note that since DSR is bipartite, each cycle has even length.

Definition 9.14 (Sign of Cycles, E-Cycles, O-Cycles, S-Cycles, Odd Intersections, [14, 40]). Let C be a cycle in a DSR graph.

1. The *sign* of C, denoted $\text{sign}(C)$, is the product of the signs of its edges. In other words, a cycle is positive (has sign +1) if it contains an even number of negative edges.
2. C is called an *e-cycle* if $(-1)^{|C|/2}\text{sign}(C) = 1$ and is called an *o-cycle* if $(-1)^{|C|/2}\text{sign}(C) = -1$. In other words, a cycle C is an e-cycle if the number of its negative (equivalently, the number of its positive) edges has the same parity as $|C|/2$.
3. Let $\{e_1, \ldots, e_{2r}\}$ denote the edges of C traversed in order. C is called an *s-cycle* if
$$\prod_{i=1}^{r} l(e_{2i-1}) = \prod_{i=1}^{r} l(e_{2i}).$$
Note that this product does not depend on the vertex of C where we start enumerating its edges.
4. Two cycles in the DSR graph are *compatibly oriented* if their orientations coincide on each undirected edge in their intersection. Two cycles of the DSR graph have *odd intersection* if they are compatibly oriented and each component of their intersection contains an odd number of edges.

To illustrate, we refer to Fig. 9.5A. The DSR has four species nodes, three reaction nodes, and two cycles: C_1: $1 \to ES \to 2 \to E \to 1$ and C_2: $1 \to ES \to 2 \to P \to 3 \to S \to 1$. Both are s-cycles and e-cycles: for example, C_2 has three negative edges, the same as half of its length. Moreover, C_1 and C_2 are compatibly oriented, and do not have odd intersection; their intersection is the path $1 \to ES \to 2$.

Likewise, Fig. 9.5B depicts the DSR graph of the network

$$2A + B \to C \to 3A, \quad A \rightleftharpoons B. \tag{9.18}$$

Here we have four cycles; C_1: $1 \to C \to 2 \to A \to 1$, C_2: $1 \to B \to 3 \to A \to 1$, C_3: $1 \to A \to 3 \to B \to 1$, and C_4: $1 \to C \to 2 \to A \to 3 \to B \to 1$. Note that C_2 and C_3 have the same edges, traversed in opposite directions. C_1

and C_4 are e-cycles, and C_2 and C_3 are o-cycles: for example, half of the length of C_2 is even (two), whereas the number of its negative edges is odd (one). None of the cycles are s-cycles: for example, the two products of alternating labels for C_1 are $1 \cdot 3 \neq 1 \cdot 2$. Cycles C_1 and C_2 have odd intersection, as do C_1 and C_4, and C_3 and C_4. For example, the latter pair intersect along the path of length three $A \to 3 \to B \to 1$.

Theorem 9.2 (The DSR Graph Theorem [40]). *Suppose $\mathcal{R}$ is a mass action CRN whose DSR graph satisfies the following property: all its e-cycles are s-cycles, and no two e-cycles have odd intersection. Then the fully open extension of $\mathcal{R}$ is injective, and therefore it does not have the capacity for MPE.*

Most networks found in applications only involve stoichiometric coefficients equal to 1, and in that case all cycles are e-cycles; therefore, the first condition in Theorem 9.2 is very often satisfied in practice. DSR e-cycles are related to feedback loops; the DSR theorem implies not only that positive feedback is needed for MPE (as in the conjecture of Thomas), but that they satisfy additional conditions. Indeed, the DSR theorem is a more powerful result [61].

All cycles are s-cycles in (Fig. 9.5A), and since the two cycles do not have odd intersection, one quickly rules out the capacity for MPE of the fully open extension of network (9.15). On the other hand, Theorem 9.2 stays silent for the open extension of network (9.18). Here both conditions of the theorem fail: C_1 is an e-cycle, but not an s-cycle, and the e-cycles C_1 and C_4 have odd intersection. In fact, one can show by methods of Section 9.5 that the open extension of Eq. (9.18) does have the capacity for MPE. We emphasize that in general, however, failure of the hypotheses in Theorem 9.2 is merely a necessary condition for noninjectivity (see Exercise 1).

The DSR graph theorem has been implemented in CoNtRol [56], which also includes a useful tool for drawing DSR graphs.

EXERCISES

1. (*Injective fully open CRN for which the DSR theorem does not apply*). Construct the DSR graph of the following network (the futile cycle with enzyme sharing):

$$E + S \rightleftharpoons ES \to E + P$$
$$E + P \rightleftharpoons EP \to E + S$$
$$E \rightleftharpoons 0,\ S \rightleftharpoons 0,\ P \rightleftharpoons 0,\ ES \rightleftharpoons 0,\ EP \rightleftharpoons 0.$$

   What conclusions does the DSR theorem allow? Show that the network is injective.
2. (*DSR of some enzymatic networks*). Study the injectivity of the fully open extensions of CRNs in Exercise 3 of Section 9.3. When the DSR theorem stays silent, use the methods of Theorem 9.1.

3. (*The sequestration network* [34, 62]). Show that if $m = 1$ or $n \geq 1$ is even, then the fully open extension of the following mass action network is injective:

$$\begin{aligned} X_1 + X_2 &\to 0 \\ X_2 + X_3 &\to 0 \\ &\vdots \\ X_{n-1} + X_n &\to 0 \\ X_1 &\to mX_n. \end{aligned} \tag{9.19}$$

Remark 9.3. One can show that all remaining possibilities for m, n lead to networks with capacity for MPE [34]. It is, however, an open question whether these equilibria are nondegenerate [62].

9.5 SUFFICIENT CONDITIONS FOR MULTISTATIONARITY: INHERITANCE OF MULTIPLE EQUILIBRIA

There has been much recent interest in studying "network motifs" in biological systems, namely small, frequently occurring subnetworks from which dynamical behaviors of the whole network can be inferred [63]. The results collected in this section fall broadly in this research direction; they study how multistationarity of a network is inherited from smaller structures, or "motifs." Specifically, we list a number of situations where MPNE and MPSE (Definitions 9.9 and 9.10) persist as we build up a network from smaller subnetworks. Here we focus on results that are shown by analytic methods, for example, the implicit function theorem [28, 29], although approaches of algebraic nature also exist in recent literature; for example, the results in [30] stem from an algebraic technique for linear elimination of species.

Using the implicit function theorem is quite natural, since certain modifications of the network result in small perturbations of the vector field which allow for local continuations of each nondegenerate positive equilibrium. For example, adding new reactions to a network without changing the stoichiometric subspace, and assigning them small rate constants results in new nullcline manifolds that are merely perturbations of the original ones. In this case, the steady-state configuration will (locally) stay the same—imagine a small perturbation of the steady-state curve in Fig. 9.4C. On the other hand, if the network is modified in a way that the stoichiometric subspace changes, then it is possible for equilibria to vanish, or become degenerate; see Exercise 1.

Throughout this section, we consider a mass action CRN $\mathcal{R}$ with the capacity for MPNE, and order its species in a vector $X = (X_1, \ldots, X_n)$. Recall that complexes are formal linear combinations of species, with nonnegative integer coefficients. If $a = (a_1, \ldots, a_n) \in \mathbb{N}^n$, we use the convenient notation $a \cdot X$ to denote the complex $a_1X_1 + \cdots + a_nX_n$. The zero complex $0x_1 + \cdots + 0x_n$ will be denoted 0. The theorem below collects a series of results about the preservation of the capacity for MPNE when $\mathcal{R}$ is being "enlarged" into a new network $\mathcal{R}'$.

Theorem 9.3. *Let $\mathcal{R}$ be a CRN that admits MPNE. If $\mathcal{R}'$ is a CRN obtained by modifying $\mathcal{R}$ in any of the following ways, then $\mathcal{R}'$ admits MPNE. If additionally $\mathcal{R}$ admits MPSE, then $\mathcal{R}'$ does as well.*

1. (Adding a dependent reaction [28, 29]) *$\mathcal{R}'$ is obtained by adding a new irreversible reaction with reaction vector in the stoichiometric subspace of $\mathcal{R}$.*
2. (Adding a trivial species [28]) *$\mathcal{R}'$ is obtained by adding into the reactions of $\mathcal{R}$ a new species Y which occurs with the same stoichiometry on both sides of each reaction in which it participates.*
3. (Adding inflows and outflows for all species [28, 29, 37, 40]) *$\mathcal{R}'$ is obtained by adding to $\mathcal{R}$ the reactions $0 \rightleftharpoons X_i$ for each $i \in \{1, \ldots, n\}$.*
4. (Adding a new species with inflow and outflow [28, 29]) *$\mathcal{R}'$ is obtained by adding into the reactions of $\mathcal{R}$ the new species Y in an arbitrary way, while also adding the new reaction $0 \rightleftharpoons Y$.*
5. (Adding new reversible reactions involving new species [28]) *$\mathcal{R}'$ is obtained by adding $m \geq 1$ new reversible reactions involving k new species such that the submatrix of the new stoichiometric matrix corresponding to the new species has rank m (this forces $k \geq m$).*
6. (Adding intermediate complexes involving new species [28]) *$\mathcal{R}'$ is obtained by replacing each of the m reactions:*

   $a_i \cdot X \to b_i \cdot X$ *with a chain* $\quad a_i \cdot X \to c_i \cdot X + \beta_i \cdot Y \to b_i \cdot X, \quad (i = 1, \ldots, m).$

   Here Y is a list of k new species whose coefficient matrix $\beta = (\beta_1 | \beta_2 | \cdots | \beta_m)$ has rank m (this implies $k \geq m$).

Remark 9.4. The proof of Theorem 9.3 relies on a continuation of each nondegenerate equilibrium of $\mathcal{R}$ into an equilibrium of $\mathcal{R}'$. With that in mind, the theorem actually says more: $\mathcal{R}'$ has at least as many nondegenerate equilibria as $\mathcal{R}$. In particular, if $\mathcal{R}$ has a positive nondegenerate equilibrium, then so does $\mathcal{R}'$.

We illustrate the applicability of these results starting with our favorite example (Eq. 9.6):

$$2A + B \to 3A, \ A \to B. \tag{9.20}$$

Recall from Sections 9.2.5 and 9.2.6 that this system has the capacity for MPNE. We now build on this network using modifications of the type shown in Theorem 9.3 to arrive at a significantly more complicated network, whose capacity for MPNE is otherwise not easy to study. For example, we can add

$A \rightleftharpoons 0$, $B \rightleftharpoons 0$ (modification 2), and $B \to A$ (modification 1); note that one can always add the reverse of an existing reaction keeping the capacity for MPNE (or MPSE). Next, we add the reversible reactions $A+C \rightleftharpoons 2D$ and $C+D \rightleftharpoons 2B$; this is allowed by modification 5, since the submatrix of the stoichiometric matrix corresponding to the new species C and D, that is, $\begin{bmatrix} -1 & -1 \\ 2 & -1 \end{bmatrix}$ has rank 2. Next we can replace $A \to 0$ by $A + E \to E$ (modification 2), and then replace this reaction and the reaction $2A+B \to 3A$ by the two chains $A+E \to C+F \to E$ and $2A+B \to F+2G \to 3A$; this is allowed by modification 6 since the matrix of coefficients $\begin{bmatrix} 1 & 1 \\ 0 & 2 \end{bmatrix}$ for the two new species F and G has rank 2. Finally, we can add $H \rightleftharpoons 0$ and add H to the reactions $A \to B$ and $0 \to B$ by $A + 2H \to B$ and $2H \to B$, respectively. We conclude that the resulting network

$$
\begin{gathered}
2A + B \to F + 2G \to 3A \\
A + 2H \to B \to A \rightleftharpoons 0 \\
H \rightleftharpoons 0 \to B \leftarrow 2H \\
A + C \rightleftharpoons 2D \\
C + D \rightleftharpoons 2B \\
A + E \to C + F \to E
\end{gathered}
$$

inherits the capacity for MPSE from network (9.20). This example may be overtly made up, but it illustrates how the simple modifications listed in Theorem 9.3 allow one to draw conclusions that are otherwise very difficult. We note that here we started with a simple network with capacity for MPNE and concluded that the (much more) complicated one keeps this property. In practice, however, one starts with a large network, and the inverse process is needed, namely finding an MPNE "subnetwork" and a series of modifications that transforms it into the large network. This is a daunting task in general, although for relatively small networks, familiarity with MPNE motifs like Eq. (9.20) and some trial and error will work in many cases. Such an example is presented in Exercise 3, which discusses an important biological example.

EXERCISES

1. *(Variations of network* (9.20)*).*

a. Show that the fully open extension of network (9.18):

$$2A + B \to C \to 3A,\ A \rightleftharpoons B,\ A \rightleftharpoons 0 \rightleftharpoons B$$

admits MPNE.

b. Show that the network

$$2A + B \to 3A,\ A \to B,\ 0 \leftrightharpoons B$$

does not have the capacity for MPE.

c. Show that the network

$$2A + B \to 3A,\ A \to B,\ 0 \leftrightharpoons A + B$$

has the capacity for MPE, but not the capacity for MPNE.

Remark 9.5. CRNs b. and c. show that adding reactions may destroy the capacity for MPE/MPNE [28]).

2. *(One-species networks [64]).* A network involving a single species can be naturally represented as a sequence of arrows on the real line. For example, $0 \leftrightharpoons A, 2A \to 3A, 4A \to 3A$ corresponds to the diagram

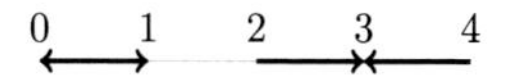

a. Show that a one-species reaction network has the capacity for MPNE if and only if it contains an arrow subsequence of the form $\to \leftarrow \to$ or $\leftarrow \to \leftarrow$. (Here the length of the arrow is not important, only its direction; the condition says that there are three reactions with distinct reactant complexes, arranged in order of their stoichiometric coefficient, that proceed left, right, left, or right, left, right.)

b. Characterize the one-species networks with capacity for MPE.

3. *(Adding enzymatic mechanisms [28]).* Let $\mathcal{R}$ be a CRN with capacity for MPNE and vector of species X as in Theorem 9.3, whose notation we adopt.

a. Suppose $\mathcal{R}'$ is obtained by replacing reaction $a \cdot X \to b \cdot X$ with the chain $cE + a \cdot X \rightleftharpoons I \to cE + b \cdot X$, where E and I are new species and $c \geq 0$. Show that $\mathcal{R}'$ has the capacity for MPNE.

b. Suppose that we add the chain in a., while also keeping the reaction $a \cdot X \to b \cdot X$. Show that this new network has the capacity for MPNE.

4. *(Multistationarity of the Huang-Ferrell-Kholodenko MAPK cascade with negative feedback [47]).*

a. Show that the following version of mass action double phosphorylation has the capacity for MPSE [28]. In the diagram, reactions $X \overset{E}{\to} Y$ are shortened representations of an enzymatic chain $E + X \rightleftharpoons EX \to E + Y$. The list of all reactions is also included.

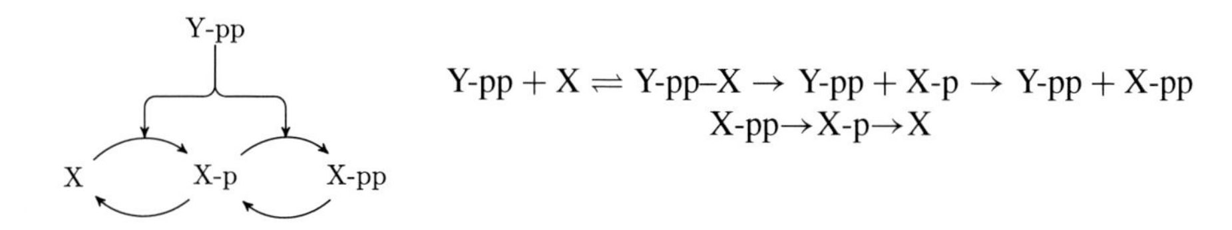

b. Show that the following network has the capacity for MPSE [28].

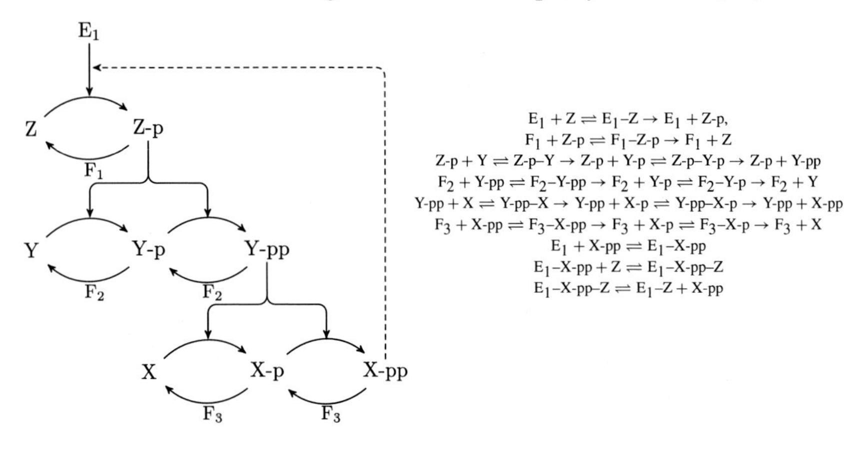

9.6 SUFFICIENT CONDITIONS FOR MULTISTATIONARITY II: THE DETERMINANT OPTIMIZATION METHOD

As we have seen (Section 9.3, Exercise 1), noninjectivity of a CRN does not imply its capacity for MPE. The following result shows that noninjectivity together with an additional condition on the stoichiometric matrix insures the capacity for MPE for fully open systems. The result is essentially Theorem 4.1 in [26], but we present it here using the matrix-theoretic framework developed in [27]. We also refer the reader to [62] for related results.

Theorem 9.4 ([26], [27]). *Let $\mathcal{R}$ be a fully open mass action CRN and let Γ and Γ_l denote the stoichiometric matrix and reactant matrix of the CRN obtained from $\mathcal{R}$ by removing all inflow reactions $0 \to X_i$. Suppose there exists a diagonal matrix D such that $\det(-\Gamma D \Gamma_l^t) < 0$ and $\Gamma D\mathbf{1} \leq 0$, where $\mathbf{1}$ denotes the column vector with all entries equal to 1. Then $\mathcal{R}$ has the capacity for MPE.*

Condition $\det(-\Gamma D \Gamma_l^t) < 0$ simply states that the CRN is not injective (Exercise 1). Verifying the hypothesis of the theorem amounts to finding a solution for a nonlinear system of inequalities in the entries of D, which can be easily implemented computationally (the webserver CoNtRol [56] includes this calculation). On the other hand, for small CRNs this can be done by hand, as we illustrate for the following example. Consider the mass action network

$$2A + B \rightleftharpoons 3A, \; A \to B, \; A \rightleftharpoons 0, \quad B \rightleftharpoons 0.$$

It is already clear that this CRN has the capacity for MPE, as easily shown by using tools from previous sections. Here we apply Theorem 9.4, and search for positive d_1, d_2, d_3, d_4 such that

$$\det\left(\begin{bmatrix} -1 & 1 & 1 & 0 \\ 1 & -1 & 0 & 1 \end{bmatrix}\begin{bmatrix} d_1 & 0 & 0 & 0 \\ 0 & d_2 & 0 & 0 \\ 0 & 0 & d_3 & 0 \\ 0 & 0 & 0 & d_4 \end{bmatrix}\begin{bmatrix} 2 & 1 \\ 1 & 0 \\ 1 & 0 \\ 0 & 1 \end{bmatrix}\right) < 0,$$

and

$$\begin{bmatrix} 1 & -1 & -1 & 0 \\ -1 & 1 & 0 & -1 \end{bmatrix}\begin{bmatrix} d_1 \\ d_2 \\ d_3 \\ d_4 \end{bmatrix} \leq 0,$$

or equivalently,

$$d_1 d_3 - 2d_1 d_4 + d_2 d_4 + d_3 d_4 \leq 0, \quad d_1 \leq d_2 + d_3, \quad d_2 \leq d_1 + d_4.$$

The first inequality can be rewritten as $-d_1d_4 + d_1d_3 + d_4(d_2 + d_3 - d_1)$, and one may notice that these inequalities are satisfied by choosing d_1 close to $d_2 + d_3$ and d_4 large enough; for example, $d_1 = 9,\ d_2 = d_3 = 5,\ d_4 = 6$.

EXERCISES

1. With the notations of Theorem 9.4, show that there exists a positive diagonal matrix D such that $\det(-\Gamma D \Gamma_l^t) < 0$ if and only if $\mathcal{R}$ is not injective.
2. Show that the following sequestration network has the capacity for MPE [62]:

$$X_1 + X_2 \to 0, \quad X_2 + X_3 \to 0, \quad X_3 \to 2X_1,$$
$$X_1 \rightleftharpoons 0,\ X_2 \rightleftharpoons 0,\ X_3 \rightleftharpoons 0.$$

3. Use the determinant optimization method to show that the following network [26] has the capacity for MPE:

$$2A + B \rightleftharpoons 3A,\ A \rightleftharpoons 0 \rightleftharpoons B.$$

9.7 RESULTS BASED ON DEFICIENCY THEORY

Network deficiency was introduced by Horn, Jackson, and Feinberg in the 1970s in a series of seminal papers [41, 42, 53]. The name "deficiency" refers to a nonnegative integer that connects the structure of a reaction network and the existence of (multiple) equilibria for the corresponding system of ODEs.

The reaction network structure enters in the definition of the deficiency via the rank of the stoichiometric matrix (denoted by s), the number of complexes (denoted by N), and the number of connected components (denoted by ℓ). For example, the network given in Fig. 9.6 consists of $N = 6$ complexes and $\ell = 2$ connected components of the reaction graph.

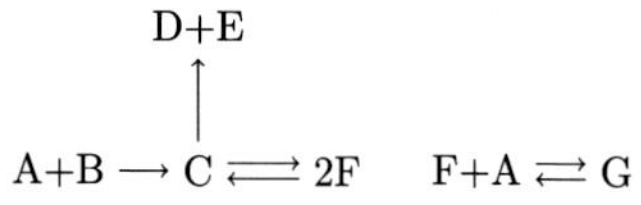

$$\begin{array}{ccc} & \text{D+E} & \\ & \uparrow & \\ \text{A+B} \longrightarrow & \text{C} \rightleftarrows 2\text{F} & \quad \text{F+A} \rightleftarrows \text{G} \end{array}$$

FIG. 9.6 A reaction network with $N = 6$ complexes (nodes) and $\ell = 2$ connected components.

Definition 9.15 (Network Deficiency). The *deficiency* δ of a CRN is defined as

$$\delta = N - \ell - s. \tag{9.21}$$

It turns out that the deficiency of a CRN is always a nonnegative integer [42]. Quite a lot is known about CRNs of deficiency zero or one. We present below results pertaining to existence and uniqueness of positive equilibria, but we note that for deficiency zero CRNs, a great deal is known about the dynamics as well [41–43]. For now, we carry on with a few deficiency computations.

It is easy to see that the CRN in Fig. 9.6 has stoichiometric subspace of dimension 4, so that its deficiency is $\delta = 6-4-2 = 0$. The network in Fig. 9.2 has $N = 7$ complexes, $\ell = 3$ connected components, and its stoichiometric subspace has dimension $s = 4$. The deficiency of the CRN is $7 - 3 - 4 = 0$.

The following version of Eq. (9.6) also has deficiency zero:

$$2A + B \rightleftharpoons 3A. \tag{9.22}$$

Indeed, it is clearly the case that $s = 1$, and so $\delta = 2 - 1 - 1 = 0$.

For a biologically relevant example, consider again the futile cycle (Eq. 9.1) and its stoichiometric matrix:

$$\begin{array}{rcl} E+S & \rightleftharpoons & ES \rightarrow E+P \\ F+P & \rightleftharpoons & FP \rightarrow F+S \end{array} \qquad \Gamma = \begin{bmatrix} -1 & 1 & 1 & 0 & 0 & 0 \\ -1 & 1 & 0 & 0 & 0 & 1 \\ 1 & -1 & -1 & 0 & 0 & 0 \\ 0 & 0 & 1 & -1 & 1 & 0 \\ 0 & 0 & 0 & -1 & 1 & 1 \\ 0 & 0 & 0 & 1 & -1 & -1 \end{bmatrix}.$$

The species ordering here was chosen (E, S, ES, P, F, FP). A simple calculation shows that $s = \text{rank}\ \Gamma = 3$, and since the futile cycle involves $N = 6$ complexes and $\ell = 2$ connected components, its deficiency is $\delta = 6-3-2 = 1$.

9.7.1 The Deficiency Zero and Deficiency One Theorems

We present two theorems exhibiting settings where a CRN cannot have the capacity for MPE. The first result is a weak version of the Deficiency Zero Theorem [43]. Although our focus is on existence and uniqueness of equilibria, the real power of the Deficiency Zero Theorem resides in its strong conclusions about dynamical properties of certain CRNs. While we omit these altogether, the reader is encouraged to survey the vast recent literature on global stability of equilibria in CRNs satisfying the hypotheses of the Deficiency Zero Theorem [65–71]. We start by setting up some terminology.

Definition 9.16 (Strongly Connected Components, Terminal Strongly Connected Components, Weakly Reversible CRNs)**.** Two complexes C_1 and C_2 are called *strongly connected* if $C_1 = C_2$ or if there exist reaction paths from C_1 to C_2, and from C_2 to C_1. A *strongly connected component* of the reaction graph is a maximal subset of nodes that are pairwise strongly connected. A strongly connected component $\mathcal{C}$ is called *terminal* if there is no reaction from a complex in $\mathcal{C}$ to a complex outside $\mathcal{C}$. A CRN is called *weakly reversible* if all connected components in its reaction graph are strongly connected. In other words, a CRN is weakly reversible if whenever there is a path following reaction arrows from a complex C_1 to a complex C_2, there is also a path from C_2 to C_1.

For example, in Fig. 9.6, C and $2F$ are strongly connected while $A + B$ and C are not, and neither are C and G. The strongly connected components of the CRN in Fig. 9.6 are $\{A+B\}$, $\{C, 2F\}$, $\{D+E\}$, $\{F+A, G\}$; out of these $\{D+E\}$ and $\{F + A, G\}$ are terminal. The network in Fig. 9.1 is weakly reversible, and so is the CRN (Eq. 9.22)—more generally, a CRN that contains only reversible reactions is clearly weakly reversible. On the other hand, the CRN in Fig. 9.6 is not weakly reversible, and neither is the futile cycle: for instance, there is a path from ES to $E + P$, but not the other way around.

Theorem 9.5. *[Equilibria in Deficiency Zero CRNs [43, Theorem 4.1]] Consider a mass action CRN with zero deficiency.*

1. *If the network is not weakly reversible, then for any choice of rate constants, the CRN has no positive equilibria.*
2. *If the network is weakly reversible, then for any choice of rate constants, there exists exactly one equilibrium in each positive stoichiometry class.*

A quick application of this theorem shows, for example, that the CRN in Fig. 9.1 has one equilibrium in each positive stoichiometry class, regardless of the choice of rate constants. The same is true for CRN (Eq. 9.22). Notice that this network is related to our favorite multistationary example (Eq. 9.6); in fact, adding inflows and outflows to Eq. (9.22) results in a CRN with capacity for MPE (Exercise 3 in Section 9.6). This is yet another illustration of the subtle connection between the structure of a network and its capacity for multistationarity. Finally, the CRN in Fig. 9.6 has no positive equilibria, no matter what rate constants it is being assigned.

Deficiency theory can be applied to draw conclusions about the existence and uniqueness of positive equilibria even if the deficiency is strictly positive. In particular, a useful result is the Deficiency One Theorem which we present next.

Theorem 9.6 (Deficiency One Theorem [43]). *Consider a CRN with mass action kinetics that satisfies the following conditions:*

1. *The deficiency of every connected component is either zero or one.*
2. *The deficiency of the overall network is the sum of the deficiencies of the connected components.*
3. *Every connected component contains exactly one terminal strongly connected component.*

If for some values of rate constants the CRN has a positive equilibrium, then (for the same rate constants) the CRN has precisely one equilibrium in each positive stoichiometry class. Moreover, if the network is weakly reversible, then for any values of rate constants, the CRN has precisely one equilibrium in each positive stoichiometry class.

Note that "Deficiency One" in the name of Theorem 9.6 refers to the deficiencies of the connected components of the network, and not to the deficiency of the network itself, which can be much higher.

9.7.2 The Deficiency One Algorithm and the Advanced Deficiency Algorithm

The Deficiency Zero and Deficiency One theorems are remarkable results, not only due to the powerful conclusions they allow, but also because of their elegant statements. While unfortunately, many biologically relevant CRNs do not fall under the scope of these results, the theorems are the basis for widely applicable algorithms that have been developed by Feinberg and his research group [72, 73]. A precise discussion of these requires machinery beyond the scope of this chapter. We merely note that the *Deficiency One Algorithm* (applicable to networks of Deficiency One) and the *Advanced Deficiency Algorithm* (for CRNs with arbitrary deficiency) translate the steady-state equations into a collection of (potentially many) linear inequality systems. If at least one of these linear systems is feasible, then *multistationarity is possible* and every solution defines a pair of steady states and corresponding rate constants. If none of the systems are feasible, then *capacity for MPE is ruled out*. Both algorithms are implemented in the Chemical Reaction Network Toolbox [58]. We also refer the reader to a strand of very interesting results related to the Deficiency One Algorithm [74, 75].

The Deficiency One and Advanced Deficiency algorithms work well for many examples of biologically relevant networks. In particular, they can be used in a systems biology context to discriminate between different reaction mechanisms [76], or to study multistable signaling motifs [77], or to analyze whole families of enzymatic networks [46].

EXERCISES

1. a. Show that the futile cycle with enzyme sharing has exactly one equilibrium in each positive stoichiometry class, for any values of the rate constants (see also Exercise 1 of Section 9.4):

$$E + S \rightleftharpoons ES \rightarrow E + P, \quad E + P \rightleftharpoons EP \rightarrow E + S.$$

b. Show that adding reversible inflow-outflow reactions for any one species or for any combination of two species results in CRNs that have exactly one equilibrium in each positive stoichiometry class.

2. Let $N \in \mathbb{N}$. Show that the following mass action CRN (the McKeithan network [78]) has exactly one equilibrium in each stoichiometry class, for any choice of rate constants:

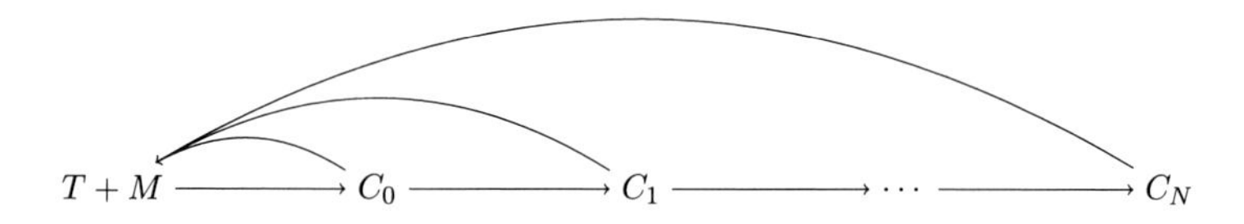

3. Consider the mass action CRN

$$2X \rightarrow Y + Z, \quad 2Y \rightarrow Z + X, \quad 2Z \rightarrow X + Y.$$

a. Set up and solve the corresponding mass action algebraic system to show that the CRN has precisely one equilibrium in each positive stoichiometry class.

b. Use deficiency theory to show the conclusion of a. *(Hint. Find a weakly reversible, deficiency zero reaction network with the same ODEs (dynamically equivalent, [79]) as the one in the exercise. An algorithm for finding weakly reversible networks, dynamically equivalent to a given CRN, is the object of [80]).*

ACKNOWLEDGMENTS

Casian Pantea was supported by NSF-DMS Grant 1517577 *Multistationarity and Oscillations in Biochemical Reaction Networks*.

REFERENCES

[1] J.J. Tyson, K.C. Chen, B. Novak, Sniffers, buzzers, toggles and blinkers: dynamics of regulatory and signaling pathways in the cell, Curr. Opin. Biol. 15 (2003) 221–231.

[2] N.I. Markevich, J.B. Hoek, B.N. Kholodenko, Signaling switches and bistability arising from multisite phosphorylation in protein kinase cascades, J. Cell Biol. 164 (2004) 353–359.

[3] G. Buzski, A. Draguhn, Neuronal oscillations in cortical networks, Science 304 (2004) 1926–1929.

[4] W. Xiong, J.E. Ferrell, A positive-feedback-based bistable "memory module" that governs a cell fate decision, Nature 426 (2003) 460–465.
[5] J.J. Tyson, K. Chen, B. Novak, Network dynamics and cell physiology, Nat. Rev. Mol. Cell Biol. 2 (2001) 908–916.
[6] A.W. Murray, Creative blocks: cell-cycle checkpoints and feedback controls, Nature 359 (1992) 599–604.
[7] K.C. Chen, A. Csikasz-Nagy, B. Gyorffy, J. Val, B. Novak, J.J. Tyson, Kinetic analysis of a molecular model of the budding yeast cell cycle, Mol. Biol. Cell. 11 (2000) 369–391.
[8] J.R. Pomerening, E.D. Sontag, J.E. Ferrell Jr., Building a cell cycle oscillator: hysteresis and bistability in the activation of CDC2, Nat. Cell Biol. 5 (2003) 346–351.
[9] D. Angeli, J.E. Ferrell Jr., E.D. Sontag, Detection of multistability, bifurcations, and hysteresis in a large class of biological positive-feedback systems, Proc. Natl Acad. Sci. USA 101 (2004) 1822–1827.
[10] R. Thomas, R. d'Ari, Biological Feedback, CRC Press, Boca Raton, FL, 1990.
[11] C. Soulé, Graphic requirements for multistationarity, ComPlexUs 1 (2003) 123–133.
[12] J.L. Gouzé, Positive and negative circuits in dynamical systems, J. Biol. Syst. 6 (1998) 11–15.
[13] U. Alon, An Introduction to Systems Biology: Design Principles of Biological Circuits, Chapman & Hall/CRC Press, London, 2006.
[14] G. Craciun, M. Feinberg, Multiple equilibria in complex chemical reaction networks: II. The species-reaction graph, SIAM J. Appl. Math. 66 (4) (2006) 1321–1338.
[15] G. Craciun, Y. Tang, M. Feinberg, Understanding bistability in complex enzyme-driven reaction networks, Proc. Natl Acad. Sci. USA 103 (23) (2006) 8697–8702.
[16] M. Banaji, G. Craciun, Graph-theoretic approaches to injectivity and multiple equilibria in systems of interacting elements, Comm. Math. Sci. 7 (4) (2009) 867–900.
[17] M. Kaufman, C. Soulé, R. Thomas, A new necessary condition on interaction graphs for multistationarity, J. Theor. Biol. 248 (2007) 675–685.
[18] G. Craciun, C. Pantea, E. Sontag, Graph-theoretic characterizations of multistability and monotonicity for biochemical reaction networks, in: H. Koeppl, D. Densmore, G. Setti, M. di Bernardo (Eds.), Design and Analysis of Biomolecular Circuits, Springer, 2011, pp. 63–73.
[19] R. Thomas, J. Richelle, Positive feedback loops and multistationarity, Discret. Appl. Math. 19 (1988) 381–396.
[20] B. Novák, J.J. Tyson, Design principles of biochemical oscillators, Nat. Rev. Mol. Cell Biol. 9 (2008) 981–991.
[21] J.E. Ferrell Jr., Self-perpetuating states in signal transduction: positive feedback, double-negative feedback and bistability, Curr. Opin. Cell Biol. 14 (2002) 140–148.
[22] J.J. Tyson, B. Novak, Regulation of the eukaryotic cell cycle: molecular antagonism, hysteresis, and irreversible transitions, J. Theor. Biol. 210 (2001) 249–263.
[23] U. Alon, Network motifs: theory and experimental approaches, Nat. Rev. Gen. 8 (2007) 450–461.
[24] T.S. Gardner, C.R. Cantor, J.J. Collins, Construction of a genetic toggle switch in *Escherichia coli*, Nature 403 (2000) 339–342.
[25] M. Banaji, Graph-theoretic conditions for injectivity of functions on rectangular domains, J. Math. Anal. Appl. 370 (2010) 302–311.
[26] G. Craciun, M. Feinberg, Multiple equilibria in complex chemical reaction networks: I. The injectivity property, SIAM J. Appl. Math. 65 (2005) 1526–1546.
[27] M. Banaji, C. Pantea, Some results on injectivity and multistationarity in chemical reaction networks, SIAM J. Appl. Dyn. Syst. 15 (2) (2016) 807–869.
[28] M. Banaji and C. Pantea, The inheritance of nondegenerate multistationarity in chemical reaction networks, SIAM J. Appl. Math. 78(2) (2018) 1105–1130.

[29] B. Joshi, A. Shiu, Atoms of multistationarity in chemical reaction networks, J. Math. Chem. 51 (1) (2013) 153–178.
[30] E. Feliu, C. Wiuf, Simplifying biochemical models with intermediate species, J. Roy. Soc. Interface 10 (2013) 20130484.
[31] M. Pérez Millán, A. Dickenstein, The structure of MESSI biological systems, arXiv:1612.08763.
[32] G. Craciun, J.W. Helton, R.J. Williams, Homotopy methods for counting reaction network equilibria, Math. Biosci. 216 (2) (2008) 140–149.
[33] C. Conradi, E. Feliu, M. Mincheva, C. Wiuf, Identifying parameter regions for multistationarity, PLoS Comput. Biol. 13 (10) (2017) e1005751.
[34] B. Joshi, A. Shiu, A survey of methods for deciding whether a reaction network is multistationary, Math. Model. Nat. Phenom. 10 (5) (2015) 47–67.
[35] S. Müller, E. Feliu, G. Regensburger, C. Conradi, A. Shiu, A. Dickenstein, Sign conditions for injectivity of generalized polynomial maps with applications to chemical reaction networks and real algebraic geometry, Found. Comp. Math. 16 (2016) 69–97.
[36] C. Wiuf, E. Feliu, Power-law kinetics and determinant criteria for the preclusion of multistationarity in networks of interacting species, SIAM J. Appl. Dyn. Syst. 12 (2013) 1685–1721.
[37] G. Craciun, M. Feinberg, Multiple equilibria in complex chemical reaction networks: extensions to entrapped species models, IEE Proc. Syst. Biol. 153 (4) (2006) 179–186.
[38] G. Shinar, M. Feinberg, Concordant chemical reaction networks, Math. Biosci. 240 (2012) 92–113.
[39] P. Donnell, M. Banaji, Local and global stability of equilibria for a class of chemical reaction networks, SIAM J. Appl. Dyn. Syst. 12 (2013) 899–920.
[40] M. Banaji, G. Craciun, Graph-theoretic approaches to injectivity and multiple equilibria in systems of interacting elements, Commun. Math. Sci. 7 (4) (2009) 867–900.
[41] M. Feinberg, Chemical Reaction Networks structure and the stability of complex isothermal reactors—I. The deficiency zero and deficiency one theorems, Chem. Eng. Sci. 42 (10) (1987) 2229–2268.
[42] M. Feinberg, Lectures on Chemical Reaction Networks, notes of lectures given at the Mathematics Research Center of the University of Wisconsin in 1979, Available from: https://crnt.osu.edu/LecturesOnReactionNetworks. (Accessed July 16, 2018).
[43] M. Feinberg, The existence and uniqueness of steady states for a class of chemical reaction networks, Arch. Ration. Mech. Anal. 132 (4) (1995) 311–370.
[44] L. Chang, M. Karin, Mammalian MAP kinase signaling cascades, Nature 410 (2001) 37–40.
[45] C. Huang, J.E. Ferrell, Ultrasensitivity in the mitogen-activated protein kinase cascade, Proc. Natl Acad. Sci. USA 93 (1996) 10078–10083.
[46] C. Conradi, A. Shiu, A global convergence result for processive multisite phosphorylation systems 77 (4) (2014) 126–155.
[47] B.N. Kholodenko, Negative feedback and ultrasensitivity can bring about oscillations in the mitogen-activated protein kinase cascades, Eur. J. Biochem. 267 (6) (2000) 1583–1588.
[48] L. Wang, E.D. Sontag, On the number of steady states in a multiple futile cycle, J. Math. Biol. 57 (2008) 29–52.
[49] C. Conradi, D. Flockerzi, Multistationarity in mass action networks with applications to ERK activation, J. Math. Biol. 65 (1) (2012) 107–156.
[50] K. Holstein, D. Flockerzi, C. Conradi, Multistationarity in sequential distributed multisite phosphorylation networks, Bull. Math. Biol. 75 (11) (2013) 2028–2058.
[51] D. Flockerzi, K. Holstein, C. Conradi, N-site phosphorylation systems with 2N-1 steady states 76 (8) (2014) 1892–1916.

[52] B. Joshi, Complete characterization by multistationarity of fully open networks with one non-flow reaction, Appl. Math. Comput. 219 (12) (2013) 6931–6945.

[53] F. Horn, R. Jackson, General mass action kinetics, Arch. Ration. Mech. Anal. 47 (1972) 81–116.

[54] P. Erdi, J. Tóth, Mathematical models of chemical reactions, in: Nonlinear Science: Theory and Applications, Manchester Univ. Press, Manchester, 1989.

[55] S. Muller, G. Regensburger, Generalized mass action systems: complex balancing equilibria and sign vectors of the stoichiometric and kinetic-order subspaces, SIAM J. Appl. Math. 72 (6) (2012) 1926–1947.

[56] M. Banaji, P. Donnell, A. Marginean, C. Pantea, CoNtRol—chemical reaction network analysis tool, 2013, Available from: https://control.math.wvu.edu/. (Accessed July 16, 2018).

[57] P. Donnell, M. Banaji, A. Marginean, C. Pantea, CoNtRol: an open source framework for the analysis of chemical reaction networks, Bioinformatics 30 (11) (2014) 1633–1634.

[58] P. Ellison, M. Feinberg, H. Ji, Chemical Reaction Network Toolbox, Available from: http://www.crnt.osu.edu/CRNTWin. (Accessed July 16, 2018).

[59] E. Feliu, C. Wiuf, Enzyme sharing as a cause of multistationarity in signaling systems, J. Roy. Soc. Interface 9 (71) (2012) 1224–1232.

[60] J. Gunawardena, Distributivity and processivity in multisite phosphorylation can be distinguished through steady-state invariants, Biophys. J. 93 (2007) 3828–3834.

[61] M. Banaji, Graph-theoretic conditions for injectivity of functions on rectangular domains, J. Math. Anal. Appl. 370 (2010) 302–311.

[62] A. Shiu, B. Felix, Z. Woodstock, Analyzing multistationarity in chemical reaction networks using the determinant optimization method, Appl. Math. Comput. 287–288 (2016) 60–73.

[63] U. Alon, Network motifs: theory and experimental approaches, Nat. Rev. Genet. 8 (2007) 450–461.

[64] B. Joshi, A. Shiu, Which small reaction networks are multistationary?, SIAM J. Appl. Dyn. Syst. 16 (2) (2016) 802–833.

[65] G. Craciun, A. Dickenstein, A. Shiu, B. Sturmfels, Toric dynamical systems, J. Symb. Comput. 44 (11) (2009) 1551–1565.

[66] D.F. Anderson, A. Shiu, The dynamics of weakly reversible population processes near facets, SIAM J. Appl. Math. 70 (6) (2010) 1840–1858.

[67] D.F. Anderson, A proof of the Global Attractor Conjecture in the single linkage class case, SIAM J. Appl. Math. 71 (4) (2011) 1487–1508.

[68] C. Pantea, On the persistence and global stability of mass action systems, SIAM J. Math. Anal. 44 (3) (2012) 1636–1673.

[69] G. Craciun, F. Nazarov, C. Pantea, Persistence and permanence of mass action and power-law dynamical systems, SIAM J. Appl. Math. 73 (2013) 305–329.

[70] M. Gopalkrishnan, E. Miller, A. Shiu, A geometric approach to the global attractor conjecture, SIAM J. Appl. Dyn. Syst. 13 (2) (2014) 758–797.

[71] G. Craciun, Toric Differential Inclusions and a Proof of the Global Attractor Conjecture, arXiv:1501.02860.

[72] M. Feinberg, Chemical reaction network structure and the stability of complex isothermal reactors—II. Multiple steady states for networks of deficiency one, Chem. Eng. Sci. 43 (1) (1988) 1–25.

[73] P.R. Ellison, The Advanced Deficiency Algorithm and Its Applications to Mechanism Discrimination, The University of Rochester, Rochester, NY, 1998.

[74] B. Boros, On the existence of the positive steady states of weakly reversible deficiency-one mass action systems, Math. Biosci. 245 (2) (2013) 157–170.

[75] B. Boros, Notes on the Deficiency-One theorem: multiple linkage classes, Math. Biosci. 235 (1) (2012) 110–122.
[76] C. Conradi, J. Saez-Rodriguez, E.D. Gilles, J. Raisch, Using chemical reaction network theory to discard a kinetic mechanism hypothesis, IEE Proc. Syst. Biol. 152 (4) (2005) 243–248.
[77] J. Saez-Rodriguez, A. Hammerle-Fickinger, O. Dalal, S. Klamt, E.D. Gilles, C. Conradi, Multistability of signal transduction motifs, IET Syst. Biol. 2 (2) (2008) 80–93.
[78] E. Sontag, Structure and stability of certain chemical networks and applications to the kinetic proofreading model of T-cell receptor signal transduction, IEEE Trans. Autom. Control 46 (7) (2001) 1028–1047.
[79] G. Craciun, C. Pantea, Identifiability of chemical reaction networks, J. Math. Chem. 44 (1) (2008) 244–259.
[80] M.D. Johnston, D. Siegel, G. Szederkényi, Computing weakly reversible linearly conjugate chemical reaction networks with minimal deficiency, Math. Biosci. 241 (1) (2013) 88–98.

Chapter 10

The Minimum Evolution Problem in Phylogenetics: Polytopes, Linear Programming, and Interpretation

Stefan Forcey*, **Gabriela Hamerlinck**[†], **Logan Keefe**[‡] **and William Sands**[§]
**Department of Mathematics, University of Akron, Akron, OH, United States, †QUBES, BioQUEST Curriculum Consortium, Boyds, MD, United States, ‡Department of Mathematics, Kent State University, Kent, OH, United States, §Department of Computational Mathematics, Science, and Engineering, Michigan State University, MI, United States*

10.1 INTRODUCTION

A *phylogenetic tree* (or *phylogeny*) is a representation of the genetic relationships between organisms, species, or genes. Mathematically, phylogenetic trees can be thought of as *partially labeled graphs*—a collection of items connected by branching edges. Phylogenies are commonly used in biology to explore the evolution, biological diversity and relatedness of species, organisms, and genes. Phylogenetic trees allow us to postulate about similar adaptations and shared genes between organisms. Determining phylogenetic relationships can be a useful step in identifying the genetic basis of an observed trait between closely related species or individuals.

10.1.1 Phylogenetic Reconstructions and Interpretation

Phylogenetic reconstructions can be inferred from morphological or genetic data. For the purposes of this chapter, we focus on the genetic data used in molecular phylogenetics. The genome carries the complete genetic material of an organism. This information is found in the deoxyribonucleic acid (DNA), which is transcribed into ribonucleic acid (RNA) where it can be processed into amino acids to form proteins in a process called translation. DNA is typically represented as a chain of nucleotide bases: adenine (A), guanine (G), thymine (T), and cytosine (C). In RNA, thymine is replaced by uracil (U). A nucleotide

Algebraic and Combinatorial Computational Biology. https://doi.org/10.1016/B978-0-12-814066-6.00010-6

sequence is thus represented by a continuous chain of repeating letters (A, C, G, and T/U). Some RNA strands encode proteins, meaning their sequence of nucleotide bases code for amino acids. There are 20 unique amino acids that, when strung together in a continuous chain, form unique proteins.

To construct a phylogenetic tree, we must first identify and align our sequences so they can be compared. These data could be DNA or RNA nucleotide sequences or an amino acid sequence. There are many programs and approaches available that will automatically align multiple sequences (see [1]). For our purposes, we assume that we have a well-defined alignment of multiple sequences and explore phylogenetic reconstructions using distance-based approaches, specifically the balanced minimum evolution (BME) and branch and bound methods.

Distance-based approaches to phylogenetic inference are one class of methods used to approximate a tree to a set of molecular data and can accommodate very large data sets. These methods use a matrix of genetic distances, which estimate the genetic dissimilarity between organisms. The distance matrix is calculated by applying a model of evolution to the multisequence alignment and can be done in a variety of molecular phylogenetic programs (see [1] for a thorough explanation). The BME method constructs a phylogenetic tree by minimizing a quantity called the tree length. In the case of error-free data, the tree and the corresponding tree length are uniquely determined by the pairwise distances, the dissimilarity matrix. The *BME method* yields an unrooted tree, without edge lengths. If wanted, edge lengths can be found by solving or approximating solutions to linear equations. Finally a root can be chosen or added based on external evidence. In this chapter, we pair the BME method with a branch and bound method. *Branch and bound* methodology is typically used to optimize the search for the most accurate phylogenetic reconstruction and we provide a demonstration of how branch and bound may be applied to the BME problem.

Phylogenetic trees contain a wealth of information about the evolution, biological diversity and relatedness of the species, organisms or genes represented in the tree. Each branch of the tree will end in a tip that represents the terminal or extant taxa that were used to construct the tree. The tips of phylogenetic tree branches are also called *leaves*. The internal branching points of the tree, called *nodes*, represent instances where genetic differentiation from the ancestral population has occurred. See Fig. 10.1 for a simple example. The lengths of the branches of a phylogenetic tree can be calibrated to estimate evolutionary time. The branch length is then indicative of the amount of genetic change between organisms. For a more detailed description of the biological interpretations of evolutionary events represented in phylogenies, see [2].

With recent technological and computing advances, researchers now have access to and have begun to utilize larger and larger genetic datasets in their phylogenetic reconstructions—into the thousands of sequences [3, 4]. Phylogenies produced from these large datasets have the potential to lead to more biological questions. For example, Fig. 10.2 shows three of the current theories of angiosperm (flowering plant) evolution.

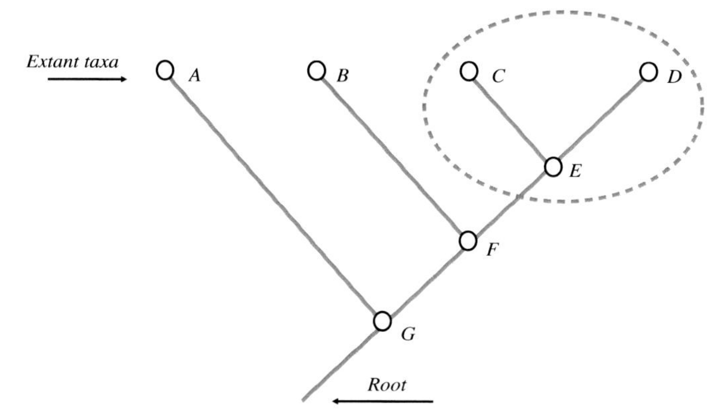

FIG. 10.1 Terminology and interpretation of phylogenetic trees. The root of a tree represents the ancestral population from which all the extant taxa are derived. The nodes in the body of the tree are branching points where differentiation (speciation) from the ancestral population has occurred. A *monophyletic group*, or *clade*, is marked with a *dashed line*. A clade with only two leaves is also known as a *cherry*. An example of a nonmonophyletic group would include taxa B and C and their common ancestor nodes, E and F.

The grouping of *Chloranthus* and *Magnolia* in Fig. 10.2A is called a *clade*, specifically a clade with two leaves which is referred to as a *cherry*. A clade will include the ancestral lineage and all the descendants of that ancestor and is also called a *monophyletic group*. The monophyly seen with *Amborella* and the plants are disputed by other studies that have shown *Amborella* and *Nymphaeales* (water lilies) are sisters and are equally distant from the rest of the flowering plants (Fig. 10.2C; [7]). The inconsistencies shown between the trees in Fig. 10.2 demonstrate a demand for new and innovative approaches to phylogenetic reconstructions as a way to approximate the true evolutionary history of organisms. The work presented in this chapter illustrates a new approach to the distance-based BME method of phylogenetic reconstructions.

Exercise 10.1. It is often helpful to present a phylogeny using cladograms (see Fig. 10.3). Redraw parts (A) and (C) of Fig. 10.2 to get the alternate versions: cladogram and unrooted tree as seen in Fig. 10.3.

10.1.2 Balanced Minimum Evolution

Utilizing molecular data collected from DNA, RNA, and amino acids allows us to measure *dissimilarities* between pairs of taxa. These dissimilarities are based on the observed differences between DNA sequences, compared positionwise, after an attempt to align them as well as possible. The dissimilarities are represented as nonnegative real numbers. If the pairwise dissimilarities obey the triangle inequality (reviewed in the next section), then they may be interpreted as distances in a metric space whose points are the taxa.

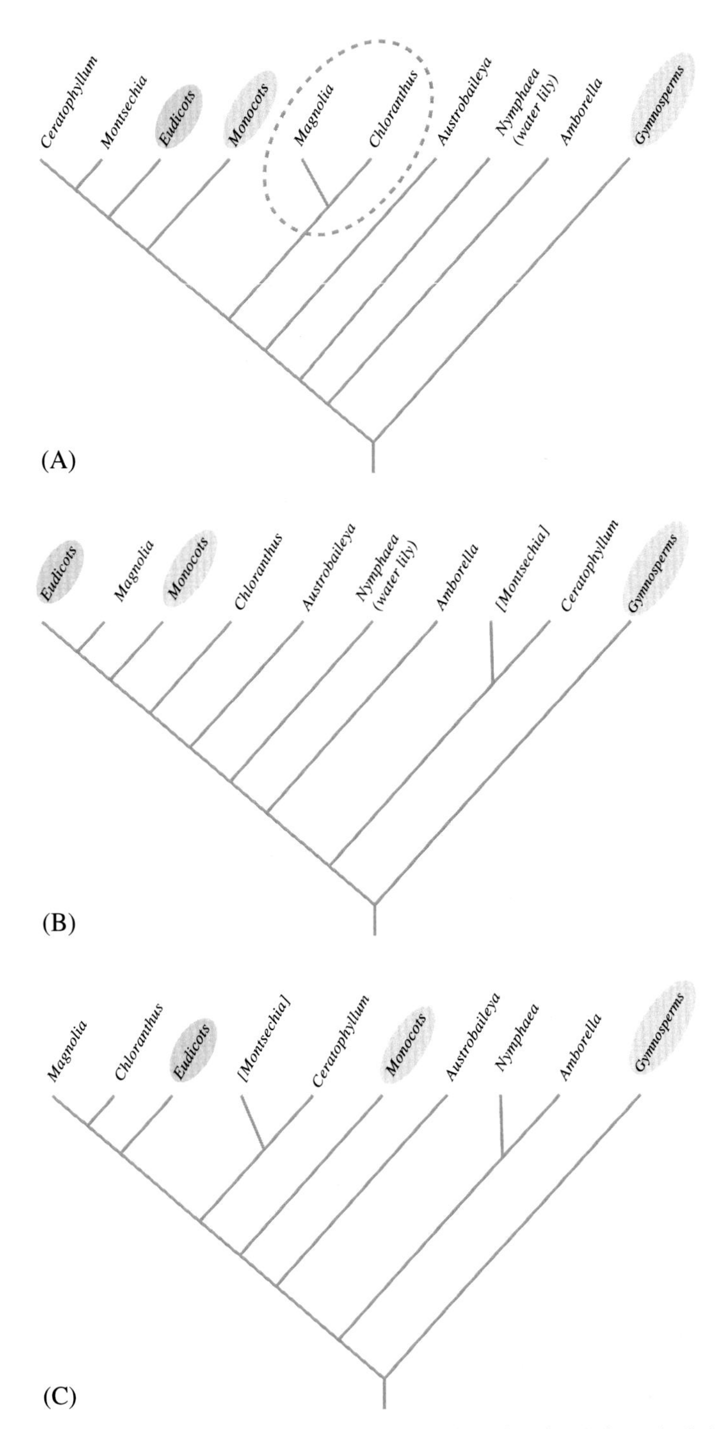

FIG. 10.2 Phylogenetic reconstructions for angiosperms (flowering plants), for each of which the gymnosperms contain the root ancestor. Top-to-bottom these show three competing phylogenies in the literature. (A) is based on results in [5], (B) is based on results in [6], and (C) is based on results in [7]. It is in [5] that the fossil *Montsechia* appears, but we add it here to extant *Ceratophyllum* for comparison.

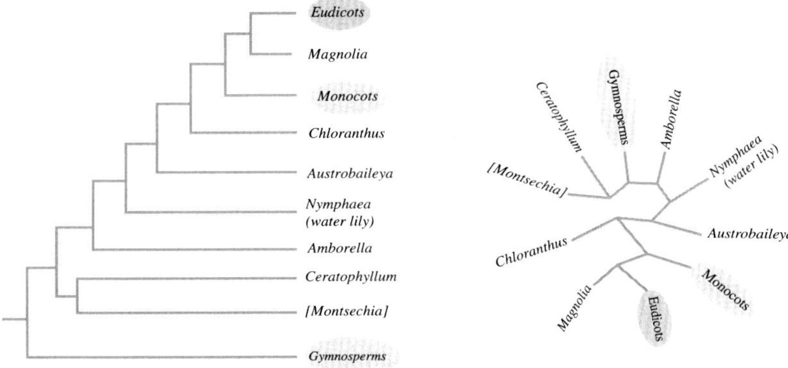

FIG. 10.3 Alternate tree diagrams, or *cladograms*, for the tree in Fig. 10.2B. The second diagram is the unrooted version.

Distance-based methods are a class of techniques that use this information to select the best tree to represent a set of aligned molecular data. The BME method, which we consider in this chapter, finds the optimal tree by solving a minimization problem designed to pick out the unique tree whose geometry is able to yield those distances (or the tree which comes closest, in a sense to be made precise). An advantage of the BME method is that it is known to be *statistically consistent*. This means that as we obtain more information related to the dissimilarity of species which indeed obey a tree structure, then our solution approaches the true, unique, tree representing that data. Furthermore, the correct tree can be recovered even in instances of missing or corrupted data, provided that the error is within bounds.

Various methods for obtaining solutions have been proposed over the course of several decades (see [8] for a survey of current methods). In [9], Saitou and Nei suggested an algorithm, known as Neighbor-Joining (NJ), which runs in polynomial time. Neighbor joining has been popular for years because of its efficiency and observed relative accuracy—it was a big improvement over earlier methods. It was not understood for some time exactly what NJ works to optimize.

In 2000, Pauplin [10] demonstrated a new method for calculating the total length of a phylogenetic tree with positive valued lengths assigned to each edge. The *length* of a tree is simply the sum of its edge lengths. However, Pauplin's method instead begins with the dissimilarities (distances between leaves) which are each found by adding the edge lengths on the path between a pair of leaves. This list of distances is treated as a vector, the *distance vector*. A second *characteristic vector* is found by considering only the numbers of edges traversed in each path from leaf to leaf, ignoring the edge lengths. We show how to calculate both vectors in the next section. Finally, the total length of the tree is recovered by finding the dot product of these two vectors. The value of this roundabout approach is that it uses both the typical given data (dissimilarities) and the combinatorial branching structure of the tree. If any alternative tree (with a different branching structure) is substituted, the calculated length is incorrect. In fact, the incorrect value will be greater than the correct value. Now assume the branching structure is unknown, but that we do know the distance vector. Then by minimizing the dot product, seen as a linear functional, we can recover the correct branching structure. This is the BME method. In [11] the authors show that NJ is actually a greedy algorithm for BME method: it seeks to minimize that same total length by making local choices. Recently, it was discovered that minimizing the tree length over the set of phylogenetic trees is equivalent to minimizing over a geometric object known as the *Balanced Minimum Evolution Polytope* [12]. Thus, the problem can be reformulated in terms of mathematical linear programming.

10.1.3 Definitions and Notation

To formulate our problem, we must introduce some definitions. We define $\mathcal{S} = [n] = \{1, \ldots, n\}$ to be a list of distinct taxa that we wish to compare. Each element in $\mathcal{S}$ is a natural number, which corresponds to an individual taxon. Let the *distance vector* $\mathbf{d}$ with $\binom{n}{2}$ components be given, where each entry d_{ij} is nonnegative and represents the so-called dissimilarity between taxa i and j, for each pair $\{i, j\} \subset \mathcal{S}$. This vector is obtained from our aligned molecular data. It is also sometimes described as a symmetric matrix, with 0's on the diagonal. If the numbers d_{ij} obey the triangle inequality $d_{xy} + d_{yz} \geq d_{xz}$ for all triples x, y, z then we say they are distances in a *metric*. If they obey the stronger four-point condition that:

$$d_{xy} + d_{wz} \leq \max(d_{xw} + d_{yz}, d_{yw} + d_{xz})$$

for all quadruples x, y, z, w then we say that the distance matrix is *additive*. For any additive $\mathbf{d}$ there is a phylogenetic tree with edge weights that realizes the values of d_{ij} as the sums of lengths on the path from leaf i to leaf j. In what follows we will use t to refer to an arbitrary phylogenetic tree without edge weights and T to refer to a phylogenetic tree with edge weights. The tree t without edge weights is often referred to as the *tree topology* shared by any weighted T created by adding edge weights to t.

Mathematically speaking, a *phylogenetic tree on* $[n]$ is a cycle-free graph with leaves labeled by the elements of $[n]$. The nonleaf vertices are not labeled, and we exclude vertices of degree 2. A *weighted* phylogenetic tree has nonnegative real numbers assigned to its edges. Let $\mathcal{T}_n$ be the set of all binary phylogenetic trees on $[n]$ *without* edge weights. Then, for each tree $t \in \mathcal{T}_n$, there is a corresponding *characteristic* vector $\mathbf{x}(t)$ with $\binom{n}{2}$ components $x_{ij}(t)$ for each pair $\{i, j\} \subset \mathcal{S}$. We define

$$x_{ij}(t) = 2^{n-2-l_{ij}(t)}, \tag{10.1}$$

where $l_{ij}(t)$ is the number of internal nodes (degree-3 vertices) in the path connecting i and j in t. The additional factor of 2^{n-2} is used to rescale Pauplin's original coordinates to be positive integers [10].

We say a vector **b** has a lexicographic ordering if the entries b_{ij} are expressed in the form

$$\mathbf{b} = \left(b_{12}, b_{13}, \ldots, b_{1n}, b_{23}, b_{24}, \ldots, b_{2n}, \ldots, b_{(n-1)n}\right).$$

We use a lexicographic ordering of the entries for vectors **d** and **x**.

Here is an example of a phylogenetic tree for $\mathcal{S} = [6]$, together with its characteristic vector.

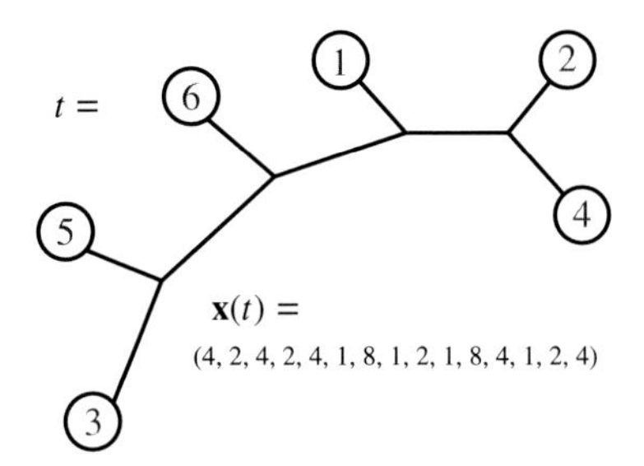

For simplicity, we choose to use a vector to express dissimilarity, rather than a matrix. The dissimilarity matrix D contains zeros along the main diagonal and is symmetric because $d_{ij} = d_{ji}$. Our vector **d** arises naturally from D in a simple way: It consists of rows from the upper triangular elements in D, excluding the main diagonal. If we are provided a binary tree T with nonnegative weights on the edges, then we can calculate **d** by adding the weights on each of the edges connecting the path from i to j in T. Once we have calculated **d**, we can determine the *rescaled length* of T using the path-length functional $\mathcal{L}\colon \mathcal{T}_n \longrightarrow \mathbb{R}$ defined by

$$\mathcal{L}(T) = \sum_{\substack{i,j \\ i<j}} d_{ij} 2^{n-2-l_{ij}(T)}. \tag{10.2}$$

Here l_{ij} does not depend on the edge lengths, just on the path lengths. As shown by Pauplin [10], $\mathcal{L}(T)$ is equal to $2^{n-2}L(T)$ where the length $L(T)$ is the sum of the edge lengths of T. Using our definition in Eq. (10.1), we can rewrite Eq. (10.2) as

$$\mathcal{L}(T) = \mathbf{d}_T \cdot \mathbf{x}(T). \tag{10.3}$$

Here is an example of a weighted phylogenetic tree for $\mathcal{S} = [6]$, together with its length and rescaled length. Note that the latter can be calculated using the dot product, and that $\mathbf{x}(T) = \mathbf{x}(t)$ from the previous example.

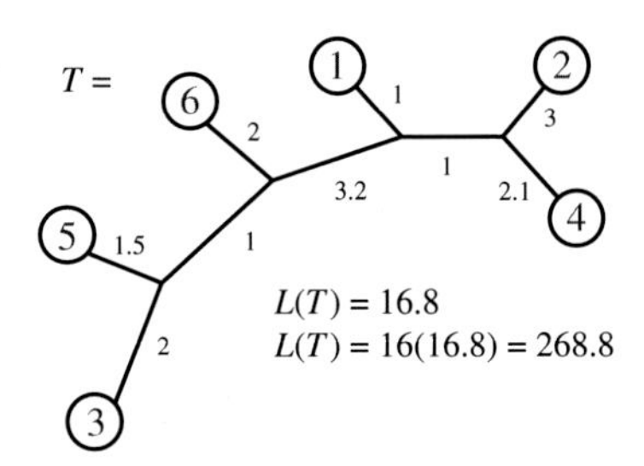

$\mathbf{d}_T =$ (5, 7.2, 4.1, 6.7, 6.2, 10.2, 5.1, 9.7, 9.2, 9.3, 3.5, 5, 8.8, 8.3, 4.5)

Since T is its own BME tree, we are left to compute a single-dot product. However, our task here is to find the BME tree represented by an arbitrary distance vector $\mathbf{d}$ using data that is potentially missing or corrupted. Therefore, we must extend this definition to handle any tree $t \in \mathcal{T}_n$ using a slight modification of Eq. (10.3). The penultimate form of our functional is

$$\mathcal{L}(t) = \mathbf{d} \cdot \mathbf{x}(t). \tag{10.4}$$

The ultimate form will come later, when we allow $\mathbf{x}$ to range over a region of Euclidean space. The primary difference between Eqs. (10.3), (10.4) is that the latter does not assume that the dissimilarity data come from a known phylogeny. For instance, we might have

$$\mathbf{d} = (5,\ 7,\ 4,\ 7,\ 6,\ 10.2,\ 5.1,\ 9.7,\ 9.2,\ 9.3,\ 3.5,\ 5,\ 8.8,\ 8.3,\ 4.5).$$

Notice that this given vector $\mathbf{d}$ does not obey the four-point condition, since $d_{1,5} + d_{2,3} > \max(d_{1,2} + d_{3,5},\ d_{1,3} + d_{2,5})$. Thus there does not exist a tree with edge weights realizing this vector exactly, even though it is just a (partly) rounded version of the example $\mathbf{d}_T$ from earlier.

However, in the case for which $\mathbf{d} = \mathbf{d}_T$ for T with positive edge lengths, the tree topology is unique and the functional $\mathcal{L}(t)$ will be minimized precisely when the minimizer t^* has the same topology as T. Thus minimizing this dot product provides a consistent way to reconstruct the tree which realizes the distance vector. Moreover, it has been shown that the method is *statistically consistent*: for any sequence of vectors $\mathbf{d}_n$ which converges to $\mathbf{d}_T$, the corresponding sequence of minimizers t_n^* approaches the topology of T.

Using Eq. (10.1), we can equivalently describe the structure of a tree t by its unique vector representation $\mathbf{x}(t)$. This allows us to minimize Eq. (10.4) over $\mathbf{x}(t)$, with the minimizer, now, being $\mathbf{x}(t^*)$. Observe that our rescaling will not affect the solution obtained through the minimization procedure. In general, the minimizer is only unique provided it does not contain edge weights that are identically equal to zero. In the latter case, we will have a finite collection of trees that minimize $\mathcal{L}(t)$ simultaneously.

Note that vectors are written several ways in this chapter. We use $(1, 2, 3)$ in the text, but for Matlab input and output this becomes [1 2 3]. For Polymake input and output we see $[1, 1, 2, 3]$ where an extra coordinate of "1" is placed in the first position. Polymake is a free software package for the analysis and computation of polytopes, see [13].

Exercise 10.2. Find the BME tree, with $n = 4$, given the distance vector $\mathbf{d} = (6, 8, 9, 12, 7, 15)$. Proceed as follows:

1. Draw all possible unrooted trees t on $n = 4$ taxa and determine the vector $\mathbf{x}(t)$ for each tree.
2. Compute the dot products $\mathbf{d} \cdot \mathbf{x}(t)$ and select the tree corresponding to the minimal dot product.
3. Then reconstruct the five edge lengths by solving six linear equations simultaneously. In general, there will be $\binom{n}{2}$ equations using the $2n - 3$ variable edge lengths. Note that if the original $\mathbf{d}$ is not additive, then the solution may not exist—and then we may be forced to choose approximate solutions.

Luckily the tree topology is the crucial ingredient we seek. Also note that in this example, the naive approach would choose the smallest distance, 6, as the first cherry. This is known as the *long-branch problem*, and in the solution you will see that one branch has an outsize length.

10.2 POLYTOPES AND RELAXATIONS

10.2.1 What Is a Polytope?

A *polytope* is the *convex hull* of finitely many points in a Euclidean space. The definition of convex hull is as follows: A set Y is said to be *convex* if for any points $a, b \in Y$, every point on the straight-line segment joining them is also in Y. The convex hull of a set of points X in Euclidean space is the smallest convex set containing X. Colloquially speaking, one way to define a polytope is as a finite set of points which have been shrink wrapped. Some examples of polytopes include polygons, cubes, tetrahedrons, pyramids, and hypercubes, also known as tesseracts. Note that a point in Euclidean space is equivalently seen as a vector from the origin, and vice versa.

Another geometric definition of a polytope utilizes *half-spaces*, which are given by *linear inequalities*. If we take finitely many linear equalities such that the set of points which obey all of them is bounded, that set of points is a polytope. We can say that the polytope is the intersection of half-spaces described by those inequalities. Any polytope given by a convex hull can also be given in this manner. Here is an example:

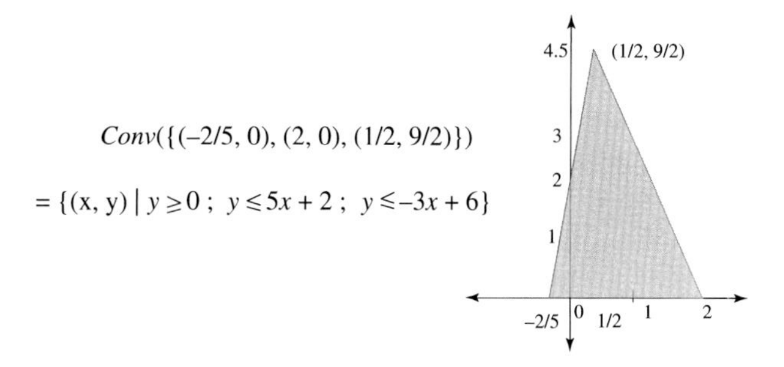

If we cannot remove a point v without changing the convex hull itself, we will call v a vertex of the polytope. For example, each corner of a triangle is a vertex. If there exists a linear inequality such that every point in the polytope satisfies it, we call the set of points that satisfy it exactly a *face* of the polytope. Each face of a polytope is itself a polytope. For example, all of the corners and edges of an octagon are faces of the octagon. The *dimension* of a polytope is the dimension of the smallest Euclidean space, which could contain it. For example, the dimension of a pentagon is 2.

A *facet* of a polytope is a face with dimension one less than that of the polytope. For example, a square is a facet of a cube. A polytope can also be described combinatorially as a *partially ordered set*, or more specifically a *lattice*. Each polytope is made up of smaller polytopes—its faces—ordered by containment. The poset of faces does not record the geometric information, such as lengths and volumes, or location of points in space. Instead, it records the combinatorial type of the polytope, and we often refer to this as the polytope itself, up to equivalence.

Many polytopes occur naturally in a sequence. For easy examples, consider the *simplices*. This sequence begins with the degenerate cases of a point and a line segment, then the triangle and the tetrahedron. In general, as we increase the dimension by one, the n-dimensional *simplex* is the convex hull of $n + 1$ points in general position, such as the $n + 1$ unit vectors of a standard basis. Another example of a sequence is the n-dimensional cubes. Our polytopes of interest occur in a similar sequence, as described in the next section, but they skip dimensions.

10.2.2 The BME Polytope

There are exactly $(2n - 5)!!$ trees that belong to $\mathcal{T}_n$ (see Table 10.1 for details). Minimizing our functional Eq. (10.4) over these trees would require us to construct the characteristic vector $\mathbf{x}(t)$ for each and every tree in $\mathcal{T}_n$, and then find its dot product with the given distance vector. A useful result from [12] states that minimizing over the set of trees in $\mathcal{T}_n$ is equivalent to minimizing over the convex hull of $\{\mathbf{x}(t) \mid t \in \mathcal{T}_n\}$. We define the *BME Polytope*, hereafter denoted as BME(n), to be the convex hull: $Conv(\{\mathbf{x}(t) \mid t \in \mathcal{T}_n\})$. For $n = 3$, since there is only one tree with three leaves, BME(3) is the single point $(1, 1, 1)$ in $\mathbb{R}^3$. For $n = 4$, the polytope BME(4) is the triangle in $\mathbb{R}^6$ pictured in Fig. 10.4.

Describing the BME polytope allows us to reformulate the BME problem as a *linear programming* problem where both the objective function and the constraint function are linear. In this context, the BME polytope represents the *feasible region* that is described by these inequalities. Next we will mention some of the facet inequalities that are shared by BME polytopes of all dimension.

TABLE 10.1 Technical Statistics for the BME Polytopes, as Seen in [15]

Number of Species	Dim. of $\mathcal{P}_n$	Vertices of $\mathcal{P}_n$	Facets of $\mathcal{P}_n$	Facet Inequalities (Classification)	Number of Facets	Number of Vertices in Facet
3	0	1	0	–	–	–
4	2	3	3	$x_{ab} \geq 1$	3	2
				$x_{ab} + x_{bc} - x_{ac} \leq 2$	3	2
5	5	15	52	$x_{ab} \geq 1$ (caterpillar)	10	6
				$x_{ab} + x_{bc} - x_{ac} \leq 4$ (intersecting-cherry)	30	6
				$x_{ab} + x_{bc} + x_{cd} + x_{df} + x_{fa} \leq 13$ (cyclic ordering)	12	5
6	9	105	90,262	$x_{ab} \geq 1$ (caterpillar)	15	24
				$x_{ab} + x_{bc} - x_{ac} \leq 8$ (intersecting-cherry)	60	30
				$x_{ab} + x_{bc} + x_{ac} \leq 16$ $(3, 3)$-split	10	9

Continued

TABLE 10.1 Technical Statistics for the BME Polytopes, as Seen in [15]—cont'd

Number of Species	Dim. of $\mathcal{P}_n$	Vertices of $\mathcal{P}_n$	Facets of $\mathcal{P}_n$	Facet Inequalities (Classification)	Number of Facets	Number of Vertices in Facet
n	$\binom{n}{2} - n$	$(2n-5)!!$	?	$x_{ab} \geq 1$ (caterpillar)	$\binom{n}{2}$	$(n-2)!$
				$x_{ab} + x_{bc} - x_{ac} \leq 2^{n-3}$ (intersecting-cherry)	$\binom{n}{2}(n-2)$	$2(2n-7)!!$
				$x_{ab} + x_{bc} + x_{ac} \leq 2^{n-2}$ $(m,3)$-split, $m > 3$	$\binom{n}{3}$	$3(2n-9)!!$
				$\sum_{i,j \in S_1} x_{ij} \leq (k-1)2^{n-3}$ (m,k)-split, $m > 2, k > 2$	$2^{n-1} - \binom{n}{2} - n - 1$	$(2m-3)!! \times (2k-3)!!$

Notes: The first four columns are found in [16, 17]. Our new and recent results from [15] are in the last three columns. The inequalities are given for any $a, b, c, \ldots \in [n]$. Note that for $n = 4$ the three facets are described twice: our inequalities are redundant.

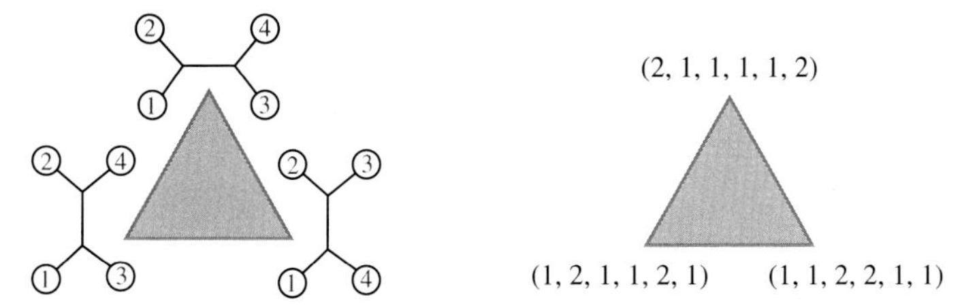

FIG. 10.4 The balanced minimum evolution polytope for $n = 4$ is the convex hull of three points in $\mathbb{R}^6$. The vertices $\mathbf{x}(t)$ on the *right* are shown for the respective trees on the *left*. *(From S. Forcey, L. Keefe, W. Sands, Facets of the balanced minimal evolution polytope, J. Math. Biol. 73 (2) (2016) 447–468.)*

Proposition 10.1 (Caterpillar Facets and Cherry Faces). *For every $i, j \in \mathcal{S}$, with $i \neq j$,*

$$1 \leq x_{ij} \leq 2^{n-3}. \tag{10.5}$$

We will begin with an example to motivate the terminology. If $n = 5$, then choosing the pair $\{1, 2\}$ gives the inequalities $x_{1,2} \geq 1$ and $x_{1,2} \leq 4$. These are obeyed by all points in BME(5), specifically by all the vertices. For vertices corresponding to trees with the cherry $\{1, 2\}$ the equality $x_{1,2} = 4$ holds. For vertices corresponding to caterpillar trees with the two cherries containing $\{1, i\}$ and $\{2, k\}$ the equality $x_{1,2} = 1$ holds.

$t =$ $\qquad\qquad$ $t =$

$\mathbf{x}(t) = (4, 1, 2, 1, 1, 2, 1, 2, 4, 2)$ $\qquad$ $\mathbf{x}(t) = (1, 4, 1, 2, 1, 4, 2, 1, 2, 2)$

This inequality provides both a lower bound and a upper bound on each of the decision variables used in the model. These inequalities suffice to guarantee that our polytope is bounded, since it is contained within the hypercube $\left[1, 2^{n-3}\right]^{\binom{n}{2}}$, in $\binom{n}{2}$-dimensional space. The right-hand side of the inequality follows immediately using the definition of x_{ij} and noting that every leaf must be separated using at least one internal node. These constraints are called the *cherry faces*. Similarly, on the left-hand side, the *caterpillar facets* follow because the distance between any two leaves in a tree is at most $n - 2$ internal nodes away.

Proposition 10.2 (Kraft Equalities). *Let $i, j \in \mathcal{S}$. Then for every $i \in \mathcal{S}$,*

$$\sum_{j \in S - \{i\}} x_{ij} = 2^{n-2}. \tag{10.6}$$

The Kraft equality is a necessary condition for a path length sequence to represent a phylogeny. These equalities are commonly encountered in information theory, specifically in *Huffman trees*, which are rooted binary trees used to represent symbols in a coding alphabet. Interestingly, Huffman trees can be described using a path length sequence [18]. Therefore, we can think of a phylogeny as a Huffman tree encoded in a binary alphabet using the taxa as symbols in the code [19, 20]. We do not provide a proof of this inequality here, but one can derive this property using an inductive edge collapsing argument and an appropriate relabeling of the taxa.

Next we see a type of facet inequality that captures sets of trees, which can have either of two cherries, where one leaf label is shared by the two. For $n = 5$, here is the set of trees that have either the cherry $\{3,4\}$ or the cherry $\{3,5\}$.

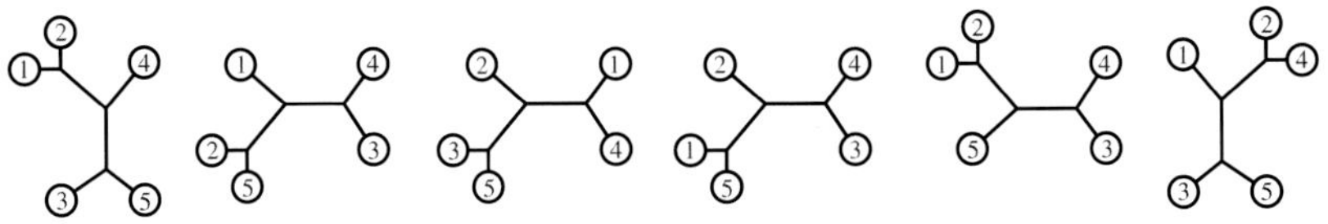

Proposition 10.3 (Intersecting-Cherry Facets). *Let $i,j,k \in S$ be distinct. Then, for any collection of phylogenetic trees with either $\{i,j\}$ or $\{j,k\}$ as cherries, we have*

$$x_{ij} + x_{jk} - x_{ik} \leq 2^{n-3}. \quad (10.7)$$

This inequality will become strict when a tree contains neither $\{i,j\}$ or $\{j,k\}$ as cherries. The proof that this inequality forms a facet of BME(n) is in Theorem 4.7 of [14].

The size of the collections of these types of facets, as well as the number of vertices each type contains, is displayed in Table 10.2. The next kind of facets has the fastest growing collection size, as n increases. A *split* of the set $[n]$ is a partition into two parts. A phylogenetic tree on $[n]$ *displays* a certain split if it has clades whose leaves are the two parts of that split.

Proposition 10.4 (Split Facets). *Consider $\pi = \{S_1, S_2\}$, a partition of S. Let $|S_1| = k \geq 3$ and $|S_2| = m \geq 3$. Then for $i,j \in S_1$*

$$\sum_{i<j} x_{ij} \leq (k-1)2^{n-3}. \quad (10.8)$$

For convenience, we will refer to types of splits using the cardinality of their partitions, for example, we say a tree exhibits a (k,m)-split. This inequality allows us to have some control on the positioning of the taxa within a subgraph of a tree t. The split inequality achieves equality for any tree that

TABLE 10.2 We Provide Some Statistics for the Inequalities Used in Our Relaxation, Sp(n), Based on [15]

Classification	Size of Collection	Vertices in Faces
Caterpillar facets	$\binom{n}{2}$	$(n-2)!$
Cherry faces	$\binom{n}{2}$	$(2n-7)!!$
Intersecting cherry facets	$\binom{n}{2}(n-2)$	$2(2n-7)!!$
Kraft equalities	n	–
Split facets	$2^{n-1}-\binom{n}{2}-n-1$	$(2m-3)!!(2k-3)!!$

Notes: The facets of first three classes of inequalities grow polynomially in n, while the facets for (k, m)-splits grow exponentially in n. The Kraft equalities appear in all faces of the polytope, so they trivially contain all of the vertices.

displays the split, and is a strict inequality for all others. In [15], the authors proved that this inequality, indeed, forms a facet of BME(n). They also showed that the number of these facets grows on the order of $\mathcal{O}(2^n)$, which will be relevant to our discussion on the performance of our proposed algorithm in later sections.

Software such as Polymake [13] is useful for finding the structure of high-dimensional polytopes. For an example, we will show the input for the two-dimensional case. From Fig. 10.4, we have the three vertices (counterclockwise from the top, from the three possible trees). In Polymake we input these vertices to model the polytope as follows (note the required extra initial coordinate 1):

```
polytope> $points=new Matrix
   ([[1,2,1,1,1,1,2],[1,1,2,1,1,2,1],[1,1,1,2,2,1,1]]);
polytope> $p=new Polytope(POINTS=>$points);
```

This creates a model of the polytope, and now we can output facts about it.

```
polytope> print $p->F_VECTOR;
```

This outputs 3 3, which simply tells us there are three vertices (the first number) and three edges in the convex hull. Also useful is:

```
polytope> print \$p->VERTICES_IN_FACETS;
```

which outputs:

{1 2}
{0 2}
{0 1}

These three pairs tell us which sets of vertices appear in a facet—here the vertices are numbered 0, 1, 2 in the order they were input, so each pair makes a facet (edge). The inequalities that create these edges geometrically are found in Table 10.1, under $n = 4$. We see that one option is to use the caterpillar inequalities involving leaf 1, which are, respectively, $x_{12} \geq 1$, $x_{13} \geq 1$, and $x_{14}\, ge1$.

Exercise 10.3. Find the vertices for the five-dimensional BME polytope. Use software (such as Polymake) to find the structure of the three types of four-dimensional facets with their inequalities for the five-dimensional BME polytope. We will start the latter process by giving the answer for a cyclic ordering facet, as described in Table 10.1 for $n = 5$.

Letting $a = 1$, $b = 2$, $c = 3$, $d = 4$, $f = 5$, the five trees which keep those five leaves in that cyclic order are the ones with vertices as follows:

$$(4, 2, 1, 1, 2, 1, 1, 2, 2, 4),\quad (2, 2, 2, 2, 1, 4, 1, 1, 4, 1),\quad (4, 1, 1, 2, 1, 1, 2, 4, 2, 2),$$

$$(1, 1, 2, 4, 4, 2, 1, 2, 1, 2),\quad (2, 1, 1, 4, 2, 2, 2, 4, 1, 1).$$

They correspond to the trees in the following picture, starting at the top and going counterclockwise, ending with the central tree (figure from [14]):

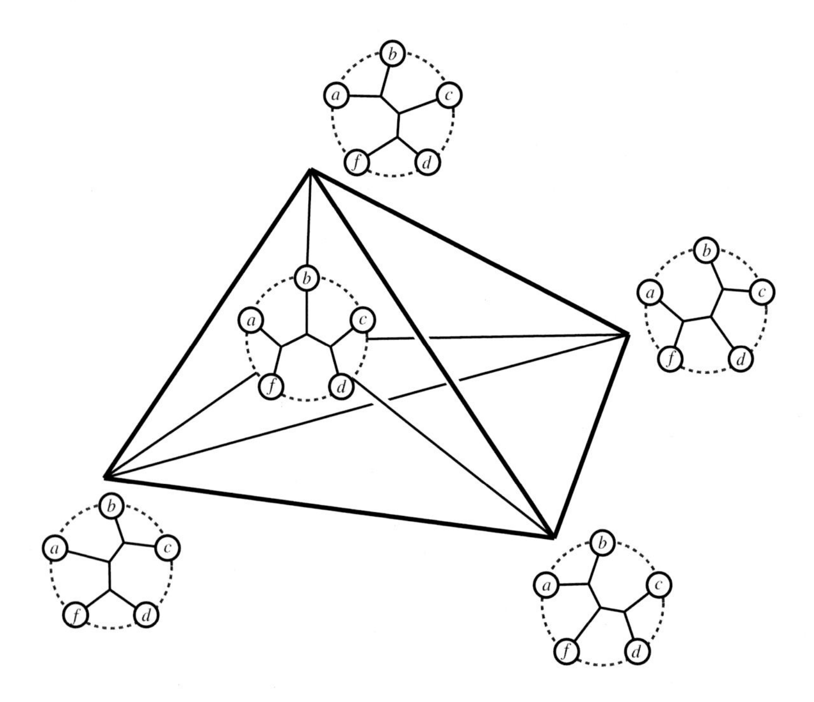

$$x_{ab} + x_{bc} + x_{cd} + x_{df} + x_{fa} \leq 13.$$

The facet formed by these trees is four-dimensional, since the polytope is five-dimensional for $n = 5$. Since their are only five vertices in the facet, the only polytope it an be is a four-dimensional simplex, that is, a pyramid on a tetrahedral base, which we picture above using a Schlegel diagram.

10.2.3 The Splitohedron

Obtaining a *complete* combinatorial or geometric description of a polytope is often difficult. By complete description, we mean that for every facet of the polytope, we would like to have a corresponding facet inequality. Furthermore, collections of facets for polytopes are often exponential or factorial in size, which are impractical for mathematical programming formulations. The BME problem has long been known to be NP-hard. In [21] it is shown that even certain approximate solutions to the BME problem are NP-hard, unless P = NP. This suggests that a complete description of the polytope is unlikely. To partly circumvent this, we can consider a relaxation of the BME polytope. A *relaxation* of a polytope P is a larger polytope R containing P. The relaxation R is specifically any polytope that can be given by a subset of a list of inequalities which define P. We can develop relaxations of the BME polytope using various combinations of the known facet inequalities. To this end, we propose several inequalities used to construct a relaxation of BME(n). We refer to the relaxations as the splitohedra, Sp(n).

We define our relaxation of the BME Polytope as the intersection of half-spaces given by Propositions 10.1–10.4. This operation forms a new polytope, which we call the *Splitohedron*, denoted as Sp(n). Some properties regarding faces of Sp(n) are provided in Table 10.2. Note that the cherry face inequalities, that is, the upper bounds $x_{ij} \leq 2^{n-3}$, are actually redundant. This can be seen since each upper bound is also implied by the pair of intersecting-cherry inequalities based on (1) the intersecting cherries $\{i, j\}$ and $\{j, k\}$ and (2) the intersecting cherries $\{i, j\}$ and $\{i, k\}$. Adding the latter two inequalities:

$$\begin{array}{rl} & x_{ij} + x_{jk} - x_{ik} \leq 2^{n-3} \\ + & x_{ij} + x_{ik} - x_{jk} \leq 2^{n-3} \\ \hline & 2x_{ij} \leq 2(2^{n-3}) \end{array}$$

yields the former upper bound. However, keeping the cherry inequalities can be a programming convenience. The first nontrivial polytope Sp(4) is the same triangle as BME(4). After that, although the facets are fewer, there are more vertices than in BME(n). From computer calculations using Polymake [13], we see that Sp(5) has 27 vertices and Sp(6) has 2335. We now state a theorem obtained in [15] relating the vertices of Sp(n) and BME(n).

Theorem 10.1. *Let t be an unrooted phylogenetic tree with n taxa. If t has at least* $\lceil \frac{n}{4} \rceil$ *cherries, then* $\mathbf{x}(t)$ *is a vertex in both BME(n) and Sp(n). For* $n \leq 11$, *the statement holds regardless of the number of cherries.*

Theorem 10.1 allows us to estimate when we begin losing information under the relaxation. As long as $n \leq 11$, the Splitohedron will contain all the vertices of BME(n). Otherwise, we begin losing some of the vertices of the BME Polytope.

Exercise 10.4. Find the vertices for the two- and five-dimensional splitohedra.

10.3 OPTIMIZING WITH LINEAR PROGRAMMING

The linearity of the half-spaces defining the Splitohedron and the underlying linear objective function suggest that we can cast our optimization problem as a linear program. Recall that, in Eq. (10.4) we defined the path-length functional that describes the length of a tree $t \in \mathcal{T}$. As previously noted, we seek a minimizer $\mathbf{x}(t^\star)$ of this functional, but we delayed defining the region containing admissible solutions. Now that we have defined our relaxation of the BME polytope, we can use it as the feasible region for our linear programming model. Our model is as follows.

Formulation (Discrete Integer Linear Programming).

$$\underset{\mathbf{x}}{\operatorname{argmin}} \quad \mathbf{d} \cdot \mathbf{x}$$

subject to:

$$\sum_{j:j \neq i} x_{ij} = 2^{n-2}, \qquad \forall i \in \mathcal{S}, \tag{10.9}$$

$$x_{ij} + x_{jk} - x_{ik} \leq 2^{n-3}, \qquad \textit{distinct } i,j,k \in \mathcal{S}, \tag{10.10}$$

$$\sum_{\substack{i,j \in S_1 \\ i<j}} x_{ij} \leq (m-1)2^{n-3}, \qquad m = |S_1| \geq 3, n-m \geq 3, \tag{10.11}$$

$$1 \leq x_{ij} \leq 2^{n-3}, \qquad \forall i,j \in \mathcal{S}, i \neq j, \tag{10.12}$$

$$x_{ij} \in \{2^y : y \in \mathbb{Z}_{\geq 0}\}, \qquad i,j \in \mathcal{S}, i \neq j. \tag{10.13}$$

Here we use *argmin* because we want to return the argument $\mathbf{x}$ that minimizes the rescaled path length $\mathcal{L}(\mathbf{x})$. This ultimate form of our functional considers $\mathbf{x}$ as a general vector *without* the dependence on t. This is a direct consequence of our relaxation, which introduces nontree realizable vectors into the feasible region. Notice that Eqs. (10.9)–(10.12) are the *Kraft Equalities*, the

Intersecting Cherry Facets, *Split Facets*, and the *Caterpillar and Cherry Faces*, respectively. These are the same inequalities we used to define our polytope Sp(n). The last constraint (10.13) states that each of the variables is a power of 2. This allows us to avoid encountering many of the potential solutions that might not belong to the BME Polytope that are present in our relaxation. It is this constraint which makes the problem difficult.

In terms of computation, the system above is *not* a polynomial sized formulation. The number of inequalities in constraint (10.11) grows according to the power set and is, therefore, $\mathcal{O}(2^n)$. While this is not an issue for smaller problems, where $n \leq 11$, larger problems consist of many inequalities and the numerical linear algebra slows the run time considerably. Therefore, larger problems require some simplifications to obtain solutions within a reasonable run time.

Exercise 10.5. Repeat Exercise 10.2 using the inequalities for the polytope BME(3), by hand or with your favorite linear programming software.

10.3.1 Discrete Integer Linear Programming: The Branch and Bound Algorithm

The problem of finding the vertex of the BME polytope which corresponds to the tree that minimizes our product belongs to a class of problems called discrete integer linear programs. That is the primary reason that we scaled the values in the solution vector to become integer powers of two. One of the most common techniques for solving this class of problems is called *branch and bound*. This process is recursive, breaking the original problem into *subproblems*, which are easier to solve. The recursive structure of this process can be visualized as traversing a rooted binary tree, where each node represents an individual linear programming problem.

To begin, the discrete valued constraints on the decision variables are relaxed. This allows us to utilize linear programming algorithms where the decision variables admit a continuum of values. The initial linear programming problem that results from this relaxation is called the *root LP*. Computing its solution allows us to determine the feasibility of the original problem. If the solution to the root LP meets our original restrictions for the decision variables, then the branch and bound routine terminates. Otherwise, we select a variable according to a *branching rule* and begin the branching process. The branching rule tells us how to divide the solution space, which results in a set of subproblems, which represent new nodes in our branch and bound tree. After separately solving each of these problems, a *selection strategy* is used to determine which nodes to explore in the branch and bound tree. If a node is not explored, we say the node was *fathomed* or *pruned*. Once a node is selected for exploration, the process is repeated.

The inequalities used in the creation of individual problems, along the path, are maintained throughout the search. Once a feasible solution satisfying the constraints on the decision variables is obtained, we can update the global bound on the objective and use it to prune subproblems, which provide a less optimal objective value. We are permitted to prune subproblems in this manner, even if the discrete constraints on the decision variables are not satisfied. We call the best current solution, the *incumbent solution*. This pruning process allows us to eliminate significant portions of the search space, which effectively reduces the algorithm's running time. Repeatedly applying this process allows us to eventually obtain the optimal solution to our original discrete programming problem.

To summarize, the steps of the general branch and bound approach are as follows:

1. Run the LP solver (such as the simplex method) on our (relaxed) polytope with our given objective function to get "answer zero": vector $\mathbf{x}_0$ and objective function value p_0.
2. If $\mathbf{x}_0$ has all coordinates powers of 2, then we say it is *complete*, and it is our final answer.
3. If $\mathbf{x}_0$ is not complete, then we create some new LP problems 1_A, 1_B, etc. by adding new inequalities one at a time, just enough to force an offending coordinate away from its disallowed value.
4. We solve each of these (as long as they are still *feasible*, i.e., as long as the intersection of the inequalities is nonempty) to get answers x_{1A}, x_{1B}, p_{1A}, p_{1B}, etc.
5. For each new answer we check whether it is complete. If complete, check if its objective function value is better than any seen so far: if it is better then save that solution as the current incumbent solution. If not complete then decide whether it *merits further branching* into more new problems; that is, whether it has an objective function value better than the current best value given by a complete answer. If it is not better, then we prune this branch—that is, we do not branch again. Otherwise we return to step (3).
6. The process ends when no more branching is indicated; and the final answer is the optimal one from among the complete answers found.

For an example, let us solve the following problem: Maximize $p = 4x + 2y$ subject to

$$y \geq 0,$$

$$y \leq 5x + 2,$$

$$y \leq -3x + 6.$$

Require all coordinates of answer $\mathbf{x} = (x, y)$ to be in the set $\{0, 1, 10\}$.

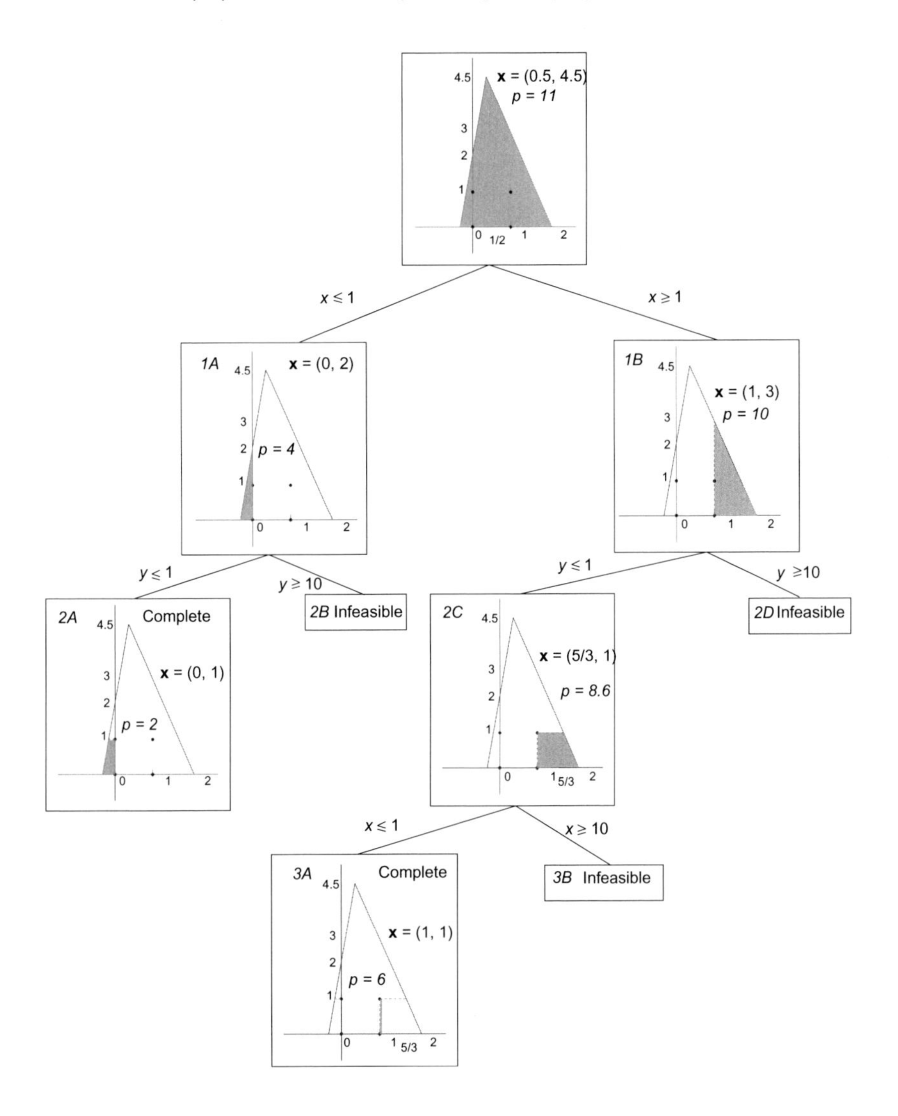

The uppermost box shows the 0-level problem with its solution. Then the branching is performed in alphabetic order on the variables x and y. Of course this is an artificial example, in which the solution is easily found and the inequalities are not very helpful! Larger examples are needed to see that the method is efficient. In this example, some of the branching paths end in an infeasible problem and some end in a complete solution. This example does not have a path that ends in a pruned solution. Next, we give as an exercise a similar problem that does have an opportunity to prune. It also has the same requirement on discrete coordinate values as our actual BME problem.

Exercise 10.6. Perform branch and bound. Maximize $p = 6.75x + 5y$ subject to

$$\begin{aligned}-x &\leq 0,\\ -y &\leq 0,\\ y &\leq 10,\\ x &\leq 8.3,\\ 79x + 18y &\leq 693.5.\end{aligned}$$

Require all coordinates of the answer (x, y) to be powers of 2. When branching, take the first alphabetical coordinate value that is not a power of 2 and introduce two branches that add the inequalities $\leq$ and $\geq$ the nearest powers of two smaller and larger than that coordinate value, respectively. This of course is an arbitrary choice of strategy for branching. In the next section we will talk about more tailored strategies for our specific BME problem.

10.3.2 Recursive Structure: Branch Selection Strategy and Fixing Values

In the next few sections, we will discuss the specific strategies we found while producing our algorithm for the BME problem. The strategies and the pseudocode are presented here and the actual code is made freely available at http://www.math.uakron.edu/~sf34/hedra.htm#splito. However, they are not guaranteed as trustworthy for any purpose, and are intended only as tools for investigation and discovery.

Designing a branch and bound algorithm is an open-ended decision process. One has a large amount of flexibility among choices for selection strategies and branching rules available for implementation. For example, if our problem required that variables belong to $\{0, 1\}$, then we could chose to branch on variables with the "most fractional" entry. For a vector $\mathbf{x}$, this is the entry i satisfying

$$\underset{i}{\operatorname{argmin}}\{|x_i - 0.5|\}. \tag{10.14}$$

In our problem, since we require that variables be powers of 2, we could modify Eq. (10.14) so that we chose the entry whose value is "farthest" from any adjoining power of 2. The expression for this is slightly more complex and takes the form

$$\underset{i}{\operatorname{argmax}}\left\{\min\{|x_i - 2^{\lfloor \log_2 x_i \rfloor}|, |x_i - 2^{\lceil \log_2 x_i \rceil}|\}\right\}. \tag{10.15}$$

The complexity of this selection strategy arises naturally from the structure of powers of 2. As the numbers in the set becomes larger, the distance between adjoining elements also increases. We can also think of Eq. (10.15) as being a bottom-up selection strategy, where the algorithm focuses on finding the cherries first. If this strategy leads to multiple maximizers, then we can choose any of them to use as a branching variable. Our strategy simply chooses the first maximizing index. Of course, many other good selection strategies may exist, perhaps based on machine-learning and examining statistical relationships between selection variables and the corresponding "gain" in the objective function (see, e.g., [22]).

Once we have selected the variable to branch on, we must determine how to subdivide the feasible region. This effect is accomplished by *bounding functions*, which introduce new inequalities to the feasible region in an effort to remove the unwanted values. Suppose we have identified the branching variable x_j. During the branching phase, we generate two new bounds

$$x_j \leq 2^{\lfloor \log_2 (x_j) \rfloor}, \tag{10.16}$$

$$x_j \geq 2^{\lceil \log_2 (x_j) \rceil}, \tag{10.17}$$

which are maintained in the constraint sets $\{A_1, \mathbf{b}_1\}$ and $\{A_2, \mathbf{b}_2\}$, respectively. Each constraint set is associated with a particular subproblem, and is passed recursively to the branch and bound algorithm, where both feasibility and validity are further examined. If the problem is infeasible, then the node is fathomed. If the problem is feasible and the solution has entries that are powers of 2, then we check the value of the objective function. If it is less than the incumbent solution (since we are minimizing), then the bound is updated. Once a valid solution has been found, its bound can be used to prune suboptimal branches in the tree.

We can also introduce *heuristic* fixing. Heuristics, on a general level, involve incorporating problem-dependent experimental information into an algorithm to eliminate unlikely candidates in the branch and bound process and promote faster convergence to the optimizer. If variables in the solution have floating point values which are "close" to a power of 2 and remain stagnant during the process, we could reduce round-off errors by fixing the values to their closest power of 2. Therefore, given a tolerance $\epsilon > 0$ and a vector $\mathbf{x}$, we select a candidate for fixing if an entry satisfies

$$\operatorname*{argmin}_i |x_i - 2^{[\log_2 (x_i)]}| < \epsilon. \tag{10.18}$$

Here, $[x]$ denotes the nearest integer function. Suppose now that we have identified a candidate variable j that satisfies Eq. (10.18). Then we set

$$x_j = 2^{[\log_2 (x_j)]} \tag{10.19}$$

in the equality constraints $\{A_{eq}, \mathbf{b}_{eq}\}$. This effectively eliminates the variable x_j as an unknown. We fix variables in two places: inside the main script, after finding the root solution, and before branching. In the first instance, we fix all cherries using equality constraints and adjust the upper bounds from 2^{n-3} to 2^{n-4}. The experimental observation here is that linear programming solvers tend to find all cherries in the given problem after solving the root LP. Therefore, it is deemed reasonable to fix their positions in the equality constraints. In the second instance, we fix any variable according to Eqs. (10.18), (10.19). Should multiple variables be eligible for fixing, we choose one closest to its rounded value. This sort of secondary fixing is commonly referred to as a *Large Neighborhood Search* (LNS) since it searches for solutions to the underlying problems in a large neighborhood of the polytope, which contains the face generated by our fixed entry. As we let $\epsilon \rightarrow 0$, the algorithm relies less on fixing and performs similarly to pure branch and bound. The size of ϵ controls how liberal we wish to be with fixing.

A noted downside of the LNS heuristic is that it sometimes can lead to infeasible situations along all branches of the branch and bound algorithm. Once the algorithm encounters universal infeasibility, the process terminates. This problem has been considered in strategies such as *Relaxation Induced Neighborhood Search* (RINS) and *Guided Dives*, which use sophisticated back-tracking processes to return to a feasible state. Authors in [23] develop these methods, exploring how each of them defines, searches, and diversifies neighborhoods to improve incumbent solutions. In practice, such techniques have allowed for fast, successful computation of often intractable optimization problems. Heuristics, in the same regard as other aspects of branch and bound, require careful experimentation. However, it should be noted that approaches utilizing heuristics can no longer guarantee that the solution obtained is the true minimizer (or maximizer) for the original problem.

Exercise 10.7. Suppose the current solution to a subproblem in the branch and bound algorithm is: $\mathbf{x} = (2, 4, 3.25, 3.42, 2, 3.68, 1.33, 1, 8, 4, 2, 2, 4, 8, 1.5)$. Use the expression (10.15) to determine the branching variable and write down the two bounds generated by branching using Eqs. (10.16), (10.17).

10.3.3 Pseudocode for the Algorithm: POLYSPLIT

Our branch and bound procedure consists of a main script to manage the entire problem, and two separate branching algorithms embedded in the main file. The purpose of the main file is to formally initialize the best current bound and evaluate the feasibility of the root LP. Before the root LP is solved, the data for the problem is collected. This data includes the given dissimilarity data $\mathbf{d}$, the sets of inequality and equality constraints, choices for parameters governing run-time and termination, and a decision to apply the LNS heuristic. Once this information has been collected, it is sent to the LP solver to identify the solution for the root and determine feasibility. If the problem has been deemed feasible, it is sent to a recursive branching function with updated constraints based on whether or not a heuristic has been applied. We provide pseudocode for the main algorithm in Fig. 10.5.

The algorithm decides to call a particular branching function depending on whether or not we use a heuristic. If we choose to use the heuristic, the algorithm

Algorithm 1: The Discrete ILP Main Algorithm PolySplit

Require: Identification of minimizer$\mathbf{x}^\star$
Input: $\mathbf{d}$, A, $\mathbf{b}$, A_{eq}, $\mathbf{b}_{eq}$, lb, ub, n, maxiter0, maxiter1,ϵ, heuristic
Output: $\mathbf{x}^\star$, $\mathcal{L}^\star$, status

1: **Initialization:** set bound $= +\infty$ and iter $= 0$
2: Solve the relaxation at the root node$\rightarrow$ $\mathbf{x}_0$, $\mathcal{L}_0$, status0
3: **if** status = infeasible **then**
4: Return $\mathbf{x}^\star$, $\mathcal{L}^\star = \emptyset$
5: **else**
6: **if** heuristic = 0 **then**
7: Call branch0 $\rightarrow \mathbf{x}^\star, \mathcal{L}^\star$, status
8: **else if** heuristic = 1 **then**
9: Find all entries s.t. $|x_{0_i} - 2^{n-3}| < \epsilon$ for $i = 1, \ldots, \binom{n}{2}$
10: Fix positions and update $A_{eq}, \mathbf{b}_{eq}$, ub
11: Call branch1 $\rightarrow \mathbf{x}^\star, \mathcal{L}^\star$, status
12: **end if**
13: **end if**

FIG. 10.5 Main algorithm PolySplit for the discrete ILP problem.

finds cherries and fixes their positions in the constraints, which are then passed as input to the branching function. It was observed in numerous test cases that LP solvers could identify cherries immediately in the root LP and that they did not change over the course of the process. Therefore, we chose to fix their values in the solution vector.

The algorithms for the branching functions are similar with a slight modification in the heuristic-based option. We provide a detailed outline in Fig. 10.6. The branching functions first evaluate the solution of the LP at the current node. If we are at the root node, then the problem is solved twice. This redundancy is not an issue and merely serves to simplify our code. If the solution at the current node is infeasible or the value of the objective function is worse than our incumbent solution, then the node is pruned. Otherwise, the algorithm checks the solution to see if it contains all powers of 2. If this holds, then the solution becomes the new incumbent and we can use the value of the objective function to prune suboptimal results later in the search. If we find that the solution produced does not contain all powers of 2, then we branch on the current solution. We can identify the branching variable, according to a devised rule and then create the two subproblems. Each subproblem receives one of the new inequalities (10.16), (10.17), respectively. Finally, we pass the information for the subproblems $\mathcal{P}_1$ and $\mathcal{P}_2$, recursively, to the branching function. Each subproblem communicates with the other through the incumbent solution. Once the branching produces suboptimal results, the algorithm terminates and reports the solution to the main algorithm.

To modify the code present in Fig. 10.6 to apply the heuristic, we need to incorporate an additional parameter, which we call "maxiter0." The algorithm

Algorithm 2: POLYSPLIT Branch0 Algorithm

Input: $\mathbf{d}$, A, $\mathbf{b}$, A_{eq}, $\mathbf{b}_{eq}$, lb, ub, $\mathbf{x}_t, \mathcal{L}_t$, ϵ, bound

Output: $\tilde{\mathbf{x}}, \tilde{\mathcal{L}}$, status, bb

1: Solve the relaxation at the current node$\rightarrow$ $\mathbf{x}_0, \mathcal{L}_0$, status0
2: **if** status0 = infeasible or $\mathcal{L}_0 >$ bound **then**
3: Return input, bb $\leftarrow$ bound
4: **else**
5: Compute $\mathcal{E} = \max_i \left\{ \min\{|x_{0_i} - 2^{\lfloor \log_2 (x_{0_i}) \rfloor}|, |x_{0_i} - 2^{\lceil \log_2 (x_{0_i}) \rceil}|\}\right\}$
6: **if** $\mathcal{E} < \epsilon$ or iter > maxiter1 **then**
7: **if** $\mathcal{L}_0 <$ bound **then**
8: $\tilde{\mathbf{x}} \leftarrow \mathbf{x}_0, \tilde{\mathcal{L}} \leftarrow \mathcal{L}_0$, bb $\leftarrow \mathcal{L}_0$
9: **else**
10: Return input, bb $\leftarrow$ bound
11: **end if**
12: Return
13: **end if**
14: Select a branching variable x_{0_j}
15: Build subproblem $\mathcal{P}_1$: Set $x_{0_j} \leq 2^{\lfloor \log_2 (x_{0_j}) \rfloor}$ in $\{A, \mathbf{b}\} \rightarrow \{A_1, \mathbf{b}_1\}$
16: Build subproblem $\mathcal{P}_2$: Set $x_{0_j} \geq 2^{\lceil \log_2 (x_{0_j}) \rceil}$ in $\{A, \mathbf{b}\} \rightarrow \{A_2, \mathbf{b}_2\}$
17: iter $\leftarrow$ iter + 2
18: Call branching routine for $\mathcal{P}_1$ $\rightarrow \mathbf{x}_1, \mathcal{L}_1$, status1, bound1
19: **if** bound1 < bound and status1 = feasible **then**
20: $\tilde{\mathbf{x}} \leftarrow \mathbf{x}_1, \tilde{\mathcal{L}} \leftarrow \mathcal{L}_1$, bound $\leftarrow$ bound1, bb $\leftarrow$ bound1, status $\leftarrow$ status1
21: **else**
22: Return $\mathcal{P}_1$ input data, bb $\leftarrow$ bound
23: **end if**
24: Call branching routine for $\mathcal{P}_2$ $\rightarrow \mathbf{x}_2, \mathcal{L}_2$, bound2, status2
25: **if** bound2 < bound and status2 = feasible **then**
26: $\tilde{\mathbf{x}} \leftarrow \mathbf{x}_2, \tilde{\mathcal{L}} \leftarrow \mathcal{L}_2$, bb $\leftarrow$ bound2, status $\leftarrow$ status2
27: **end if**
28: **end if**

FIG. 10.6 Branching algorithm for pure branch and bound.

is designed to work with pure branch and bound until the number of iterations reaches maxiter0. At this point, the algorithm has not found the solution, so instead, we begin applying the heuristic, according to Eqs. (10.18), (10.19). This process works in conjunction with pure branch and bound until we (1) reach the solution, (2) encounter an infeasibility, or (3) reach maxiter1 and the algorithm terminates. This adjustment is reflected in our code in Fig. 10.7.

It is important to note that for our purposes, we use a single value for each of the tolerances in our branching algorithms presented in Figs. 10.6 and 10.7. In principle, we could have defined different tolerances, say ϵ_0, ϵ_1, and ϵ_2, associated with the stopping criterion, branching variable selection, and the LNS search, respectively. Another area to explore would be variations in the iterations

Algorithm 3: PolySplit Branch1 Algorithm

Input: $\mathbf{d}$, A, $\mathbf{b}$, A_{eq}, $\mathbf{b}_{eq}$, lb, ub, $\mathbf{x}_t, \mathcal{L}_t$, ϵ, bound
Output: $\tilde{\mathbf{x}}, \tilde{\mathcal{L}}$, status, bb

1: Solve the relaxation at the current node$\rightarrow$ $\mathbf{x}_0, \mathcal{L}_0$, status0
2: **if** status0 $=$ infeasible or $\mathcal{L}_0 >$ bound **then**
3: Return input, bb$\leftarrow$ bound
4: **else**
5: Compute $\mathcal{E} = \max_i \{ \min\{|x_{0_i} - 2^{\lfloor \log_2 (x_{0_i}) \rfloor}|, |x_{0_i} - 2^{\lceil \log_2 (x_{0_i}) \rceil}|\}\}$
6: **if** $\mathcal{E} < \epsilon$ or iter $>$ maxiter1 **then**
7: **if** $\mathcal{L}_0 <$ bound **then**
8: $\tilde{\mathbf{x}} \leftarrow \mathbf{x}_0, \tilde{\mathcal{L}} \leftarrow \mathcal{L}_0$, bb $\leftarrow \mathcal{L}_0$
9: **else**
10: Return input, bb$\leftarrow$ bound
11: **end if**
12: Return
13: **end if**
14: **if** iter $>$ maxiter0 **then**
15: Find an entry $k = \operatorname{argmin}_i |x_{0_i} - 2^{[\log_2 (x_{0_i})]}| < \epsilon$ for $i = 1, \ldots, \binom{n}{2}$
16: Set $x_{0_k} = 2^{[\log_2 (x_{0_k})]}$ and update $A_{eq}, \mathbf{b}_{eq}$
17: **end if**
18: Select a branching variable x_{0_j}
19: Build subproblem$\mathcal{P}_1$: Set $x_{0_j} \leq 2^{\lfloor \log_2 (x_{0_j}) \rfloor}$ in $\{A, \mathbf{b}\} \rightarrow \{A_1, \mathbf{b}_1\}$
20: Build subproblem$\mathcal{P}_2$: Set $x_{0_j} \geq 2^{\lceil \log_2 (x_{0_j}) \rceil}$ in $\{A, \mathbf{b}\} \rightarrow \{A_2, \mathbf{b}_2\}$
21: iter $\leftarrow$ iter $+ 2$
22: Call branching routine for$\mathcal{P}_1 \rightarrow \mathbf{x}_1, \mathcal{L}_1$, status1, bound1
23: **if** bound1 $<$ bound and status1 $=$ feasible**then**
24: $\tilde{\mathbf{x}} \leftarrow \mathbf{x}_1, \tilde{\mathcal{L}} \leftarrow \mathcal{L}_1$, bound $\leftarrow$ bound1, bb $\leftarrow$ bound1, status$\leftarrow$ status1
25: **else**
26: Return $\mathcal{P}_1$ input data, bb$\leftarrow$ bound
27: **end if**
28: Call branching routine for$\mathcal{P}_2 \rightarrow \mathbf{x}_2, \mathcal{L}_2$, bound2, status2
29: **if** bound2 $<$ bound and status2 $=$ feasible **then**
30: $\tilde{\mathbf{x}} \leftarrow \mathbf{x}_2, \tilde{\mathcal{L}} \leftarrow \mathcal{L}_2$, bb $\leftarrow$ bound2, status$\leftarrow$ status2
31: **end if**
32: **end if**

FIG. 10.7 Branching algorithm using an LNS heuristic. Once we specify a tolerance, the algorithm determines if a decision variable can be fixed before branching.

before the heuristic is applied. For instance, does the algorithm exhibit better performance and solution quality when the heuristic is applied earlier?

Once the algorithm terminates, it is possible to draw the phylogenetic tree simply by passing the solution to the distance function, which determines the topological distances l_{ij} of the phylogenetic tree. Using our definition for x_{ij} in Eq. (10.1) and inverting the exponential piece via logarithms, we obtain a formula for the l_{ij}'s in terms of the x_{ij}'s. The ability to draw the tree is conditional on the tree realizability of the solution $\mathbf{x}^\star$. We have observed experimentally that the above algorithms seem to always yield a result that is realizable as a tree, although a formal proof of this is yet to be complete.

Exercise 10.8. The Matlab code for PolySplit is available from the Encyclopedia of Combinatorial Polytope Sequences, at www.math.uakron.edu/~sf34/hedra.htm#splito. Sample input is given there as well, for instance, the distance vector for a tree with nine leaves:

$$\mathbf{d} = (4,5,3,2,6,7,8,8,3,3,4,4,5,6,6,4,5,3,4,5,5,3,5,6,7,7,\\ 6,7,8,8,3,4,4,3,3,2).$$

Find the optimal tree using the code. The output will be the powers-of-2 vector. Use it to redraw the tree. The Matlab files are also found in [24], available from etd.ohiolink.edu.

10.4 NEIGHBOR JOINING AND EDGE WALKING

The most frequently used method for distance-based phylogeny reconstruction is Neighbor Joining (NJ), developed in [25]. As shown in [11], NJ is a greedy algorithm for finding the BME tree. However, NJ loses accuracy quickly at the point of considering seven or eight taxa (leaves) as seen in [26]. For eight taxa, the theoretical accuracy is between 69% and 72% for trees with more than two cherries, but drops to 62% for caterpillar trees. The conjecture in [26] is that the accuracy will continue to drop quickly, especially for caterpillar trees with more taxa. Our algorithm, on the other hand, experimentally shows 100% accuracy up to 11 taxa, even in the caterpillar case, with and without noise.

10.4.1 NNI and SPR Moves, and FastME 2.0

In an ideal situation, where the entire polytope is known via its facet inequalities, linear programming via the simplex method has a nice geometrical interpretation. The pivot moves and row reductions in the simplex method correspond to moving from vertex to vertex of the polytope along edges—each time choosing the edge which most improves the objective function.

In the nonideal situation, working with a relaxation of the polytope is the best we can do if we want to use the inequalities. If, alternatively, we know some subset of the edges, we can attempt a solution that uses any known edge in order to move from a current vertex to an adjacent vertex which improves the objective. This requires having combinatorial knowledge of the edges—that is, knowing how to find at least some of the adjacent phylogenetic trees to any given tree.

The best current published improvement on NJ is the algorithm by Desper and Gascuel known as FastME2.0 [27]. This is an improved version of the original FastME [28], which piggy-backed on NJ by performing nearest-neighbor interchanges (NNI) on the current tree candidate. The FastMe2.0 algorithm improves this by also doing subtree-prune-and-regraftings (SPR). As shown in [17], both operations correspond to edges of the BME polytope. There are many more edges in general. Thus, FastMe uses edge walking on a subset of the edges of BME(n), which is essentially the bottom-up analog of our top-down use of facet inequalities. In [29] the FastMe algorithm is shown to improve on NJ by between 3.5% and 7% for 24 taxa, and as much as 21.3% for 96 taxa. While it is difficult to make a conclusive statement without further development of our PolySplit algorithm for greater numbers of taxa, it appears that our algorithm has the potential to improve even more, based on our 100% accuracy rate with up to 11 taxa. In order to test higher numbers, we plan to develop heuristics for selectively decreasing the number of facets from our list. The number of split facets, for instance, grows like 2^n, so we need to use dynamically chosen subsets of these facets in order to manage the run-time for $n > 12$.

10.5 SUMMARY

There are many phylogenetic questions that remain unanswered in biology and we have addressed only one method of tree reconstruction in this chapter. We have, however, demonstrated a new algorithm for finding the BME tree. This approach, accompanied by the general introduction to the branch and bound method of linear programming using the discrete integer set of powers-of-2, can be used to explore these unanswered questions. We hope that the approaches in this chapter will provide a new and exciting method of phylogenetic reconstruction, capable of accounting for the very large datasets biologists are able to generate.

Additionally, there are many other methods that can be used to approximate a phylogenetic tree. Maximum likelihood (ML) and Bayesian methods are commonly used by biologists. ML methods assume a probabilistic model of mutations and choose the tree with the maximum probability. These probabilistic methods can be combined with BME: the ML method is used to build the dissimilarity (distance) matrix and then BME is used to construct the tree. Alternative linear programming approaches to BME, such as that in [30], have also been tested on large sets of taxa. A natural extension of the work presented in this chapter would be to compare the various linear programming methods to identify areas where the approaches could be combined to further optimize reconstructions.

REFERENCES

[1] M. Salemi, P. Lemey, A.-M. Vandamme, The Phylogenetic Handbook: A Practical Approach to Phylogenetic Analysis and Hypothesis Testing, Cambridge University Press, Cambridge, MA, 2009.

[2] D. Baum, Reading a phylogenetic tree: the meaning of monophyletic groups, Nat. Educ. 1 (1) (2008) 190.

[3] D.R. Maddison, K.S. Schulz, W.P. Maddison, The tree of life web project, in: Z.Q. Zhang, W.A. Shear (Eds.), Linnaeus Tercentenary: Progress in Invertebrate Taxonomy. Zootaxa 1668:1–766, 2007, pp. 19–40.
[4] F.D. Ciccarelli, T. Doerks, C. von Mering, C.J. Creevey, B. Snel, P. Bork, Toward automatic reconstruction of a highly resolved tree of life, Science 311 (5765) (2006) 1283–1287.
[5] B. Gomez, V. Daviero-Gomez, C. Coiffard, C. Martín-Closas, D.L. Dilcher, Montsechia, an ancient aquatic angiosperm, Proc. Natl Acad. Sci. 112 (35) (2015) 10985–10988.
[6] C.M. Morton, Newly sequenced nuclear gene (Xdh) for inferring angiosperm phylogeny, Ann. Mo. Bot. Gard. 98 (1) (2011) 63–89.
[7] T.J. Barkman, G. Chenery, J.R. McNeal, J. Lyons-Weiler, W.J. Ellisens, G. Moore, A.D. Wolfe, C.W. dePamphilis, Independent and combined analyses of sequences from all three genomic compartments converge on the root of flowering plant phylogeny, Proc. Natl Acad. Sci. 97 (24) (2000) 13166–13171.
[8] D. Catanzaro, The minimal evolution problem: overview and classification, Networks 53 (2) (2007) 112–125.
[9] N. Saitou, M. Nei, The neighbor joining method: a new method for reconstructing phylogenetic trees, Mol. Biol. Evol. 4 (1987) 406–425.
[10] Y. Pauplin, Direct calculation of a tree length using a distance matrix, J. Mol. Evol. 51 (1) (2000) 41–47.
[11] O. Gascuel, M. Steel, Neighbor-joining revealed, Mol. Biol. Evol. 23 (2006) 1997–2000.
[12] D. Haws, T. Hodge, R. Yoshida, Optimality of the neighbor joining algorithm and faces of the balanced minimum evolution polytope, Bull. Math. Biol. 73 (11) (2011) 2627–2648.
[13] E. Gawrilow, M. Joswig, Polymake: a framework for analyzing convex polytopes, in: G. Kalai, G.M. Ziegler (Eds.), Polytopes—Combinatorics and Computation, Birkhäuser, Basel, 2000, pp. 43–74.
[14] S. Forcey, L. Keefe, W. Sands, Facets of the balanced minimal evolution polytope, J. Math. Biol. 73 (2) (2016) 447–468.
[15] S. Forcey, L. Keefe, W. Sands, Split facets of balanced minimal evolution polytopes and the permutoassociahedron, Bull. Math. Biol. 79 (5) (2016) 975–994.
[16] P. Huggins, Polytopes in Computational Biology (Ph.D. Dissertation), U.C. Berkeley, 2008.
[17] D.C. Haws, T.L. Hodge, R. Yoshida, Optimality of the neighbor joining algorithm and faces of the balanced minimum evolution polytope, Bull. Math. Biol. 73 (11) (2011) 2627–2648.
[18] W. Rytter, Trees with minimum weighted path length, in: D. Mehta, S. Sahni (Eds.), Handbook of Data Structures and Applications, chap. 10, Chapman and Hall/CRC, London, 2004, pp. 1–22.
[19] D. Catanzaro, M. Labbé, R. Pesenti, J.-J. Salazar-González, The balanced minimal evolution problem, INFORMS J. Comput. 24 (2) (2012) 276–294.
[20] B. Fortz, O. Oliveira, C. Requejo, Compact mixed integer linear programming models to the minimum weighted tree reconstruction problem, Eur. J. Oper. Res. 256 (2017) 242–251.
[21] S. Fiorini, G. Joret, Approximating the balanced minimum evolution problem, Oper. Res. Lett. 40 (1) (2012) 31–35.
[22] T. Achterberg, T. Koch, A. Martin, Branching rules revisited, Oper. Res. Lett. (33) (2005) 42–54.
[23] E. Danna, E. Rothberg, C. Le Pape, Exploring relaxation induced neighborhoods to improve MIP solutions, Math. Program. 102 (1) (2005) 71–90.
[24] W. Sands, Thesis: Phylogenetic Inference Using a Discrete-Integer Linear Programming Model, etd.ohiolink.edu, 2017.

[25] N. Saitou, M. Nei, The neighbor joining method: a new method for reconstructing phylogenetic trees, Mol. Biol. Evol. 4 (4) (1987) 406–425.
[26] K. Eickmeyer, P. Huggins, L. Pachter, R. Yoshida, On the optimality of the neighbor-joining algorithm, Algorithms Mol. Biol. 3 (5) (2008), Available from: http://www.almob.org/content/3/1/5.
[27] V. Lefort, R. Desper, O. Gascuel, FastME 2.0: a comprehensive, accurate, and fast distance-based phylogeny inference program, Mol. Biol. Evol. (2015) https://doi.org/10.1093/molbev/msv150.
[28] R. Desper, O. Gascuel, Fast and accurate phylogeny reconstruction algorithms based on the minimum-evolution principle, J. Comput. Biol. 9 (5) (2002) 687–705.
[29] R. Desper, O. Gascuel, Fast and accurate phylogeny reconstruction algorithms based on the minimum-evolution principle, J. Comp. Biol. 9 (5) (2002) 687–705.
[30] D. Catanzaro, R. Aringhieri, M.D. Summa, R. Pesenti, A branch-price-and-cut algorithm for the minimum evolution problem, Eur. J. Oper. Res. 244 (3) (2015) 753–765.

Chapter 11

Data Clustering and Self-Organizing Maps in Biology

Olcay Akman*, Timothy Comar†, Daniel Hrozencik‡ and Josselyn Gonzales*

*Illinois State University, Normal, IL, United States, †Benedictine University, Lisle, IL, United States, ‡Chicago State University, Chicago, IL, United States

11.1 CLUSTERING: AN INTRODUCTION

In his 1936 article, "The use of multiple measurements in taxonomic problems," statistician and biologist Ronald Fisher published a data set that looked at 50 samples from each of three species of Iris flower: *Iris setosa*, *Iris virginica*, and *Iris versicolor* [1]. Each sample consisted of the length and width of the flower sepal and the length and width of the petals, where all four measurement components are in centimeters. Fig. 11.1 is a collection of scatter plots that compares the data of the three species according to plots of the different measurement components. We can clearly see from the plots that the points seem to separate into three fairly distinct groups despite the overlap between *I. versicolor* and *I. virginica*. The existence of discriminating features between the three species allowed Fisher to develop a model capable of classifying measurement observations into the correct species of Iris flower.

Charu C. Aggarwal, a researcher for IBM and author of a number of data mining books, describes the problem of data clustering with the following statement: "Given a set of data points, partition them into a set of groups which are as similar as possible" [2]. In other words, the purpose of data clustering is to take a set of data and separate it according to its natural data structures. These separations, or partitions, of a data set are called *clusters*.

Although the definition of a cluster can be stated in a number of different ways, we attempt to provide the most general definition. A cluster is a collection of objects which are very similar to each other. Hence, objects belonging to different clusters are not as similar to each other as are objects belonging to the same cluster. (Note that other terms for cluster in the literature are *group* and *class*.)

Algebraic and Combinatorial Computational Biology. https://doi.org/10.1016/B978-0-12-814066-6.00011-8

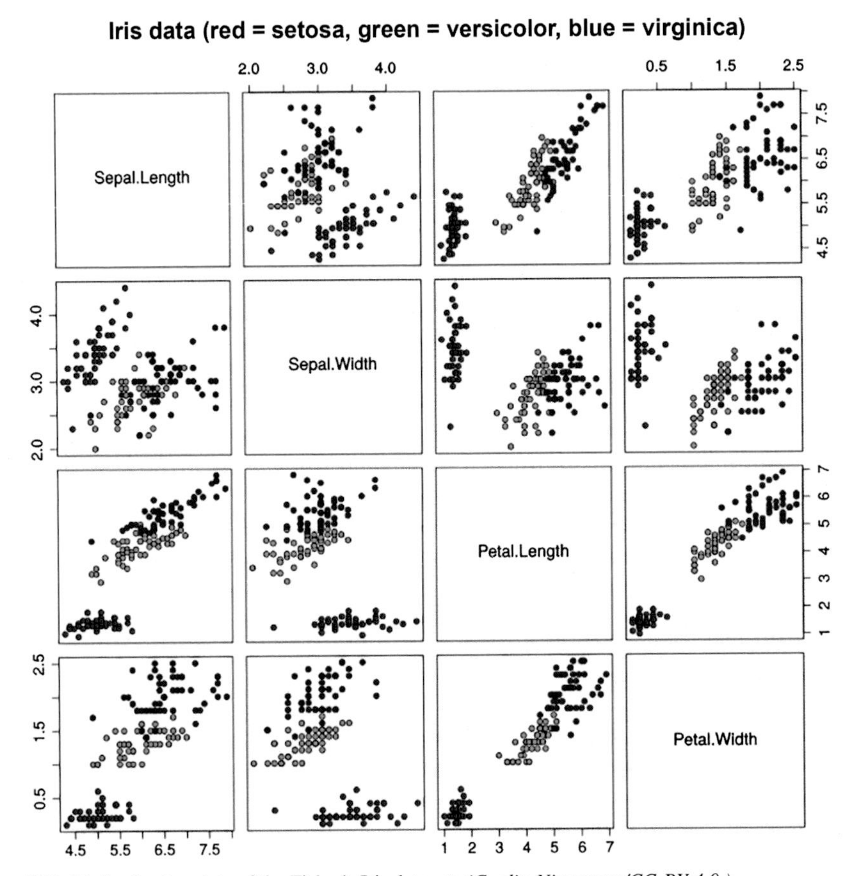

FIG. 11.1 Scatter plots of the Fisher's Iris data set. *(Credit: Nicoguaro/CC-BY-4.0.)*

A more rigorous definition of a cluster, of course, comes with a more rigorous definition of similarity using a proximity measure. One could, for example, use a distance function, that is, the closer two data points are, the more similar they are and the more likely they are to be clustered together. One could also consider the relative densities of regions within a space to determine clusters.

The concept of a cluster is very natural, especially in fields like morphology. For example, each scatter plot in Fig. 11.1 shows that there is an obvious tendency in the data to separate into three clusters. For a closer look, Fig. 11.2 focuses on the scatter plot of sepal width versus sepal length. These two features of the data, while discriminating do not completely allow us to distinguish between the three Iris species. Considering all four features together, however, as in Fig. 11.1 provides us with greater insight. Thus, we can expect a clustering of the Iris data to fairly accurately separate it according to species.

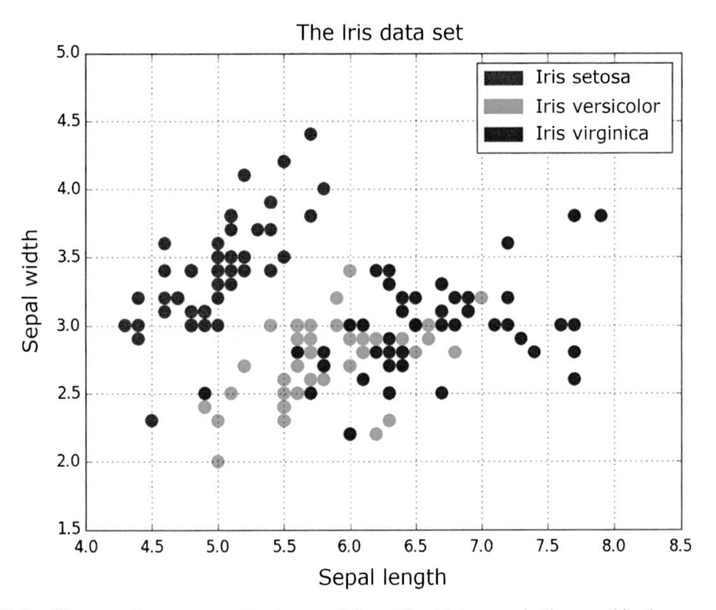

FIG. 11.2 Here, we focus on a particular panel from Fig. 11.1, namely the panel in the *second row and first column*. We see that *I. setosa* has distinctive sepal length and width features that distinguish it from the other two species. On the other hand, the overlap between *I. versicolor* and *I. virginica* indicates that the sepal length and width, all by themselves, may not be enough to distinguish these species. This is a prime of example of need for sophisticated clustering methods where multivariable traits are used to accomplish the goal of discriminating variables based on specific attributes.

Cluster analysis algorithms are *unsupervised classification* algorithms, meaning it classifies data objects into classes based on the natural features of the data. By contrast, a *supervised classification* algorithm classifies data objects based on predefined classes or labels.

Because clustering algorithms are unsupervised, they are relatively quick and easy to run. Additionally, the user does not necessarily need to have a priori knowledge about the data set in order to perform a cluster analysis. In fact, in any type of clustering algorithm, the clustering of the data set is entirely dependent on the proximity measure and similarity criterion and not on any user-predefined labels. In that sense, cluster analysis is more than just simple segmentation of objects and more than just the results of some database query.

Cluster analysis has countless applications in fields such as engineering, medical sciences, earth sciences, economics, and many others. Clustering helps researchers gain meaningful insight into the distribution of the data with which they are working. It usually serves as a preliminary step for other statistical algorithms.

In general, there are two types of clustering: hard clustering and fuzzy clustering. In *hard clustering*, each object belongs to only one cluster, while in *fuzzy clustering*, an object can belong to one or more clusters with particular probabilities. The differences between hard and fuzzy clustering are few but significant. Whereas in hard clustering, each object is assigned a value of either 0 or 1, in fuzzy clustering, each object is assigned a value within the interval [0, 1], according to some probability function.

Hard clustering further breaks down into partitional and hierarchical clustering. *Partitional clustering* seeks to partition the data set into a finite number of disjoint, nonempty sets. In *hierarchical clustering*, the resulting partitions form a nested, tree-like structure. We will go into further detail about these types of clustering in Section 11.3.

11.2 CLUSTERING: A BASIC PROCEDURE

Although the process of data clustering varies according to the particular problem being addressed, Xu [3] and Jain [4] each provide a basic outline of the procedure for performing a typical cluster analysis. In most cases, the best clustering of the data comes from performing these steps multiple times. Here we provide a summary of those steps:

1. data representation,
2. clustering algorithm selection,
3. cluster validation, and
4. result interpretation.

11.2.1 Data Representation

Before attempting to cluster a data set, it is important to recognize its patterns and features. What kind of cluster structure do we expect to find in the data set? This step requires the user to identify and select the features that are relevant to the cluster analysis problem. This is referred to as *feature selection*, which involves effectively choosing significant and distinctive features of the data in a way that results in both an easy-to-understand design process and a cost-effective clustering algorithm. Feature selection and extraction is especially crucial to the efficient clustering of high dimensional data. One example of a feature extraction method is principal component analysis, which is a dimensionality-reduction tool for numerical data [5]. Given a data set consisting of d data vector components or features, principal components are the uncorrelated linear combinations of the d features and have the largest possible variances. If a small number p of principal components (relative to d) explain the majority of the total variance of all the principal components, then these p principal components can be used instead of the original d features without a loss of too much information [6].

One other factor that must be taken into consideration when selecting the best features on which to focus on is the type of the data making up the data set, that is, the data types involved. Many types of data exist, and most of these are specific to a particular research area or field of study. Hence, it may sometimes be necessary to standardize the data. *Standardizing* transforms the data so that it is dimensionless and, therefore, easier to use and interpret. However, such transformations may result in a loss of original information, so it is important that the chosen transformation technique preserves as much of the information contained in the data as possible. In fact, each area of study tends to have its own conventions for the standardization of data to address this very concern. See [5] for commonly used standardization techniques.

11.2.2 Clustering Algorithm Selection

Once the important characteristics of the data are determined, the natural next step is to design or select a clustering algorithm that works best with the data. Designing a clustering algorithm mainly involves determining a distance or similarity measure and a similarity criterion. The resulting clusters, of course, are entirely dependent upon these choices.

There is no universal clustering algorithm. Again, many fields of study have their own conventions when it comes to clustering. However, certain parameters must be met in order for the distance/similarity measure to provide any meaningful information. It is important to note the inverse relationship between distance and similarity despite the interchangeable use of the terms. To state the relationship more explicitly, the smaller the distance, the more similar two objects are and vice versa.

In order for d to be a distance measure applied to a data set $\{x_1, \ldots, x_n\}$, the following conditions must hold for indices $1 \leq i, j, k \leq n$:

(i) $d(x_i, x_j) = d(x_j, x_i)$,
(ii) $d(x_i, x_j) \geq 0$,
(iii) $d(x_i, x_k) \leq d(x_i, x_j) + d(x_j, x_k)$, and
(iv) $d(x_i, x_j) = 0$ if and only if $x_i = x_j$.

Condition (i) is symmetry, that is, the distance between any two objects will remain the same no matter in what order the measurement is taken. Condition (ii) requires that all distances be nonnegative for there would be no applicable meaning otherwise. Condition (iii) describes the triangle inequality, which essentially states that the shortest path between two objects is always the most direct path. Finally, condition (iv) states that the distance between an object and itself is always zero, making this the only time distance is nonpositive.

There are many different kinds of distance measures. The most well-known and commonly used for numerical data is the Euclidean distance in two-dimensional space. With data objects $x_i = (x_{i1}, x_{i2})$ and $x_j = (x_{j1}, x_{j2})$, the *Euclidean distance* d between any two points x_i and x_j is given by

$$d(x_i, x_j) = \sqrt{(x_{i1} - x_{j1})^2 + (x_{i2} - x_{j2})^2}.$$

This definition can be extended to p-dimensional space where $p \geq 1$. The generalized Euclidean distance in p-dimensions between two points $x_i = (x_{i1}, \ldots, x_{ip})$ and $x_j = (x_{j1}, \ldots, x_{jp})$ is given by

$$d(x_i, x_j) = \left(\sum_{k=1}^{p} (x_{ik} - x_{jk})^2 \right)^{1/2}.$$

For numerical data, other examples of distance functions are the Manhattan distance (also known as taxicab distance), maximum distance, and average distance. See [3, 5, 7] for more examples of distance functions.

Once a measure is selected, a similarity criterion is then stipulated by which each data object is assigned to a cluster. How similar must two objects be to belong to the same cluster? What conditions place two objects in different clusters?

11.2.3 Cluster Validation

The term *cluster validation* is used to design the procedure of evaluating the goodness of clustering algorithm results. It is an important step to avoid falling into the trap of finding patterns in a random data, as well as, in situations where the efficacy of clustering algorithms are compared. This step is arguably the most challenging one in the clustering process. As mentioned before, the resulting clusters of any clustering algorithm are almost entirely dependent on the measure and similarity criterion decided on and, therefore, are subjective. Hence, an objective validation process is required to prove that the number of clusters is optimal and that the clusters themselves are meaningful.

Generally, clustering validation statistics can be categorized into three classes (see [8–10]).

Internal cluster validation uses the internal information of the clustering process to evaluate the goodness of a clustering structure without reference to external information. It can be also used for estimating the number of clusters, which remains an important factor in the analysis.

External cluster validation is used in comparing the results of a cluster analysis to an externally known result, such as externally provided class labels. For instance, in Fisher's Iris data set, each species is labeled relative to the measurements obtained from each unit. Since we know the "true" cluster number in advance, this approach is mainly used for selecting the appropriate clustering algorithm for a specific data set.

Relative cluster validation evaluates the clustering structure by varying different parameter values for the same algorithm (e.g., varying the number of clusters k). This is generally used for determining the optimal number of clusters. Self-organizing maps (SOMs) are a good example where this type of validation is implemented. We will discuss these in Section 11.5.

For an in-depth discussion about test statistics and validity indices that are commonly used, see [5, 11].

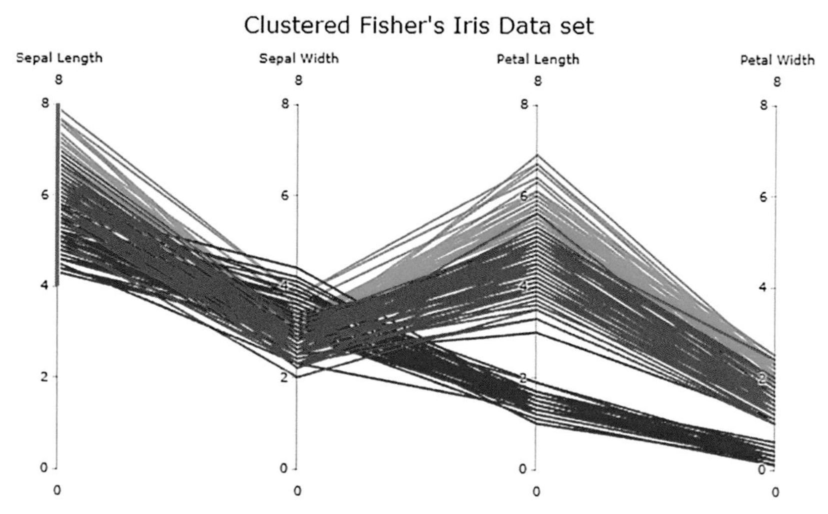

FIG. 11.3 A clustered Fisher's Iris data set plotted in parallel coordinates. The vertical axes, respectively, correspond to sepal length, sepal width, petal length, and petal width. Each *color* is a different cluster and each individual *line* connecting the four axes represents an iris flower. In the interactive version of the plot, the *thick, pink line* on the sepal length axis activates the color-coding of the highlighted data points. Flowers of the same color, that is, data points belonging to the same cluster, sit near each other in this coordinate system. As such, parallel coordinates are useful in data and cluster visualization of relatively low-dimensional data.

11.2.4 Interpretation of Results

Recall that the purpose of clustering a data set is to separate the elements, or data objects, in a way that is reflective of the natural data structures of the data set. By doing so, one can gain an understanding of the data that otherwise would not have been clear. It is important to note that data clustering does not automatically provide solutions to whatever research problem one is trying to solve. In fact, in many cases, the first attempt at clustering a data set results in a clustering that may not be the most effective, so multiple clusterings must be done.

Visualization of the clustered data plays an important role in the interpretation of the clustering. In some cases, even after reducing the dimension of the data set through some feature extraction method, the clustered data set may still be too high-dimensional to visualize easily. Several visualization techniques have been developed to address this very concern. One example of a two-dimensional representation of a high-dimensional data set involves the use of parallel coordinates [12] (example in Fig. 11.3). In a *parallel coordinate system*, each dimension is represented by a vertical axes that is parallel to the other dimensions. For exceptionally large data sets, however, parallel coordinates may lead to hard-to-read data. Furthermore, the scaling of each dimension may have a significant effect on the perceived distance between data points. *Star coordinates*, as developed by Kandogan [13], are another useful two-dimensional coordinate system in which each dimension extends from a common origin and is initially placed at equal angles from each of the other dimensions. The placement of each point, then, is the result of a spiral-shaped path corresponding to the components of that data point. A well-designed, interactive star coordinate generation tool for the Iris data set can be found at https://star-coordinates.com/.

11.3 TYPES OF CLUSTERING

The two main types of clustering are known as hard and fuzzy clustering. Hard clustering requires that each data object in a data set be clustered into one and only one cluster. This type of clustering includes partitional and hierarchical clustering, which further breaks down into *divisive* and *agglomerative* clustering. In fuzzy clustering, each data object may belong to one cluster or more in which its presence in a cluster corresponds to some probability.

11.3.1 Hard Clustering

A clustered data set in both hard and fuzzy clustering may be represented by a $k \times n$ matrix U. Sticking with the notation from Gan [5], the matrix looks as follows:

$$U = \begin{bmatrix} u_{11} & u_{12} & \cdots & u_{1n} \\ u_{21} & u_{22} & \cdots & u_{2n} \\ \vdots & \vdots & \ddots & \vdots \\ u_{k1} & u_{k2} & \cdots & u_{kn} \end{bmatrix}, \tag{11.1}$$

where k is the resulting number of clusters and n is the number of data points in the original data set. Each entry in U is denoted by u_{ji} where $j \in \{1, \ldots, k\}$ and $i \in \{1, \ldots, n\}$. In hard clustering, u_{ji} may only take a value of either 0 or 1. If data point i is in cluster j, then $u_{ji} = 1$. Otherwise, $u_{ji} = 0$. In other words, each row describes a cluster, and the 1-entries tell us which data points it contains. Each column corresponds to a data point, and a 1-entry denotes the cluster to which it pertains.

There are two natural conditions that U must satisfy in hard clustering:

(1) $\sum_{j=1}^{k} u_{ji} = 1$, and
(2) $\sum_{i=1}^{n} u_{ji} > 0$.

Condition (1) states that a data object i may only belong to one cluster. Simply put, only one entry may take the value of 1 within any particular column. To meet condition (2), there must be no empty clusters, that is, each row must contain an entry of value 1. In fuzzy clustering, condition (1) is different, and it will be discussed in Section 11.4.

As mentioned, partitional and hierarchical clustering algorithms fall under hard clustering. The main difference between these two types of clustering is the resulting structure of the clustered data set. Whereas partitional clustering results in a structure consisting of discrete partitions, hierarchical clustering results in a tree-like, nested structure.

11.3.1.1 Partitional Clustering

In this type of clustering, the goal is to find the optimal partitioning of a data set according to some criterion function. Partitional clustering algorithms tend to be very efficient (relative to other clustering algorithms) when applied to big data sets [5]. One of the most common algorithms that is used in partitional clustering is the *k-means algorithm*, which involves some objective function that quantifies the quality of the clustering. The optimal partitioning of the data set is the partitioning that minimizes this objective function. The k-means algorithm is only for numerical data sets and typically uses the Euclidean distance.

Suppose the data set to be clustered has n elements and we wish to cluster them into k clusters. Let C_i denote the ith cluster for $1 \leq i \leq k$. The standard k-means algorithm uses an error function defined as

$$E = \sum_{i=1}^{k} \sum_{\mathbf{x} \in C_i} d(\mathbf{x}, \mu(C_i))$$

where $\mu(C_i)$ is the centroid of the ith cluster (i.e., the centroid is the arithmetic mean of all data points in C_i) and d is the Euclidean distance. By constantly taking the distances between each data point $\mathbf{x}$ and the centroid of cluster to which it belongs, the objective of this algorithm is to find a partitioning of the data set that minimizes E.

There are two parts to any k-means clustering algorithm: initialization and iteration. In the *initialization* phase, the number of desired clusters is determined, that is, a value for k is chosen, and the data set is split into k groups. In the *iteration* phase, the distances between each data point $\mathbf{x}$ and the k centroids are calculated, and the minimum of these distances is chosen. That is, for each $\mathbf{x}$, we find the cluster j in which the minimum is achieved,

$$j = \arg\min_{1 \leq i \leq k} d(\mathbf{x}, \mu(C_i)),$$

and then place $\mathbf{x}$ into the jth cluster. If the minimum is achieved in more than one cluster, then $\mathbf{x}$ may be randomly placed into one of them. After all points are placed in a cluster, the k centroids are recalculated and the iteration step is repeated. The clustering is complete when E is minimized or, equivalently, when there is no more significant change in the centroids or the cluster membership of each data point from one iteration to the next.

Ultimately, the clustering result of a k-means algorithm is entirely dependent on the desired number of clusters k and on the choice of initial k groups. One way to select the optimal k value is to try multiple values. Typically, as k increases, the average distance of all the points in the data set D to their respective centroid decreases. At some value of k, however, this average distance no longer changes significantly. This is the best value of k to use. To initialize the k clusters, k data points can be randomly chosen as centroids, and the rest of the points are placed into one of these clusters. A more computationally efficient way of initializing the clusters requires some a priori knowledge of the data set either through observation or by performing a separate clustering analysis. One could also choose the initial k centroids by choosing points that are relatively equally dispersed throughout the data set. In this way, there is no danger of mistakenly separating points that should be clustered together. An example of a k-means clustering for $k = 3$ is shown in Fig. 11.4. Note that a suboptimal choice in initial centroids may result in a higher misclassification rate.

There are a number of variants of k-means algorithms. They are relatively simple to run and their complexity depends on the number of iterations, number of clusters k, the number of data points in D, and the dimension of these data points. They work relatively well with big or high-dimensional data sets. However, because they only work with numerical data sets, their applications are limited.

Another example of partitional clustering is SOMs. We will discuss SOMs in detail in Section 11.4.

11.3.1.2 Hierarchical Clustering

Hierarchical clustering results in a clustering structure consisting of nested partitions. In an *agglomerative clustering algorithm*, the clustering begins with singleton sets of each point. That is, each data point is its own cluster. At each time step, the most similar cluster pairs are combined according to the chosen similarity measure, and this step is repeated either until all data points are included in a single cluster or until some predetermined criteria are met. If the nesting occurs in the other direction, that is, the clustering begins with one large cluster and breaks down into smaller clusters, it is known as a *divisive clustering algorithm*.

For either type of hierarchical clustering, the data set X is partitioned into Q sets $\{H_1, \ldots, H_Q\}$. That is, if subsets C_i and C_j satisfy that $C_i \in H_m, C_j \in H_l$, and $m > l$, then either $C_i \subset C_j$ or $C_i \cap C_j = \emptyset$ for all $i \neq j, m, l = 1, \ldots, Q$ [3]. In other words, for two subsets of any hierarchical partitions, either one subset contains the other entirely or they are completely disjoint. *Dendograms*, as in Fig. 11.5, provide an easy-to-understand way to represent the nested structure of the clustering.

Hierarchical clustering works especially well with smaller data sets. Agglomerative algorithms, because they not only have to determine the best way to pair the clusters at each iteration but also when the clustering is complete, become more computationally expensive as more data points are considered. In some ways, divisive algorithms are more efficient. At each time step, the algorithm only needs to split each cluster into two in a way that satisfies some criteria, for example, a minimization of the sum of squares error. The clustering is complete when all clusters are split into singleton sets. For further reading please see [14, 15].

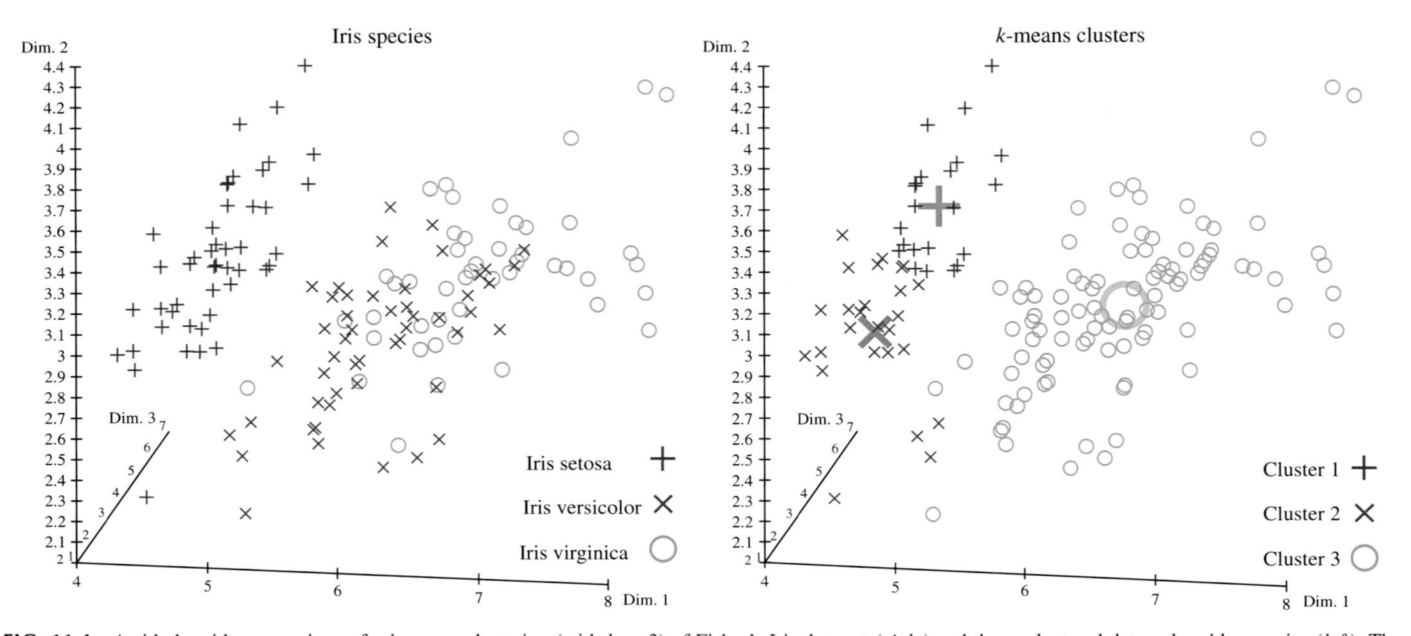

FIG. 11.4 A side-by-side comparison of a k-means clustering (with $k = 3$) of Fisher's Iris data set (*right*) and the unclustered data colored by species (*left*). The *blue* ×, *red* +, and *green* ○ in the right plot are the initial centroids of each cluster. The misclassification of observations is most likely due to the initial three centroids chosen. *(Credit: Chire/PD-self.)*

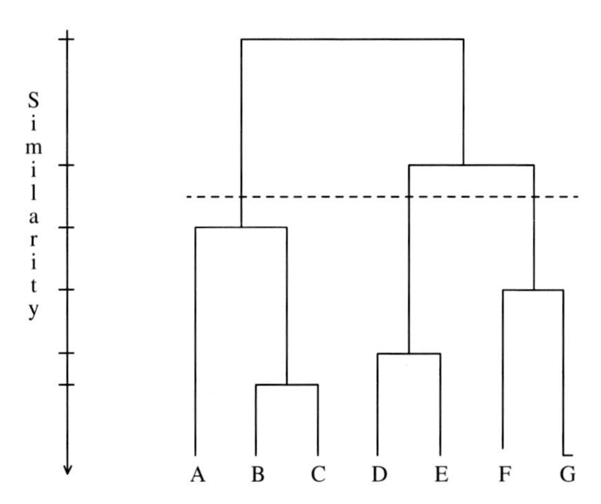

FIG. 11.5 The dendogram representation of the results of a hierarchical agglomerative clustering algorithm. The *dashed line* serves as the stopping point of this particular algorithm and produces three clusters. One clusters contains A, B, and C. Another contains D and E. The final cluster contains F and G. An algorithm with a stricter similarity criterion would result in fewer clusters at the stopping point. Likewise, a looser criterion would result in more clusters. *(Credit: Henriquerocha/PD-self.)*

11.3.2 Exercises

Exercise 11.1. The following exercise illustrating k-means clustering is borrowed from http://mnemstudio.org. Consider the following two-dimensional data set for seven individuals.

Subject	A	B
1	1.0	1.0
2	1.5	2.0
3	3.0	4.0
4	5.0	7.0
5	3.5	5.0
6	4.5	5.0
7	3.5	4.5

1. Using Euclidean distance, identify the two individuals that are the furthest apart. These two individuals initialize the two clusters and their A and B values will serve as cluster centroids.
2. For the first step, choose the next subject and determine to which cluster it belongs by comparing its distance to each centroid. Once the subject is placed into a cluster, recalculate the centroid of that cluster.
3. Place the next subject into the cluster it is closest to and calculate the new centroid. Repeat this for the rest of the data set.

4. Check whether each subject belongs to the correct cluster by comparing its distance from both current centroids. If it is closer to its cluster centroid, the subject is in the correct cluster. If it is closer to the centroid of the opposite cluster, move the subject to the correct cluster. Recalculate the centroids.
5. Repeat Step 4 until no more relocation is necessary. What is the final clustering?

Exercise 11.2. Consider the data set given following.

1. Create a scatter plot by placing V_1 on the y-axis and V_2 on the x-axis.
2. Create a *proximity matrix* by calculating the Euclidean distance between each pair of points in the data set.

Case	V_1	V_2
1	1	1
2	2	1
3	4	5
4	7	7
5	5	7

3. Reduce the size of the data set by replacing the closest pair of points with their midpoint.
4. Create the new similarity matrix for this new data set.
5. Repeat this process until you exhaust all points.
6. Identify each cluster and list the points that belong to them.
7. Build a *dendogram* that represents the clusters and the distances of observations.

11.4 FUZZY CLUSTERING

Fuzzy clustering makes use of fuzzy sets (defined by Zadeh [16]), which are sets whose elements have degrees of membership. More formally, supposed X is a set of data points. A *fuzzy set* A is formed if there exists a function $f_A\colon X \to [0,1]$ such that each element $a \in A$ is of the form $f_A(x) = a$, for some $x \in X$. That is, each data point in X is assigned a value between 0 and 1 which describes its degree of membership or the probability of its placement in the set A. Fuzzy clustering, then, results in data objects belonging to one or more clusters and their membership in a particular cluster corresponding to some probability.

The results of a fuzzy clustering can be represented by the same $k \times n$ matrix U defined in Eq. (11.1). Although the conditions for fuzzy clustering are similar to the conditions for hard clustering, we state them explicitly here again. Recall that for hard clustering, each entry of U is of the form $u_{ji} \in \{0, 1\}$ where $j \in \{1, \ldots, k\}$ and $i \in \{1, \ldots, n\}$ index the cluster and data point, respectively.

Each row corresponds to clusters each column to data points. The following conditions must still hold [5]:

(1) $\sum_{j=1}^{k} u_{ji} = 1$, and
(2) $\sum_{i=1}^{n} u_{ji} > 0$.

For fuzzy clustering we modify the requirement that u_{ji} is either 0 or 1 to one that allows $u_{ji} \in [0, 1]$. Next, condition (1) now requires that for each data object, the sum of its degrees of membership across all clusters be equal to 1. Condition (2), as before, requires there to be no empty clusters.

One example of a fuzzy clustering algorithm is the *fuzzy k-means algorithm* (sometimes referred to as the *c-means algorithm* in the literature). Similar to its hard clustering counterpart, the goal of a fuzzy k-means algorithm is to minimize some objective function. Suppose we have a data set $D = \{\mathbf{x}_1, \ldots, \mathbf{x}_n\}$ and let $q \in [0, 1]$. Here, q is known as the *fuzzifier*, which determines the fuzziness of the resulting clusters. The larger the q value, the smaller the membership values u_{ji} and, thus, the fuzzier the clustering. The objective function is defined as

$$E_q = \sum_{i=1}^{n} \sum_{j=1}^{k} u_{ji}^{q} d^2(\mathbf{x}_i, V_j),$$

where d is an inner product metric function and the V_js are the centroids of the initial clustering of the data. This initial clustering of D is, of course, allowed to overlap as long as all points are included in at least one cluster. At any iteration, the degree of membership of the data point $\mathbf{x}_i$ in the cluster j is

$$u_{ji} = \left(\frac{d^2(\mathbf{x}_i, V_j)}{\sum_{l=1}^{k} d^2(\mathbf{x}_i, V_l)} \right)^{\frac{1}{1-q}}.$$

Each centroid V_j is recalculated in the following way:

$$V_j = \frac{\sum_{i=1}^{n} u_{ji}^{q} \mathbf{x}_i}{\sum_{i=1}^{n} u_{ji}^{q}}.$$

After each iteration, the membership matrices of consecutive times steps are compared. If

$$\max_{ji} |u_{ji}^{old} - u_{ji}^{new}| < \varepsilon,$$

where ε 0 is some predefined criterion for stability, then the fuzzy k-means algorithm is complete. Otherwise, the membership matrix, using new centroids, is calculated.

Although fuzzy k-means algorithms tend to be more time-consuming and complex than their hard clustering counterparts, fuzzy clustering has important applications due to its flexibility in grouping data that has its basis in uncertain parameters, such as consumer behavior and market segmentation. See [17] for further reading on fuzzy clustering in pattern recognition.

11.5 SELF-ORGANIZING MAPS

SOMs, also known as *Kohonen maps*, form a class of clustering algorithms that make use of artificial neural networks to generate a low-dimensional representation of a high-dimensional data set and to reflect the structure of the data set in a visual way [18]. As with all clustering algorithms, SOMs are unsupervised classification algorithms. This technique requires the input of a numerical data set and, through an iterative algorithm, outputs a one- or two-dimensional map representing the data. This output map eases the visualization of the underlying structures of the data set as it preserves the similarity relationships between the original data objects and, as such, results in a clustering of the data. As a clustering tool, SOMs are not as computationally expensive as other algorithms and result in fast clustering, which is why they are often used in combination with other well-known clustering algorithms [3].

SOMs involve *competitive learning algorithms*. An algorithm is considered competitive learning when, during each iteration, the elements of the artificial neural network, in a sense, compete against each other for the chance to respond to the input. In any competitive learning system, there are input nodes and output nodes. The input nodes are the input data objects and the output nodes are a set of units which are each assigned a weight vector either randomly or using a priori knowledge of the data set. More clearly, let the input nodes be denoted by $i \in \{1, \ldots, n\}$ and the output nodes by $j \in \{1, \ldots, k\}$, where n is the dimension of the data set and k corresponds to the number of clusters. Then each output node j is weighted by a vector $\mathbf{w_j} \in \mathbb{R}^n$ whose components w_{ij} are each connected to the ith component of the input vector $\mathbf{x} \in \mathbb{R}^n$. That is, for an input vector $\mathbf{x}$, each of its components x_i is connected to every output node j by some weighted connection denoted by w_{ij}. An iterative algorithm then compares the input vector $\mathbf{x}$ to every weight vector $\mathbf{w_j}$ and seeks out the index J for which the similarity between $\mathbf{x}$ and $\mathbf{w_j}$ is maximized or, equivalently, for which the distance between the two is minimized. For node J and for some neighborhood around J, an *activation* (or *update*) takes place when the weights of each node are updated in a way that makes them more similar to $\mathbf{x}$. For the next time step, a new input vector is presented and the nodes adjust accordingly. As the algorithm progresses, the size of the updating neighborhood around J decreases until the neighborhood contains only the node J. The algorithm is complete when the positions of the nodes satisfy some predetermined condition of stability.

In SOMs, the output nodes are called *neurons* and the weighted connections between the components of the input vector and the neurons are called *synapses*. That is, for an input vector $\mathbf{x}$, each of its components x_i is connected to every neuron j by some weighted synapse denoted by w_{ij}. A visual example of an

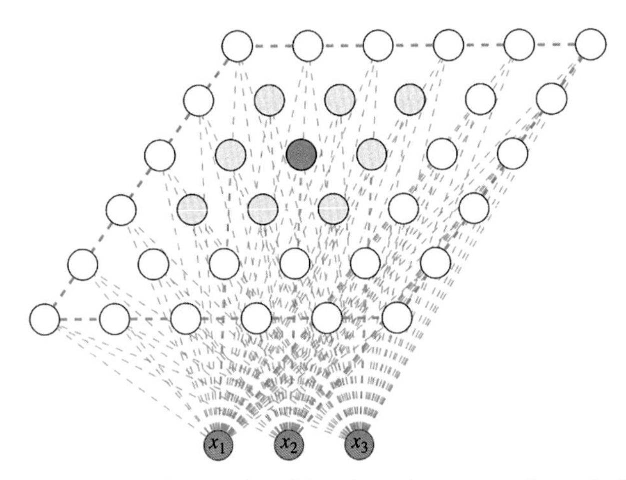

FIG. 11.6 The SOM shown here consists of three data points $\mathbf{x}_1, \mathbf{x}_2$, and $\mathbf{x}_3$ and a 6×6-neuron grid. The *dotted lines* represent the weighted connections between the data point and the neurons. After the presentation of one of the data points to the grid, the winning neuron (*dark gray*) and its nearest neighbors (*light gray*) are updated. *(Credit: Martin Thoma/CC0 1.0.)*

SOM setup is shown in Fig. 11.6. For each input vector $\mathbf{x}$, the goal is to identify the "winning" weight vector, that is, the weight vector that is most similar to the input. If d is a distance measure, then the winning vector is $\mathbf{w}_J$ whose index is defined as

$$J := \arg\min_j \{d(\mathbf{x}, \mathbf{w}_j)\}.$$

Again, if the minimum is achieved for multiple weight vectors W_j, then one of them may be chosen randomly.

Once a winning vector is determined, an activation takes place for all weight vectors within some predetermined neighborhood of J. Let the neighborhood around J be denoted by $N(J)$. The activation function of a weight vector at iteration $t + 1$ is

$$\mathbf{w}_j(t+1) = \begin{cases} \mathbf{w}_j(t) + \eta(t)(\mathbf{x}(t) - \mathbf{w}_j(t)) & \text{for } j \in N(J) \\ \mathbf{w}_j(t) & \text{for } j \notin N(J), \end{cases}$$

where $\eta(t)$ is called the learning rate. It is a monotonically decreasing function selected to ensure that a more fine-tuned learning is taking place toward the end of the algorithm. Commonly used learning functions are linear, inversely related to the total number of iterations performed by the algorithm, exponential, or power series.

After the update of the appropriate weights is complete, the next iteration of the algorithm begins with a new input vector $\mathbf{x}$ presented to the updated neurons. As the algorithm proceeds, the size of the neighborhood around the winning neuron decreases until the only neuron affected by the update is the winning neuron itself. Additionally, the learning rate $\eta(t)$ also decreases for each time step, which means the effect the update has on the winning neuron and its neighbors lessens over time. The algorithm is considered complete when the change in the position of the neurons occurring between time steps is below some predetermined positive number.

There are several decisions that must be made before performing an SOM. To begin with, the number of desired clusters must be determined. This can be done using various methods. The weight vectors of each neuron must then be initialized either by randomly assigning each entry a value between 0 and 1 or by assigning values using prior knowledge of the structure of the data set. The latter, if done correctly, can increase the efficiency of the SOM. For example, if it is obvious that some data objects are extremely different from each other, it may be more beneficial to initialize the weight vectors in a way that reflects this structure so as not to waste computational energy.

Next, we determine the best type of neighborhood or neighborhood function to use. Common neighborhoods are circular, square, or hexagonal. Defining a *neighborhood function* allows us to specify how much we want each neuron within the neighborhood $N(J)$ to learn. If we define the neighborhood function by

$$h_{Jj} = \begin{cases} \eta(t) & \text{for} \quad j \in N(J) \\ 0 & \text{for} \quad j \notin N(J) \end{cases},$$

then the updating function can simply be written as

$$\mathbf{w_j}(t+1) = \mathbf{w_j}(t) + h_{Jj}(\mathbf{x}(t) - \mathbf{w_j}(t)).$$

This neighborhood function specifies that each neuron within $N(J)$ is affected in the same exact way by the learning function $\eta(t)$. If we want the neurons closest to J to be affected more than those near the border of $N(J)$, we can multiply the learning function by a factor that utilizes the distance from J, as seen in the commonly used Gaussian neighborhood function

$$h_{Jj} = \eta(t) \exp\left(\frac{-||r_J - r_j||^2}{2\sigma^2(t)}\right),$$

where r_J and r_j are the positions of the Jth and jth neuron, respectively, and $\sigma(t)$ is some decreasing kernel width function which provides the radius of the neighborhood at time step t. For more details about kernel width functions, see [19].

Finally, the last item to be set is the condition for stability of the SOM, which determines the stopping point of the algorithm. This condition simply requires that the change in position of the neurons from one time step to the next remains less than some small, positive number.

Once an SOM is complete, the resulting map allows for an easy-to-understand visual of the structure of the data set. Furthermore, the map also tells us that each input vector $\mathbf{x}$ belongs to the Jth cluster.

11.6 SOM APPLICATIONS TO BIOLOGICAL DATA

We provide here two examples of clustering applications to biological data and discuss their respective interpretations for SOM, as described in the previous section. The first example considers the effect that type of tide may have on the composition of water in bays and creeks on the coast of South Carolina. This example is based upon an analysis performed in [20]. The data and a description of the data can also be found at [20]. The second example clusters grasshopper size data first by sex and then by age [21]. For a more an in-depth analysis of this grasshopper data, see [22].

All the Kohonen maps shown in these examples were created using Databionic Emergent Self-Organizing Maps (ESOMs) Tools, which may be found at http://databionic-esom.sourceforge.net. All settings were kept at default, so the clustering was done using Euclidean distance, the Gaussian neighborhood function with a radius that decreases linearly, and a linear learning function.

11.6.1 Example 1: Water Composition Data

The geographical map in Fig. 11.7 represents a region in South Carolina from which 3486 water samples were collected to study the relationship between composition of water and tidal stage. The data consist of chlorine, dissolved oxygen, pH, temperature, conductivity, and dissolved organic matter

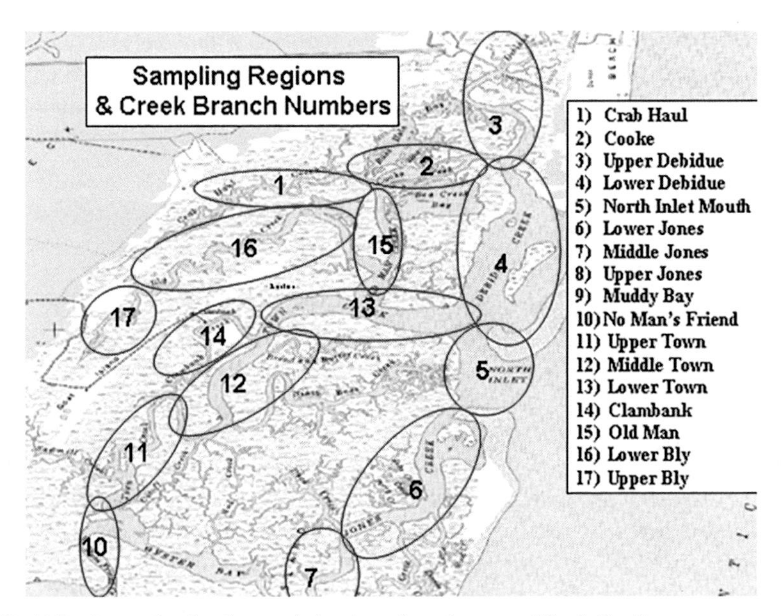

FIG. 11.7 A map showing the sampled regions along the coast of South Carolina.

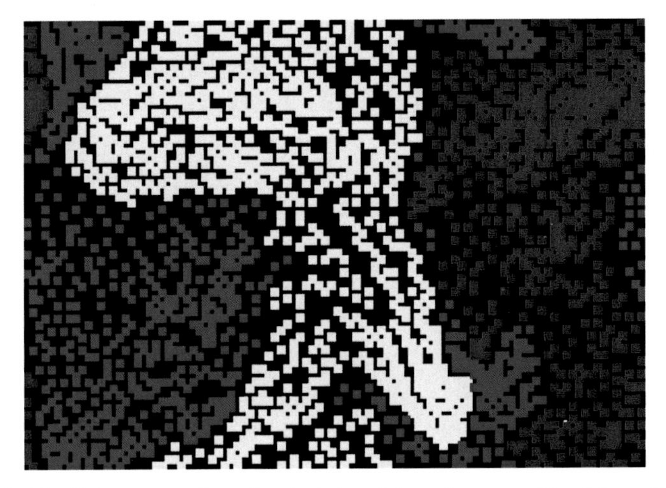

FIG. 11.8 The Kohonen map shows the clustering of the water composition data vectors, where the observations corresponding to low tide are colored *yellow* and those corresponding to medium tide are *red and blue*. While the low tide observations remain grouped together in the mapping, the medium tide observations do not. The separation of the medium tide points into two clusters implies that there must be some other discriminating factor at play. Further analysis shows that the points colored *blue* are water samples taken from the northernmost sampled regions while the points colored *red* were taken from the southernmost regions.

measurements. The objective of the cluster analysis is to examine whether the type of tide (low or medium) is a discriminating factor in water composition. The Kohonen map presented in Fig. 11.8 shows a clustering based on type of tide and location.

In general, during a low tide, there tends to be more fresh water in the sampled regions than salt water. Similarly, during a higher tide, more salt water is present than fresh water. The yellow data points in Fig. 11.8 represent the composition of all water samples after a low tide. The fact that these points remain grouped together in the Kohonen map implies a relatively uniform composition of the water samples in these regions. The blue and red data points are the water composition data vectors after a medium tide. These points did not remain grouped together which implies there is a difference in water composition. Upon closer inspection of the data, we find that the blue points correspond to the samples taken from the northern regions and the red points correspond to the samples taken from the southern regions. Fig. 11.9 shows which regions were colored blue and red. Because the clustering according to tide differentiates the northern and southern sampled regions, a medium tide must affect one of these regions more that the other.

Adding creeks 6, 7, 12, 14, 16, or 17 to either location group causes interaction between the two. During all tides, the regions outlined in blue in Fig. 11.9 are closer to fresh water while the regions outlined in red are closer to

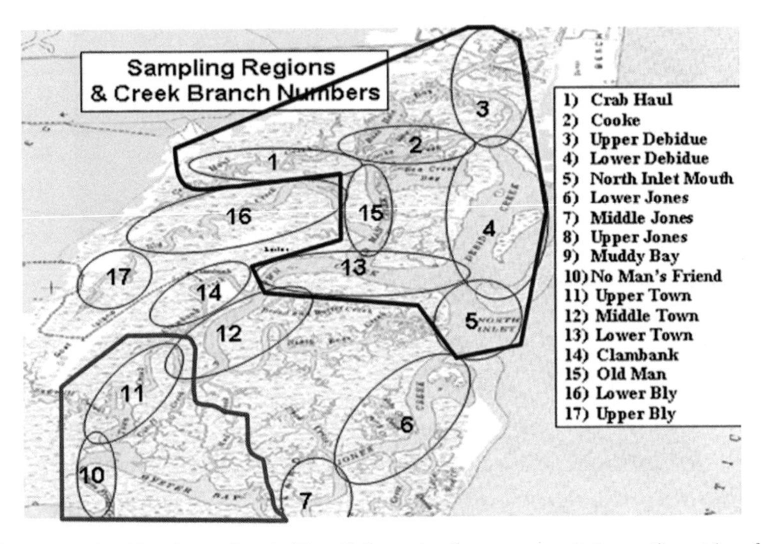

FIG. 11.9 The *blue* data points in Fig. 11.8 are the data vectors of observations taken from the northernmost sampled regions, shown here outlined in *blue*. The *red* data points in the map correspond to the southern regions outlined in *red*.

salt water. During the low tide, when the salt water recedes, the creeks that are in the blue region show dominant fresh water characteristics, much more so than the creeks in the red region. That is because the fresh water creeks feed into the blue region directly while the red region is only impacted after the fresh water is mixed with salt water from the creeks in between. As seen in Fig. 11.10, the type of tide that the creeks are exposed to is a discriminating factor in water composition.

In summary, the type of tide and the source location of observations serve as distinguishing features in a cluster analysis of the water sample composition data, as made clear by the general lack of misclassification. That is, a single observation vector consisting of the same six measurements also provides information about the source location of the sample and the type of tide that was recently experienced in the area.

11.6.2 Example 2: Grasshopper Size Data

The grasshopper data used in this section consists of the following length measurements for each individual: head, thorax, abdomen, front wing, and the hind wing. Each of these anatomical regions are labeled in Fig. 11.11. The Kohonen maps illustrating the clustering of this data are depicted in Figs. 11.12 and 11.13.

The two panels in Fig. 11.12 show a clustering of the grasshopper data where the data is colored according to sex. The blue and purple data points

FIG. 11.10 The distinct clusters shown here indicate that the northern creeks show characteristics consistent with fresh water attributes that are substantially different than the southern creeks. The *white* points in between are the nonclustered points whose attributes depend on the type of tide that they are exposed to.

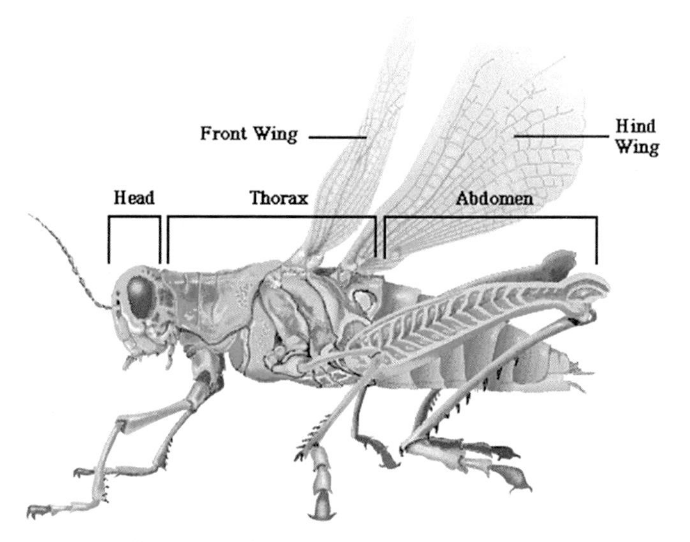

FIG. 11.11 The anatomical regions of the grasshopper considered in the data set.

correspond to the male and female grasshoppers, respectively. In general, female grasshoppers are bigger than male grasshoppers, so we expect size to be distinguishing feature. Indeed, the clusters in Fig. 11.12 are fairly distinct. Any overlap in data is due either to relatively small females or relatively large males.

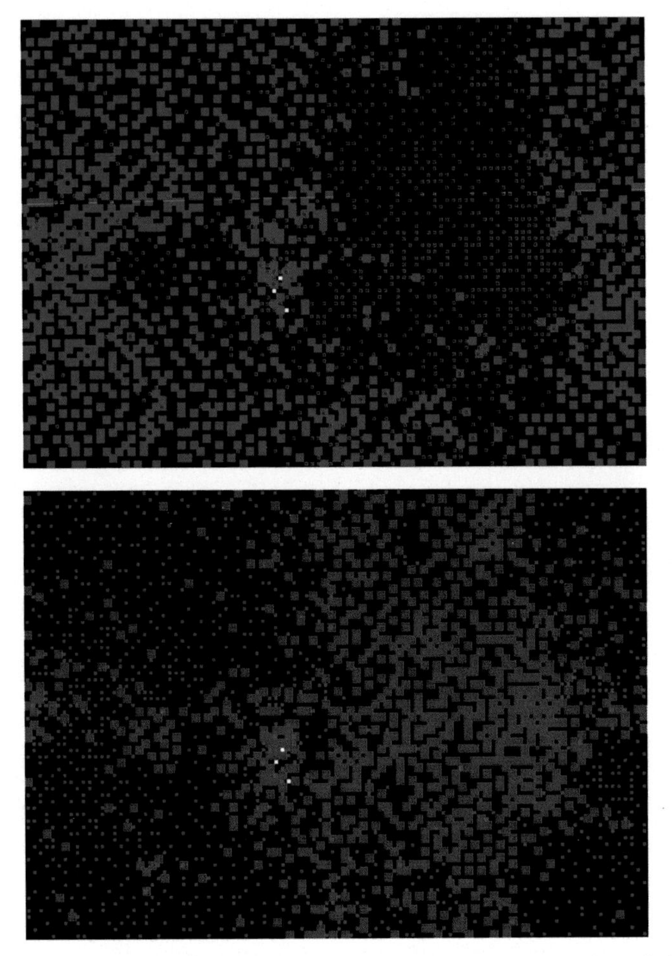

FIG. 11.12 The observation vectors corresponding to male grasshoppers are shown in *blue*. The female grasshoppers are shown in *purple*. Although both panels represent the same map, the points are, respectively, enlarged to allow for a better visualization of the clustering in each panel. *White* points are empty cells, that is, missing observations.

Fig. 11.13 shows a clustering of the same data, this time colored by life stage. The life stages being considered are nymphs, or young grasshoppers (shown in blue) and adult grasshoppers (red). As expected, the size of a grasshopper is a distinguishing factor in the classification of life stage. The Kohonen map results in two very clear clusters. Overlapping points are either large nymphs or small adults.

From Figs. 11.12 and 11.13, we are able to conclude that size is a distinguishing feature in the cluster analysis of grasshopper data. In other words,

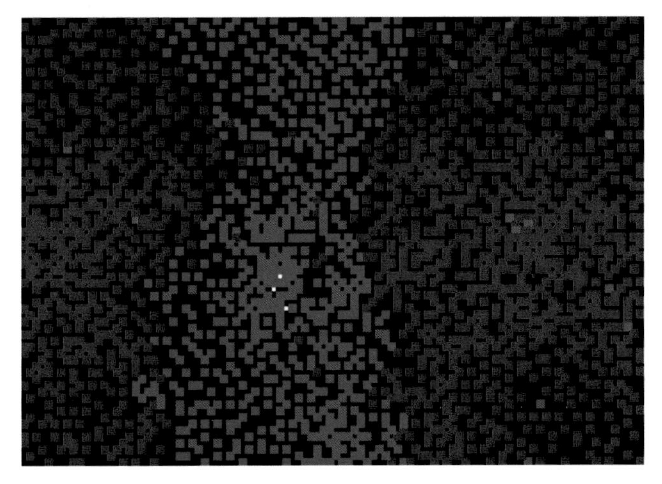

FIG. 11.13 This map shows the grasshopper data clustered by life stage: young grasshoppers, or nymphs (shown in *blue*) and adult grasshoppers (*red*). *White points*, as before, correspond to missing data.

the size measurements of a grasshopper provide us with enough information to predict with little error the sex and life stage of the grasshopper.

11.6.3 Group Project: Iris Data Set

Fisher's Iris data set can be found at https://en.wikipedia.org/wiki/Iris_flower_data_set#Data_set. Use the Databionic ESOMs Tools to cluster the data according to species. Discuss the results.

REFERENCES

[1] R.A. Fisher, The use of multiple measurements in taxonomic problems, Ann. Eugenics 7 (7) (1936) 179–188.

[2] Charu C. Aggarwal, Chandan K. Reddy, eds. Data Clustering: Algorithms and Applications, CRC Press, Boca Raton, FL, 2013.

[3] R. Xu, D.C. Wunsch, Clustering, IEEE Press Series on Computational Intelligence, John Wiley & Sons Inc, Hoboken, NJ, 2009.

[4] A.K. Jain, M.N. Murty, P.J. Flynn, Data clustering: a review, ACM Comput. Surv. 31 (3) (1999) 264–323.

[5] G. Gan, C. Ma, J. Wu, Data Clustering: Theory, Algorithms, and Applications, SIAM, Philadelphia, PA, 2007.

[6] R.A. Johnson, D.W. Wichern, Applied Multivariate Statistical Analysis, Prentice Hall, Inc., Upper Saddle River, NJ, 1998.

[7] B. Everitt, Cluster Analysis, fifth ed., Wiley Series in Probability and Statistics, John Wiley & Sons Ltd, Chichester, 2011.

[8] M. Charrad, N. Ghazzali, V. Boitequ, A. Niknafs, NbClust: an R package for determining the relevant number of clusters in a data set, J. Stat. Softw. 61 (6) (2014) 1–36.

[9] G. Brock, V. Pihur, S. Datta, S. Datta, clValid: an R package for cluster validation, J. Stat. Softw. 25 (4) (2008) 1–22.

[10] S. Theodoridis, K. Koutroumbas, Pattern Recognition, fourth ed., Elsevier Inc., Amsterdam, 2009.

[11] A.K. Jain, R.C. Dubes, Algorithms for Clustering Data, Prentice Hall, Englewood Cliffs, NJ, 1988.

[12] A. Inselberg, Multidimensional detective, in: Proceedings 1997 IEEE Symposium on Information Visualization, 1997, pp. 100–107.

[13] E. Kandogan, Star coordinates: A multi-dimensional visualization technique with uniform treatment of dimensions. In Proceedings of the IEEE Information Visualization Symposium, vol. 650, p. 22, Citeseer, 2000.

[14] L. Kaufman, P.J. Rousseeuw, Finding Groups in Data: An Introduction to Cluster Analysis, John Wiley and Sons, Inc., New York, NY, 1990.

[15] T.J. Hastie, R.J. Tibshirani, J.H. Friedman, The Elements of Statistical Learning: Data Mining, Inference, and Prediction, Springer, New York, NY, 2009.

[16] L.A. Zadeh, Fuzzy sets, Inf. Control. 8 (3) (1965) 338–353.

[17] J.C. Bezdek, Pattern Recognition With Fuzzy Objective Function Algorithms, Plenum Press, New York, NY, 1987.

[18] T. Kohonen, Self-Organizing Maps, Springer Series in Information Sciences, vol. 30, third ed., Springer, Berlin, 2001.

[19] J. Lampinen, T. Kostiainen, Self-Organizing Maps in data analysis-notes on overfitting and overinterpretation. In ESANN, 2000, pp. 239–244.

[20] O. Akman, K. Carr, E. Koepfler, Kohonen network modeling of tidal stage impact on spatial variability of estuarine water quality in North inlet-South Carolina, Far East J. Ocean Res. 1 (2-3) (2007) 63–80.

[21] M.P. Walker, C.J. Lewis, D.W. Whitman, Effects of males on the fecundity and fertility of female *Romalea microptera* grasshoppers, J. Orthop. Res. 8 (1999) 277.

[22] O. Akman, D. Whitman, Analysis of body size and fecundity in a grasshopper, J. Orthop. Res. 17 (2) (2008) 249–257.

Chapter 12

Toward Revealing Protein Function: Identifying Biologically Relevant Clusters With Graph Spectral Methods

Robin Davies*, Urmi Ghosh-Dastidar†, Jeff Knisley‡ and Widodo Samyono§

**Biomedical Sciences, Jefferson College of Health Sciences, Roanoke, VA, United States, †Department of Mathematics, New York City College of Technology, Brooklyn, NY, United States, ‡Department of Mathematics and Statistics, East Tennessee State University, Johnson City, TN, United States, §Department of Mathematics, Jarvis Christian College, Charles A. Meyer Science and Mathematics Center, Hawkins, TX, United States*

12.1 INTRODUCTION TO PROTEINS

12.1.1 Protein Structures

Proteins are the most diverse of the biological macromolecules that make up living cells. Many cellular proteins are enzymes, the catalysts of cellular reactions. Others are membrane transport proteins, which allow vital food molecules such as glucose to enter the cell. Still others are motor proteins which move materials around the cell, or cytoskeletal proteins which give cells their specific shapes.

In all of these examples and more, each protein has a particular function, which is determined by its three-dimensional (3D) shape. The shape of each protein is the result of a complex series of molecular interactions acting upon the protein chain. Before we consider these interactions, however, we will introduce protein structure.

Proteins are composed of amino acid monomers bound together into a polymer, the polypeptide chain. The covalent bonds, which bind the monomers together, are called peptide bonds. There are 20 different amino acids, each with an amino group, a carboxyl group, a hydrogen atom, and an R group,

Algebraic and Combinatorial Computational Biology. https://doi.org/10.1016/B978-0-12-814066-6.00012-X

all bound to the same central carbon called the alpha carbon. As there are 20 different amino acids, there are 20 different R groups. The names, three-letter abbreviations, and one-letter abbreviations of the amino acids are given in Table 12.1.

Fig. 12.1 shows a cartoon of a polypeptide chain on the left (A) and an unbound amino acid on the right (B).

Each individual protein has a specific sequence of amino acid monomers that makes up its polypeptide chain. The order in which the amino acids appear in the chain is called the primary structure of the protein, seen schematically in part A of Fig. 12.1. Depending upon the amino acid sequence of a protein (the primary structure), the protein chain may form helical structures (α helices) or sheets (β sheets or pleated sheets). (See part A of Fig. 12.2.) These helices and sheets constitute elements of the final 3D shape and are called the secondary structure of the protein.

Also depending upon the amino acid sequence is the tertiary structure, which is the 3D configuration of an individual protein chain and consists of the way in which the elements of the secondary structure and the rest of the protein chain are spatially related. (See part B of Fig. 12.2.) Finally, in those proteins which consist of more than one chain of amino acids, there is the quaternary structure, which is the manner in which the individual folded chains of the complex protein interact with each other in 3D space. (See part C of Fig. 12.2.)

The primary structure of a protein is determined by the genetic information in the DNA. The section of DNA, which contains the information for producing a protein, is called a gene. When a cell needs to make a particular protein, it makes a copy of the gene sequence in a related molecule called RNA. The RNA is used by the protein-synthesizing machinery, the ribosome, to direct the formation of the protein, linking the correct amino acids together in the correct sequence or primary structure. Once the amino acids are all linked together according to the sequence, one might suppose that the formation of the correct 3D shape of the protein was a foregone conclusion. Alas, it is not that simple.

As the growing chain of amino acids leaves the ribosome, it is in danger of folding inappropriately. Hydrophobic amino acids will bind to each other in the aqueous environment of the cell, and, if allowed to bind in an uncontrolled manner, may form inactive protein aggregates rather than correctly folded proteins. To prevent this wasteful or possibly dangerous situation, proteins called chaperones bind to the nascent protein chain and keep it from folding incorrectly. The chaperones allow the correct folding of the protein to occur. Other chaperones can bind unfolded or misfolded proteins and assist them in folding correctly. Any proteins which fail to achieve their correctly folded state are marked for disposal and are broken down into their constituent amino acids for recycling into new proteins.

TABLE 12.1 The 20 Standard Amino Acids With Their Three-Letter and One-Letter Designations

Amino acid	Alanine	Arginine	Asparagine	Aspartic acid	Cysteine	Glutamic acid	Glutamine
Three-letter	Ala	Arg	Asn	Asp	Cys	Glu	Gln
One-letter	A	R	N	D	C	E	Q
Amino acid	Glycine	Histidine	Isoleucine	Leucine	Lysine	Methionine	Phenylalanine
Three-letter	Gly	His	Ile	Leu	Lys	Met	Phe
One-letter	G	H	I	L	K	M	F
Amino acid	Proline	Serine	Threonine	Tryptophan	Tyrosine	Valine	
Three-letter	Pro	Ser	Thr	Trp	Tyr	Val	
One-letter	P	S	T	W	Y	V	

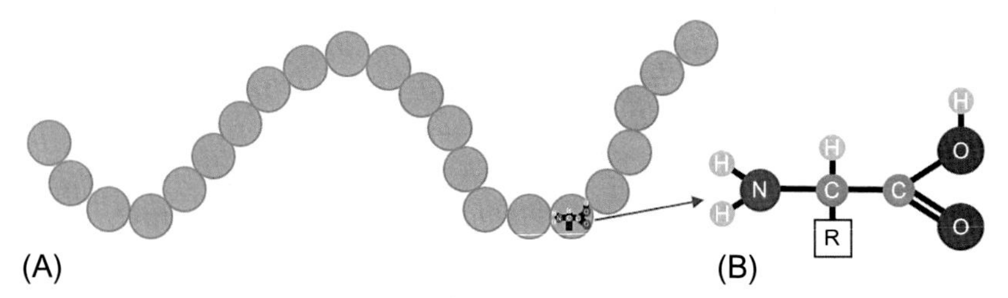

FIG. 12.1 A polypeptide chain is a sequence of amino acids.

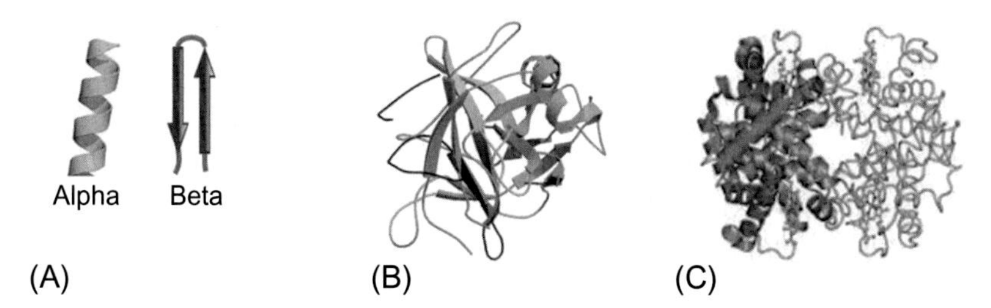

FIG. 12.2 The hierarchy of protein structure. Part A represents the secondary structure. Part B is the tertiary structure. Part C is the quaternary structure, the relationship among subunits of a multipeptide protein. *(Tertiary images from H.M. Berman, J. Westbrook, Z. Feng, G. Gilliland, T.N. Bhat, H. Weissig, L.N. Shindyalov, P.E. Bourne, Protein Data Bank, Nucleic Acids Res. 28 (2000) 235–242, https://doi.org/doi:10.1093/nar/28.1.235. rendered using Jmol, an open-source Java viewer for chemical structures in 3D (http://www.jmol.org/).)*

12.1.2 Experimental Determination of Protein Structure

The traditional method for determining protein 3D structure is X-ray crystallography. In order to use this technique, one must first isolate and purify the protein and then allow it to form a crystal. The crystal will then be placed in an X-ray beam and the regular array of the crystal will result in the X-ray beam being scattered in a specific pattern. The pattern, when analyzed, will reveal the position of the atoms in the protein and result in a 3D structure being derived from the scattering pattern. According to PDB-101, the informational webpages of the Protein Data Bank (PDB), the majority of the protein structures in the PDB were determined using X-ray crystallography [1].

X-ray crystallography works well if the protein has an inflexible structure which forms well-ordered crystals. However, many proteins do not fit these criteria. For these, other tools in the protein structure determination arsenal such as nuclear magnetic resonance (NMR) spectroscopy may be used. In preparation for NMR, proteins must be purified, but they remain in an aqueous solution. This means that flexible proteins (and any other proteins that do not readily form crystals) can be studied in conditions which more nearly reproduce the aqueous environment of the cell. Thus far, NMR has proven to be useful for smaller proteins; larger proteins produce complex datasets which current technology is unable to resolve. See "Appendix" for further explanations of protein structure determinations using linear algebra techniques.

12.1.3 Isofunctional Families

In summation, DNA encodes RNA which is subsequently translated into proteins, where a protein is a sequence of amino acids that form a polypeptide chain with individual side chains (R groups) projecting from the main chain (also known as the peptide backbone). Because each amino acid in the sequence is 1 of the 20 standard amino acids in Table 12.1, we often write the primary sequence of a protein simply as a list (more precisely, string) of the one-letter designations of the corresponding amino acids. Table 12.2 lists the primary sequence of one of the proteins in the serine protease family. Fig. 12.3 shows the secondary structures associated with the primary sequence in Table 12.2.

The tertiary structure of most proteins determined to date have been obtained experimentally using X-ray crystallography. A result from X-ray crystallography is a prediction of the 3D coordinates of the atoms (other than hydrogen) in the protein crystal. Because proteins are dynamic, this crystal structure of a protein is at best a snapshot of it at a single moment in time. An important representation of this crystal structure is a 3D image with cartoon representations of the secondary structures. An example of this for the protein from Table 12.2 is part B in Fig. 12.2.

TABLE 12.2 Primary Sequence for the Serine Protease RCSB PDB 1AB9 Chain B

IVNGEEAVPGSWPWQVSLQDKTGFHFCGGSLINENWVVTAAHCGVTTS
DVVVAGEFDQGSSSEKIQKLKIAKVFKNSKYNSLTINNDITLLKLSTA
ASFSQTVSAVCLPSASDDFAAGTTCVTTGWGLTRY

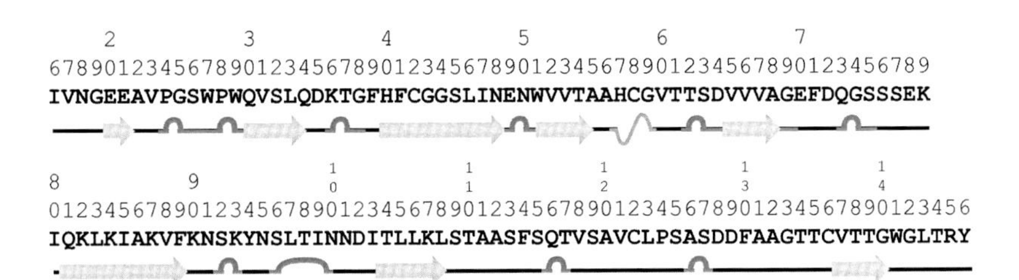

FIG. 12.3 Secondary sequence for RCSB PDB 1AB9. Indexing for the section being illustrated begins at 16. There is one alpha helix, which is from index 55 to index 61. Turns, where the protein bends significantly, are indicated by *arches*. Beta strands are individual strands in beta sheets and are denoted by *arrows*.

The tertiary structure of a protein is important to its function, including how it behaves dynamically and how it interacts with other biological macromolecules. Such functional properties include binding sites, which are regions in the tertiary structure where a protein binds to another molecule, as well as enzyme active sites and other similar local functions. Often, a separate goal in the study of proteins is to relate functional properties of a protein to its primary structure, which is known as annotation. Such annotations can be used to study a protein's role in a number of biological and biomedical contexts.

In summary, the tertiary structure of a protein is of central importance, and as a consequence, the problem of predicting the tertiary structure of a protein from its primary sequence is among one of the most important problems in biology. Unfortunately, the protein folding problem, as this problem is called, is also one of the most difficult problems scientists have ever faced. Moreover, experimental investigation of a protein is very expensive and may reveal little about the protein's annotation. In fact, most proteins can only be annotated computationally and that is also at great expense, if even possible, given the difficulty of the protein folding problem [2].

As a result, alternative methods for studying the tertiary structure of a protein are highly valued. One such approach is to study a given protein as a member of a protein family, which is a group of proteins that share a common evolutionary origin [3]. Each member of a protein family plays approximately the same role across several different species of organisms. Thus, proteins in a given family are assumed to be closely related, and yet different proteins in a given family often have different functions and annotations. In such cases, it may be possible to group or cluster the family into subsets with similar functions, in which case a subset is called an *isofunctional subfamily* [4]. Because proteins in a family tend to have similar (though not identical) primary structures, protein annotation may be greatly assisted by the fact that proteins in an isofunctional subfamily share similar functions that are not common to the whole family.

To summarize, the proteins in a protein family have similar primary sequences because they are coded by the same or similar genes across species, and as we will see next, this similarity serves as an important concept in the study of protein function.

Exercise 12.1. What are the normal functions of the serine proteases? In what disease processes are serine proteases involved? Utilizing internet resources, investigate the roles of serine proteases in health and disease.

12.1.4 Sequence Motifs and Logos

Two proteins in the same family are said to be *homologous* to each other, where homology refers to the existence of shared ancestry between two structures. For example, the forelimbs of vertebrates are homologous, even though they may take the form of arms, legs, wings, or fins across different species of vertebrates. Likewise, while homologous proteins have similar primary, secondary, and tertiary structures, they may have significantly different functionalities due to local variations in structure.

A sequence motif is a tool for representing the individual variation within primary structures for a protein family. Because homologous proteins need not have primary sequences of the same length—one protein may have additions or deletions with respect to another in the family—the proteins in a family must first be aligned, where a sequence alignment is a prediction of where each protein in a family has insertions or deletions with respect to the remainder of the family.

An alignment produces a well-defined concept of position in the primary sequence for the family, after which a sequence motif is the frequency of each of the 20 residues for each position.

There are many ways to represent a sequence motif. One approach is a position frequency matrix, which is a table whose rows correspond to positions in the motif, whose columns correspond to the 20 amino acids, and whose coefficients are the individual counts at each position. Often, a position frequency matrix is normalized by dividing each coefficient by the number of proteins in the protein family, resulting in a table of relative frequencies of the amino acids at each position. A position matrix may also be scored using some scoring scheme to create a position weight matrix.

A position frequency matrix can subsequently be visualized using a *sequence logo*, also known as a consensus logo, in which the positions are placed on a horizontal axis and then at each position, the one-letter designation for the most frequent amino acid is drawn with a height proportional to the negative base 2 logarithm of its relative frequency at that position (the negative base 2 logarithm of a probability is its information in units of bits). The next most frequent amino acid is drawn below, again with height proportional to frequency, and so on, resulting in a stack of letters at each position. The result is a graphical view of the sequence motif of the family. Fig. 12.4 is the consensus logo for the 15 proteins in Table 12.3.

The rest of the chapter is organized as follows: In Section 12.2, we present a survey of some of the most common clustering methods. In Section 12.3, we use an example to show how our methods can be used to identify

FIG. 12.4 A sequence logo for Table 12.3. Because a sequence logo can become unreadable, we tend to produce several logos for each alignment. This logo corresponds to indexes 16–52 of the alignment in Table 12.3. *(From weblogo.berkeley.edu. See [5] for details.)*

TABLE 12.3 Fifteen Proteins in the Serine Protease Family Alignment Using *FASTA* Notation in Which a *dash* Indicates an Insertion to Allow Subsequent Alignment

	——1——2——3——4——5– 1234567890123456789012345678901234567890123456789012
1MCT-A	————IVGGYTCAANSIPYQVSLNS——GSHFCGGSLIN
1BIT	————IVGGYECKAYSQAHQVSLNS——GYHFCGGSLVN
2CGA-A	CGVPAIQPVLSGLSRIVNGEEAVPGSWPWQVSLQDKT —GFHFCGGSLIN
1ELG	————VVGGTEAQRNSWPSQISLQYRS-GSSWAHTCGGTLIR
1FUJ-A	————IVGGHEAQPHSRPYMASLQMRG–NPGSHFCGGTLIH
1TRY	————IVGGTSASAGDFPFIVSISRN——GGPWCGGSLLN
1DST	————ILGGREAEAHARPYMASVQLN——GAHLCGGVLVA
1ETR-H	————IVEGQDAEVGLSPWQVMLFRKS-P–QELLCGASLIS
1ELT	————VVGGRVAQPNSWPWQISLQYKS-GSSYYHTCGGSLIR
1LMW-B	————IIGGEFTTIENQPWFAAIYRRHRGGSVTYVCGGSLMS
1HNE-E	————IVGGRRARPHAWPFMVSLQL——RGGHFCGATLIA
1PPF-E	————IVGGRRARPHAWPFMVSLQLR——GGHFCGATLIA
1SGT	————VVGGTRAAQGEFPFMVRLSM———GCGGALYA
3RP2-A	————IIGGVESIPHSRPYMAHLDIVTEKG-LRVICGGFLIS
1HYL-A	————IINGYEAYTGLFPYQAGLDITLQDQ-RRVWCGGSLID

Notes: The RCSB PDB number-chain (if more than one chain) is followed by the first 52 out of 288 total positions in the protein family sequence motif. There are approximately 1500 proteins in this serine protease family.

11 isofunctional subfamilies in the Serine Protease family. The appendix, which can be found at the chapter website https://github.com/appmathdoc/AlgebraicAndCombinatorialComputationalBiologyChapter12, elaborates on the value of linear algebra for determining protein structure and also contains some Python resources for clustering.

12.2 CLUSTERING OF DATA

12.2.1 An Overview

We can illustrate many of the ideas in clustering by considering points in the plane where similarity corresponds to Euclidean distance, thus allowing us to visualize clusters and to illustrate how clustering methods partition the data into

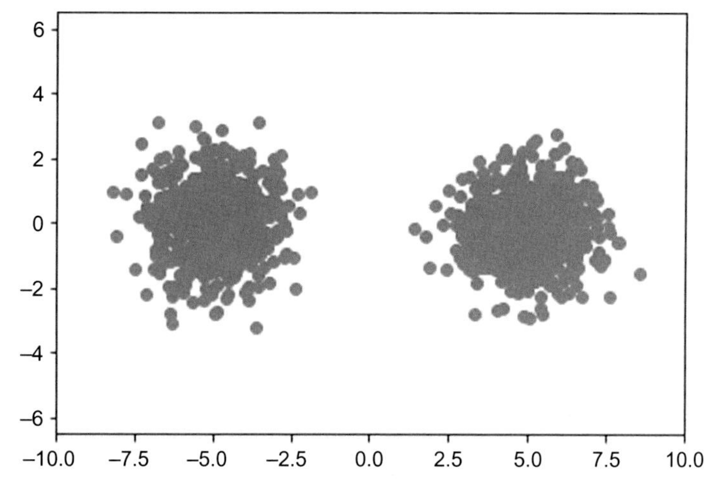

FIG. 12.5 Two-dimensional data separated into two separate "blobs."

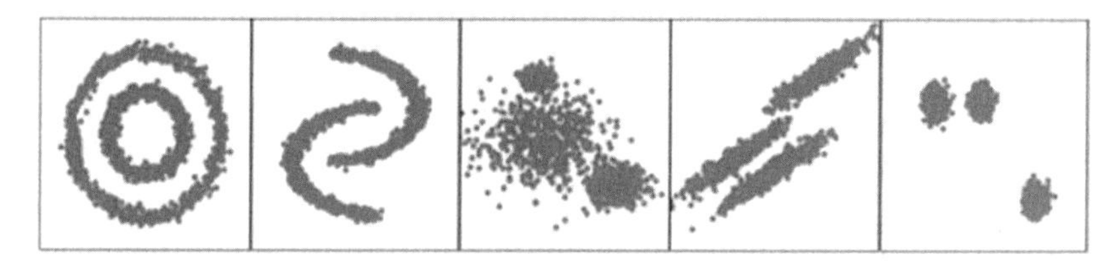

FIG. 12.6 Examples of data intuitively forming either two or three clusters.

k clusters. In this context, our intuition readily identifies that widely separated "blobs" of data correspond to clusters, as illustrated in Fig. 12.5.

However, there is more to the story, even in our simple 2D context. Fig. 12.6 also has some natural concepts of "clusters." In the "noisy concentric circles," the "outer ring" is one cluster while the "inner ring" is another. The last two plots intuitively correspond to $k = 3$ clusters, but notice in particular that in all five of the plots in Fig. 12.6, one can choose two points in at least one cluster that are more distant from each other than they are from points in another cluster.

That is, clustering does not mean trying to force every point in a cluster to be similar to every other, but instead, it means forcing all the points in one cluster to be dissimilar to all the points in every other cluster. Clustering, in fact, is based on how different points in a data set can be.

Let us explain this more rigorously. A data set is a sample from an overall population, where a population is a possibly infinite set that is often defined as a theoretical abstraction. Statistical methods, models, and tests are typically developed as methods for inferring properties of a population from a set of data. Outliers and confounding factors can limit the applicability of a statistical method or model, thus requiring some initial step to identify these features and ameliorate their impact on the data. Large variance within a population may imply distinct subpopulations, thus requiring techniques like stratified sampling to ensure that the information inferred from the data is accurate and unbiased [6]. If the data are collected without stratification, then mixture models can be applied to subpopulations so as to obtain accurate and unbiased information about the entire population [7].

A good example of a statistical population is an isofunctional subfamily—proteins with similar structures and function. However, an entire protein family is not necessarily a statistical population, or more accurately, there is no reason to believe that every isofunctional subfamily necessarily provides information about the larger family of proteins. Protein families are based on shared ancestry, not shared functionality. Consequently, the most appropriate methods for identifying the isofunctional subfamilies within a protein family are clustering methods.

Clustering methods are based on differences between individual elements of an overall set. To the degree that data points inside a cluster are similar, two data points drawn from different clusters should be dissimilar. More specifically for our purposes, a cluster—i.e., an isofunctional family—should constitute a statistical population that can be studied using statistical models, tests, and methods; and correspondingly, samples from one cluster should provide comparatively little or no information about another cluster.

Given that it is appropriate to cluster a data set, there are three requirements that must be addressed in any clustering of that data:

1. A quantitative measure of how similar two data points are.
2. Criteria for determining how many clusters the data should be clustered into.
3. A method for separating the data into the given number of clusters.

The three requirements are also somewhat independent of each other. Common similarity measures include Euclidean distance, correlation, cosine similarity, and even non-Euclidean distance metrics. However, the methods used to separate the data into clusters are typically based only on a matrix of pairwise similarity comparisons; and not on how those outcomes are produced. Likewise, choosing the number of clusters is largely independent of the other two requirements.

In what follows, our focus will be mainly on the clustering methods (item 3). The similarity measure used (item 1) will be based on the primary, secondary, and tertiary structure of proteins, as will the number of clusters (item 2).

12.2.2 Clustering Methods

There are several methods for clustering data into k clusters for some given k, although they tend to be combinations of four different approaches—cluster centers, linkages, densities, and graph theory (networks). All of them are illustrated here with 2D data sets in the xy-plane, although they apply to any finite-dimensional vector space.

1. *K-means*: The k-means method clusters data using cluster centers. The method begins with k initial guesses for the centers, after which it repeats the following until the centers become fixed.
   a. Divide the data into k subsets, where the ith subset is the collection of data points that are closest to the ith center.
   b. Replace the set of k centers with the means of the k subsets, respectively. Fig. 12.7 illustrates this approach graphically in our 2D context.
2. *Agglomerative methods (linkages)*: These methods, which include Ward's linkage, dendrograms, and hierarchical clustering in general, begin with every point in the data set as an individual cluster. In each iteration, existing clusters become subsets of new clustering based on a linkage criteria (i.e., a means of combining existing clusters into larger clusters), continuing until the entire data set is agglomerated into one cluster. The results can be tracked backwards to identify k clusters for any given k. Fig. 12.8 illustrates agglomerative methods for a small data set in the xy-plane.
3. *Density-based methods*: The closer points in a data set are together, the more of them occur in a fixed volume. The number of points per unit volume is known as the data's local density in that volume. Once the data's density has been estimated, then clusters correspond to high-density regions that are surrounded by areas of low density.

   The DBScan method accomplishes this objective using hyperspheres with radius ε for some sufficiently small $\varepsilon > 0$. The neighborhood of a point p in the data is the set of all points in the dataset within distance ε of p. A point q in the data is reachable from a point p if there is a sequence of points $p = p_0, p_1, p_2, \ldots, p_n = q$ where p_i is in the neighborhood of p_{i-1} for each $i = 1, \ldots, n$. A cluster is a subset in which any point in the subset is reachable from any other point in the subset. Points in the data whose neighborhoods are empty (or too small, depending on the implementation) are identified as "noise" or "outliers." Fig. 12.9 illustrates several density-based methods.
4. *Network methods*: Network-based methods use a similarity measure to construct a graph theoretic (i.e., network) representation of the data, after which graph algorithms or eigenvalue methods are used to identify clusters. The graph theoretic representation is typically a nearest neighbor network, which like the density methods, defines a concept of neighborhood for each data point. If a data point q is a neighbor of a data point p, then we say that p is adjacent to q and write $p \sim q$. Algorithms such as "max-flow, min-cut" are then applied to the network representation, as illustrated in Fig. 12.10.

Each type of method has its advantages and its challenges. The k-means algorithm, for example, can be applied to very large data sets in which each data point is in $\mathbb{R}^n$ for large n, especially if the data are presented to the k-means algorithm in batches, which are subsets of the data implemented so that each data point is used by an algorithm the same number of times. This is, in fact, known as the minibatch algorithm, which is illustrated in Fig. 12.11. Batches

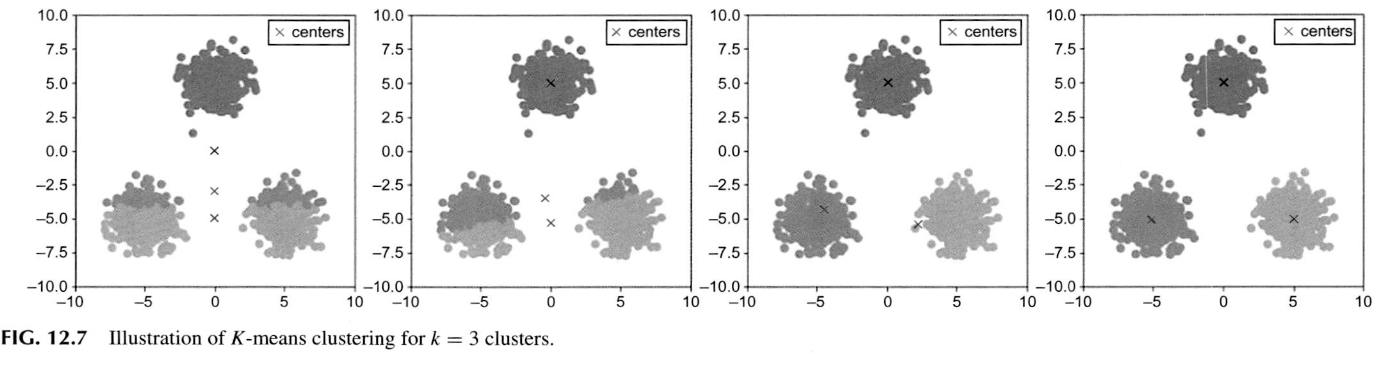

FIG. 12.7 Illustration of K-means clustering for $k = 3$ clusters.

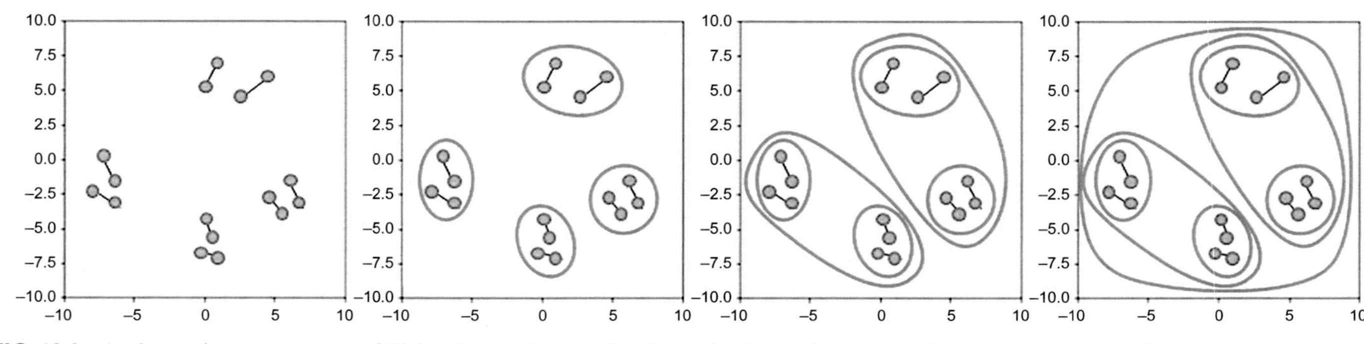

FIG. 12.8 Agglomeration uses a concept of "linkage" to combine smaller clusters into larger clusters (resulting in a smaller number of clusters).

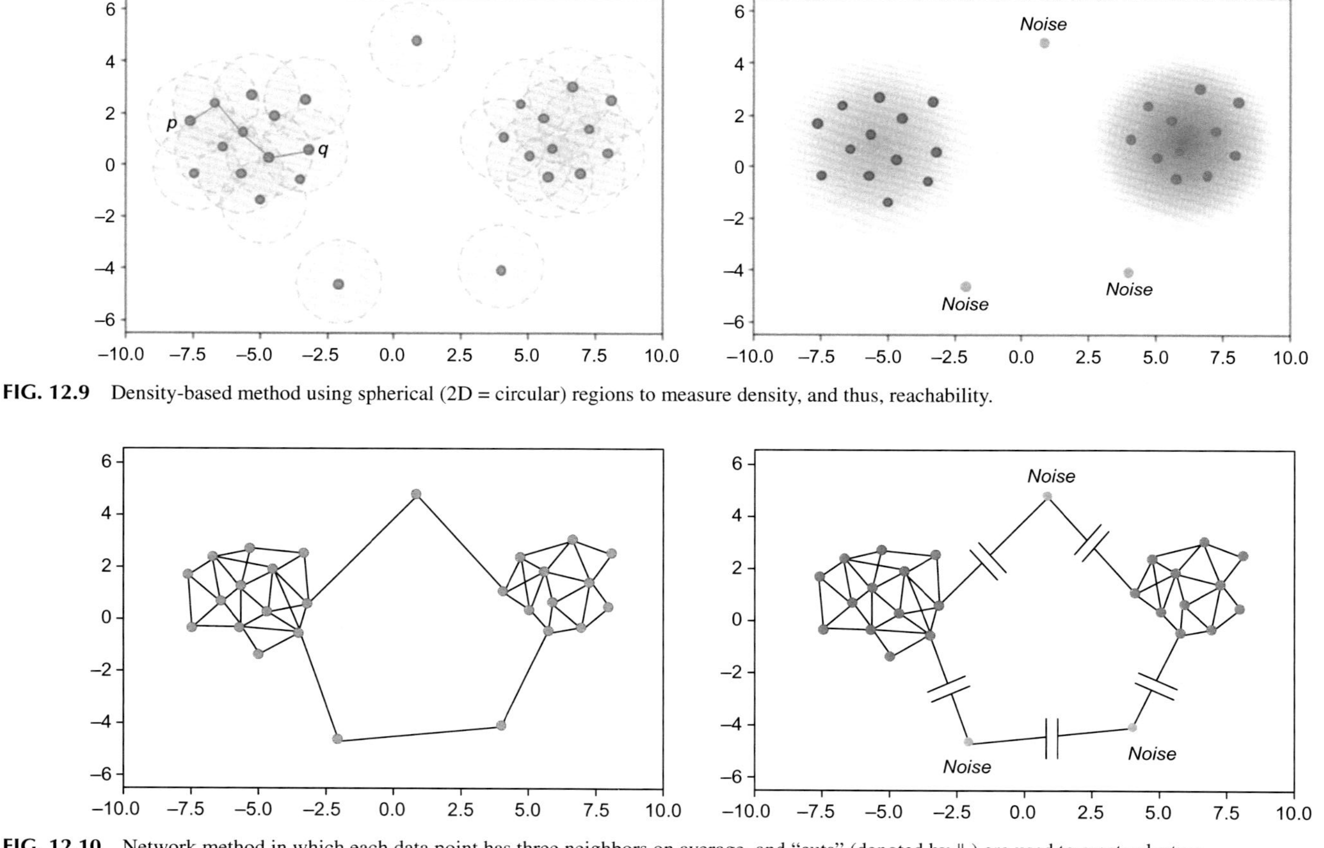

FIG. 12.9 Density-based method using spherical (2D = circular) regions to measure density, and thus, reachability.

FIG. 12.10 Network method in which each data point has three neighbors on average, and "cuts" (denoted by ∥) are used to create clusters.

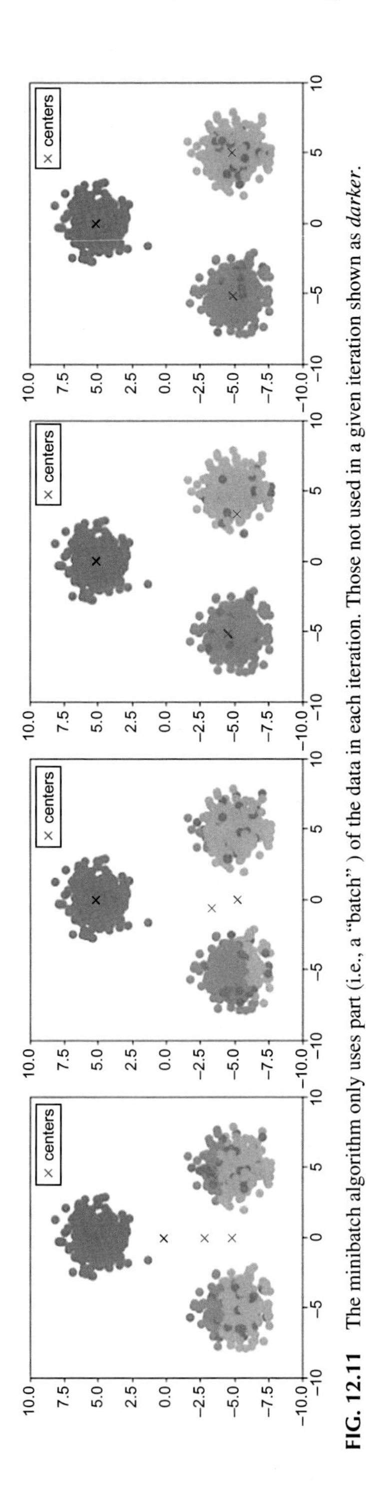

FIG. 12.11 The minibatch algorithm only uses part (i.e., a "batch") of the data in each iteration. Those not used in a given iteration shown as *darker*.

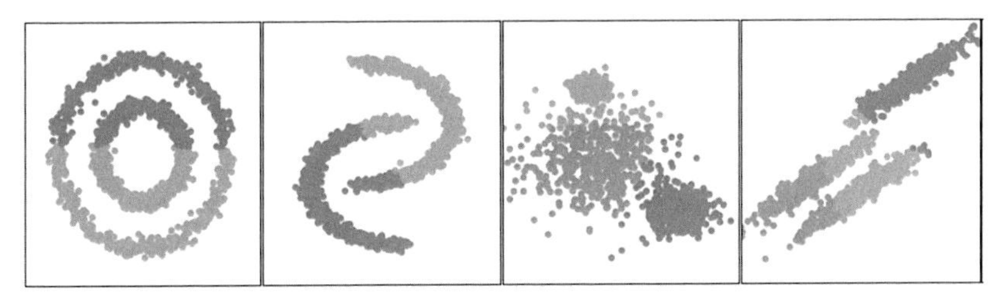

FIG. 12.12 The minibatch algorithm assuming $k = 2$ clusters in the left two plots and $k = 3$ clusters in the right two plots. *Colors* indicate cluster membership. *(From scikit learn (http://scikit-learn.org/stable/modules/clustering.html).)*

can be created in many different ways—the data may have been collected in batches, or the data are streaming and a batch is the data arriving in a given time interval. Given a large data set, however, batches are usually created either by random sampling or random partitioning. Regardless, the motivation for using batches is the same—modern data sets are too large to be analyzed in whole within a reasonable amount of time.

However, if the clusters applied by the data cannot be separated by hyperplanes (e.g., by lines in the xy-plane), then k-means does not work well. We say that such clusters are *nonconvex*. Fig. 12.12 illustrates the results of the k-means minibatch algorithm for several nonconvex data sets.

The k-means algorithm also does not identify outliers (i.e., "noise"), and it can be very sensitive to the choice of k. Clustering protein tertiary structures, much like clustering spatial data in general, tends both to be nonconvex and have outliers [8].

Agglomeration, density, and network methods are superior to k-means for nonconvex data, but they also have their challenges [9]. Agglomeration methods do not identify outliers and if the implementation at each iteration considers all possible ways that smaller clusters can be combined to produce larger clusters, then agglomeration takes a large amount of time (i.e., is computationally expensive) for larger data sets [10].

Density and network methods can identify outliers, but like agglomeration, they can be computationally expensive, even prohibitively so [8]. In addition, density methods tend to be based on the concept of distance as the similarity measure, although we often use similarity scoring functions that are more correlation-like than distance-like. Agglomeration and network methods are arguably (though a bit subjectively) less sensitive to the criterion used to measure similarity than are k-means and density-based methods, although the latter in particular if carefully implemented may have all the advantages of the other three [11].

Nonetheless, as we will see in the next section, network methods, and in particular, spectral methods, may be the best choice for clustering protein families into isofunctional subfamilies. At the very least, the next section will explain why network methods, and in particular network spectral methods, are our choice for the study of protein functionality and tertiary structure.

12.2.3 Network Spectral Methods

Network models are based on two important concepts we have already encountered—a measure of similarity and the concept of a network, or equivalently, a graph. Spectral clustering methods are based on the eigenvalues and eigenvectors of matrices that are implied by or associated with graphs and networks. Before motivating why and how network spectral methods are used for clustering data, let us first revisit these two important concepts to put them on a firm foundation.

In order to make the concept of similarity more rigorous, let us first suppose that each data point in a data set X is a vector in $\mathbb{R}^n$. It follows that we can write X as a matrix with m rows, one row for each element of the data set X. Typically, a row is called an observation, and a column is called either a feature or a factor. A data set which can be written in such a matrix form is often said to be a *structured data set.*

Given a data set X, a similarity score $S(p, q)$ is a symmetric function mapping each pair of observations to a real number no greater than 1. We interpret $S(p, q) > S(p, r)$ to mean that observation p is more similar to observation q than it is to observation r, and we interpret $S(p, q) = 1$ to mean that p and q are as similar as is possible with this scoring method (although it does not necessarily mean that p and q are the same).

For example, if

$$p = (p_1, \ldots, p_n) \quad \text{and} \quad q = (q_1, \ldots q_n)$$

are observations (i.e., rows in the data set) and if $\bar{p}$ and $\bar{q}$ are the means of p and q, respectively, then a possible similarity score for p and q is given by $S(p, q) = r_{pq}$, where

$$r_{pq} = \frac{\sum_{j=1}^{n} (p_j - \bar{p}) (q_j - \bar{q})}{\sqrt{\left(\sum_{j=1}^{n} (p_j - \bar{p})^2\right) \left(\sum_{j=1}^{n} (q_j - \bar{q})^2\right)}}$$

is the Pearson correlation coefficient for p and q. Cosine similarity is also a frequently used measure of similarity and is given by

$$S_{\cos}(p, q) = \frac{\sum_{j=1}^{n} p_j q_j}{\sqrt{\left(\sum_{j=1}^{n} p_j^2\right) \left(\sum_{j=1}^{n} q_j^2\right)}}.$$

Cosine similarity is the same as Pearson correlation if the means of the observations are 0.

Moreover, a similarity scoring function can also be constructed from a distance metric. For example, suppose that $\|\cdot\|$ is a vector space norm on $\mathbb{R}^n$. Then $\|p - q\|$ is the distance between p and q. For $\beta > 0$ and $r > 0$, define

$$S(p, q) = e^{-\beta \|p-q\|^r}.$$

Then $S(p,q)$ is a similarity score because the exponential of a negative number is less than 1 and $S(p,q) = 1$ only if $\|p-q\| = 0$, which is true only if $p = q$. Typically, β and r are called tuning parameters because they can be adjusted to produce the desired concept of similarity in a given application.

Similarity scoring functions lead to network representations via undirected graphs. Specifically, a simple, undirected graph G is a pair of sets (V, E) where V is the set of vertices of G and E is a collection of two element subsets of V, known as the edges of G. If $\{u, v\} \in E$, then we say that u and v are adjacent and write $u \sim v$. Since E is itself a set, an edge can occur at most once in a graph—that is, there are no multiple edges in a simple, undirected graph. Also, a vertex cannot be adjacent to itself and because subsets of points have no order, the edges have no direction. For our purposes, edges are models of similarities, which is to say that $u \sim v$ only if u and v as points in a dataset are sufficiently similar [12].

Given a similarity scoring function, a network representation of a structured dataset X can be constructed in several different ways. One particularly simple k nearest neighbors method is to let data points in X be the vertices of the graph (=network) and then for each vertex, if necessary, add edges corresponding to nearest neighbors until that vertex has at least k neighbors. That is, we iterate over some order of the vertices, and in each iteration, if the vertex being considered has fewer than k neighbors, then we add edges to its closest neighbors until it has at least k neighbors. The result is that all the vertices will have at least k neighbors, although some may have more than k neighbors. Examples for $k = 2$ and $k = 4$ are shown in Fig. 12.13. There are also more sophisticated methods for creating a network representation of a graph, and often, the method for creating a network representation for the data can be derived from the nature of the application itself—as will be the case for us later in this chapter.

Exercise 12.2. Fig. 12.14 is a set of data points in the xy-plane with integer coordinates. Construct a nearest neighbor graph in which each vertex has at least three neighbors. Then construct a nearest neighbor graph in which each vertex has at least five neighbors.

Exercise 12.3. Cluster the data in Fig. 12.14 by partitioning the network graphs from Exercise 12.2 into two separate networks in a way that removes the *fewest* number of edges.

12.2.4 Spectral Clustering

Given a set of n data points, a spectral clustering method partitions the graph into k clusters based on some similarity measure. The basic idea is to construct a similarity graph from the initial data set where each vertex represents a data

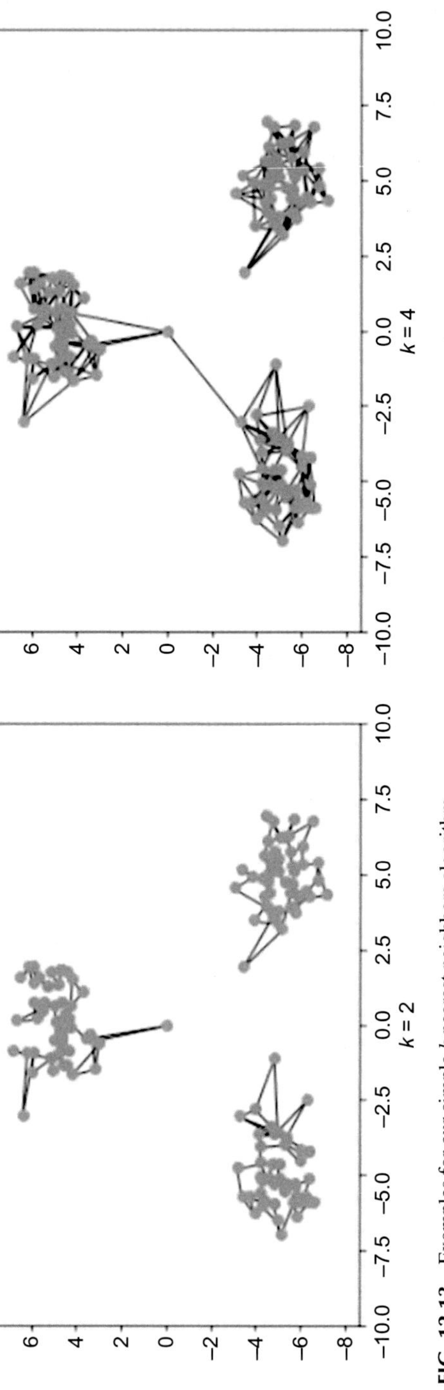

FIG. 12.13 Examples for our simple k nearest neighbors algorithm.

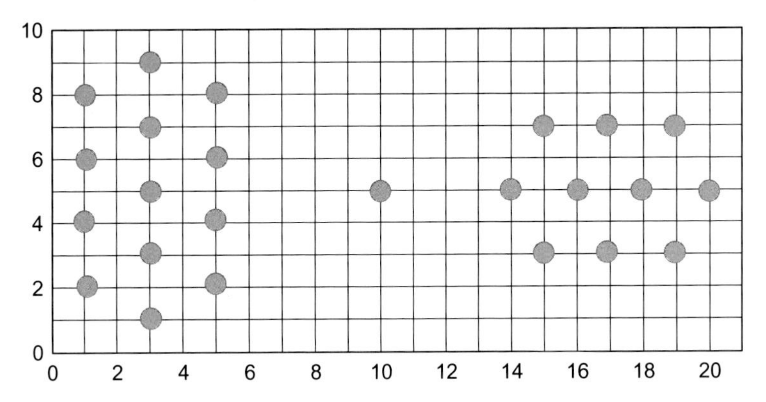

FIG. 12.14 Points have integer coordinates.

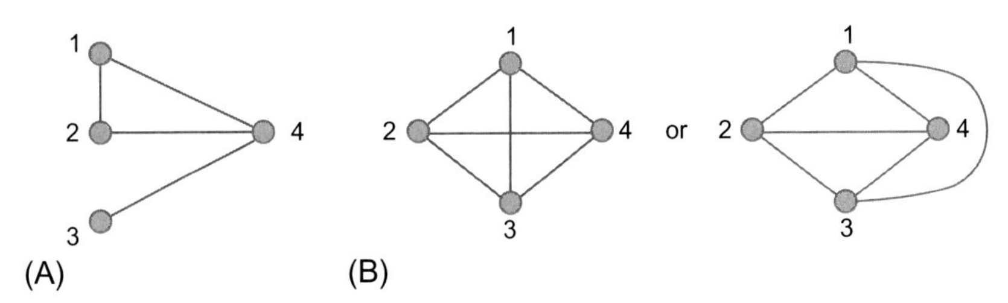

FIG. 12.15 Two examples of unweighted undirected graphs. Both representations in B are of the same graph. It is how a graph is connected that is important rather than how it is drawn.

object, and each unweighted or weighted edge represents the similarity between two objects (see Fig. 12.15). This clustering technique often outperforms traditional approaches [13–19]. We next describe this technique in detail.

A similarity graph is an unweighted or weighted undirected graph with *adjacency matrix* A. The adjacency matrix of an unweighted graph $G = \{V, E\}$ is denoted $A = [A_{ij}]$ where

$$A_{ij} = \begin{cases} 1 & \text{if } \{i,j\} \in E, \\ 0 & \text{otherwise.} \end{cases}$$

Edges $\{i,j\}$ may also have weights w_{ij}, as shown in Fig. 12.16, in which case the adjacency matrix is

$$A_{ij} = \begin{cases} w_{ij} & \text{if } \{i,j\} \in E, \\ 0 & \text{otherwise.} \end{cases}$$

For example, the adjacency matrices of graphs A and B in Fig. 12.15 are, respectively,

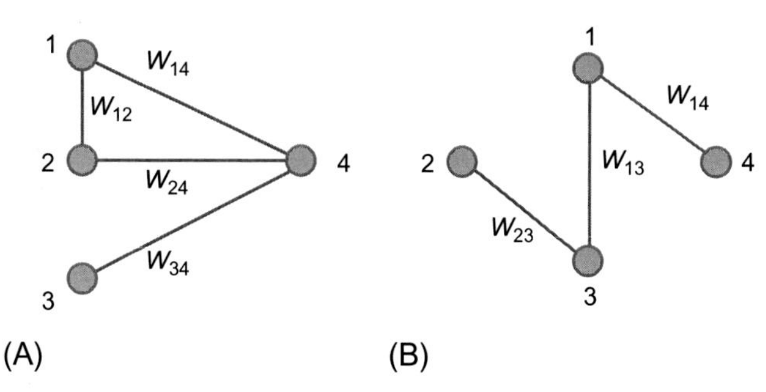

FIG. 12.16 Two examples of weighted undirected graphs.

$$A = \begin{bmatrix} 0 & 1 & 0 & 1 \\ 1 & 0 & 0 & 1 \\ 0 & 0 & 0 & 1 \\ 1 & 1 & 1 & 0 \end{bmatrix} \quad \text{and} \quad B = \begin{bmatrix} 0 & 1 & 1 & 1 \\ 1 & 0 & 1 & 1 \\ 1 & 1 & 0 & 1 \\ 1 & 1 & 1 & 0 \end{bmatrix}.$$

In a weighted, undirected graph, the weights must satisfy $w_{ij} = w_{ji}$. For example, the adjacency matrices of graphs A and B in Fig. 12.16 are

$$A = \begin{bmatrix} 0 & w_{12} & 0 & w_{14} \\ w_{12} & 0 & 0 & w_{24} \\ 0 & 0 & 0 & w_{34} \\ w_{14} & w_{24} & w_{34} & 0 \end{bmatrix} \quad \text{and} \quad B = \begin{bmatrix} 0 & 0 & w_{13} & w_{14} \\ 0 & 0 & w_{23} & 0 \\ w_{13} & w_{23} & 0 & 0 \\ w_{14} & 0 & 0 & 0 \end{bmatrix}.$$

The *degree* matrix D of an unweighted, undirected graph is the diagonal matrix of the degrees of the vertices, where the degree $\deg(v_i)$ of vertex i is the number of its neighbors. More generally, if the edges E have weights, then the *weighted* degree of a vertex $i \in V$ is

$$d_i = \sum_{\{i,j\} \in E} w_{ij}$$

and the corresponding *weighted* degree matrix D_{w} is the diagonal matrix of the weighted degrees of the graph. For example, the weighted degrees of the vertices of graph A in Fig. 12.16 are

$$d_1 = w_{12} + w_{14}, \quad d_2 = w_{12} + w_{24}, \quad d_3 = w_{34},$$
$$\text{and } d_4 = w_{14} + w_{24} + w_{34}$$

and the weighted degree matrix of A in Fig. 12.16 is the diagonal matrix of these weighted degrees.

The Laplacian matrix L of a graph is a matrix that denotes the difference between the degree matrix D and the adjacency matrix A:

$$L = D - A.$$

The Laplacian matrix is motivated by the concept of the *Laplacian* of a topological manifold, and not surprisingly, it reveals a great deal of information about the structure of a graph [15, 19–21].

Specifically, $L = [L_{ij}]$, where for unweighted and weighted graphs, respectively,

$$L_{(\text{unweighted})ij} = \begin{cases} \deg(v_i) & \text{if } i = j, \\ -1 & \text{if } \{i,j\} \in E, \\ 0 & \text{otherwise,} \end{cases} \quad L_{(\text{weighted})ij} = \begin{cases} d_i & \text{if } i = j, \\ -w_{ij} & \text{if } \{i,j\} \in E, \\ 0 & \text{otherwise.} \end{cases}$$

For example, the Laplacian of graph A in Fig. 12.15 is

$$L = \begin{bmatrix} 2 & 0 & 0 & 0 \\ 0 & 2 & 0 & 0 \\ 0 & 0 & 1 & 0 \\ 0 & 0 & 0 & 3 \end{bmatrix} - \begin{bmatrix} 0 & 1 & 0 & 1 \\ 1 & 0 & 0 & 1 \\ 0 & 0 & 0 & 1 \\ 1 & 1 & 1 & 0 \end{bmatrix} = \begin{bmatrix} 2 & -1 & 0 & -1 \\ -1 & 2 & 0 & -1 \\ 0 & 0 & 1 & -1 \\ -1 & -1 & -1 & 3 \end{bmatrix}.$$

Similarly, the Laplacian for B in Fig. 12.15 is

$$L = \begin{bmatrix} 0 & -1 & -1 & -1 \\ -1 & 0 & -1 & -1 \\ -1 & -1 & 0 & -1 \\ -1 & -1 & -1 & 0 \end{bmatrix}$$

and the Laplacian L_w for the weighted graph A in Fig. 12.16 is

$$L_w = \begin{bmatrix} w_{12} + w_{14} & -w_{12} & 0 & -w_{14} \\ -w_{12} & w_{12} + w_{24} & 0 & -w_{24} \\ 0 & 0 & w_{34} & -w_{34} \\ -w_{14} & -w_{24} & -w_{34} & \sum_{i=1}^{3} w_{i4} \end{bmatrix}.$$

The Laplacian matrix L of a graph has several important properties. These properties are from linear algebra, and a linear algebra background—is assumed from here forward [22] (i.e., just put citation at the end of this sentence). For example, the Laplacian is symmetric and also positive semidefinite. This means that its eigenvalues are nonnegative. Moreover, each row and column of the Laplacian matrix sums to zero, which in vector form means that

$$L\mathbf{1} = \mathbf{0} \quad \text{or} \quad L\mathbf{1} = 0\mathbf{1}$$

if $\mathbf{1}$ is the vector of all ones (the "ones" vector). That is, zero is an eigenvalue of L with eigenvector $\mathbf{1}$.

Typically, we write the eigenvalues of L as

$$0 = \lambda_1 \leq \lambda_2 \leq \cdots \leq \lambda_n,$$

where n is the number of vertices of the graph, which is also the number of rows of L. A nonzero eigenvector of L corresponding to the eigenvalue λ_2 is called a *Fiedler* vector of L, which we denote as f. It is necessary that $f \neq 0$. The Fiedler vector partitions a graph into two separate clusters based on the sign of the coefficients of f (clustering into $k = 2$ clusters is called a *partition* of the data).

That is, a basic spectral partitioning algorithm for a given data set is as follows [15]:

1. Use a similarity function to score the data.
2. Define a network representation in which two vertices are adjacent if they are sufficiently similar.
3. Compute the Laplacian matrix of the network.
4. Use a Fiedler vector $f = (f_1, \ldots, f_n)$ to partition the vertices (i.e., data points) for which $f_i \neq 0$ into two sets

$$\{i : f_i < 0\} \quad \text{and} \quad \{j : f_j > 0\}.$$

We do not assign vertices i for which $f_i = 0$ to either cluster. Instead, we interpret such vertices as *outliers* which do not belong in either cluster.

To obtain $k > 2$ clusters, we use more eigenvectors of L, or we iteratively examine each cluster to determine if it should be further partitioned until the desired number of clusters is obtained. We will use this latter approach as it is both more intuitive and does not require us to determine the number of clusters beforehand.

Exercise 12.4. What are the Laplacian matrices for the graphs in Fig. 12.17?

12.2.5 Spectral Clustering With Outliers

As already mentioned, this algorithm either assigns a data point to a cluster or identifies it as an outlier. We briefly discuss the motivation for doing so before moving on to our discussion of clustering of protein families into isofunctional families.

A graph G is said to have at least r *components* if its vertices V can be partitioned into

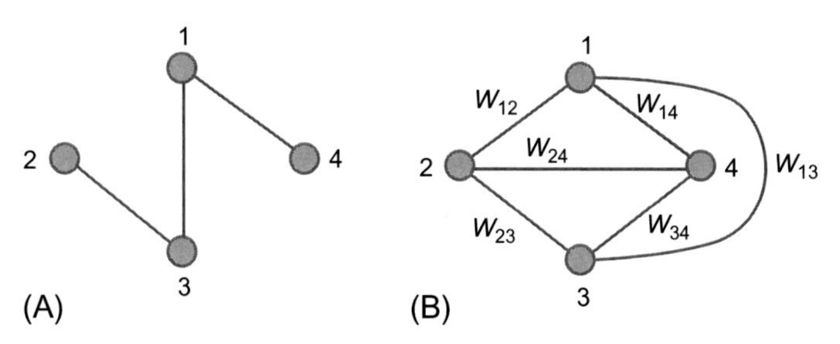

FIG. 12.17 A is unweighted, while B is weighted.

$$V = V_1 \cup V_2 \cup \cdots \cup V_r,$$

where there are no edges between V_i and V_j for all $i \neq j$. G is said to be a *connected graph* if it has only one component.

If G is an undirected graph with edges having nonnegative weights, then for the ith component, define a "characteristic vector" $\mathbf{c}_i$ which has coefficients of 1 for the vertices corresponding to the ith component and 0 elsewhere. Each $\mathbf{c}_i$ is an eigenvector of the Laplacian corresponding to the zero eigenvalue, and indeed, the multiplicity of the zero eigenvalue is equal to the number of connected components [15, 23]. If G has two components G_1 and G_2 with n_1 and n_2 vertices, respectively, then $\mathbf{c}_1 + \mathbf{c}_2$ is the ones vector while

$$f = \frac{1}{n_1}\mathbf{c}_1 - \frac{1}{n_2}\mathbf{c}_2$$

is an eigenvector corresponding to the zero eigenvalue that is orthogonal to $\mathbf{1}$. It is, in fact, the *Fiedler vector* of a graph with two components, and clearly, it is negative on one component and positive on the other.

The Laplacian is a matrix of a linear transformation, and a key concept in linear algebra is that linear transformations on finite vector spaces are *continuous*. Thus, small variations in the coefficients of a matrix only produce small variations in the eigenvectors of that matrix, and this concept can be used to prove that if a single edge is added between two sufficiently large components of a graph, then the signs of the corresponding Fiedler vector do not change, although the coefficients may change in small, individual ways. Consequently, the signs of the coefficients of f will be conserved as long as only relatively few edges between G_1 and G_2 are added to the graph. This is the basic idea that leads to the use of the Fiedler vector to partition a graph.

To illustrate why vertices corresponding to zero coefficients of f are considered outliers, let us first consider the special case shown in Fig. 12.18 in which the two subgraphs G_1 and G_2 are identical (i.e., isomorphic) and there is a single vertex v not in either G_1 or G_2. If there is a single edge from a vertex u in G_1 to v and there is an edge between the vertex in G_2 corresponding to u and v, then there is no more reason to assign v to a cluster containing G_1 but not G_2 than there is in assigning it to a cluster containing G_2 but not G_1. That is, v does not belong in either cluster, and thus, we say that v is an *outlier* to the clustering problem (or is a vertex that is *unassociated* with either cluster).

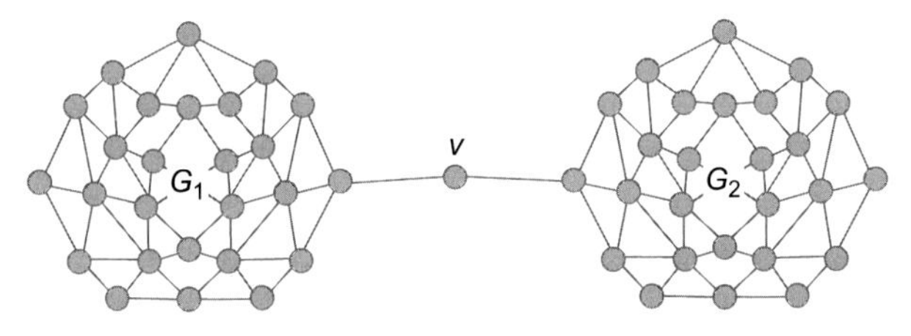

FIG. 12.18 Special case: G_1 and G_2 are identical (mathematically, isomorphic).

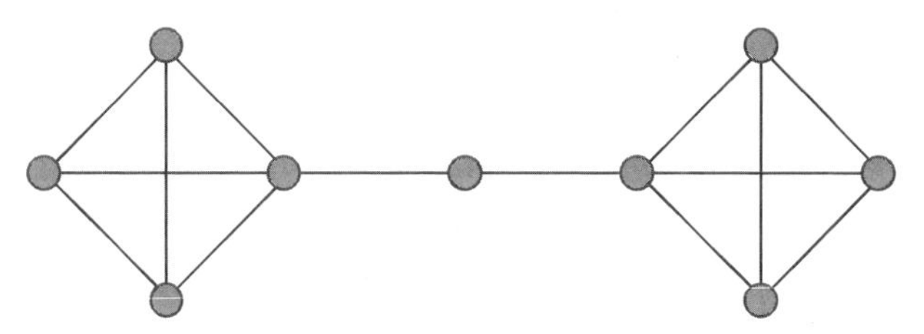

FIG. 12.19 A graph with 11 vertices.

Moreover, because of the symmetry of the graph in Fig. 12.18, the Fiedler vector must also be symmetric. That can only occur if the coefficient corresponding to v in the Fiedler vector is zero.

Thus, the Fiedler vector, the eigenvector corresponding to the second smallest Laplacian eigenvalue (for a connected graph), is known to provide a measure of graph connectivity and can be used to obtain graph partitions such that the nodes within the same cluster are more strongly connected than with the nodes belonging to the different clusters. The second smallest eigenvalue is called the *algebraic connectivity* of the graph. The smaller this value, the easier it is to cut the graph into unconnected components; a larger value indicates greater bonding strength and thus, it would be harder to partition the graph into disconnected components [23, p. 16].

Clustering is a means of separating a large, messy dataset into samples from different populations. If the data are a mixture of "apples" and "oranges," then the differences between them can separate them from each other, even if there is not initially any concept of how an "apple" differs from an "orange." And if there are few pineapples, cantaloupes, watermelons, and other assorted "fruits" mixed in, then they are outliers to both. Clustering allows the apples and oranges to be separated into groups while also preventing those "fruits" that are neither from contaminating the results. Specifically, as we will see next, clustering allows us to separate isofunctional families from each other inside a protein family while also allowing isolated proteins to remain unassociated from those subfamilies.

Exercise 12.5. Compute the Laplacian matrix of the graph in Fig. 12.19. Then find a Fiedler eigenvector of its Laplacian (we recommend doing so with software). What does the Fiedler vector say about how the graph should be clustered?

12.3 CLUSTERING TO IDENTIFY ISOFUNCTIONAL FAMILIES

12.3.1 Similarity Scores Based on Tertiary Structure

It was already emphasized that the functional properties of a protein are defined both by its chemical properties and its tertiary structure. Thus, if two proteins have both similar primary and tertiary structures, then they can be expected to have similar functional properties [4]. Consequently, isofunctional subfamilies of a protein family are predicted by clusters of a protein family if similarity is based on primary and tertiary structures.

However, it is difficult at best to determine if two proteins have a similar tertiary structure. Identical proteins can nonetheless have different 3D coordinates because each X-ray crystallography experiment has its own "lab frame," which is to say that a protein can occur with different rotations and translations—that is, different orientations—within the coordinate system defined as part of an experiment. This issue can be readily addressed by finding rotations through angles θ, ϕ, ψ, respectively, and translations by dx, dy, dz, respectively, that map the coordinate system of one protein into the coordinate system of another.

While that is a good start, it is not enough. Proteins are not static, so a final conformation is not a rigid 3D geometric structure but rather a specific topological shape. Consequently, even after the lab frame for one crystal structure of a protein is rotated into the lab frame for another of the same protein, the 3D coordinates are still unlikely to match perfectly.

Proteins in the same functional subfamily need not even have primary structures of the same length, and often they do not have identical secondary structures either. Nonetheless, there is hope. The backbone of a protein tends to be somewhat rigid, and every amino acid has a first carbon (the alpha carbon, often called the C-alpha of the amino acid). All alpha carbons in identical proteins in the same coordinate system should have the same positions, and correspondingly, two tertiary structures in the same coordinate system should have C-alpha carbons which are close to each other [2].

This is the concept behind Template Modeling scores (TM-scores). The similarity score between two proteins P and Q is calculated as follows [2]:

1. The primary structures of the two proteins are aligned to produce the two subsets of residues, P_{aligned} and Q_{aligned}, respectively, for which P and Q are aligned with the sequence motif for the entire family. For example, Table 12.4 shows that the proteins 1MCT-A and 1BIT are aligned for vertices 16–35 and for vertices 42–52. The sets P_{aligned} and Q_{aligned} are the same size, which we denote L_{aligned}. For example, $L_{\text{aligned}} = 31$ for 1MCT-A and 1BIT as shown in Table 12.4.

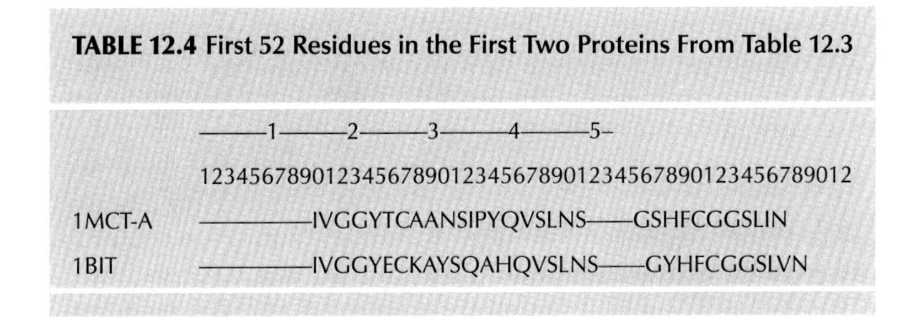

TABLE 12.4 First 52 Residues in the First Two Proteins From Table 12.3

	——1——2——3——4——5–
	1234567890123456789012345678901234567890123456789012
1MCT-A	——————IVGGYTCAANSIPYQVSLNS——GSHFCGGSLIN
1BIT	——————IVGGYECKAYSQAHQVSLNS——GYHFCGGSLVN

2. Let L_{target} be the length of the longer of the primary sequences of P and Q, respectively. For example, the proteins 1MCT-A and 1BIT are lengths 223 and 237, respectively, so that $L_{\text{target}} = 237$ for this specific pair of proteins.
3. For each possible set of rotations and translations of the 3D coordinates of P_{aligned}, define the function

$$f(\theta, \phi, \psi, dx, dy, dz) = \frac{1}{L_{\text{target}}} \sum_{i=1}^{L_{\text{aligned}}} \frac{1}{1 + (d_i/d_M)^2},$$

where d_i is the distance between the ith residue of P_{aligned} in the new coordinates and Q_{aligned}, respectively. In general, the distances d_i tend to range from 0 to 10 or so angstroms. The scaling factor $d_M = 1.24\left(L_{\text{target}} - 15\right)^{1/3} - 1.8$ is an empirically derived "maximum possible distance." Specifically, maximum distances are observed to satisfy a power law distribution with respect to L_{target}, and d_M represents a good approximation to that power law distribution [24].
4. The TM-score is the maximum of the function f over all possible coordinate systems. That is,

$$\text{TM-score} = \max_{\theta,\phi,\psi,dx,dy,dz} \left(\frac{1}{L_{\text{target}}} \sum_{i=1}^{L_{\text{aligned}}} \frac{1}{1 + (d_i/d_M)^2} \right).$$

If P and Q are identical, then $L_{\text{target}} = L_{\text{aligned}}$ and there are some choices of angles and translations for which $d_i = 0$ for all i. Consequently, the TM-score for identical proteins is 1, even if they are in different coordinate systems to begin with. If there is some coordinate system in which $d_i \approx 0$ for all i, then the TM-score is close to 1 if the proteins are well aligned (i.e., if $L_{\text{target}} \approx L_{\text{aligned}}$). Thus, a pair of similar proteins has a TM-score that is close to 1.

Since the TM-score cannot exceed 1, and two proteins are maximally similar if their TM-score is 1, so the TM-score is indeed a similarity score. It is also independent of the length of the proteins, at least in theory. The TM-score is based on proteins assuming rigid geometric configurations, and thus in practice, it depends on how much a protein's geometry can vary within its topological shape. Thus, when clustering based on TM-scores, outliers are typical and scores are often lower than would have been expected theoretically [4].

The paper [4] by de Lima and colleagues is a thorough investigation of this approach. Protein families are clustered into isofunctional subfamilies via spectral clustering using similarities based on TM-scores and alignment scores. They consider four protein families, including serine proteases. They also augment TM-scoring with additional similarity measures. In the paper, they use agglomerative clustering (specifically, hierarchical clustering) to suggest two relevant clustering numbers for the serine proteases: 4 clusters and 11 clusters.

Given the intended audience for this chapter includes undergraduates and given that the results in [4] require large amounts of time and resources, we

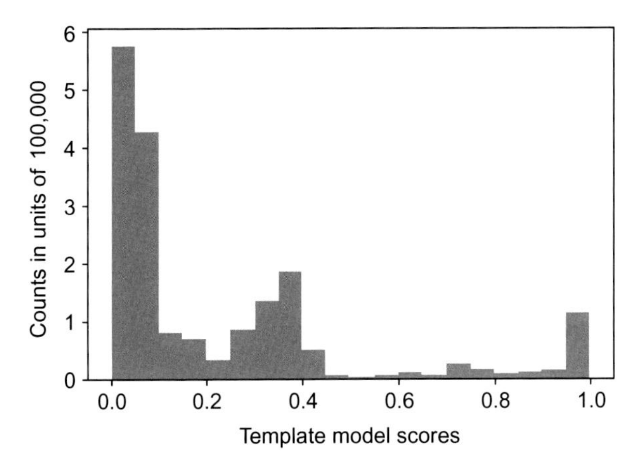

FIG. 12.20 Template model scores for serine protease family.

limit ourselves here only to the serine protease family. In addition, we focus only on a single measure—TM-scores—rather than on the multiple scoring approaches used in [4], again due primarily to the intended audience for this chapter. Instead, we augment their approach to clustering using outlier removal as previously discussed, and we use successive partitioning, which means successively dividing a set into two clusters. The use of successive partitioning is again primarily due to the intended audience, as it requires only the interpretation of only one eigenvector of the Laplacian for each partition. Even given these simplifications with respect to [4], we demonstrate here that spectral clustering including outlier removal produces quality clusters, even though based solely on TM-scores. We begin by noting how the scores are distributed, which is shown in Fig. 12.20.

It is widely accepted that TM-scores below 0.2 imply completely unrelated proteins, whereas scores above 0.5 imply closely related proteins [25]. Thus, we create a graph model in which two proteins are adjacent if their TM-score similarity is more than 0.5. The result presented in Fig. 12.21 is a graph with several components, or more specifically, a number of small components with one large component, which we often call the "giant component."

The existence of the giant component is expected not only in the comparison of proteins but also in real world network models in general [26]. Relationships modeled by networks—tertiary structure similarity in this case—tend to be the result of dynamic processes, and such processes tend to create relationships both on a large scale—the giant component—as well as on smaller scales.

Indeed, in Fig. 12.21, the giant component (center) contains 1408 proteins, whereas no other of the smaller components has more than 50 proteins; there are a huge number of components with only 2 or 3 proteins, and the average number of proteins in a smaller component is 8. Fig. 12.22 shows the giant component itself, which appears to be a "loosely" connected collection of 10–13 clusters, where a cluster is a large collection of vertices that appear to be almost a single

FIG. 12.21 The network model of TM-score similarity of the serine protease family has a giant component (*center*) and a large number of "very small" components.

FIG. 12.22 The giant component for the serine protease family. Depending on the interpretation of the representation, there appear to be somewhere between 10 and 13 clusters in this network.

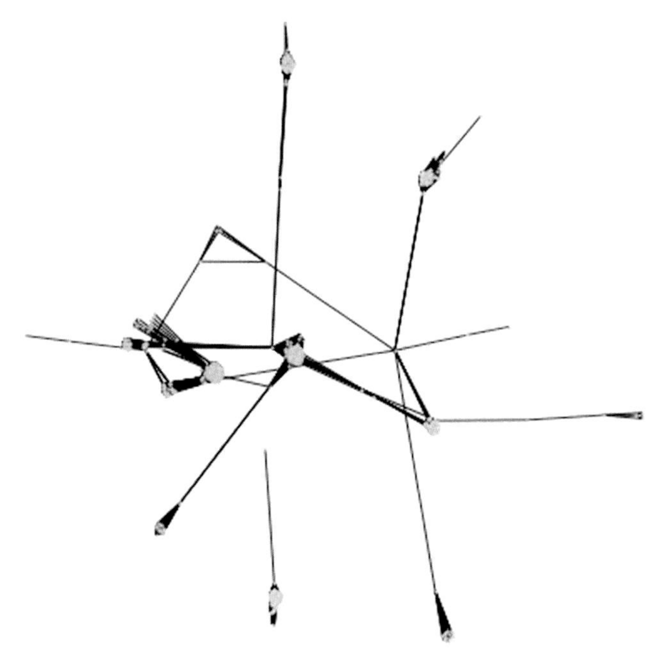

FIG. 12.23 The Fiedler eigenvector splits off the "pink" cluster (compare with Fig. 12.22). The pink cluster is south of the central cluster.

node. As suggested in [4], the use of 11 clusters is well justified. To further identify outliers, we partition the giant component into two clusters, which is sometimes called a "cut" of a network. The result should be that some subset of the 11 apparent clusters are separated from the remaining ones. Outliers are those proteins that do not tend to associate with one partition or the other.

Clustering in [4] is done in a single step, but here we produce the clusters iteratively, beginning first by computing the Fiedler eigenvector of the Laplacian matrix of the giant component. Coefficients of the eigenvector that are close to 0 correspond to proteins that are outliers, and otherwise the sign of the coefficient identifies the partition to which the protein belongs. Fig. 12.23 illustrates the first iteration of this process, in which one cluster is split away from the network in Fig. 12.22. Fig. 12.24 shows the final clustering into 11 clusters.

The quality of the results is to be judged on the degree to which the clusters group proteins into isofunctional subfamilies within which all the proteins are closely related. Indeed, in some of the clusters, the primary sequences of the proteins are all identical—e.g., the "pink" cluster, which is the cluster split off in Fig. 12.23. Even if not identical, the proteins within a cluster are highly similar. For example, the "purple" cluster, which is the cluster immediately southeast of the central cluster, has the logo given in Fig. 12.25.

FIG. 12.24 The final clustering of the serine protease family into 11 isofunctional subfamilies.

FIG. 12.25 The first 30 positions for the consensus logo for the *purple* cluster, which is southeast of the central cluster.

Likewise, the "orange" cluster, which is the cluster northeast of the central cluster, has the sequence logo shown in Fig. 12.26. The "pink," "purple," and "orange" clusters contain 113, 113, and 88 proteins, respectively.

The "yellow" cluster, which is the cluster in the center, is large, containing 351 proteins, but as its sequence logo in Fig. 12.27 illustrates, the proteins are nonetheless highly similar in structure—and thus, also, in function.

Nonetheless, the largest cluster is the "green" cluster, which is the cluster just west of the central cluster, and it shows that we still have some work to do.

FIG. 12.26 The first 30 positions for the consensus logo for the "orange" cluster, which is the cluster that is northeast of the central cluster.

FIG. 12.27 The first 30 positions for the consensus logo for the *yellow* cluster, which is the cluster in the center.

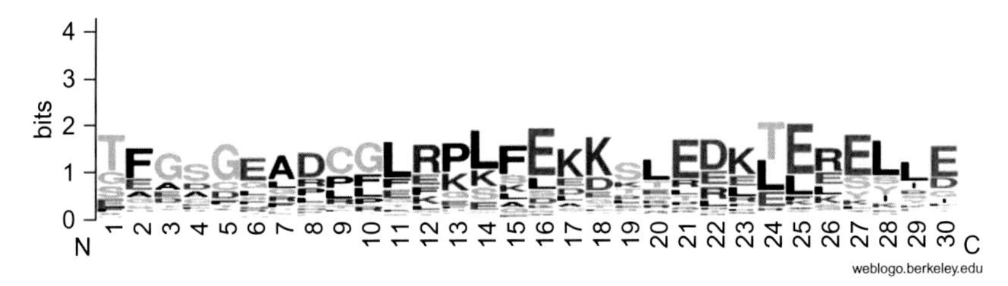

FIG. 12.28 The first 30 positions for the consensus logo for the "green" cluster, which is the largest cluster just west of the central cluster.

Although the proteins it contains are not similar to proteins in other clusters, they also are not close to each other, as Fig. 12.28 illustrates.

Indeed, the logo in Fig. 12.28 is essentially what we would expect if amino acids occurred randomly in every position in this family. That is, the largest cluster is analogous to the residuals in a regression fit. It represents those proteins in the family that are not outliers and yet are also not members of an identifiable isofunctional family. Further statistical analysis of the "green" cluster might reveal more about how these represent randomly distributed proteins within the given isofunctional family, but that is beyond the scope of this chapter.

Although sequence logos are excellent for intuitively assessing the quality of a cluster, they are not rigorous measures of how well a cluster predicts an isofunctional subfamily. In general, at this point we would apply rigorous statistical and information-theoretic measures, such as mutual information scoring [10]. However, what we have seen in this chapter is that clustering can be used to differentiate between different isofunctional subfamilies, or more generally, between different statistical populations.

Suggested projects:

1. Interesting: Very often, a data set is *labeled*, which means that observations are in one "class" or another. For example, observations in the Wisconsin Breast cancer dataset are labeled either "benign" or "malignant." (This data set is included both in the Python library scikit-learn and in the statistical software tool R. It can also be found online at https://www.kaggle.com/uciml/breast-cancer-wisconsin-data as well as in the UCI machine learning repository). Can you partition (cluster into $k = 2$ clusters) a labeled data set *without using the labels* so that most—if not all—of the observations in a given cluster share the same label? That is, can you produce a topological structure for the data that predicts the labeling?
2. Challenging: There were several other protein families in [4]. This chapter and their paper can assist in clustering another protein family.

The code and related files used to produce the results in this chapter are available at https://github.com/appmathdoc/AlgebraicAndCombinatorialComputationalBiologyChapter12.

APPENDIX

Linear algebra is one of the important mathematical tools in biomolecular modeling for protein structure determinations [27]. There are three main methods for determining protein structure using linear algebra: NMR spectroscopy, X-ray crystallography, and homology modeling. In this appendix we only discuss the first method. More information can be found in [28].

Let n be the number of atoms in a given protein and $x_1, \ldots, x_n$ be the coordinate vectors for the atoms, where $x_i = (x_{i1}, x_{i2}, x_{i3})^T$ and x_{i1}, x_{i2}, and x_{i3} are the first, second, and third coordinates of atom i. The distance d_{ij} between atoms i and j is given by $d_{ij} = \|x_i - x_j\|$, where $\| \cdot \|$ is the standard Euclidean norm. As a result, the coordinate and distance matrices for the protein are, respectively

$$X^o = \{x_{ij} : i = 1, \ldots, n, j = 1, 2, 3\} \quad \text{and}$$
$$D^o = \{d_{ij} : i, j = 1, \ldots, n\}.$$

If the coordinate matrix X^o is known, then the distance D^o follows immediately. On the other hand, if the distance matrix D^o is known, then a coordinate matrix X^o can be obtained from D^o. If D^o is only partially known, then X^o can be approximated from D^o, although the computation is not straightforward. Finding or approximating a coordinate matrix given a distance matrix is known mathematically as the *distance geometry problem* [27, 29, 30].

Borrowing from [27] to illustrate the distance geometry problem, we consider a simple case in which a complete set of exact distances

$$D^o = \{d_{ij} : i, j = 1, \ldots, n\}$$

is given for all *ij*-pairs of atoms in a protein. A coordinate matrix is an $n \times 3$ real matrix of the form

$$X^o = \left[x_{ij} : i = 1, \ldots, n, j = 1, 2, 3\right]_{i,j}^{n,3} = [x_i]_{i=1}^{n} \quad \text{where } x_i = (x_{i1}, x_{i2}, x_{i3}),$$

and it must satisfy $\|x_i - x_j\| = d_{ij}$ for all $i, j = 1, \ldots, n$, and

$$\|x_i\|^2 - 2x_i^T x_j + \|x_j\|^2 = d_{ij}^2, \; i, j = 1, \ldots, n.$$

Since X^o is unique only up to translation or rotation, so we set a reference system so the origin is located at the last atom, $x_n = (0, 0, 0)^T$. It then follows that

$$d_{in}^2 - 2x_i^T x_j + d_{jn}^2 = d_{ij}^2, \quad i, j = 1, \ldots, n-1$$

and correspondingly, we define a new set of coordinate and distance matrices,

$$X = \{x_{ij} : i = 1, \ldots, n-1, j = 1, 2, 3\}$$

$$D = \left\{(d_{i,n}^2 - d_{i,j}^2 - d_{j,n}^2)/2 : i, j = 1, \ldots, n-1\right\}.$$

Therefore, $XX^T = D$, and since X is an $n \times 3$ matrix, the distance matrix D must be of maximum rank 3 (it could be 1 or 2) [31, Theorem 2.1]. Solving for X in $XX^T = D$ yields a solution X^o to the distance geometry problem for distance matrix D^o.

One approach to solving $XX^T = D$ for X is to use the singular value decomposition [32, 33]. An alternative is the geometric buildup algorithm (GBA) by Dong and Wu [29, 30]. Since D is rank 3, the SVD requires at least $O(n^2)$ floating-point operations [33], but the GBA is more efficient with $O(n)$ floating-point operations [29]. In addition, the GBA also requires a smaller number of distances and is easier to extend to problems with sparse sets of distances [30]. On the other hand, while the SVD method requires all of the distances, it may be preferred over the GBA if the distances contain some errors or are not consistent [34]. The solution from the GBA for such a case may or may not be a good approximation, depending on the choice of the initial four atoms and hence the distances it uses to build the structure. For more details see [27].

REFERENCES

[1] H.M. Berman, J. Westbrook, Z. Feng, G. Gilliland, T.N. Bhat, H. Weissig, L.N. Shindyalov, P.E. Bourne, Protein Data Bank, Nucleic Acids Res. 28 (2000) 235–242, https://doi.org/10.1093/nar/28.1.235.

[2] P. Radivojac, W.T. Clark, T.R. Oron, A.M. Schnoes, T. Wittkop, A. Sokolov, K. Graim, C. Funk, K. Verspoor, A. Ben-Hur, et al., A large-scale evaluation of computational protein function prediction, Nat. Methods 10 (3) (2013) 221–227.

[3] A. Bateman, L. Coin, R. Durbin, R.D. Finn, V. Hollich, S. Griffiths-Jones, A. Khanna, M. Marshall, S. Moxon, E.L.L. Sonnhammer, et al., The Pfam protein families database, Nucleic Acids Res. 32 (suppl_1) (2004) D138–D141.

[4] E.B. de Lima, W.M. Júnior, R.C. de Melo-Minardi, Isofunctional Protein Subfamily Detection Using Data Integration and Spectral Clustering, PLoS Comput. Biol. 12 (6) (2016) e1005001.

[5] G.E. Crooks, G. Hon, J.-M. Chandonia, S.E. Brenner, WebLogo: a sequence logo generator, Genome Res. 14 (6) (2004) 1188–1190.

[6] D.R. Bellhouse, Systematic sampling, in: Handbook of Statistics, vol. 6, Elsevier, 1988 pp. 125–145.

[7] G. McLachlan, D. Peel, Finite Mixture Models, John Wiley & Sons, New York, NY, 2004.

[8] J. Sander, M. Ester, H.-P. Kriegel, X. Xu, Density-based clustering in spatial databases: the algorithm GDBScan and its applications, Data Min. Knowl. Disc. 2 (2) (1998) 169–194.

[9] A.K. Jain, M.N. Murty, P.J. Flynn, Data clustering: a review, ACM Comput. Surv. 31 (3) (1999) 264–323.

[10] R. Xu, D. Wunsch, Survey of clustering algorithms, IEEE Trans. Neural Netw. 16 (3) (2005) 645–678.

[11] B.-R. Dai, I.-C. Lin, Efficient map/reduce-based DBScan algorithm with optimized data partition, in: Cloud Computing (CLOUD), 2012 IEEE 5th International Conference on, IEEE, 2012, pp. 59–66.

[12] B.H. Junker, F. Schreiber, Analysis of Biological Networks, vol. 2, John Wiley & Sons, New York, NY, 2011.

[13] U. Brandes, M. Gaertler, D. Wagner, Experiments on Graph Clustering Algorithms, Springer, New York, NY, 2003.

[14] D. Hamad, P. Biela, Introduction to spectral clustering, in: Information and Communication Technologies: From Theory to Applications, 2008. ICTTA 2008. 3rd International Conference on, IEEE, 2008, pp. 1–6.

[15] U. Von Luxburg, A tutorial on spectral clustering, Stat. Comput. 17 (4) (2007) 395–416.

[16] U. Von Luxburg, M. Belkin, O. Bousquet, Consistency of spectral clustering, Ann. Statist. (2008) 555–586.

[17] M. Masum, Vertex Weighted Spectral Clustering (Electronic Theses and Dissertation), 2017, Available from: https://dc.etsu.edu/etd/3266/, Accessed 21 October 2017.

[18] C.H.Q. Ding, X. He, H. Zha, M. Gu, H.D. Simon, A min-max cut algorithm for graph partitioning and data clustering, in: Data Mining, 2001. ICDM 2001, Proceedings IEEE International Conference on, IEEE, 2001, pp. 107–114.

[19] P. Zumstein, Comparison of spectral methods through the adjacency matrix and the Laplacian of a graph, TH Diploma, ETH Zürich, 2005.

[20] S.K. Butler, Eigenvalues and Structures of Graphs, University of California, San Diego, CA, 2008.

[21] A.J. Seary, W.D. Richards, Partitioning networks by eigenvectors, in: Proceedings of the International Conference on Social Networks, vol. 1, 1995, pp. 47–58.

[22] G. Strang, Introduction to Linear Algebra, vol. 3, Wellesley-Cambridge Press, Wellesley, MA, 1993.

[23] W. Ellens, Effective Resistance and Other Graph Measures for Network Robustness (Ph.D. thesis, Master thesis), Leiden University, 2011.

[24] Y. Zhang, J. Skolnick, Scoring function for automated assessment of protein structure template quality, Proteins Struct. Funct. Bioinf. 57 (4) (2004) 702–710.

[25] J. Xu, Y. Zhang, How significant is a protein structure similarity with TM-score = 0.5?, Bioinformatics 26 (7) (2010) 889–895.

[26] F.R.K. Chung, L. Lu, Complex Graphs and Networks, vol. 107, American Mathematical Society, 2006.

[27] Z. Wu, Linear algebra in biomolecular modeling, in: L. Hogben (Ed.), Handbook of Linear Algebra, second ed., Chapman and Hall/CRC, New York, NY, 2016.

[28] T.E. Creighton, Proteins: Structures and Molecular Properties, Freeman and Company, Boca Raton, FL, 2013.

[29] Q. Dong, Z. Wu, A linear-time algorithm for solving the molecular distance geometry problem with exact inter-atomic distances, J. Glob. Optim. 22 (1–4) (2002) 365–375.

[30] Q. Dong, Z. Wu, A geometric buildup algorithm for solving the molecular distance geometry problem with sparse distance data, J. Glob. Optim. 26 (3) (2003) 321–333.

[31] L.M. Blumenthal, Theory and Applications of Distance Geometry, Clarendon Press, Oxford, 1953.

[32] G.M. Crippen, T.F. Havel, Distance Geometry and Molecular Conformation, John Wiley & Sons, New York, NY, 1988.

[33] G. Golub, C. Van Loan, MatrixComputation, Johns Hopkins University Press, Baltimore, MD, 1989.

[34] T.F. Havel, Distance geometry: theory, algorithms, and chemical applications, in: Encyclopedia of Computational Chemistry, John Wiley & Sons, New York, NY, 1998, pp. 1–20.

Index

Note: Page numbers followed by *f* indicate figures and *t* indicate tables.

D

E

F

G

H

I

J

K

L

T

U

V

W

X